DUXBURY

Introduction to Regression Modeling

Bovas Abraham
University of Waterloo
Johannes Ledolter
University of Iowa

THOMSON
BROOKS/COLE

Australia • Canada • Mexico • Singapore • Spain
United Kingdom • United States

THOMSON

BROOKS/COLE

Acquisitions Editor: *Carolyn Crockett*
Assistant Editor: *Ann Day*
Editorial Assistant: *Rhonda Letts*
Technology Project Manager: *Burke Taft*
Marketing Manager: *Tom Ziolkowski*
Marketing Assistant: *Jessica Bothwell*
Advertising Project Manager: *Nathaniel Bergson-Michelson*
Project Manager, Editorial Production: *Belinda Krohmer*
Art Director: *Rob Hugel*
Print/Media Buyer: *Karen Hunt*
Permissions Editor: *Kiely Sexton*

Printed in the United States of America

1 2 3 4 5 6 7 09 08 07 06 05

Library of Congress Control Number: 2004109839

ISBN 0-534-42075-3

Production Service: *Matrix Productions*
Copy Editor: *Dan Hays*
Illustrator: *International Typesetting and Composition*
Cover Designer: *Carolyn Deacy*
Cover Image: *© Alberto Incrocci/Getty Images*
Cover Printer: *Quebecor World, Taunton*
Compositor: *International Typesetting and Composition*
Printer: *Quebecor World, Taunton*

Thomson Higher Education
10 Davis Drive
Belmont, CA 94002-3098
USA

Asia (including India)
Thomson Learning
5 Shenton Way #01-01
UIC Building
Singapore 068808

Australia/New Zealand
Thomson Learning Australia
102 Dodds Street
Southbank, Victoria 3006
Australia

Canada
Thomson Nelson
1120 Birchmount Road
Toronto, Ontario M1K 5G4
Canada

UK/Europe/Middle East/Africa
Thomson Learning
High Holborn House
50/51 Bedford Row
London WC1R 4LR
United Kingdom

Latin America
Thomson Learning
Seneca, 53
Colonia Polanco
11560 Mexico D.F.
Mexico

Spain (includes Portugal)
Thomson Paraninfo
Calle Magallanes, 25
28015 Madrid, Spain

Contents

Preface

Regression modeling is one of the most useful statistical techniques. It is an activity that leads to a functional relationship between a response and a set of explanatory variables. Such modeling has many different purposes. A regression model indicates which explanatory variables have an effect on the response, telling us which explanatory variables to change in order to affect the response. The functional relationship allows us to estimate the response for given values of the explanatory variables and to infer the response for values of the explanatory variables that were not studied directly. The regression model allows us to ask "what—if" type questions; for example, what happens to sales if we keep the product's price the same, but increase the advertising by ten percent. The prediction of future values of a response is another important purpose of regression; a model relating sales at time t to sales at previous periods allows us to make predictions of future sales. Furthermore, regression models may indicate that a variable that is difficult and expensive to measure is well explained by other variables that are easy and cheap to obtain. Such information is useful as then the cheaper measurements can be used as proxies for the more expensive ones.

This book is intended as a regression text for undergraduate and graduate students in statistics, business, engineering, and the physical/biological sciences desiring a solid introduction to this area. Our book is a blend of theory and interesting applications. We explain in detail the theory behind regression, using results from matrix algebra and adopting a data-driven approach that also emphasizes regression applications. The book includes several case studies from a wide range of application areas, and it covers the analysis of observational data as well as of data that arise from designed experiments. Special emphasis is given to the difficulties when working with observational data, such as problems arising from multicollinearity and "messy" data situations that violate some of the usual regression assumptions.

The book goes beyond the typical linear regression model as it also covers nonlinear models, regression models with time series errors, and logistic and Poisson regression models. These topics are of great importance as the response data in many application areas are categorical involving small counts, and because observations often arise in the form of time series.

Prerequisites for this book are an introduction to linear algebra including matrix operations, a first course in statistics, and knowledge of probability distributions and introductory statistical inference including confidence intervals and probability values.

WHY ANOTHER REGRESSION BOOK AND HOW THIS TEXT DIFFERS FROM COMPETING BOOKS

People may ask why we have written yet another regression book as several other successful regression texts are already available on the market. We have engaged in this project because we believe that we can do better. Time will tell whether we have been successful.

Students taking a course on regression must understand the purpose of regression. Chapter 1 addresses this issue. Many regression books do not address this topic in such detail as we have done in this book. We believe that students should know how, when and where regression models work. Hence we have included several case studies that illustrate the process of regression modeling, emphasize its benefits, but also warn of its pitfalls and problems. We believe that the data-driven approach used in our book will teach students practical modeling skills.

We also believe that most students want to know why some things work and why others do not. The text not only teaches students the use of regression, but also provides a rigorous coverage of the theory behind regression. This gives students the theoretical foundation that is needed for subsequent courses and further self-study. Geometric interpretations complement the algebraic results whenever possible.

EXERCISES, CASE STUDIES AND PROJECTS

Our book is unique because of the many excellent exercises, data sets, and project suggestions that are drawn together from several different areas of application such as engineering, business, social sciences, and the physical sciences. Projects in this book address questions that will interest readers from diverse fields of study. You may be interested whether and how it is possible to predict the price of fine French wine, and how to predict the winner of the next U.S. presidential election. You may want to know how to explain fuel efficiencies of automobiles, and whether race places a role in death penalty sentencing. You may want to model the scholastic achievement of U.S. students and determine the driving factors behind scholastic success. You may want to establish whether certain additives in building materials affect the ambient air quality, infer the effect of advertisement on the sales of a product, or learn about the factors that contributed to the survival of the Donner party that attempted to cross the Sierra Nevada mountain range in the 1840's on their westward trek. This book will interest you if you want to find the answers to these questions.

One learns statistics and regression modeling best by solving exercises that emphasize theoretical concepts, by analyzing real data sets, and by working on projects that require one to identify a problem of interest and to collect data that are relevant to the problem's solution. Suggestions for project topics and guidelines for dealing with projects are provided in Chapter 8. Successful projects involve the application of the studied regression techniques to solve real problems. Many of our exercises are based on real situations giving the students ample chance to practice on meaningful problems. Most exercises have been pre-tested in various classes on regression, time series modeling (Chapter 10), and statistical methods for business applications (Chapter 11 on logistic regression).

HOW TO USE THIS TEXT

Some parts of the book require more mathematical/statistical background than others. However, the more theoretical portions of this book can be omitted without compromising the main ideas. We suggest omitting the following sections when teaching students with weaker mathematics/statistics background: Most of the material in Chapter 3, except for elementary matrix algebra, can be skipped. In Chapter 4, one can omit the geometric approach in section 4.2.1, many of the derivations in sections 4.2.2 and 4.3, the discussion involving the geometric approach in section 4.4.1, the derivations in section 4.4.2, the joint confidence regions in section 4.4.3, generalized least squares in section 4.6, and the appendix. In Chapter 6, one can skip section 6.2.2 on added variable plots, the derivations in sections 6.4.1, section 6.5 on transformations, and the appendix.

The complete book (all 12 chapters) can be covered comfortably in two terms. If only one term or only one semester is available, we suggest covering Chapters 1 through 8 on the standard linear regression model, followed by a brief introduction to one of the additional topics in Chapters 9 through 12. For students with weaker background in mathematics/statistics we recommend that the more theoretical sections listed previously are omitted. For a target audience that wishes to concentrate on practical modeling we recommend that additional emphasis be put on the case studies in Chapter 8.

Chapters 1–7 of the book are based on notes that have been used many times at the University of Waterloo in an advanced undergraduate course on "Applied Linear Models." Students from Actuarial Science, Computer Science, Math-Business and Math-Accounting, Operations Research, Statistics, and Systems Design Engineering take this course. Materials from all chapters of the book have been class-tested in several courses at the University of Iowa as well.

SUPPLEMENTARY MATERIALS: SOLUTIONS TO EXERCISES, DATA FILES, AND COMPUTER PROGRAMS

Several supplements for this book are available. Files containing the data sets used throughout the text can be found on the enclosed data disk and on the webpage http://statistics.brookscole.com/abraham_ledolter/. The files are in ASCII format, as well as in the format of frequently-used statistical software packages.

One cannot learn and understand regression without using statistical computer software. Our text is not tied to a specific computer program, but we discuss computer output from several commonly used packages such as Minitab, R, S-Plus, SPSS, and SAS. Most packages are menu-driven and knowing one helps you understand how to use the others. Since not all programs are alike, we encourage you to try and experiment with several.

An instructor's manual and a student solutions manual are also available. They provide solutions to many of the exercises, as well as helpful hints on how to access and work with statistical computer software. Furthermore, brief answers to selected exercises are listed in the back of the text.

ACKNOWLEDGEMENTS

We gratefully acknowledge the contributions and the teaching of George Box, Norman Draper, Arthur Goldberger, and George Tiao from whom we have learned immensely over the past years.

We would like to thank the various professional societies, companies and publishers who have given permission to reproduce their materials in our text. Permission credit is acknowledged at various places in the book.

We also like to acknowledge the comments of our students at the University of Iowa, the University of Waterloo, and the Vienna University of Economics and Business Administration who have used various parts of our book.

We like to acknowledge the help of our editor, Carolyn Crockett, and her efficient staff. Carolyn was always there when we needed help and encouragement. She helped us along by providing useful and insightful reviews at just the right time when help was needed. We are also grateful to the many anonymous reviewers at Duxbury who have looked at various versions of this manuscript. Bovas Abraham would like to thank Mary Lou Dufton for typing parts of an earlier version of the manuscript.

Finally, we would like to acknowledge the help of our families for their understanding and assistance in completing this project. Writing a book is always more time-consuming than anticipated, and our families gave of their times freely. Thank you.

BACKGROUND OF THE AUTHORS

Bovas Abraham is the former Director of the Institute for Improvement in Quality and Productivity and also a professor in the Department of Statistics and Actuarial Science at the University of Waterloo. Bovas received his Ph.D. from the University of Wisconsin, Madison, USA. He has held visiting positions at the University of Wisconsin, the University of Iowa, and the University of Western Australia. He is the author of the book *Statistical Methods for Forecasting* (with Johannes Ledolter) published by Wiley in 1983, and the editor of the volume *Quality Improvement Through Statistical Methods* published by Birkhauser in 1998.

Johannes Ledolter is the John F. Murray Professor of Management Sciences at the University of Iowa, and a Professor at the Vienna University of Economics and Business Administration. His graduate degrees are in Statistics (MS and Ph.D. from the University of Wisconsin, and MS from the University of Vienna). He has held visiting positions at Princeton University and Yale University. He is the author of four books: *Statistical Methods for Forecasting* (with Bovas Abraham) published by Wiley in 1983, *Statistics for Engineers and Physical Scientists* (2nd edition, with Robert V. Hogg) published by Macmillan in 1991, *Statistical Quality Control* (with Claude W. Burrill) published by Wiley in 1999, and *Achieving Quality Through Continual Improvement* (with Claude W. Burrill) published by Wiley in 1999.

1 Introduction to Regression Models

1.1 INTRODUCTION

Regression modeling is an activity that leads to a mathematical description of a process in terms of a set of associated variables. The values of one variable frequently depend on the levels of several others. For example, the yield of a certain production process may depend on temperature, pressure, catalyst, and the rate of throughput. The number or the rate of defectives of a process may depend on the speed of the production line. The number of defective seals on toothpaste tubes may depend on the temperature and the pressure of the sealing process. The volume of a tree is related to the diameter of the tree at breast height, the height of the tree, and the taper of the tree. The fuel efficiency of an automobile depends, among others, on the weight of the car and characteristics of its body and engine. Employee efficiency may be related to the performance on employment tests, years of training, and educational background. The salaries of managers, athletes, and college teachers may depend on their seniority, the size of the market, and their performance. Many additional examples can be given, and in Exercise 1.1 we ask you to comment on several other relationships in detail.

The supply of a product depends on the price customers are willing to pay; one can expect that more products are brought to market when the price is high. Economists refer to this relationship as the **production function**. Similarly, the demand for a product depends on the price of the item, the price of the competition, and the amount spent on its advertisement. Economists refer to this relationship as the **demand function**. One can expect lower sales if the price is high, increased sales if the price of the competition is higher, and increased sales if more money is spent on promotion. However, price and advertising may also interact. Advertising may be more effective if the price is low; furthermore, the effect of the competition's price on sales may depend on one's own price. Also, seasonal components may have an impact on sales during a certain period because sales of a summer item during winter months will be low in northern states, irrespective of the product's price.

In all these situations we are interested in obtaining a "model" or a "law" (i.e., a mathematical description) for the relationship among the variables. Regression analysis deals with modeling the functional relationship between a **response variable** and one or more **explanatory variables**. In some instances one has a fairly good idea about the form of these models. Often the laws from physics or chemistry tell us how a response is related to the explanatory variables. These laws may involve complicated mathematical equations that contain functions such as logarithms and exponentials. In some instances, the constants in the equations are also known, but more often the constants need to be determined empirically by "fitting" the models to data. In many social science applications, theoretical models are absent, and one must develop empirical models that describe the main features of the relationship entirely from data.

Let us consider a few illustrative examples in detail.

1.2 EXAMPLES

1.2.1 PAYOUT OF AN INVESTMENT

Consider the payout of a principal P that you invest for a certain number of years (length of maturity) T, at an annual interest rate of $100R$ percent. We know from simple actuarial mathematics that the payout is given by

$$\text{Payout} = f(P, R, T) = P(1 + R)^T \tag{1.1}$$

provided that interest is compounded annually. With continuous compounding the resulting payout is slightly different. In this case, it can be calculated from $\text{Payout} = Pe^{RT}$, where e is Euler's number ($e = 2.71828\ldots$).

This first example illustrates a **deterministic relationship**. Each investment of principal P at rate R and maturity T leads to the exact same payout—nothing more and nothing less. We are very familiar with this law, and we would not need any data (or regression methods) to arrive at this particular model. However, assume for a moment that one was unfamiliar with the theory but had data on the payouts of different investments P, with different interest rates and maturities. Since the relationship is deterministic, payouts from identical investments would be identical and would not provide any additional information. Given this information, one would—after some trial and error and carefully constructed plots of the information—"see" the underlying functional relationship. This model would "fit" the data perfectly.

We have actually used the previous relationship to generate payouts for different principals, interest rates, and maturities, and we ask you in Exercise 1.2 to document the approach you use to find the model. You will experience firsthand the value of good theory; good theory will avoid much trial and error. Note that for payouts from continuous compounding, a plot of the logarithm of payout against the product of interest rate and length of maturity (RT) will show points falling on a line with slope one and intercept $\log(P)$.

1.2.2 PERIOD OF OSCILLATION OF A PENDULUM

Consider the period of oscillation (let us call it μ) of a pendulum of length L. It is a well-known fact from physics that the period of oscillation is proportional to the square root of the pendulum's length L, $\mu = \beta L^{1/2}$. However, the value of the proportionality factor β may be unknown.

In this example, we are given the functional form of the relationship, but we are missing information on the key constant, the proportionality factor β. In statistics we refer to unknown constants as **parameters**. The values of the parameters are usually determined by collecting data and using the resulting data to estimate the parameters.

The situation is also more complicated than in the first example because there is **measurement error**. Although the length of the pendulum is easy to measure, the determination of the period of oscillation is subject to variability. This means that sometimes our measurement of the "true" period of oscillation is too high and sometimes too low. However, for a calibrated measurement system we can expect that there is no bias (i.e., on average there is no error). If measured oscillation periods are plotted against the square roots of varying pendulum lengths, then the points will not line up exactly on a straight line through the origin, and there will be some scatter.

Mathematically, we characterize the relationship between the true period of oscillation μ and the length of the pendulum L as $\mu = \beta L^{1/2}$. However, the measured oscillation period OP is the sum of the true period (which we sometimes call the **signal**) and the measurement error ε (which we sometimes call the **noise**). Typically, we use a symmetric distribution about zero for the measurement error since the error is supposed to reflect only unbiased variability; if there were some bias in the measurement error, then such bias could be incorporated into the signal component of the model. Combining these two components (the signal and the noise) leads to the model

$$\text{OP} = \mu + \varepsilon = \beta L^{1/2} + \varepsilon \tag{1.2}$$

This model is similar to the one in Example 1.2.1 because we use theory (in this case, physics) to suggest the functional form of the relationship. However, in contrast to the previous example, we do not know certain constants (parameters) of the function. These parameters need to be estimated from empirical information. Furthermore, we have to deal with measurement variability, which leads to variability (or scatter) around the function (here, a line through the origin). We include a stochastic component ε in the model in order to capture this measurement variability.

1.2.3 SALARY OF COLLEGE TEACHERS

The third example represents a situation in which there is no theory about the functional form of the relationship and there is considerable variability in the measurements. In this situation, the data must perform "double duty," namely to determine the functional form of the model and the values of the parameters in these functions. Moreover, the modeling must be carried out in the presence of considerable variability. We refer to such models as **empirical models** (in contrast to the theory-based models discussed in Examples 1.2.1 and 1.2.2), and we refer to the process of constructing such models as **empirical model building**. Examples of this type arise in the social sciences, economics, and business, where one usually has little *a priori* theory of what the functions should look like.

Consider building a model that explains the annual salary of a college professor. We probably agree that salary should be related to experience (the more experience, the higher the salary), teaching performance (better teachers are paid more), performance on research (significant papers and books increase the salary), and whether the job includes administrative duties (administrators usually get paid more). However, we are lacking a theory that tells us the functional form of the model. Although we know that salary should increase with years of experience, we do not know whether the function should be linear in years, quadratic, or whether an even more complicated function of the number of years should be used. The same applies to the other variables.

Moreover, we notice considerable variability in salary because professors with virtually identical background often are paid vastly different salaries. So there may be additional factors that one has overlooked. Feel free to brainstorm and add to this initial list of variables. For example, salary may also depend on gender and racial factors (use of these factors would be illegal), the year the professor was hired, whether the professor is easy to get along with, whether the professor has had a relationship with the dean's spouse or had made an inappropriate remark at last year's holiday party, and so on. Knowing these factors may improve the fit of the model to the data. However, even after factoring all these variables into the model, substantial random variation will still exist.

Another aspect that makes the modeling within the social science context so difficult is problems with measuring the variables. Consider, for example, the teaching performance of an instructor. Although student ratings from end-of-the-semester questionnaires could be used as an indicator of teaching performance, one could argue that these ratings are only a poor proxy. Demanding teachers, difficult subject matter, and lectures held in large classes are known to lower these ratings, thus biasing the measure. Assessment of research performance is another good case in point. One could use the number of publications and books and use this as a proxy for research. However, such a simple-minded count does not incorporate the quality of the publications. Even if one decides to somehow incorporate publication quality, one notices very quickly that reasonable people differ in their judgments. Of course, not being able to accurately measure the factors that we believe to have an effect on the response affects the results of the empirical modeling.

In summary, we find that empirical modeling faces many difficulties: little or no theory on how the variables fit together, often considerable variability in the

TABLE 1.1 HARDNESS DATA [DATA FILE: hardness]

Run	1	2	3	4	5	6	7	8	9	10	11	12	13	14
x = Temperature	30	30	30	30	40	40	40	50	50	50	60	60	60	60
y = Hardness	55.8	59.1	54.8	54.6	43.1	42.2	45.2	31.6	30.9	30.8	17.5	20.5	17.2	16.9

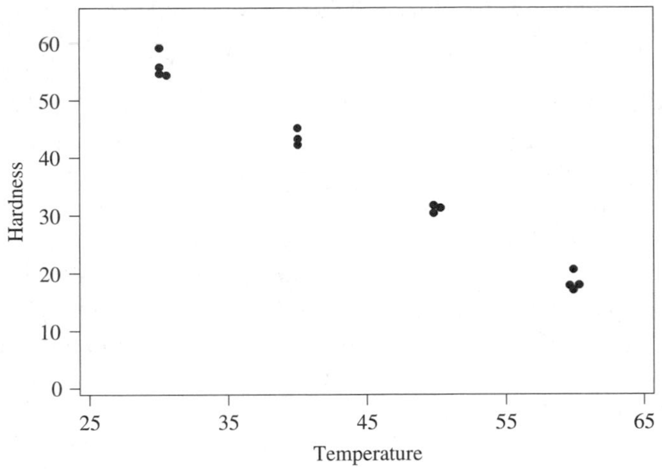

FIGURE 1.1 Scatter plot of hardness against quench bath temperature

response, and difficulties in obtaining appropriate measures for the variables that go into the model.

1.2.4 HARDNESS DATA

The quench bath temperature in a heat treatment operation was thought to affect the Rockwell hardness of a certain coil spring. An experiment was run in which several springs were treated under four temperatures: 30, 40, 50 and 60°C. The springs used in this experiment were selected from springs that had been produced under very similar conditions; all springs came from the same batch of material. Table 1.1 lists the (coded) hardness measurements and the temperatures at which the springs were treated.

We are interested in understanding how quench bath temperature affects hardness. Knowing this relationship is useful because it allows us to select the temperature that achieves a specified level of hardness.

Hardness is the dependent (or response) variable, and we denote it by y. Quench bath temperature is the independent (predictor, explanatory) variable that is supposed to help us predict the hardness; we denote it by x. For each experiment (coil spring—also called run or case) i, we have available a temperature that we select and control (the value x_i) and a measurement on the resulting hardness that we determine from the manufactured part (the value y_i). A scatter plot of hardness (y_i) against quench bath temperature (x_i) is shown in Figure 1.1.

We want to build a model (i.e., a mathematical relationship) to describe y in terms of x. Note that y cannot be a function of x alone since we have observed different y's (55.8, 59.1, 54.8, and 54.6) for the same $x = 30$. Furthermore, since no theoretical information is available to us to construct the model, we have to study the relationship empirically. The scatter plot of y against x indicates that y is approximately linear in x.

The scatter plot suggests the following model:

$$y(\text{hardness}) = \beta_0 + \beta_1 x(\text{temperature}) + \varepsilon \tag{1.3}$$

where β_0 and β_1 are the constants (parameters), and ε is the random disturbance (or error) that models the deviations from the straight line. The model is the sum of two components, the deterministic part (or signal) $\mu = \beta_0 + \beta_1 x$ and the random part ε. The deterministic part $\mu = \beta_0 + \beta_1 x$ is a linear function of x with parameters β_0 and β_1. More important, it is linear in the parameters β_0 and β_1, and hence we refer to this model as a **linear** model. The random component ε models the variability in the measurements around the regression line. This variability may come from the measurement error when determining the response y and/or changes in other variables (other than temperature) that affect the response but are not measured explicitly.

In order to emphasize that the model applies to each considered (and potential) experiment, we introduce subscripts. The temperature and the hardness from the ith experiment are written as (x_i, y_i). With these subscripts, our model can be expressed as

$$y_i = \beta_0 + \beta_1 x_i + \varepsilon_i, \quad \text{where } i = 1, 2, \ldots, n \tag{1.4}$$

We complete the model specification by making the following assumptions about the random component ε:

$$\begin{gathered} E(\varepsilon_i) = 0,\ V(\varepsilon_i) = \sigma^2 \text{ for all } i = 1, 2, \ldots, n \\ \varepsilon_i \text{ and } \varepsilon_j \text{ are independent random variables for } i \neq j \end{gathered} \tag{1.5}$$

In this example, we treat x_i as deterministic. The experimenter selects the temperature and knows exactly the temperature of the quench bath. There is no uncertainty about this value. In later sections of this book (Section 2.9), we consider the case when the values of the explanatory variable are random. For example, the observed temperature may only be a "noisy" reading of the true temperature.

Our assumptions about the error ε and the deterministic nature of the explanatory variable x imply that the response y_i is a random variable, with mean $E(y_i) = \mu_i = \beta_0 + \beta_1 x_i$ and variance $V(\varepsilon_i) = \sigma^2$. Furthermore, y_i and y_j are independent for $i \neq j$.

The mean, $E(y_i) = \mu_i = \beta_0 + \beta_1 x_i$, is a linear function of x. The intercept β_0 represents $E(y)$ when $x = 0$. If the value $x = 0$ is uninteresting or impossible, the intercept is a rather meaningless quantity. The slope parameter β_1 represents the change in $E(y)$ if x is increased by one unit. For positive β_1, the mean $E(y)$

TABLE 1.2 UFFI DATA [DATA FILE: uffi]

$y = CH_2O$	$x =$ Air Tightness	$z =$ UFFI Present
31.33	0	0
28.57	1	0
39.95	1	0
44.98	4	0
39.55	4	0
38.29	5	0
50.58	7	0
48.71	7	0
51.52	8	0
62.52	8	0
60.79	8	0
56.67	9	0
43.58	1	1
43.30	2	1
46.16	2	1
47.66	4	1
55.31	4	1
63.32	5	1
59.65	5	1
62.74	6	1
60.33	6	1
53.13	7	1
56.83	9	1
70.34	10	1

increases for increasing x (and decreases for decreasing x). For negative β_1, the mean $E(y)$ decreases for increasing x and increases for decreasing x.

Our assumption in Eq. (1.5) implies that $V(y) = \sigma^2$ is the same for each x. This states that if we repeat experiments at a value of x (as is the case in this example), we should see roughly the same scatter at each of the considered x's. Figure 1.1 shows that the variability in hardness at the four levels of temperature—$x = 30$, 40, 50, and 60—is about the same.

1.2.5 UREA FORMALDEHYDE FOAM INSULATION

Data were collected to check whether the presence of urea formaldehyde foam insulation (UFFI) has an effect on the ambient formaldehyde concentration (CH_2O) inside the house. Twelve homes with and 12 homes without UFFI were studied, and the average weekly CH_2O concentration (in parts per billion) was measured. It was thought that the CH_2O concentration was also influenced by the amount of air that can move through the house via windows, cracks, chimneys, etc. A measure of "air tightness," on a scale of 0 to 10, was determined for each home.

The data are shown in Table 1.2. CH_2O concentration is the response variable (y) that we try to explain through two explanatory variables: the air tightness

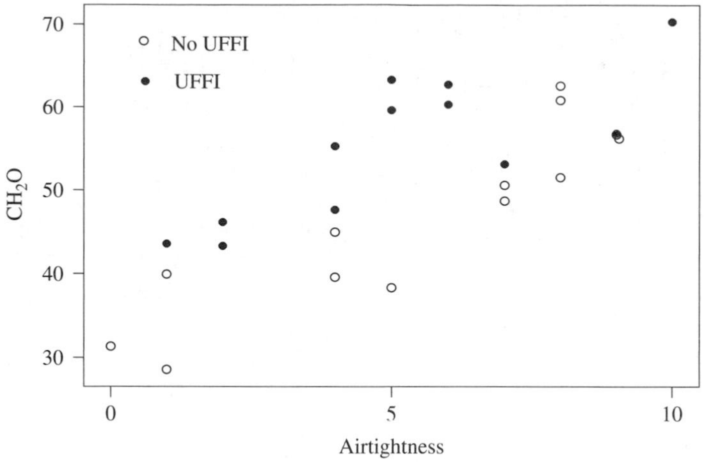

FIGURE 1.2 Scatter plot of CH_2O against air tightness for homes with and without urea formaldehyde foam insulation (UFFI)

of the home (x) and the absence/presence of UFFI (z). A scatter plot of CH_2O against air tightness for homes with and without UFFI is shown in Figure 1.2. The absence/presence of UFFI is expressed through an indicator variable. If insulation is present, then UFFI $= 1$; if it is absent, then UFFI $= 0$. The points in the scatter plot are labeled with solid and open circles, depending on whether or not UFFI is present. The plot shows strong evidence that CH_2O concentrations increase with increasing air tightness of the home.

It is important to emphasize that the data-generating mechanism in this example differs from that in the previous one. In the previous example, we were able to set the quench bath temperature at one of the four levels (30, 40, 50, and 60°C), conduct the experiment, and then measure the hardness of the spring. We refer to this as a **controlled** experiment, one in which the experimenter sets the values of the explanatory variable. In the current example, we select 12 houses with UFFI present and 12 houses in which it is not and measure the CH_2O concentration (the response y) as well as the air tightness (the explanatory x variable). It is not possible to preselect (or control) the air tightness value; the x values become available only after the houses are chosen. These data come from an **observational study**.

The basic objective of this particular observational study is to determine whether differences in the CH_2O concentrations can be attributed to the presence of insulation. Note, however, that we want to take into account the effect of air tightness. This can be achieved by considering the following model. Let

$$y = \beta_0 + \beta_1 x + \beta_2 z + \varepsilon \tag{1.6}$$

where

- y is the CH_2O concentration,
- x is the air tightness of the house,
- z is 1 or 0, depending on whether or not UFFI is present,

- ε is the error component that measures the random component, and
- β_0, β_1, and β_2 are constants (parameters) to be estimated.

CH_2O concentration is the response variable (y). It is the sum of a deterministic component ($\beta_0 + \beta_1 x + \beta_2 z$) and a random component ε. The random component ε is again modeled by a random variable with $E(\varepsilon) = 0$ and $V(\varepsilon) = \sigma^2$; it describes the variation in the CH_2O concentration among homes with identical values for x and z. Large variation in CH_2O concentration y among homes with the same insulation and tightness is characterized by large values of σ^2. The variability arises because of measurement errors (it is difficult to measure CH_2O accurately) and because of other aspects of the house (beyond air tightness and the presence of UFFI insulation) that have an influence on the response but are not part of the available information.

The deterministic component, $\beta_0 + \beta_1 x + \beta_2 z$, is the sum of three parts. The intercept β_0 measures the average CH_2O concentration for completely airtight houses ($x = 0$) without UFFI insulation ($z = 0$). The parameter β_2 can be explained as follows: Consider two houses with the same value for air tightness (x), the first house with UFFI ($z = 1$) and the second house without it ($z = 0$). Then $\beta_2 = E(y \mid \text{house } 1) - E(y \mid \text{house } 2)$ represents the difference in the average CH_2O concentrations for two identical houses (as far as air tightness is concerned) with and without UFFI. This is exactly the quantity we are interested in. If $\beta_2 = 0$, we cannot link the formaldehyde concentration to the presence of UFFI.

Similarly, β_1 is the expected change in CH_2O concentrations that is due to a unit change in air tightness in homes with (or without) UFFI. Model (1.6) assumes that this change is the same for homes with and without UFFI. This is a consequence of the additive structure of the model: The contributions of the two explanatory variables, $\beta_1 x$ and $\beta_2 z$, get added. However, additivity does not have to be the rule. The more general model that involves the product of x and z,

$$y = \beta_0 + \beta_1 x + \beta_2 z + \beta_3 xz + \varepsilon \qquad (1.7)$$

allows air tightness to affect the two types of homes differently. For a house without UFFI, $E(y) = \beta_0 + \beta_1 x$, and β_1 expresses the effect on the CH_2O concentrations of a unit change in air tightness. For a house with UFFI, $E(y) = (\beta_0 + \beta_2) + (\beta_1 + \beta_3)x$, and $(\beta_1 + \beta_3)$ expresses the effect of a unit change in air tightness. The effect is now different by the factor β_3.

1.2.6 ORAL CONTRACEPTIVE DATA

An experiment was conducted to determine the effects of five different oral contraceptives (OCs) on high-density lipocholesterol (HDLC), a substance found in blood serum. It is believed that high levels of this substance (the "good" cholesterol) help delay the onset of certain heart diseases. In the experiment, 50 women were randomly divided into five equal-sized groups; 10 women were assigned to each OC group. An initial baseline HDLC measurement was taken on each subject before oral contraceptives were started. After having used the respective

TABLE 1.3 ORAL CONTRACEPTIVE DATA [DATA FILE: contraceptive]

OC1 y = Final HDLC	OC1 z = Initial HDLC	OC2 y	OC2 z	OC3 y	OC3 z	OC4 y	OC4 z	OC5 y	OC5 z
43	49	58	56	100	102	50	57	41	37
61	73	46	49	52	64	50	55	58	60
45	55	66	64	49	60	52	64	58	39
46	55	59	63	51	51	58	49	69	60
59	63	71	90	48	59	65	78	68	71
57	53	64	56	51	57	71	63	64	63
56	51	53	46	40	63	52	62	46	51
68	74	50	64	52	62	49	50	56	64
46	58	68	75	44	61	49	60	51	45
47	41	35	58	50	58	58	59	57	58

drug for 6 months, a second HDLC measurement was made. The objective of the experiment was to study whether the five oral contraceptives differ in their effect on HDLC. The data are shown in Table 1.3. A scatter plot of final HDLC against the initial readings, ignoring the information on the respective treatment groups, is shown in Figure 1.3a. Figure 1.3b repeats this graph for groups 1, 2, and 5, using different plotting symbols to denote the three OC groups. Such a graph can highlight potential differences among the groups. (In order to keep the graph simple, only three groups are shown in Figure 1.3b).

Let y_i be the final HDLC measurement on subject $i (i = 1, 2, \ldots, 50)$ and let z_i be the initial HDLC reading. Furthermore, define five indicator variables $x_1, \ldots, x_5$ so that

$$\begin{aligned} x_{ik} &= 1 \text{ if subject } i \text{ is a participant in the } k\text{th OC group} \\ &= 0 \text{ otherwise} \end{aligned}$$

Here, we need two subscripts because there are five x variables. The first index in this double-subscript notation refers to the subject or case i; the second subscript refers to the explanatory variable (OC group) that is being considered. The following model relates the final HDLC measurement to six explanatory variables: the initial HDLC reading (z) and the five indicator variables ($x_1, \ldots, x_5$). For subject i,

$$y_i = \alpha z_i + \beta_1 x_{i1} + \beta_2 x_{i2} + \cdots + \beta_5 x_{i5} + \varepsilon_i \tag{1.8}$$

The usual assumption on the random component specifies that $E(\varepsilon_i) = 0$ and $V(\varepsilon_i) = \sigma^2$ for all i, and that ε_i and ε_j, for two different subjects $i \neq j$, are independent.

The deterministic component of the model, $E(y_i) = \alpha z_i + \beta_1 x_{i1} + \beta_2 x_{i2} + \cdots + \beta_5 x_{i5}$, represents five parallel lines in a graph of $E(y_i)$ against the initial HDLC, z_i. The six parameters can be interpreted as follows: The parameter α represents the common slope. The coefficients $\beta_1, \beta_2, \ldots, \beta_5$ represent the intercepts of the five lines and measure the effectiveness of the five OC treatment groups. Their comparison is of primary interest because there is no difference among the five drugs when $\beta_1 = \beta_2 = \cdots = \beta_5$.

FIGURE 1.3
Scatter plots of final HDLC against initial HDLC

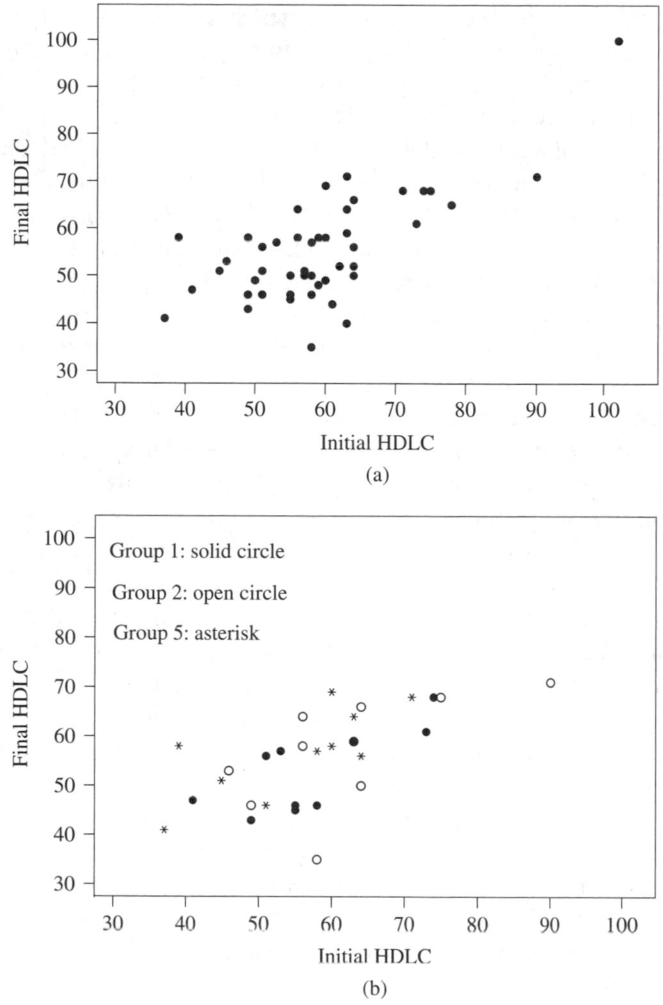

Consider two subjects (subjects i and j), both from the same OC group. Since the five indicator variables are the same on these two subjects ($x_{i1} = x_{j1}, \ldots, x_{i5} = x_{j5}$), the model implies $E(y_i) - E(y_j) = \alpha(z_i - z_j)$. The parameter α represents the expected difference in the final HDLC of two subjects who take the same drug but whose initial HDLC measurements differ by one unit. Next, consider two subjects with identical initial HDLC measurements but from different OC groups. Assume that the first woman is from group r, whereas the second is from group s. Then $E(y_i) - E(y_j) = \beta_r - \beta_s$, representing the expected difference in their final HDLC measurements.

1.2.7 GAS CONSUMPTION DATA

Let us give another illustration of empirical model building. Assume that we are interested in modeling the fuel efficiency of automobiles. First, we need to decide how to measure fuel efficiency. A typical measure of fuel efficiency used by the Environmental Protection Agency (EPA) and car manufacturers is "miles/gallon." It expresses how many miles a car can travel on 1 gallon of fuel. However, there is an alternative way to express fuel efficiency considering gallons per 100 traveled miles, "gallons/100 miles." It expresses the amount of fuel that is needed to travel 100 miles. The second measure is the scaled reciprocal of the first: [gallons/100 miles] = 100/[miles/gallon]. In Chapter 6, we discuss how to intelligently choose among these two measures. Assume for the time being, that we have settled on the second measure, [gallons/100 miles].

Next, we need to think about characteristics of the car that can be expected to have an impact on fuel efficiency. Weight of the car is probably the first variable that comes to mind. Weight should have the biggest impact, as we know from physics that we need a certain force to push an object, and that force is related to the fuel input. Heavy cars require more force and, hence, more fuel. Size (displacement) of the engine probably matters also. So does, most likely, the number of cylinders, horsepower, the presence of an automatic transmission, acceleration from 0 to 60 mph, the wind resistance of the car, and so on. However, how many explanatory variables should be in the model, and in what functional form should fuel consumption be related to the explanatory variables? Theory does not help much, except that physics seems to imply that [gallons/100 miles] should be related linearly to weight. However, how the other variables enter into the model and whether there should be interaction effects (e.g., whether changes in weight affect fuel efficiency differently depending on whether the car has a small or large engine) are open questions.

Assume, for the sake of this introductory discussion, that we have settled on the following three explanatory variables: x_1 = weight, x_2 = engine displacement, and x_3 = number of cylinders. Table 1.4 lists the fuel efficiency and the characteristics of a sample of 38 cars. We assume that the data are a representative sample (**random sample**) from a larger population. You can always replicate this study by going to recent issues of *Consumer Reports* and selecting another random sample. If you have ample time, you can select all given cars and study the population. The fact that we are dealing with a **random** sample is very important because we want to extend any conclusions from the analysis of these 38 cars to the larger population at hand. Our results should not be restricted to just this one set of 38 cars, but our conclusions on fuel efficiency should apply more generally to the population from which this sample was taken. If our set of 38 cars is not a representative sample, then it is questionable whether the inference can be extended to the population.

Note that fuel consumption in Table 1.4 is given in "miles/gallon" and "gallons/100 miles." Convince yourself that the entries in the second column are obtained through the simple transformation, [gallons/100 miles] = 100/[miles/gallon]. In addition to data on weight, engine displacement, and number of cylinders, the table includes several other variables that we will use in later chapters.

TABLE 1.4 GAS CONSUMPTION DATA [DATA FILE: gasconsumption]

Miles/ gallon	Gallons/ 100 miles	Weight, 1000 lb	Displacement (cubic inches)	No. of Cylinders	Horsepower	Acceleration (sec)	Engine Type: V(0), straight(1)
16.9	5.917	4.360	350	8	155	14.9	1
15.5	6.452	4.054	351	8	142	14.3	1
19.2	5.208	3.605	267	8	125	15.0	1
18.5	5.405	3.940	360	8	150	13.0	1
30.0	3.333	2.155	98	4	68	16.5	0
27.5	3.636	2.560	134	4	95	14.2	0
27.2	3.676	2.300	119	4	97	14.7	0
30.9	3.236	2.230	105	4	75	14.5	0
20.3	4.926	2.830	131	5	103	15.9	0
17.0	5.882	3.140	163	6	125	13.6	0
21.6	4.630	2.795	121	4	115	15.7	0
16.2	6.173	3.410	163	6	133	15.8	0
20.6	4.854	3.380	231	6	105	15.8	0
20.8	4.808	3.070	200	6	85	16.7	0
18.6	5.376	3.620	225	6	110	18.7	0
18.1	5.525	3.410	258	6	120	15.1	0
17.0	5.882	3.840	305	8	130	15.4	1
17.6	5.682	3.725	302	8	129	13.4	1
16.5	6.061	3.955	351	8	138	13.2	1
18.2	5.495	3.830	318	8	135	15.2	1
26.5	3.774	2.585	140	4	88	14.4	0
21.9	4.566	2.910	171	6	109	16.6	1
34.1	2.933	1.975	86	4	65	15.2	0
35.1	2.849	1.915	98	4	80	14.4	0
27.4	3.650	2.670	121	4	80	15.0	0
31.5	3.175	1.990	89	4	71	14.9	0
29.5	3.390	2.135	98	4	68	16.6	0
28.4	3.521	2.670	151	4	90	16.0	0
28.8	3.472	2.595	173	6	115	11.3	1
26.8	3.731	2.700	173	6	115	12.9	1
33.5	2.985	2.556	151	4	90	13.2	0
34.2	2.924	2.200	105	4	70	13.2	0
31.8	3.145	2.020	85	4	65	19.2	0
37.3	2.681	2.130	91	4	69	14.7	0
30.5	3.279	2.190	97	4	78	14.1	0
22.0	4.545	2.815	146	6	97	14.5	0
21.5	4.651	2.600	121	4	110	12.8	0
31.9	3.135	1.925	89	4	71	14.0	0

The first car on this list has weight 4,360 pounds (i.e., the value for variable x_1 for the first car is $x_{11} = 4.360$), cubic displacement of 350 in.^{3} (i.e., the value for x_2 for the first car is $x_{12} = 350$), eight cylinders (i.e., the value for x_3 for the first car is $x_{13} = 8$), and gets 16.9 miles to the gallon. The value of the response y, fuel efficiency measured in gallons/100 miles, is $y_1 = 100/16.9 = 5.917$; the car needs 5.917 gallons to travel 100 miles. The second car of our data set measures at $x_{21} = 4.054$, $x_{22} = 351$, $x_{23} = 8$, and $y_2 = 100/15.5 = 6.452$ (i.e., weight 4,054 pounds, 351 in.^{3} displacement, eight cylinders, and 6.452 gallons/100 miles). The last car (car 38) measures at $x_{38,1} = 1.925$, $x_{38,2} = 89$, $x_{38,3} = 4$, and $y_{38} = 100/31.9 = 3.135$ (i.e., weight 1,925 pounds, 89 in.^{3} displacement, four cylinders, and 3.135 gallons/100 miles).

Observe the notation that we use throughout this book. For the ith unit (in this case, the car), the values of the explanatory variables $x_1, x_2, \ldots, x_p$ (here, $p = 3$) and the response y are denoted by $x_{i1}, x_{i2}, \ldots, x_{ip}$, and y_i. Usually, there are several explanatory variables, not just one. Hence, we must use a double-index notation for x_{ij}, where the first index $i = 1, 2, \ldots, n$ refers to the case, and the second index $j = 1, 2, \ldots, p$ refers to the explanatory variable. For example, $x_{52} = 98$ is the value of the second explanatory variable (displacement, x_2) of the fifth car. Since we are dealing with a single response variable y, there is only one index (for case) in y_i.

A reasonable starting model relates fuel efficiency (gallons/100 miles) to the explanatory variables in a linear fashion. That is,

$$y = \mu + \varepsilon = \beta_0 + \beta_1 x_1 + \beta_2 x_2 + \beta_3 x_3 + \varepsilon \tag{1.9}$$

As before, the dependent variable is the sum of a random component, ε, and a deterministic component, $\mu = \beta_0 + \beta_1 x_1 + \beta_2 x_2 + \beta_3 x_3$, which is linear in the parameters β_0, β_1, β_2, and β_3.

Cars with the same weight, same engine displacement, and the same number of cylinders can have different gas consumption. This variability is described by ε, which is taken as a random variable with $E(\varepsilon) = 0$ and $V(\varepsilon) = \sigma^2$. If we consider cars with the same weight, same engine displacement, and same number of cylinders, then the average deviation from the mean value in gas consumption of these "alike" cars is zero. The variance σ^2 provides a measure of the variability around the mean value. Furthermore, we assume that $E(\varepsilon) = 0$ and $V(\varepsilon) = \sigma^2$ is the same for all groups of cars with identical values on x_1, x_2, and x_3. The variability is there because of measurement variability in determining the gas consumption. However, it also arises because of the presence of other characteristics of the car that affect fuel consumption but are not part of the data set. Cars may differ with respect to such omitted variables. If the omitted factors affect fuel consumption, then the fuel consumption of cars that are identical on the measured factors will be different.

The deterministic component $\mu = \beta_0 + \beta_1 x_1 + \beta_2 x_2 + \beta_3 x_3$ is linear in the parameters β_0, β_1, β_2, and β_3. We expect a positive value for the coefficient β_1 because a heavier car (with fixed engine displacement and number of cylinders) needs more fuel. Similarly, we expect a positive coefficient β_2 because a larger engine on a car of fixed weight and number of cylinders should require more fuel. We also expect a positive coefficient for β_3 because more cylinders on a car of fixed weight and engine displacement should require more fuel.

In order to understand the deterministic component μ more fully, consider two cars i and j with identical engine displacement and number of cylinders.

Since $x_{i2} = x_{j2}$ and $x_{i3} = x_{j3}$, the difference

$$E(y_i) - E(y_j) = \beta_1(x_{i1} - x_{j1})$$

Thus, β_1 represents the difference in the mean values of y (the mean difference in the gas consumption) of two cars whose weights (x_1) differ by one unit but that have the same engine displacement (x_2) and the same number of cylinders (x_3). Similarly, β_2 represents the difference in the mean values of y of two cars whose engine displacements (x_2) differ by one unit but that have the same weight (x_1) and the same number of cylinders (x_3). The parameter β_3 represents the difference in the mean values of y of two cars whose number of cylinders (x_3) differ by one unit but that have the same weight (x_1) and the same engine displacement (x_2).

In the modeling context, one often is not certain whether the variables under consideration are important or not. For instance, we might be interested in the question whether or not x_3 (number of cylinders) is necessary to predict y (gas consumption) once we have included the weight x_1 and the engine displacement x_2 in the model. Thus, we are interested in a test of the hypothesis that $\beta_3 = 0$, given that x_1 and x_2 are in the model. Such tests may lead to the exclusion of certain variables from the model. On the other hand, other variables such as horsepower x_4 may be important and should be included. Then the model needs to be extended so that its predictive capability is increased.

The model in Eq. (1.9) is quite simple and should provide a useful starting point for our modeling. Of course, we do not know the values of the model coefficients, nor do we know whether the functional representation is appropriate. For that we need data. One must keep in mind that there are only 38 observations and that one cannot consider models that contain too many unknown parameters. A reasonable strategy starts with simple **parsimonious** models such as the one specified here and then checks whether this representation is capable of explaining the main features of the data. A parsimonious model is simple in its structure and economical in terms of the number of unknown parameters that need to be estimated from data, yet capable of representing the key aspects of the relationship. We will say more on model building and model checking in subsequent chapters. The introduction in this chapter is only meant to raise these issues.

1.3 A GENERAL MODEL

In all of our examples, we have looked at situations in which a single response variable y is modeled as

$$y = \mu + \varepsilon \tag{1.10a}$$

The deterministic component μ is written as

$$\mu = \beta_0 + \beta_1 x_1 + \beta_2 x_2 + \cdots + \beta_p x_p \tag{1.10b}$$

where $x_1, x_2, \ldots, x_p$ are p explanatory variables. We assume that the explanatory variables are "fixed"—that is, measured without error. The parameter

$\beta_i (i = 1, 2, \ldots, p)$ is interpreted as the change in μ when changing x_i by one unit while keeping **all** other explanatory variables the same.

The random component ε is a random variable with zero mean, $E(\varepsilon) = 0$, and variance $V(\varepsilon) = \sigma^2$ that is constant for all cases and that does not depend on the values of $x_1, x_2, \ldots, x_p$. Furthermore, the errors for different cases, ε_i and ε_j, are assumed independent. Since the response y is the sum of a deterministic and a random component, we find that $E(y) = \mu$ and $V(y) = \sigma^2$.

We refer to the model in Eq. (1.10) as **linear in the parameters**. To explain the idea of linearity more fully, consider the following four models with deterministic components:

$$\begin{aligned}
&\text{i. } \mu = \beta_0 + \beta_1 x \\
&\text{ii. } \mu = \beta_0 + \beta_1 x_1 + \beta_2 x_2 \\
&\text{iii. } \mu = \beta_0 + \beta_1 x + \beta_2 x^2 \\
&\text{iv. } \mu = \beta_0 + \beta_1 \exp(\beta_2 x)
\end{aligned} \tag{1.11}$$

Models (i)–(iii) are linear in the parameters since the derivatives of μ with respect to the parameters β_i, $\partial\mu/\partial\beta_i$, do not depend on the parameters. Model (iv) is nonlinear in the parameters since the derivatives $\partial\mu/\partial\beta_1 = \exp(\beta_2 x)$ and $\partial\mu/\partial\beta_2 = \beta_1 x \exp(\beta_2 x)$ depend on the parameters.

The model in Eqs. (1.10a) and (1.10b) can be extended in many different ways. First, the functional relationship may be nonlinear, and we may consider a model such as that in Eq. (1.11iv) to describe the nonlinear pattern. Second, we may suppose that $V(y) = \sigma^2(x)$ is a function of the explanatory variables. Third, responses for different cases may not be independent. For example, we may model observations (e.g., on weight) that are taken on the same subject over time. Measurements on the same subject taken close together in time are clearly related, and the assumption of independence among the errors is violated. Fourth, several different response variables may be measured on each subject, and we may want to model these responses simultaneously. Many of these extensions will be discussed in later chapters of this book.

1.4 IMPORTANT REASONS FOR MODELING

Statistical modeling, as discussed in this text, is an activity that leads to a mathematical description of a process in terms of the variables of the process. Once a satisfactory model has been found, it can be used for several different purposes.

i. Usually, the model leads to a simple description of the main features of the data at hand. We learn which of the explanatory variables have an effect on the response. This tells us which explanatory variables we have to change in order to affect the response. If a variable does not affect a response, then there may be little reason to measure or control it. Not having to keep track of something that is not needed can lead to significant savings.

ii. The functional relationship between the response and the explanatory variables allows us to estimate the response for given values of the explanatory variables. It makes it possible to infer the response for values of the explanatory variables that were not studied directly. It also allows us to ask "what if"-type questions. For example, a model for sales can give us answers to questions of the following form: "What happens to sales if we keep our price the same, but increase the amount of advertising by 10%?" or "What happens to the gross national product if interest rates decrease by one percentage point?" Knowledge of the relationship also allows us to control the response variable at certain desired levels. Of course, the quality of answers to such questions depends on the quality of the models that are being used.

iii. Prediction of future events is another important application. We may have a good model for sales over time and want to know the likely sales for the next several future periods. We may have developed a good model relating sales at time t to sales at previous periods. Assuming that there is some stability over time, we can use such a model for making predictions of future sales.

In some situations, the models seen here are well grounded in theory. However, often theory is lacking and the models are purely descriptive of the data that one has collected. When a model lacks a solid theoretical foundation, it is questionable whether it is possible to extrapolate the results to new cases that are different from the ones occurring in the studied data set. For example, one would be very reluctant to extrapolate the findings in Example 1.2.4 and predict hardness for springs that were subjected to temperatures that are much higher than 60°C.

iv. A regression analysis may show that a variable that is difficult and expensive to measure can be explained to a large extent by variables that are easy and cheap to obtain. This is important information because we can substitute the cheaper measurements for the more expensive ones. It may be quite expensive to determine someone's body fat because this requires that the whole body be immersed in water. It may be expensive to obtain a person's bone density. However, variables such as height, weight, and thickness of thighs or biceps are easy and cheap to obtain. If there is a good model that can explain the expensively measured variable through the variables that are easy and cheap to obtain, then one can save money and effort by using the latter variables as proxies.

1.5 DATA PLOTS AND EMPIRICAL MODELING

Good graphical displays are very helpful in building models. Let us use the data in Table 1.4 to illustrate the general approach. Note that with one response and p explanatory variables, each case (in this situation, each car) represents a point in $(p + 1)$ dimensional space. Most empirical modeling starts with plots of the data in a lower dimensional space. Typically, one starts with pairwise (two-dimensional) scatter plots of the response against each of the explanatory variables. The scatter plot of fuel consumption (gallons/100 miles) against weight of the car in Figure 1.4a illustrates that heavier cars require more fuel. It also shows that the relationship between fuel consumption and weight is well approximated by a linear function. This is true at least over the observed weight range from approximately 2,000 to 4,000 pounds. How the function looks for very light and very heavy cars is difficult to tell because such cars are not in our group of considered cars; extrapolation beyond the observed range on weight is certainly a very tricky task.

Knowing that this relationship is linear simplifies the interpretation of the relationship because each additional 100 pounds of weight increases fuel efficiency by the same amount, irrespective of whether we talk about a car weighing 2,000 or 3,500 pounds. For a quadratic relationship the interpretation would not be as straightforward because the change in fuel consumption implied by a change in weight from 2,000 to 2,100 pounds would be different than the one implied by a change in weight from 3,500 to 3,600 pounds.

Another notable aspect of the data and the graph in Figure 1.4a is that the observations do not lie on the line exactly. This is because of variability. Our model recognizes this by allowing for a random component. On average, the fuel efficiency can be represented by a simple straight-line model, but individual observations (the fuel consumption of individual cars) vary around that line. This variation can result from many sources. First, it can be pure measurement error. Measuring the fuel consumption on the very same car for a second time may result in a different number. Second, there is variation in fuel consumption among cars taken from the very same model line. Despite being from the same model line and having the same weight, cars are not identical. Third, cars of identical weight may come from different model lines with very different characteristics. It is not just weight that affects the fuel consumption; other characteristics may have an effect. Engine sizes may be different and the shapes may not be the same. One could make the model more complicated by incorporating these other factors into it. Although this would reduce the variability in fuel consumption, one should not make the function so complicated that it passes through every single point. Such an approach would ignore the natural variability in measurements and attach too much importance to random variation. Henri Poincare, in *The Foundations of Science* [Science Press, New York, 1913 (reprinted 1929), p. 169] expresses this very well when he writes,

> Pass to an example of a more scientific character. I wish to determine an experimental law. This law, when I know it, can be represented by a curve. I make a certain number of isolated observations; each of these will be represented by a point. When I have obtained these different points, I draw a curve between them, striving to pass as near to them as possible and yet preserve for my curve a regular form, without angular points, or inflections too accentuated, or brusque variation of the radius of curvature. This curve will represent for me the probable law, and I assume not only that it will tell me the values of the function intermediate

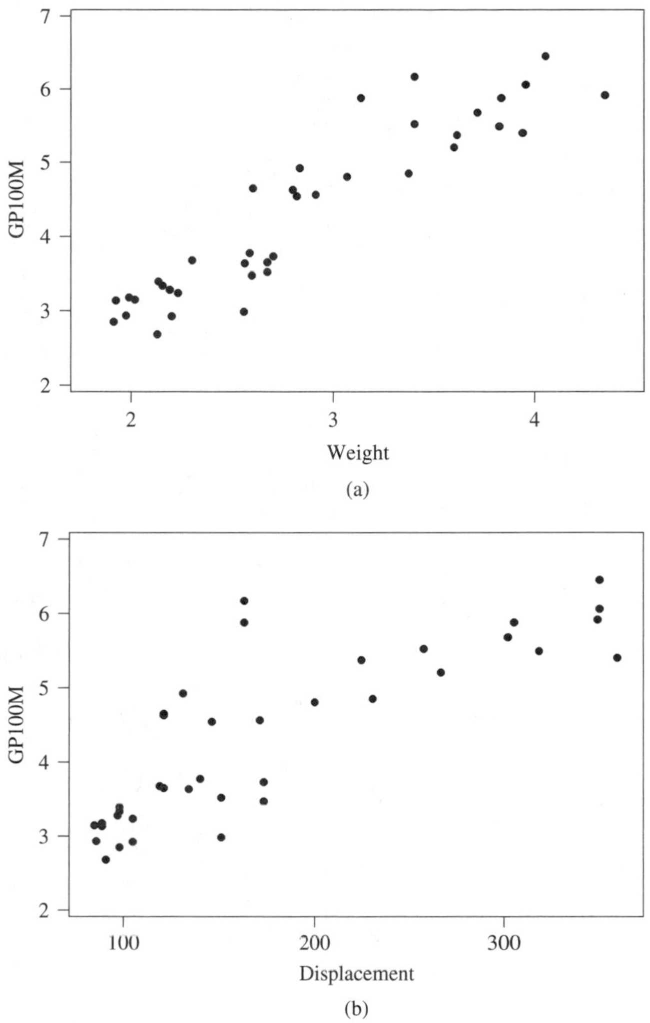

FIGURE 1.4
(a) Pairwise scatter plot of *y* (gallons/100 miles) against weight. (b) Pairwise scatter plot of *y* (gallons/100 miles) against displacement. (c) Pairwise scatter plot of *y* (gallons/100 miles) against number of cylinders. (d) Three-dimensional plot of *y* (gallons/100 miles) against weight and displacement

between those which have been observed, but also that it will give me the observed values themselves more exactly than direct observation. This is why I make it pass near the points, and not through the points themselves.

Here, we have described a two-dimensional representation of fuel consumption y and weight x_1. Similar scatter plots can be carried out for fuel consumption (y) and displacement x_2 and also fuel consumption (y) and number of cylinders x_3. The graphs shown in Figures 1.4b and 1.4c indicate linear relationships, even though the strengths of these relationships differ.

We notice from Figure 1.4 that each pairwise scatter plot exhibits linearity. Is this enough evidence to conclude that the model with the three explanatory variables should be linear also? The answer to this question is "**no**" in general. Although the linear model, $y = \beta_0 + \beta_1 x_1 + \beta_2 x_2 + \beta_3 x_3 + \varepsilon$, may provide a good starting point, the model may miss more complicated associations.

FIGURE 1.4
(Continued)

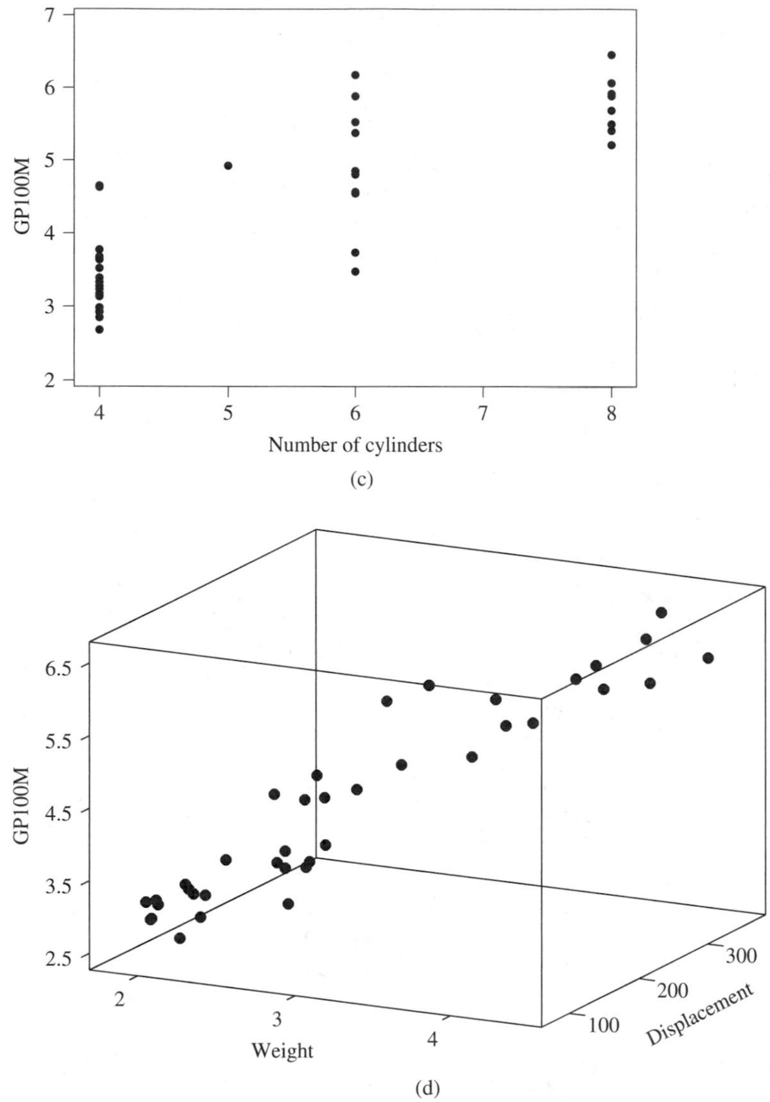

Two-dimensional displays are unable to capture the **joint** relationships among the response and more than one explanatory variable. In order to bring out the joint relationships between a response and two explanatory variables (e.g., weight x_1 and displacement x_2), one needs to look at a three-dimensional graph of fuel consumption y on both x_1 and x_2. This is done in Figure 1.4d. One notices that in such graphs it becomes considerably more difficult to recognize patterns, especially when working with relatively small data sets. However, at least from this graph it appears that the relationship can be approximated by a plane. The equation of a plane is given by $y(\text{gallons/100 miles}) = \beta_0 + \beta_1 x_1 + \beta_2 x_2$. This function implies that for a fixed value of x_2, a change in x_1 by one unit changes the response y by β_1 units. Similarly, for a fixed value of x_1, a change in x_2 by one unit changes the response y by β_2 units. The effects of changes in x_1 and x_2 are additive. Additivity is a special feature of this particular representation. It is a convenient simplification but need not be true in general. For some relationships the effects of a change in one explanatory variable depend on the value of a second explanatory variable. One says that the explanatory variables **interact** in how they affect the response y.

Up to now, we have incorporated the effects of x_1 and x_2. What about the effect of the third explanatory variable x_3? It is not possible to display all four variables in a four-dimensional graph. However, judging from the pairwise scatterplots and the three-dimensional representations (y, x_1, x_2), (y, x_1, x_3), and (y, x_2, x_3), our linear model in x_1, x_2, and x_3 may provide a sensible starting point.

1.6 AN ITERATIVE MODEL BUILDING APPROACH

An understanding of relationships can be gained in several different ways. One can start from a well-developed theory and use the data mostly for the estimation of unknown parameters and for checking whether the theory is consistent with the empirical information. Of course, any inconsistencies between theory and data should lead to a refinement of the model and a subsequent check whether the revised theory is consistent with the data.

Another approach, and one that is typically used in the social sciences, is to start from the data and use an empirical modeling approach to derive a model that provides a reasonable characterization of the relationship. Such a model may in fact lead to a new theory. Of course, theories must be rechecked against new data, and in cases of inconsistencies with the new information, new models must be developed, estimated, and checked again. Notice that good model building is a continual activity. It does not matter much whether one starts from theory or from data; what matters is that this process continues toward convergence.

A useful strategy for building such models is given in Figure 1.5. Initially, a tentative model is specified from relevant data and/or available theory. In some cases, theory will suggest certain models. In other situations, theory may not

FIGURE 1.5 A model building system

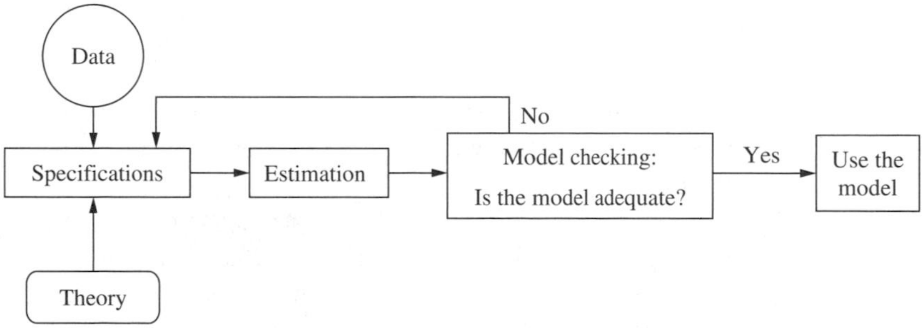

exist or may be incomplete, and data must be used to specify a reasonable initial model; exploratory data analysis and innovative ways of displaying information graphically are essential. The tentatively entertained model usually contains unknown parameters that need to be estimated. Model fitting procedures, such as **least squares** or **maximum likelihood**, have been developed for this purpose. This is discussed further in the next chapter.

Finally, the modeler must check the adequacy of the fitted model. Inadequacy can occur in several ways. For example, the model may miss important variables, it may include inappropriate and unnecessary variables that make the interpretation of the model difficult, and the model may misspecify the functional form. If the model is inadequate, then it needs to be changed, and the iterative cycle of "model specification—parameter estimation—model checking" must be repeated. One needs to do this until a satisfactory model is obtained.

1.7 SOME COMMENTS ON DATA

This discussion shows that good data are an essential component of any model building. However, not all data are alike, and we should spend some time discussing various types of data. We should distinguish between **data arising from designed experiments** and data from **observational studies**.

Many data sets in the physical sciences are the result of designed studies that are carefully planned and executed. For example, an engineer studying the impact of pressure and temperature on the yield of a production process may manufacture several products under varying levels of pressure and temperature. He or she may select three different pressures and four different settings for temperature and conduct one or several experiments at each of the $(3) \times (4) = 12$ different factor-level combinations. A good experimenter will suspect that other factors may have an impact on the yield but may not know for sure which ones. It could be the purity of the raw materials, environmental conditions in the plant during the manufacture, and so on. In order to minimize the effects of these uncontrolled factors, the investigator will randomize the arrangement of the experimental runs. He or she will do this to minimize the effects of unknown time trends. For example, one certainly would not want to run all experiments with the lowest temperature on one day and all experiments using the high temperature on another. If the process is sensitive to daily fluctuations in plant conditions, an observed difference in the results of the two days may not be due to temperature but due to the different conditions in the plant. Good experimenters will be careful when changing the two factors of interest, keeping other factors as uniform as possible. What is important is that the experimenter is actively involved in all aspects of obtaining the data.

Observational data are different because the investigator has no way of impacting the process that generates the data. The data are taken just as the data-generating process is providing them. Observational data are often referred to as "happenstance" data because they just happen to be available for analysis.

Economic and social science information is usually collected through a **census** (i.e., every single event is recorded) or through **surveys**. The problem with many social science data sets is that several things may have gone wrong during the data-gathering process, and the analyst has no chance to recover from these problems. A survey may not be representative of the population that one wants to study. Data definitions may not match exactly the factors that one wants to measure, and the gathered data may be poor proxies at best. Data quality may be poor because there may not have been enough time for careful data collection and processing. There may be missing data. The data that come along may not be "rich" enough to separate the effects of competing factors.

Consider the following example as an illustration. Assume that you want to explain college success as measured by student grade point average. Your admission office provides the student ACT scores (on tests taken prior to admission), and you have survey data on the number of study hours per week. Does this information allow you to develop a good model for college success? Yes, to a certain degree. However, there are also significant problems. First, college GPA is quite a narrow definition of college success. GPA figures are readily available, but one needs to discuss whether this is the information one really wants. Second, the range of ACT scores may be not wide enough to find a major impact of ACT scores on college GPA. Most good universities do not accept marginal students with low ACT scores. As a consequence, the range of ACT scores at your institution will be narrow, and you will not see much effect over that limited range. Third, study hours are self-reported, and students may have a tendency to make themselves look better than they really are. Fourth, ACT scores and study hours tend to be correlated. Students with a high ACT scores tend to have good study skills; it will be rare to find someone with a very high ACT score who does not study. The correlation between the two explanatory variables, ACT score and study hours, makes it difficult to separate the effects on college GPA of ACT scores and study hours.

EXERCISES

1.1. Consider the following relationships. Comment on the type of relationships that can be expected, supporting your discussion with simple graphs. In which of these cases can you run experiments that can help you learn about the relationship between the response y and the explanatory variables x?

a. Tensile strength of an alloy may be related to hardness and density of the stock.

b. Tool life may depend on the hardness of the material to be machined and the depth of the cut.

c. The weight of the coating of electrolytic tin plate may be affected by the current, acidity, rate of travel of the strip, and distance from the anode.

d. The diameter of a condenser coil may be affected by the thickness of the coil, number of turns, and tension in the winding.

e. The moisture content of lumber may depend on the speed of drying, the drying temperature, and the dimension of the pieces.

f. The performance of a foundry may be affected by atmospheric conditions.

g. The life of a light bulb may vary with the quality of the filament; the tile finish may depend on the temperature of firing.

1.2. Consider the payout of the following 18 investments (data file **payout**). The investments vary according to the invested principal P, the monthly interest rate R, and the length of maturity T (in months). The data in the following table were generated from the deterministic continuous compounding model, $Payout = Pe^{RT}$. No uncertainty was added to the equation.

It is reasonable to assume that the payout increases with the principal, the interest rate, and the maturity. However, without theory, the form of the relationship is not obvious. An empirical model building strategy that does not utilize available theory will be inefficient and may never find the hidden model. Construct scatter plots of the response (payout) on the explanatory variables (principal, interest rates, and maturity), and you will see what we mean. It is quite difficult to "see" the correct relationship.

Plot the logarithm of payout against the product of interest rate and maturity, and label the points on the scatter plot according to the invested principal (1,000, 1,500, and 2,000). What do you see, and how does this help you arrive at the correct model?

Principal	Interest Rate	Time (Months)	Payout
1,000	0.001	12	1,012.1
1,000	0.002	24	1,049.2
1,000	0.003	12	1,036.7
1,000	0.001	36	1,036.7
1,000	0.002	12	1,024.3
1,000	0.003	36	1,114.0
1,500	0.001	36	1,555.0
1,500	0.002	24	1,573.8
1,500	0.003	24	1,612.0
1,500	0.010	12	1,691.2
1,500	0.010	36	2,150.0
1,500	0.010	12	1,691.2
1,200	0.015	12	1,436.7
1,200	0.015	36	2,059.2
1,200	0.015	36	2,059.2
1,200	0.005	24	1,353.0
1,200	0.005	12	1,274.2
1,200	0.005	36	1,436.7

1.3. Look ahead in the book and read the problem descriptions of several exercises in Chapters 2 and 4–8. Find examples where the data originate from a designed experiment. Find examples where the data are the result of observational studies.

1.4. List other examples of designed experiments and observational studies.

1.5. Look ahead in the text and consider the data sets in Exercises 2.8, 2.9, 2.16–2.18, 2.21, 2.24, and 2.25 of Chapter 2. Construct pairwise scatter plots of the response variable against the explanatory variable(s). Discuss whether a linear model gives an appropriate description of the relationship. Speculate on the reasons for the variability of the response around the fitted line.

1.6. Experiment with three-dimensional displays of the information. For example, consider the data generated in Exercise 1.2. Consider the logarithm of payout as the response and the logarithm of the principal and the product of interest rate and maturity as the two explanatory variables. Discuss whether it is easy to spot three-dimensional relationships from such graphs. As a second example, consider the silkworm data in Exercise 4.15 of Chapter 4.

1.7. Explain the statement that a nonlinear model in the explanatory variables may turn out to be linear in the parameters. Give examples. For instance, is the quadratic model in x a model that is linear in the parameters? Explain.

1.8. Give examples of regression models that are nonlinear in the regression parameters.

1.9. Can you think of situations in which the variability in the response depends on the level of the explanatory variables? For example, consider sales that increase over time. Why is it reasonable to expect more variability in sales when the sales are high? Discuss.

1.10. Causality and correlation. Assume that a certain data set exhibits a strong association among two variables. Does this imply that there is a causal link? Can you think of examples where two variables are correlated but not causally related? What about the annual number of storks and the annual number of human births? Assume that your data come from a time period of increasing prosperity, such as the one immediately following World War II. Prosperity may impact the storks, and it may also affect couples' decisions to have families. Hence, you may see a strong (positive) correlation between the number of storks and the number of human births. However, you know that there is no causal effect. Can you describe the underlying principle of this example? Give other examples?

1.11. Collect the following information. Obtain average test scores for the elementary schools in your state (region). Obtain data on the proportion of children on subsidized lunch. Construct scatter plots. Do you think that there is a causal link between test scores and the proportion of children on subsidized lunch? If not, how do you explain the results you see. What if you had data on average income or the educational level of parents in these districts? Do you expect similar results?

1.12. Salary raises are usually expressed in percentage terms. This means that two people with the same percentage raise, but different previous salaries, will get different monetary (dollar) raises. Assume that the relative raise (R = RelativeRaise = PercentageRaise/100) is strictly proportional to performance (that is, RelativeRaise = βPerformance).
a. A plot of RelativeRaise against Performance exhibits a perfect linear relationship through the origin. Would a plot of AbsoluteRaise against Performance also exhibit a perfect linear association? Would a regression of AbsoluteRaise against Performance lead to the desired slope parameter β?
b. Consider the logarithmic transformation of the ratio (CurrentSalary/ PreviousSalary). What if you were to plot the logarithm of this ratio against the performance? How would this help you?

1.13. Consider Example 1.2.6 in which we studied the effectiveness of five oral contraceptives. We used model (1.8),

$$y_i = \alpha z_i + \beta_1 x_{i1} + \cdots + \beta_5 x_{i5} + \varepsilon_i$$

where z_i and y_i are the HDLC readings at the beginning and after 6 months, and $x_1, \ldots, x_5$ are indicators for the five treatment (contraceptive) groups.

How would you convince someone that this is an appropriate specification? In order to address this question, you may want to look at five separate scatter plots of y against z, one for each contraceptive group. Make sure these graphs are made with identical scales on both axes. Have the statistical software of your choice draw in the "best fitting" straight lines. Your model in Eq. (1.8) puts certain requirements on the slopes of these five graphs. What are these requirements?

How would you explain a model in which $\alpha = 1$? In this case, the emphasis is on changes in the HDLC, and the question becomes whether the magnitudes of the changes are related to the contraceptives. How would you analyze the data under this scenario?

2 Simple Linear Regression

2.1 THE MODEL

In this chapter, we consider the linear regression model with a single predictor (regressor) variable. The model is given as

$$y = \mu + \epsilon, \text{ where } \mu = \beta_0 + \beta_1 x \tag{2.1}$$

It is commonly referred to as the **simple linear regression model** because only one predictor variable is involved. Suppose we have n pairs of observations (x_i, y_i) $i = 1, 2, \ldots, n$. Then we can characterize these observations as

$$y_i = \beta_0 + \beta_1 x_i + \epsilon_i, \qquad i = 1, 2, \ldots, n$$

For the hardness data in Example 1.2.4 of Chapter 1, we have

$$\begin{aligned} 55.8 &= \beta_0 + \beta_1 .30 + \epsilon_1 \\ 59.1 &= \beta_0 + \beta_1 .30 + \epsilon_2 \\ \vdots\ \ &\qquad \vdots \\ 16.9 &= \beta_0 + \beta_1 .60 + \epsilon_{14} \end{aligned} \tag{2.2}$$

2.1.1 IMPORTANT ASSUMPTIONS

The standard analysis is based on the following assumptions about the regressor variable x and the random errors ϵ_i, $i = 1, \ldots, n$:

1. The regressor variable is under the experimenter's control, who can set the values $x_1, \ldots, x_n$. This means that x_i, $i = 1, 2, \ldots, n$, can be taken as constants; they are not random variables.
2. $E(\epsilon_i) = 0$, $i = 1, 2, \ldots, n$.
   This implies that $\mu_i = E(y_i) = \beta_0 + \beta_1 x_i$, $i = 1, 2, \ldots, n$.
3. $V(\epsilon_i) = \sigma^2$ is constant for all $i = 1, 2, \ldots, n$.
   This implies that the variances $V(y_i) = \sigma^2$ are all the same. All observations have the same precision.

4. Different errors ϵ_i and ϵ_j, and hence different responses y_i and y_j, are independent. This implies that $\text{Cov}(\epsilon_i, \epsilon_j) = 0$, for $i \neq j$.

The model implies that the response variable observations y_i are drawn from probability distributions with means $\mu_i = E(y_i) = \beta_0 + \beta_1 x_i$ and constant variance σ^2. In addition, any two observations y_i and y_j, for $i \neq j$, are independent.

2.1.2 OBJECTIVES OF THE ANALYSIS

Given a set of observations, the following questions usually arise:

1. Can we establish a relationship between y and x?
2. Can we predict y from x? To what extent can we predict y from x?
3. Can we control y by using x?

In order to answer these questions within the context of the simple linear regression model with mean $\mu = \beta_0 + \beta_1 x$, we need to estimate β_0, β_1, and σ^2 from available data (x_i, y_i), $i = 1, 2, \ldots, n$. The slope β_1 is of particular interest, because a zero slope indicates the absence of a linear association.

2.2 ESTIMATION OF PARAMETERS

2.2.1 MAXIMUM LIKELIHOOD ESTIMATION

This is a common method of estimating the parameters. Maximum likelihood estimation selects the estimates of the parameters such that the likelihood function is maximized. The likelihood function of the parameters β_0, β_1, σ^2 is the joint probability density function of $y_1, y_2, \ldots, y_n$, viewed as a function of the parameters. One looks for values of the parameters that give us the greatest probability of observing the data at hand.

A probability distribution for y must be specified if one wants to use this approach. In addition to the assumptions made earlier, we assume that ϵ_i has a normal distribution with mean zero and variance σ^2. This in turn implies that y_i has a normal distribution with mean $\mu_i = \beta_0 + \beta_1 x_i$ and variance σ^2. We write $\epsilon_i \sim N(0, \sigma^2)$ and $y_i \sim N(\beta_0 + \beta_1 x_i, \sigma^2)$.

The probability density function for the ith response y_i is

$$p(y_i | \beta_0, \beta_1, \sigma^2) = \frac{1}{\sqrt{2\pi}\sigma} \exp\left[-\frac{1}{2\sigma^2}(y_i - \beta_0 - \beta_1 x_i)^2\right] \tag{2.3}$$

and the joint probability density function of $y_1, y_2, \ldots, y_n$ is

$$p(y_1, y_2, \ldots, y_n | \beta_0, \beta_1, \sigma^2) = \left(\frac{1}{\sqrt{2\pi}}\right)^n \sigma^{-n} \exp\left[-\frac{1}{2\sigma^2}\sum_{i=1}^{n}(y_i - \beta_0 - \beta_1 x_i)^2\right]$$

Treating this as a function of the parameters leads us to the likelihood function $L(\beta_0, \beta_1, \sigma^2 | y_1, y_2, \ldots, y_n)$, and its logarithm

$$l(\beta_0, \beta_1, \sigma^2) = lnL(\beta_0, \beta_1, \sigma^2) = K - nln\sigma - \frac{1}{2\sigma^2}\sum_{i=1}^{n}(y_i - \beta_0 - \beta_1 x_i)^2 \quad (2.4)$$

Here, $K = (-n/2)ln(2\pi)$ is a constant that does not depend on the parameters. The maximum likelihood estimators (MLEs) of $\beta_0, \beta_1, \sigma^2$ maximize $l(\beta_0, \beta_1, \sigma^2)$. Maximizing the log-likelihood $l(\beta_0, \beta_1, \sigma^2)$ with respect to β_0 and β_1 is equivalent to minimizing $\sum_{i=1}^{n}(y_i - \beta_0 - \beta_1 x_i)^2$. The method of estimating β_0 and β_1 by minimizing $S(\beta_0, \beta_1) = \sum_{i=1}^{n}(y_i - \beta_0 - \beta_1 x_i)^2$ is referred to as the **method of least squares**.

2.2.2 LEAST SQUARES ESTIMATION

This discussion shows that maximum likelihood estimation, with the assumption of a normal distribution, leads to least squares estimation. However, least squares can be motivated in its own right, without having to refer to a normal distribution. One wants to obtain a line $\mu_i = \beta_0 + \beta_1 x_i$ that is "closest" to the points (x_i, y_i). The errors $\epsilon_i = y_i - \mu_i = y_i - \beta_0 - \beta_1 x_i$ $(i = 1, 2, \ldots, n)$ should be as small as possible. One approach to achieve this is to minimize the function

$$S(\beta_0, \beta_1) = \sum_{i=1}^{n}\epsilon_i^2 = \sum(y_i - \mu_i)^2 = \sum(y_i - \beta_0 - \beta_1 x_i)^2 \quad (2.5)$$

with respect to β_0 and β_1. This approach uses the squared distance as a measure of closeness. Note that other measures could be used, such as the absolute value of the difference, or some other power of the absolute difference. We use a symmetric loss function where positive and negative differences are treated the same. One could also think of nonsymmetric loss functions where over- and underpredictions are weighted differently. The squared error loss is the function that arises from the maximum likelihood procedure.

Taking derivatives with respect to β_0 and β_1, and setting the derivatives to zero,

$$\frac{\partial S(\beta_0, \beta_1)}{\partial \beta_0} = -2\sum(y_i - \beta_0 - \beta_1 x_i) = 0$$

and

$$\frac{\partial S(\beta_0, \beta_1)}{\partial \beta_1} = -2\sum(y_i - \beta_0 - \beta_1 x_i)x_i = 0$$

leads to the two equations:

$$\begin{aligned} n\beta_0 + \left(\sum x_i\right)\beta_1 &= \sum y_i \\ \left(\sum x_i\right)\beta_0 + \left(\sum x_i^2\right)\beta_1 &= \sum x_i y_i \end{aligned} \quad (2.6)$$

These are referred to as the **normal equations**. Suppose that $\hat{\beta}_0$ and $\hat{\beta}_1$ denote the solutions for β_0 and β_1 in the two-equation system (2.6). Simple algebra shows that these solutions are given by

$$\hat{\beta}_1 = \frac{\sum x_i y_i - \frac{(\sum x_i)(\sum y_i)}{n}}{\sum x_i^2 - \frac{(\sum x_i)^2}{n}} = \frac{\sum (x_i - \bar{x})(y_i - \bar{y})}{\sum (x_i - \bar{x})^2} = \frac{s_{xy}}{s_{xx}} \tag{2.7}$$

$$\hat{\beta}_0 = \bar{y} - \hat{\beta}_1 \bar{x}, \quad \text{where } \bar{y} = \frac{\sum y_i}{n} \quad \text{and} \quad \bar{x} = \frac{\sum x_i}{n}$$

They are called the **least squares estimates** (LSEs) of β_0 and β_1, respectively.

2.3 FITTED VALUES, RESIDUALS, AND THE ESTIMATE OF σ^2

The expression $\hat{\mu}_i = \hat{\beta}_0 + \hat{\beta}_1 x_i$ is called the **fitted value** that corresponds to the ith observation with x_i as the value for the explanatory variable. It is the value that is implied by the fitted model. Some textbooks refer to it as $\hat{y}_i$.

The difference between y_i and $\hat{\mu}_i$, $y_i - \hat{\mu}_i = e_i$, is referred to as the **residual**. It is the vertical distance between the observation y_i and the estimated line $\hat{\mu}_i$ evaluated at x_i.

The simple linear regression model, sample data, and the fitted line are illustrated in Figure 2.1. The broken line represents the mean $E(y_i) = \mu_i = \beta_0 + \beta_1 x_i$. The data are generated from distributions with densities sketched on the graph. The resulting data are used to determine the LSEs $\hat{\beta}_0$ and $\hat{\beta}_1$. The solid line on the graph represents the estimated line $\hat{\mu}_i = \hat{\beta}_0 + \hat{\beta}_1 x_i$. Imagine repeating this experiment with another set of n observations y_i at these specified x's. Due to the random component ϵ_i in the model, the observations will be different, and the estimates and the fitted line would change.

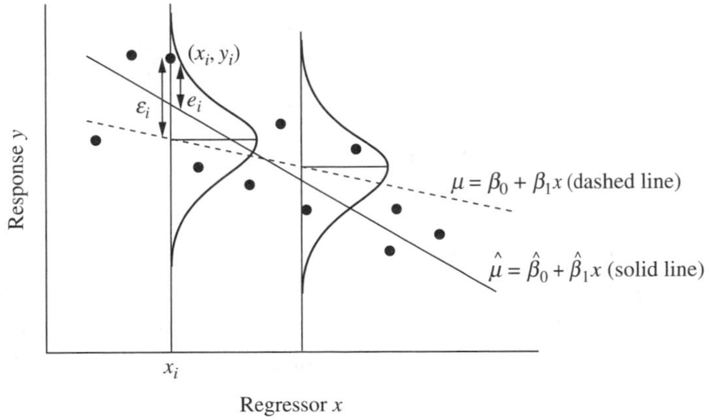

FIGURE 2.1 Mean Response and Estimated Regression Line: Simple Linear Regression

2.3.1 CONSEQUENCES OF THE LEAST SQUARES FIT

Least squares estimates set the derivatives of $S(\beta_0, \beta_1)$ equal to zero. The equations, evaluated at the least squares estimates,

$$\frac{\partial S(\beta_0, \beta_1)}{\partial \beta_0} = -2\sum[y_i - (\hat{\beta}_0 + \hat{\beta}_1 x_i)] = 0$$

and

$$\frac{\partial S(\beta_0, \beta_1)}{\partial \beta_1} = -2\sum[y_i - (\hat{\beta}_0 + \hat{\beta}_1 x_i)]x_i = 0$$

imply certain restrictions:

i. $\sum_{i=1}^{n} e_i = 0$. This can be seen from the derivative with respect to β_0.
ii. $\sum_{i=1}^{n} e_i x_i = 0$. This follows from the derivative with respect to β_1.
iii. $\sum_{i=1}^{n} \hat{\mu}_i e_i = 0$. This is because

$$\sum \hat{\mu}_i e_i = \sum(\hat{\beta}_0 + \hat{\beta}_1 x_i)e_i = \hat{\beta}_0 \sum e_i + \hat{\beta}_1 \sum x_i e_i = 0$$

due to the results in (i) and (ii).
iv. $(\bar{x}, \bar{y})$ is a point on the line $\hat{\mu} = \hat{\beta}_0 + \hat{\beta}_1 x$. Evaluating the fitted model at $\bar{x}$ leads to $\hat{\beta}_0 + \hat{\beta}_1 \bar{x} = (\bar{y} - \hat{\beta}_1 \bar{x}) + \hat{\beta}_1 \bar{x} = \bar{y}$.
v. $S(\hat{\beta}_0, \hat{\beta}_1) = \sum_{i=1}^{n} e_i^2$ is the minimum of $S(\beta_0, \beta_1)$.

2.3.2 ESTIMATION OF σ^2

Minimization of the log-likelihood function $l(\beta_0, \beta_1, \sigma^2)$ in Eq. (2.4) with respect to σ^2 leads to the MLE

$$\hat{\sigma}^2 = \frac{S(\hat{\beta}_0, \hat{\beta}_1)}{n} \tag{2.8}$$

The numerator $S(\hat{\beta}_0, \hat{\beta}_1) = \sum_{i=1}^{n}(y_i - (\hat{\beta}_0 + \hat{\beta}_1 x))^2 = \sum_{i=1}^{n} e_i^2$ is called the **residual sum of squares**; it is the minimum of $S(\beta_0, \beta_1)$.

The LSE of σ^2 is slightly different. It is obtained as

$$s^2 = \frac{S(\hat{\beta}_0, \hat{\beta}_1)}{n-2} \tag{2.9}$$

It is also called the **mean square error** (MSE). The only difference between the estimates in Eqs. (2.8) and (2.9) is in the denominator. The MLE divides by n, whereas the LSE divides by $n - 2$.

The residual sum of squares, $S(\hat{\beta}_0, \hat{\beta}_1) = \sum_{i=1}^{n} e_i^2$ consists of n squared residuals. However, the minimization of $S(\beta_0, \beta_1)$ has introduced two constraints among these n residuals; see (i) and (ii) given previously. Hence, only $n - 2$ residuals are needed for its computation. The remaining two residuals can always be calculated from $\sum e_i = \sum e_i x_i = 0$. One says that the residual sum of squares has $n - 2$ "degrees of freedom." The number of degrees of freedom symbolizes the number of independent components that are needed to determine the sum of squares.

The difference between the ML and the LS estimate of σ^2 is small, especially if n is reasonably large. In practice, one prefers the LSE s^2 because it is an **unbiased** estimate of σ^2; this is discussed further in Chapter 4.

2.3.3 LEAST SQUARES CALCULATIONS FOR THE HARDNESS EXAMPLE

Here we have $n = 14$, $\sum_{i=1}^{14} x_i = 630$, $\sum_{i=1}^{14} y_i = 520.2$, $\sum_{i=1}^{14} x_i y_i = 20{,}940$, $\sum_{i=1}^{14} x_i^2 = 30{,}300$:

$$\hat{\beta}_1 = \frac{20{,}940 - (630 \times 520.2/14)}{30{,}300 - (630^2/14)} = -1.266$$

$$\hat{\beta}_0 = \frac{520.2}{14} - (-1.266)\frac{630}{14} = 94.123$$

$$s^2 = \frac{1}{12}\sum_{i=1}^{14} e_i^2 = 2.235$$

The slope estimate $\hat{\beta}_1$ is negative. It implies lower than average hardness for items produced under higher than average temperatures. Is the estimate $\hat{\beta}_1 = -1.266$ extreme enough to claim that the unknown (population) slope β_1 is different from zero? For the answer to this question, one needs to understand the sampling properties of the estimators. In other words, if the true slope were zero and if we repeated the experiment many times at the same given temperature values, what would be the natural variability in the estimates $\hat{\beta}_1$? Would the one observed estimate $\hat{\beta}_1 = -1.266$ appear like an extreme realization from this sampling distribution? If our estimate is large compared to the sampling distribution that can be expected, then the estimate suggests that β_1 is different from zero.

2.4 PROPERTIES OF LEAST SQUARES ESTIMATES

Let us write the LSE of β_1 in Eq. (2.7) in slightly more convenient form:

$$\hat{\beta}_1 = \frac{\sum_{i=1}^{n}(x_i - \bar{x})(y_i - \bar{y})}{\sum_{i=1}^{n}(x_i - \bar{x})^2} = \frac{\sum_{i=1}^{n}(x_i - \bar{x})y_i - \bar{y}\sum_{i=1}^{n}(x_i - \bar{x})}{\sum_{i=1}^{n}(x_i - \bar{x})^2}$$

$$= \frac{\sum_{i=1}^{n}(x_i - \bar{x})y_i}{\sum_{i=1}^{n}(x_i - \bar{x})^2} = \sum_{i=1}^{n} c_i y_i$$

where $c_i = (x_i - \bar{x})/s_{xx}$ are constants; $s_{xx} = \sum_{i=1}^{n}(x_i - \bar{x})^2$.

The constants c_i have several interesting properties:

$$\text{i. } \sum_{i=1}^n c_i = \sum_{i=1}^n (x_i - \bar{x})/s_{xx} = 0$$
$$\text{ii. } \sum_{i=1}^n c_i x_i = \sum_{i=1}^n x_i(x_i - \bar{x})/s_{xx} = 1 \tag{2.10}$$
$$\text{iii. } \sum_{i=1}^n c_i^2 = \sum_{i=1}^n (x_i - \bar{x})^2/s_{xx}^2 = 1/s_{xx}$$

These results can be used to derive the expected values and the variances of the LSEs $\hat{\beta}_0$ and $\hat{\beta}_1$.

2.4.1 EXPECTED VALUES OF LEAST SQUARES ESTIMATES

1. $$E(\hat{\beta}_1) = E\left(\sum_{i=1}^n c_i y_i\right) = \sum_{i=1}^n c_i E(y_i) = \sum c_i(\beta_0 + \beta_1 x_i)$$
$$= \beta_0 \sum c_i + \beta_1 \sum c_i x_i = 0 + \beta_1 \times 1 = \beta_1 \tag{2.11}$$

   Since $E(\hat{\beta}_1) = \beta_1$, we say that $\hat{\beta}_1$ is an unbiased estimator of β_1. This implies that when the experiment is repeated a large number of times, the average of the estimates $\hat{\beta}_1$ [i.e., $E(\hat{\beta}_1)$] coincides with the true value β_1.
2. Similarly,
$$E(\hat{\beta}_0) = E(\bar{y} - \hat{\beta}_1\bar{x}) = E(\bar{y}) - \bar{x}E(\hat{\beta}_1) = E(\bar{y}) - \beta_1\bar{x}$$
   However, $E(\bar{y}) = E\left(\sum y_i/n\right) = \left[\sum_{i=1}^n (\beta_0 + \beta_1 x_i)\right]/n = \beta_0 + \beta_1\bar{x}$.
   Hence,
$$E(\hat{\beta}_0) = \beta_0 + \beta_1\bar{x} - \beta_1\bar{x} = \beta_0 \tag{2.12}$$
   Thus, $\hat{\beta}_0$ is also unbiased for β_0.
3. The LSE of $\mu_0 = \beta_0 + \beta_1 x_0$ is given by $\hat{\mu}_0 = \hat{\beta}_0 + \hat{\beta}_1 x_0$ and $E(\hat{\mu}_0) = \beta_0 + \beta_1 x_0 = \mu_0$. Hence, $\hat{\mu}_0$ is unbiased for μ_0.
4. It can also be shown that s^2 is an unbiased estimator of σ^2. That is,
$$E(s^2) = \sigma^2 \tag{2.13}$$
   This result will be proved in Chapter 4 for the general regression model.

2.4.2 VARIANCES OF LEAST SQUARES ESTIMATES

1. $V(\hat{\beta}_1) = V(\sum_{i=1}^n c_i y_i) = \sum c_i^2 V(y_i) = \sum c_i^2 \sigma^2$ since the y_i's are independent and $V(y_i) = \sigma^2$ is constant.
   Hence, from Eq. (2.10)
$$V(\hat{\beta}_1) = \sigma^2/s_{xx} \tag{2.14}$$
2. In order to obtain the variance of $\hat{\beta}_0$, we write the estimator $\hat{\beta}_0$ as follows:
$$\hat{\beta}_0 = \bar{y} - \hat{\beta}_1\bar{x} = \sum_{i=1}^n (y_i/n) - \bar{x}\sum_{i=1}^n (x_i - \bar{x})y_i/s_{xx}$$
$$= \sum_{i=1}^n k_i y_i, \quad \text{where } k_i = \frac{1}{n} - \frac{\bar{x}(x_i - \bar{x})}{s_{xx}}$$

Then,

$$V(\hat{\beta}_0) = \sum_{i=1}^{n} k_i^2 \sigma^2 = \sigma^2 \left[\frac{1}{n} + \frac{\bar{x}^2}{s_{xx}} \right] \tag{2.15}$$

Simple algebra shows that $\sum k_i^2$ equals the second factor in the previous expression.

3. For the variance of $V(\hat{\mu}_0)$, we write

$$\begin{aligned}\hat{\mu}_0 &= \hat{\beta}_0 + \hat{\beta}_1 x_0 = \bar{y} - \hat{\beta}_1 \bar{x} + \hat{\beta}_1 x_0 = \bar{y} + \hat{\beta}_1 (x_0 - \bar{x}) \\ &= \sum_{i=1}^{n} \left\{ \frac{y_i}{n} + (x_0 - \bar{x}) \frac{(x_i - \bar{x}) y_i}{s_{xx}} \right\} \\ &= \sum_{i=1}^{n} \left\{ \frac{1}{n} + \frac{(x_0 - \bar{x})(x_i - \bar{x})}{s_{xx}} \right\} y_i \\ &= \sum_{i=1}^{n} d_i y_i, \text{ where } d_i = \left\{ \frac{1}{n} + \frac{(x_0 - \bar{x})(x_i - \bar{x})}{s_{xx}} \right\}\end{aligned}$$

Then,

$$V(\hat{\mu}_0) = \sum_{i=1}^{n} d_i^2 \sigma^2 = \sigma^2 \left[\frac{1}{n} + \frac{(x_0 - \bar{x})^2}{s_{xx}} \right] \tag{2.16}$$

Simple algebra shows that $\sum d_i^2$ equals the second factor in the previous expression.

2.5 INFERENCES ABOUT THE REGRESSION PARAMETERS

The objective of most statistical modeling is to say something about the parameters of the population from which the data were taken (sampled). Of course, the more data, the smaller the uncertainty about the estimates. This fact is reflected in the variances of the LSEs; the denominators in the variances in Eqs. (2.14)–(2.16) increase with the sample size.

The uncertainty in the estimates can be expressed through confidence intervals, and for that one needs to make assumptions about the distribution of the errors. In the following discussion, we assume that the errors, and hence the observations, are normally distributed. That is,

$$y_i = \beta_0 + \beta_1 x_i + \epsilon_i, \quad \text{where } \epsilon_i \sim N(0, \sigma^2)$$

2.5.1 INFERENCE ABOUT β_1

The question whether or not the slope β_1 is zero is of particular interest. The slope β_1 expresses the effect on $E(y)$ of a unit change in the x variable.

Linear combinations of normal random variables are again normally distributed. The estimator $\hat{\beta}_1 = \sum_{i=1}^{n} c_i y_i$, where $c_i = (x_i - \bar{x})/s_{xx}$, is a linear combination of normal random variables and hence itself a normal random variable. This result is shown in Chapter 3 for a more general situation. The mean and the variance were obtained before. We find that

$$\hat{\beta}_1 \sim N\left(\beta_1, \frac{\sigma^2}{s_{xx}}\right)$$

or, after standardization,

$$\frac{\hat{\beta}_1 - \beta_1}{\sigma/\sqrt{s_{xx}}} \sim N(0, 1)$$

The factor σ^2 in the variance is unknown and must be estimated. For inferences about β_1 we replace the unknown σ^2 by its LSE s^2 in Eq. (2.9). We consider the ratio

$$T = \frac{(\hat{\beta}_1 - \beta_1)}{s/\sqrt{s_{xx}}} = \frac{(\hat{\beta}_1 - \beta_1)}{\sigma/\sqrt{s_{xx}}} \bigg/ \sqrt{\frac{(n-2)s^2}{\sigma^2(n-2)}} \tag{2.17}$$

The last identity (which you can check easily) appears unnecessary. However, the motivation for writing it in this form is that it facilitates the derivation of the sampling distribution. It can be shown that

i. The first term $Z = \dfrac{\hat{\beta}_1 - \beta_1}{\sigma/\sqrt{s_{xx}}}$ follows a standard normal distribution.
ii. $\dfrac{(n-2)s^2}{\sigma^2}$ follows a chi-square distributon with $n-2$ degrees of freedom, χ^2_{n-2} (see the appendix in Chapter 4 for the proof in the general case).
iii. s^2 and $\hat{\beta}_1$ are independent (this is proved for the general case in Chapter 4).
iv. If $Z \sim N(0, 1)$, $U \sim \chi^2_v$, and Z and U are independent, then it follows that $Z/\sqrt{U/v}$ has a Student t distribution with v degrees of freedom. We denote this distribution as $t(v)$.

From the results in (i)–(iv), it follows that

$$T = \frac{\hat{\beta}_1 - \beta_1}{s/\sqrt{s_{xx}}} \sim t(n-2) \tag{2.18}$$

Standardization of $\hat{\beta}_1 - \beta_1$ by the standard deviation of $\hat{\beta}_1$, $\sigma/\sqrt{s_{xx}}$, leads to a standard normal distribution. Standardization by the **estimated** standard deviation, $s/\sqrt{s_{xx}}$, leads to a t distribution. The estimated standard deviation of $\hat{\beta}_1$ is also referred to as the **standard error** of the estimate $\hat{\beta}_1$, and we sometimes write it as s.e.$(\hat{\beta}_1) = s/\sqrt{s_{xx}}$. The standard error tells us about the variability of the sampling distribution of $\hat{\beta}_1$; that is, the extent to which an estimate can differ from the true (population) value.

Confidence Interval for β_1

Let us use $t(1-\alpha/2; n-2)$ to denote the $100(1-\alpha/2)$ percentile of a t distribution with $n-2$ degrees of freedom. Since the t distribution is symmetric, we

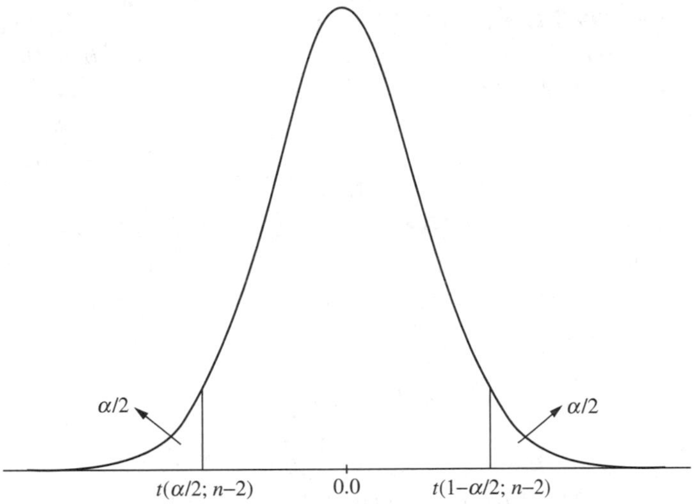

FIGURE 2.2
***t* Distribution and Confidence Intervals**

have that the $100(\alpha/2)$ percentile is given by $t(\alpha/2; n-2) = -t(1-\alpha/2; n-2)$ (Figure 2.2).

For example, the 97.5th and the 2.5th percentiles of the $t(12)$ distribution are given by $t(0.975; 12) = 2.18$ and $t(0.025; 12) = -2.18$, respectively.

The sampling distribution result in Eq. (2.18) implies

$$P\left[-t\left(1-\frac{\alpha}{2}; n-2\right) < \frac{\hat{\beta}_1 - \beta_1}{s/\sqrt{S_{xx}}} < t\left(1-\frac{\alpha}{2}; n-2\right)\right] = 1-\alpha$$

or

$$P\left[\hat{\beta}_1 - t\left(1-\frac{\alpha}{2}; n-2\right)\frac{s}{\sqrt{S_{xx}}} < \beta_1 < \hat{\beta}_1 + t\left(1-\frac{\alpha}{2}; n-2\right)\frac{s}{\sqrt{S_{xx}}}\right] = 1-\alpha$$

Hence, a $100(1-\alpha)$ percent confidence interval for β_1 is defined by the previous equation, and it is given by

$$\left[\hat{\beta}_1 - t\left(1-\frac{\alpha}{2}; n-2\right)\frac{s}{\sqrt{S_{xx}}},\ \hat{\beta}_1 + t\left(1-\frac{\alpha}{2}; n-2\right)\frac{s}{\sqrt{S_{xx}}}\right] \tag{2.19}$$

Note the form of this interval. You get it by starting with the point estimate $\hat{\beta}_1$ and by adding and subtracting a certain multiple of its standard error. That is,

$$\text{Estimate} \pm (t \text{ value})(\text{standard error of estimate})$$

where the standard error of the estimate is the estimated standard deviation of the sampling distribution of $\hat{\beta}_1$, given by $s/\sqrt{S_{xx}}$. For a 95% confidence interval and $\alpha = 0.05$, one needs to look up the 97.5th percentile $t(0.975; n-2)$ and the 2.5th percentile, $t(0.025; n-2) = -t(0.975; n-2)$.

Testing a Hypothesis about β_1

When testing $H_0: \beta_1 = 0$ against the alternative $H_1: \beta_1 \neq 0$, one assesses the magnitude of the t ratio

$$t_0 = \frac{\hat{\beta}_1}{\text{s.e.}(\hat{\beta}_1)} = \frac{\hat{\beta}_1}{s/\sqrt{s_{xx}}} \tag{2.20}$$

The t ratio is the standardized difference of the estimate $\hat{\beta}_1$ from the null hypothesis (which in this case is zero). The issue is whether the observed t ratio is large enough in absolute value so that one can also claim that the population parameter β_1 is different from zero. A comment on notation: The subscript zero in the observed t ratio $t_0 = \hat{\beta}_1/\text{s.e.}(\hat{\beta}_1)$ makes reference to the zero constraint in the null hypothesis $\beta_1 = 0$. We also write this t ratio as $t_0(\hat{\beta}_1)$ or simply as $t(\hat{\beta}_1)$.

Under the null hypothesis ($\beta_1 = 0$), the t ratio, $T = \hat{\beta}_1/\text{s.e.}(\hat{\beta}_1)$, follows a $t(n-2)$ distribution. Hence, one can calculate the probability

$$p = P[|T| \geq |t_0|] = 2P[T \geq |t_0|] \tag{2.21}$$

This is referred to as the p value, or the **probability value**. If this p value is small (smaller than a selected significance level, usually 0.05), then it is unlikely that the observed t ratio has come from the null hypothesis. In such case, one would not believe in the null hypothesis and reject the hypothesis that $\beta_1 = 0$. On the other hand, if this p value is large (larger than the significance level), one would conclude that the observed t value may have originated from the null distribution. In this case, one has no reason to reject H_0.

2.5.2 INFERENCE ABOUT $\mu_0 = \beta_0 + \beta_1 x_0$

We saw that $\hat{\mu}_0 = \sum_{i=1}^{n} d_i y_i$, where $d_i = \frac{1}{n} + \frac{(x_0 - \bar{x})(x_i - \bar{x})}{s_{xx}}$. Since $\hat{\mu}_0$ is a linear combination of normal random variables, $\hat{\mu}_0$ is normal. Earlier we derived the mean $E(\hat{\mu}_0) = \mu_0 = \beta_0 + \beta_1 x_0$ and the variance

$$V(\hat{\mu}_0) = \left[\frac{1}{n} + \frac{(x_0 - \bar{x})^2}{s_{xx}}\right]\sigma^2$$

Hence, the standardized random variable

$$\frac{\hat{\mu}_0 - \mu_0}{\sigma\left[\frac{1}{n} + \frac{(x_0-\bar{x})^2}{s_{xx}}\right]^{1/2}} \sim N(0, 1)$$

Substitution of the estimate s for σ changes the sampling distribution from a normal to a $t(n-2)$ distribution. As before, it can be shown that

$$T = \frac{\hat{\mu}_0 - \mu_0}{s\left[\frac{1}{n} + \frac{(x_0-\bar{x})^2}{s_{xx}}\right]^{1/2}} = \frac{(\hat{\beta}_0 + \hat{\beta}_1 x_0) - (\beta_0 + \beta_1 x)}{s\left[\frac{1}{n} + \frac{(x_0-\bar{x})^2}{s_{xx}}\right]^{1/2}} \sim t(n-2) \tag{2.22}$$

Using a t distribution with $n-2$ degrees of freedom (d.f.), a $100(1-\alpha)$ percent confidence interval for μ_0 is given by

$$\underbrace{(\hat{\beta}_0+\hat{\beta}_1 x_0)}_{\text{estimate}} \pm \underbrace{t\left(1-\frac{\alpha}{2}; n-2\right)}_{t\text{ value}} \underbrace{s\left[\frac{1}{n}+\frac{(x_0-\bar{x})^2}{s_{xx}}\right]^{1/2}}_{\text{s.e.}(\hat{\beta}_0+\hat{\beta}_1 x_0)} \tag{2.23}$$

Recall our rule about the construction of such intervals. Start with the point estimate and add/subtract a multiple of the estimated standard deviation of the point estimate, which is also referred to as the standard error of the estimate.

For the special case when $x_0=0$, μ_0 simplifies to $\mu_0=\beta_0$ and we can obtain a $100(1-\alpha)$ percent confidence interval for β_0 by setting $x_0=0$ in the previous interval Eq. (2.23). This turns out to be

$$\hat{\beta}_0 \pm t\left(1-\frac{\alpha}{2}; n-2\right) s\left[\frac{1}{n}+\frac{\bar{x}^2}{s_{xx}}\right]^{1/2} \tag{2.24}$$

2.5.3 HARDNESS EXAMPLE CONTINUED

$\hat{\beta}_0=94.134$, $\hat{\beta}_1=-1.266$, $s_{xx}=\sum(x_i-\bar{x})^2=1{,}950$, $n=14$, $s^2=2.235$. The relevant degrees of freedom are $n-2=12$. For a 95% confidence interval for β_1, we need $t(0.975, 12)=2.18$; and $\dfrac{s}{\sqrt{s_{xx}}}=\sqrt{\dfrac{2.235}{1{,}950}}=0.034$. The 95% confidence interval for β_1 is

$$\hat{\beta}_1 \pm t(0.975; 12)\frac{s}{\sqrt{s_{xx}}}$$
$$-1.266 \pm (2.18)(0.034)$$
$$-1.266 \pm 0.072$$

The confidence interval for β_1 extends from -1.338 to -1.194. Since "zero" is not in this interval, the data provide substantial evidence to reject $\beta_1=0$. Temperature appears to have a significant effect on hardness. Since $\hat{\beta}_1$ is negative, the hardness decreases as temperature increases.

Formally, one can test the null hypothesis $\beta_1=0$ by calculating the t ratio

$$t_0=\frac{\hat{\beta}_1}{\text{s.e.}(\hat{\beta}_1)}=\frac{-1.266}{0.034}=-37.4$$

and its probability value, $P(|T|>37.4)\approx 0.0001$. Since this is extremely small, there is overwhelming evidence against the hypothesis $\beta_1=0$. Temperature has a major impact on hardness.

A 95% confidence interval for β_0 uses the standard error

$$\text{s.e.}(\hat{\beta}_0)=\sqrt{s^2\left(\frac{1}{n}+\frac{\bar{x}^2}{s_{xx}}\right)}=1.575$$

The interval is given by

$$\hat{\beta}_0 \pm t(0.975;\, 12)\text{s.e.}(\hat{\beta}_0)$$
$$94.134 \pm (2.18)(1.575) \text{ or } 94.134 \pm 3.434$$
$$90.700 < \beta_0 < 97.578$$

The 95% confidence interval for the mean response $\mu_0 = \beta_0 + \beta_1 x_0$ when $x_0 = 55$ is centered at

$$\hat{\mu}_0 = 94.134 + (-1.266)(55) = 24.504$$

The standard error is

$$\text{s.e.}(\hat{\mu}_0) = \sqrt{s^2\left(\frac{1}{n} + \frac{(x_0 - \bar{x})^2}{S_{xx}}\right)} = \sqrt{2.235\left(\frac{1}{14} + \frac{100}{1950}\right)}$$
$$= \sqrt{0.2742} = 0.524$$

The 95% confidence interval for the mean response at $x_0 = 55$ is

$$24.504 \pm (2.18)(0.524),\ 24.504 \pm 1.142, \text{ or}$$
$$23.362 < \mu_0 < 25.646$$

We are 95% confident that our interval from 23.362 to 25.646 covers the average hardness for parts produced with temperature set at 55 degrees.

2.6 PREDICTION

We consider now the prediction of a **single** observation y, resulting from a **new** case with level x_p on the regressor variable. For illustration, in the hardness example one may be interested in the next run with temperature 55, and one may wish to predict the resulting hardness. Here, the emphasis is on a single observation and not on the mean (average) response for a given x_p.

The new observation y_p is the result of a new trial that is independent of the trials on which the regression analysis is based. However, we continue to assume that the model used for the sample data is also appropriate for the new observation.

The distinction between drawing inferences about the mean response $\mu_p = \beta_0 + \beta_1 x_p$ and predicting a new observation y_p must be emphasized. In the former case, we discuss the mean of the probability distribution of all responses at x_p. In the latter case, we are concerned about an individual outcome from this probability distribution (Figure 2.3).

The new observation can be written as

$$y_p = \beta_0 + \beta_1 x_p + \epsilon_p$$

where ϵ_p is a future unknown random error, and x_p is assumed known. Initially, let us assume that β_0 and β_1 are known. Then the "best" prediction of y_p is obtained

FIGURE 2.3
Prediction: Simple Linear Regression

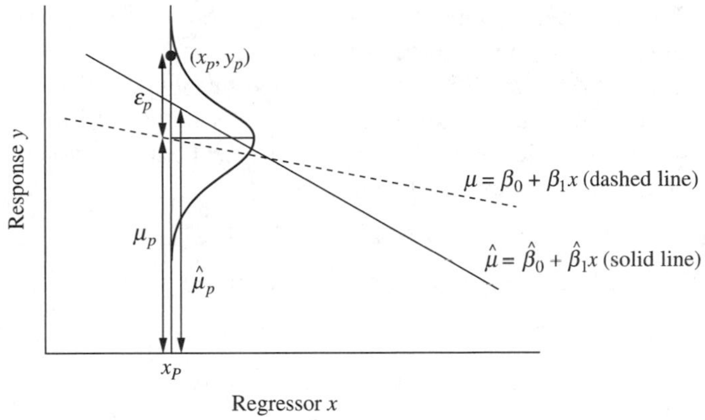

by replacing ϵ_p with its expected value, namely zero. If β_0 and β_1 are known, the prediction is given by

$$\hat{y}_p = \beta_0 + \beta_1 x_p$$

and the prediction error by

$$y_p - \hat{y}_p = \epsilon_p$$

The variance of the prediction error is

$$V(y_p - \hat{y}_p) = V(\epsilon_p) = \sigma^2 \tag{2.25}$$

In this case, the uncertainty about the prediction comes only through the random error ϵ_p.

Next, suppose that β_0 and β_1 are unknown and that they are estimated by $\hat{\beta}_0$ and $\hat{\beta}_1$. Then the best prediction is obtained by

$$\hat{y}_p = \hat{\beta}_0 + \hat{\beta}_1 x_p$$

and the prediction error is

$$y_p - \hat{y}_p = (\beta_0 + \beta_1 x_p) - (\hat{\beta}_0 + \hat{\beta}_1 x_p) + \epsilon_p = \mu_p - \hat{\mu}_p + \epsilon_p \tag{2.26}$$

The prediction error is the sum of two components: the new random error ϵ_p, and the error in estimating the mean response at x_p, $\mu_p - \hat{\mu}_p$. The LSEs $\hat{\beta}_0$ and $\hat{\beta}_1$, and hence the estimation error $\hat{\mu}_p - \mu_p$, are independent of ϵ_p since ϵ_p is a future random error that is unrelated to the data at hand. Hence, using the variance $V(\hat{\mu}_p)$ in Eq. (2.16), we find that the variance of the forecast error is

$$\begin{aligned} V(y_p - \hat{y}_p) &= V(\hat{\mu}_p) + V(\epsilon_p) \\ &= \sigma^2\left[\frac{1}{n} + \frac{(x_p - \bar{x})^2}{s_{xx}}\right] + \sigma^2 \\ &= \left[1 + \frac{1}{n} + \frac{(x_p - \bar{x})^2}{s_{xx}}\right]\sigma^2 \end{aligned} \tag{2.27}$$

Our uncertainty about the prediction $\hat{y}_p$ comes from two sources: (i) the random future error ϵ_p and (ii) the estimation of β_0 and β_1.

So far, we have discussed the properties (expectation and variance) of the prediction error. For prediction intervals we need to study the distribution of the error. Since the prediction error $y_p - \hat{y}_p$ is a linear combination of normal random variables, it is also normal. The mean

$$E(y_p - \hat{y}_p) = \mu_p - E(\hat{\mu}_p) + E(\epsilon_p) = \mu_p - \mu_p + 0 = 0$$

implying an unbiased forecast. The variance is given in Eq. (2.27). Therefore,

$$\frac{y_p - \hat{y}_p}{\sigma\left[1 + \frac{1}{n} + \frac{(x_p - \bar{x})^2}{s_{xx}}\right]^{1/2}} \sim N(0, 1)$$

Replacing σ with its LSE s leads us to the ratio

$$T = \frac{y_p - \hat{y}_p}{\text{s.e.}(y_p - \hat{y}_p)} \tag{2.28}$$

where $\text{s.e.}(y_p - \hat{y}_p) = s\left(1 + \dfrac{1}{n} + \dfrac{(x_p - \bar{x})^2}{s_{xx}}\right)^{1/2}$. Following similar arguments as before, it can be shown that T in Eq. (2.28) has a Student t distribution with $n - 2$ degrees of freedom. Hence,

$$P\left[\hat{y}_p - t\left(1 - \frac{\alpha}{2}; n-2\right)\text{s.e.}(y_p - \hat{y}_p) < y_p < \hat{y}_p + t\left(1 - \frac{\alpha}{2}; n-2\right)\text{s.e.}(y_p - \hat{y}_p)\right] = 1 - \alpha$$

and a $100(1-\alpha)$ percent **prediction interval** for y_p is given by

$$\hat{y}_p \pm t\left(1 - \frac{\alpha}{2}; n-2\right)s\left[1 + \frac{1}{n} + \frac{(x_p - \bar{x})^2}{s_{xx}}\right]^{1/2} \tag{2.29}$$

This interval is usually much wider than the confidence interval for the mean response μ_p at $x = x_p$. This is because of the random error ϵ_p reflecting the fact that individual observations vary around the mean level μ_p.

2.6.1 HARDNESS EXAMPLE CONTINUED

Suppose we are interested in predicting the hardness in the forthcoming run with temperature 55. Our prediction is $\hat{y}_p = 94.134 + (-1.266) \times (55) = 24.504$. The prediction error variance is estimated as

$$V(y_p - \hat{y}_p) = \left[1 + \frac{1}{14} + \frac{(55-45)^2}{1950}\right]2.235 = 2.5093$$

and a 95% prediction interval is given by

$$\hat{y}_p \pm t(0.975; 12)\sqrt{2.5093}$$
$$24.504 \pm (2.18)(1.584), \text{ or } 24.504 \pm 3.453$$

We are 95% confident that the interval from 21.051 to 27.957 will cover the hardness of the next run at temperature 55 degrees.

Note that this interval is considerably wider than the interval for the mean response μ_0 in Section 2.5. This is because of the additional uncertainty that comes through ϵ_p.

2.7 ANALYSIS OF VARIANCE APPROACH TO REGRESSION

In this section, we develop an approach for assessing the strength of the linear regression relationship. This approach can be extended quite easily to the more general regression models discussed in subsequent chapters.

Variability among the y_i's is usually measured by their deviations from the mean, $y_i - \bar{y}$. Thus, a measure of the total variation about the mean is provided by the **total sum of squares** (SST):

$$\text{SST} = \sum_{i=1}^{n} (y_i - \bar{y})^2$$

If SST $= 0$, all observations are the same. The greater is SST, the greater is the variation among the y observations. The standard deviation of the y's is obtained through

$$s_y = \sqrt{\text{SST}/(n-1)}$$

The objective of the analysis of variance is to partition the total variation SST into two parts: (i) the variation that is accounted for by the model and (ii) the variation that is left unexplained by the model. We can write the deviation of the response observation from its mean as

$$y_i - \bar{y} = (\hat{\mu}_i - \bar{y}) + (y_i - \hat{\mu}_i), \qquad i = 1, 2, \ldots, n$$

where $\hat{\mu}_i = \hat{\beta}_0 + \hat{\beta}_1 x_i$ is the estimated (fitted) mean (Figure 2.4). The total sum

FIGURE 2.4 Decomposition of the Variation: Simple Linear Regression

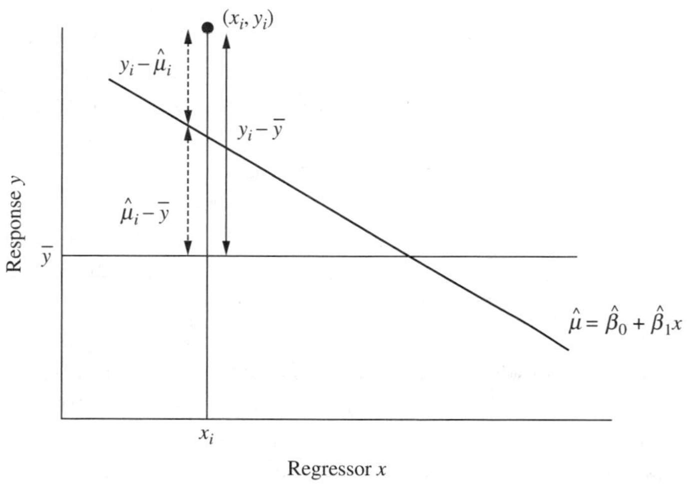

TABLE 2.1 THE ANALYSIS OF VARIANCE TABLE

Source	df	SS	MS	F
Regression (Model)	1	$SSR = \sum(\hat{\mu}_i - \bar{y})^2$	$MSR = \frac{SSR}{1}$	$\frac{SSR/1}{s^2}$
Residual	$n-2$	$SSE = \sum(y_i - \hat{\mu}_i)^2$	$MSE = \frac{SSE}{n-2} = s^2$	
Total (corrected)	$n-1$	$SST = \sum(y_i - \bar{y})^2$		

of squares can be written as

$$\begin{aligned} SST = \sum(y_i - \bar{y})^2 &= \sum(\hat{\mu}_i - \bar{y})^2 + \sum(y_i - \hat{\mu}_i)^2 + 2\sum(\hat{\mu}_i - \bar{y})(y_i - \hat{\mu}_i) \\ &= \sum(\hat{\mu}_i - \bar{y})^2 + \sum(y_i - \hat{\mu}_i)^2 \\ &= SSR + SSE \end{aligned} \tag{2.30}$$

since the cross-product term $\sum(\hat{\mu}_i - \bar{y})(y_i - \hat{\mu}_i) = \sum e_i(\hat{\mu}_i - \bar{y}) = \sum e_i\hat{\mu}_i - \bar{y}\sum e_i = 0$; this follows from properties (i) and (iii) in Section 2.3.

The difference $(y_i - \hat{\mu}_i) = e_i$ is the residual, and it reflects the component in the response that could not be explained by the regression model. The second term in Eq. (2.30), $SSE = \sum(y_i - \hat{\mu}_i)^2 = \sum e_i^2$, is known as the **residual (error) sum of squares**. It measures the variability in the response that is unexplained by the regression model. The first component in Eq. (2.30), $SSR = \sum(\hat{\mu}_i - \bar{y})^2$, is referred to as the **regression sum of squares**. The $\hat{\mu}_i$ are the fitted values of the response variable that are implied by the model. SSR measures the variability in the response variable that is accounted for by the model. SSR can also be written in the following equivalent way:

$$\begin{aligned} SSR = \sum(\hat{\mu}_i - \bar{y})^2 = \sum(\hat{\beta}_0 + \hat{\beta}_1 x_i - \bar{y})^2 &= \sum(\bar{y} - \hat{\beta}_1\bar{x} + \hat{\beta}_1 x_i - \bar{y})^2 \\ &= \hat{\beta}_1^2\sum(x_i - \bar{x})^2 \end{aligned} \tag{2.31}$$

This expression will be useful later on.

Equation (2.30) shows that the SST can be partitioned into these two components: SST = SSR + SSE. The first component, SSR, expresses the variability that is explained by the model; the second component, SSE, is the variability that could **not** be explained. The decomposition of the SST is usually displayed in a table, the so-called analysis of variance (ANOVA) table (Table 2.1).

Column 2 in the ANOVA table contains the **degrees of freedom** of the sum of squares contributions. The degrees of freedom are the number of independent components that are needed to calculate the respective sum of squares.

The total sum of squares, $SST = \sum(y_i - \bar{y})^2$, is the sum of n squared components. However, since $\sum(y_i - \bar{y}) = 0$, only $n - 1$ components are needed for its calculation. The remaining one can always be calculated from $(y_n - \bar{y}) = -\sum_{i=1}^{n-1}(y_i - \bar{y})$. Hence, SST has $n - 1$ degrees of freedom.

$SSE = \sum e_i^2$ is the sum of the n squared residuals. However, there are two restrictions among the residuals, coming from the two normal equations

$[\sum e_i = \sum e_i x_i = 0]$. Hence, only $n - 2$ residuals are needed to calculate SSE because the remaining two can be computed from the restrictions. One says that SSE has $n - 2$ degrees of freedom: the number of observations minus the number of estimated regression coefficients β_0 and β_1.

This leaves $\text{SSR} = \hat{\beta}_1^2 \sum (x_i - \bar{x})^2$. Only one (linear) function of the responses, $\hat{\beta}_1 = \sum c_i y_i$, is needed for its calculation. Hence, the degrees of freedom for SSR is one. Observe that also the degrees of freedom add up: d.f. (SST) = d.f. (SSR) + d.f. (SSE).

The sums of squares in column 3 are divided by their degrees of freedom in column 2. The resulting ratios are called the **mean squares**: MSR = SSR/1 is the mean square due to regression; $s^2 = \text{MSE} = \text{SSE}/(n - 2)$ is the mean square due to residual; it is also called the mean square error (see our discussion in Section 2.3).

The last column in the ANOVA table contains the F ratio:

$$F = \text{MSR/MSE} = \text{SSR}/s^2 \tag{2.32}$$

It will soon become clear why this is called the F ratio.

In Eq. (2.13) we mentioned that $E(s^2) = \sigma^2$; $\text{MSE} = s^2$ is an unbiased estimate of $V(\epsilon_i) = \sigma^2$. One can show that the expectation of $E(\text{MSR})$ is given by

$$\begin{aligned} E(\text{MSR}) = E(\text{SSR}) &= E\left[\hat{\beta}_1^2 \sum (x_i - \bar{x})^2\right] = \left[\sum (x_i - \bar{x})^2\right] E(\hat{\beta}_1^2) \\ &= \left[\sum (x_i - \bar{x})^2\right]\left[V(\hat{\beta}_1) + \left[E(\hat{\beta}_1)\right]^2\right] \\ &= \left[\sum (x_i - \bar{x})^2\right]\left[\frac{\sigma^2}{\sum (x_i - \bar{x})^2} + \beta_1^2\right] \\ &= \sigma^2 + \beta_1^2 \sum (x_i - \bar{x})^2 \end{aligned} \tag{2.33}$$

When $\beta_1 = 0$, then also $E(\text{MSR}) = \sigma^2$. On the other hand, when $\beta_1 \neq 0$, $E(\text{MSR})$ is greater than σ^2 since the term $\beta_1^2 \sum (x_i - \bar{x})^2$ is always positive. Thus, a test whether $\beta_1 = 0$ can be constructed by comparing the MSR and the mean square due to residuals MSE. A MSR substantially larger than s^2 (the mean square of residuals) suggests that $\beta_1 \neq 0$. This is the basic idea behind the **analysis of variance test** which is discussed next.

Let us consider the ratio in the last column of the ANOVA table. We note the following:

i. The variance of $\hat{\beta}_1$, $V(\hat{\beta}_1) = \sigma^2/\sum (x_i - x)^2$ was derived in Eq. (2.14). Since $\dfrac{(\hat{\beta}_1 - \beta_1)}{\sigma/\{\sum (x_i - \bar{x})^2\}^{1/2}} \sim N(0, 1)$, it follows that its square $\dfrac{(\hat{\beta}_1 - \beta_1)^2}{\sigma^2} \sum (x_i - \bar{x})^2 \sim \chi_1^2$. Hence, for $\beta_1 = 0$, we have that $\dfrac{\hat{\beta}_1^2 \sum (x_i - \bar{x})^2}{\sigma^2} = \dfrac{\text{SSR}}{\sigma^2} \sim \chi_1^2$.

ii. $\frac{\text{SSE}}{\sigma^2} = \frac{(n-2)s^2}{\sigma^2} \sim \chi^2_{n-2}$ (This is shown in the appendix in Chapter 4).

iii. SSR and SSE are independent. (This is proved in the appendix in Chapter 4).

These facts imply the following result for the ratio:

$$F = \frac{\frac{\text{SSR}}{\sigma^2}/1}{\frac{\text{SSE}}{\sigma^2}/(n-2)} = \frac{\text{SSR}}{s^2}$$

If $\beta_1 = 0$ (i.e., there is no linear relationship between y and x), the F ratio (Eq. 2.32) in the last column of the ANOVA table follows an F distribution with 1 and $n-2$ d.f. We write $F \sim F(1, n-2)$. The degrees of freedom are easy to remember because they stand in the d.f. column next to the respective sum of squares. For $\beta_1 \neq 0$, we expect larger values for F. For testing the hypothesis that $\beta_1 = 0$, we calculate the probability value

$$p = P(F \geq f_0)$$

where f_0 is the observed value of the F statistic. If the p value is small (smaller than a preselected significance level, usually 0.05), then there is evidence against the hypothesis $\beta_1 = 0$. If the p value is large, then there is no evidence to reject the null hypothesis that $\beta_1 = 0$.

2.7.1 COEFFICIENT OF DETERMINATION: R^2

We now discuss a descriptive measure that is commonly used in practice to describe the degree of linear association between y and x. Consider the identity

$$\text{SST} = \text{SSR} + \text{SSE}$$

The ratio

$$R^2 = \frac{\text{SSR}}{\text{SST}} = 1 - \frac{\text{SSE}}{\text{SST}} \tag{2.34}$$

is used to assess the "fit" of a regression model. It expresses the proportion of the total variation of the response around the mean that is explained by the regression model.

R^2 must always be between 0 and 1: $0 \leq R^2 \leq 1$. $R^2 = 0$ indicates that none of the variability in the y is explained by the regression model. $\text{SSE} = 0$ and $R^2 = 1$ indicate that all observations fall on the fitted line exactly.

Given the observations $y_1, y_2, \ldots, y_n$, SST is a certain fixed quantity. However, SSR (and SSE) change with the choice of the model. Models with larger SSR (smaller SSE) and larger R^2 are usually preferable to models with smaller SSR (larger SSE) and smaller R^2. However, a large R^2 does not necessarily imply that a particular model fits the data well. Also, a small R^2 does not imply that the model is a poor fit. Thus one should use R^2 with caution. The coefficient of determination R^2 does not capture the essential information as to whether a given relation is useful in a particular application. We will discuss this more fully in Chapter 4.

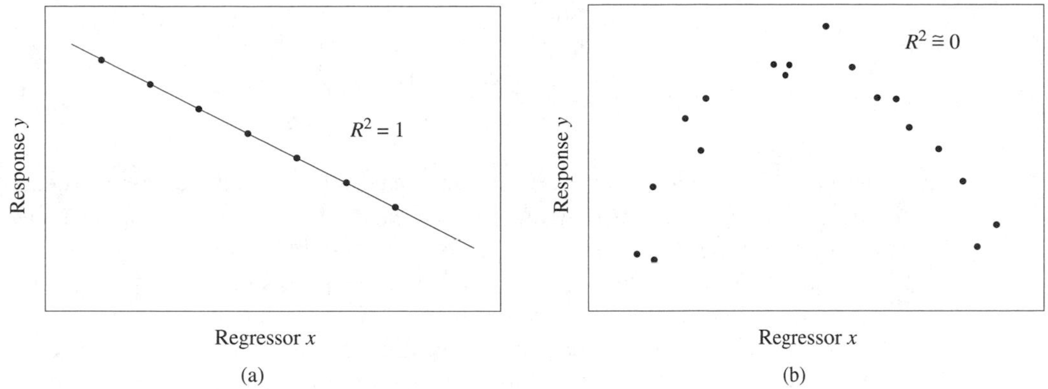

FIGURE 2.5 R^2 for Different Situations

It should also be emphasized that R^2 is a measure of the **linear** association between y and x. A small R^2 does not always imply a poor relationship between y and x. As indicated in Figure 2.5, the relation between y and x may be quadratic and R^2 could be a small value.

There is a simple relationship between the R^2 in simple linear regression and the correlation coefficient between y and x. R^2 is the square of the sample correlation coefficient r. You can see this by writing

$$\begin{aligned} R^2 &= \frac{\text{SSR}}{\text{SST}} = \frac{\hat{\beta}_1^2 \sum(x_i - \bar{x})^2}{\sum(y_i - \bar{y})^2} \\ &= \frac{\left[\sum(x_i - \bar{x})(y_i - \bar{y})\right]^2}{\left[\sum(x_i - \bar{x})^2\right]^2} \frac{\sum(x_i - \bar{x})^2}{\sum(y_i - \bar{y})^2} = \left[\frac{\sum(x_i - \bar{x})(y_i - \bar{y})}{\sqrt{\sum(x_i - \bar{x})^2}\sqrt{\sum(y_i - \bar{y})^2)}}\right]^2 \\ &= r^2 \end{aligned} \tag{2.35}$$

2.7.2 HARDNESS EXAMPLE CONTINUED

ANOVA TABLE FOR HARDNESS EXAMPLE

Source	df	SS	MS	F
Regression (Model)	1	3,126.134	3,126.134	1,398.7
Residual	12	26.820	2.235	
Total (corrected)	13	3,152.954		

Here, the F statistic has 1 and 12 df. From the F tables we find that

$$p = P(F \geq 1,398.7) < 0.0001$$

is tiny. Thus, there is considerable evidence in the data against the hypothesis $\beta_1 = 0$. We can safely reject $\beta_1 = 0$. There is a strong relationship between hardness and temperature.

Note

$$f_0 = \frac{\text{SSR}}{s^2} = \frac{\hat{\beta}_1^2 \sum(x_i - \bar{x})^2}{s^2} = \left[\frac{\hat{\beta}_1}{s/s_{xx}}\right]^2 = t_0^2 \tag{2.36}$$

where t_0 is the t ratio in Eq. (2.20) in Section 2.5. The F statistic obtained here is the square of the t statistic that was used earlier for testing $\beta_1 = 0$. We know that in general the square of a $t(n-2)$ random variable follows an $F(1, n-2)$ distribution; see Exercise 2.2. Hence, the F test discussed here and the t ratio discussed earlier represent two equivalent tests of the null hypothesis $\beta_1 = 0$.

The coefficient of determination in this example is $R^2 = 3,126.134/3,152.954 = 0.991$. Thus, 99.1% of the total sum of squares is explained by the regression model. Temperature is an important predictor of hardness.

2.8 ANOTHER EXAMPLE

This example addresses the variation in achievement test scores among Iowa elementary schools. The test scores are the average "core" scores from the Iowa Test of Basic Skills, a commonly used standardized test for elementary schools in Iowa. The core score includes vocabulary, reading comprehension, spelling, capitalization, punctuation, usage, expression, and math concepts and problems. The data set in Table 2.2 contains the average fourth-grade tests scores of all elementary schools in the six largest Iowa communities.

Average test scores vary from school to school. The average test score of a school depends on, among other factors, the natural ability of the children attending the school, the quality of the educational programs offered by the school, and the support the children get from their parents. These explanatory factors tend to be related to the economic situation that surrounds each school. Causal links are complicated, but one can expect that richer communities tend to have more resources, children of economically secure parents have more opportunities, and well-to-do parents select their residences such that their children can go to "better" schools. We use the percentage of students in the federal free and reduced-price breakfast and lunch program as an economic indicator; it serves as a proxy for "poverty." Students qualify for this program if they come from families with incomes at or below 130% of the poverty level. The information on $n = 133$ schools is taken from an article in the *Des Moines Register*, November 2000. Poverty and test scores are from the 1999–2000 school year.

A scatter plot of test scores against poverty is shown in Figure 2.6. One notices that average test scores of schools with a high percentage of children in the subsidized lunch program are considerably lower than those of schools with small percentages. The relationship between test scores and the proportion of children on subsidized lunch is roughly linear, which leads us to the simple linear regression model, $y = \beta_0 + \beta_1 x + \epsilon$, that we study in this chapter.

TABLE 2.2 IOWA TEST OF BASIC SKILLS DATA [DATA FILE: iowatest]

School	Poverty	Test Scores	City
Coralville Cen.	20	65	Iowa City
Hills	42	35	Iowa City
Hoover	10	84	Iowa City
Horn	5	83	Iowa City
Kirkwood	34	49	Iowa City
Lemme	17	69	Iowa City
Lincoln	3	88	Iowa City
Longfellow	24	63	Iowa City
Lucas	21	65	Iowa City
Mann	34	58	Iowa City
Penn	24	52	Iowa City
Roosevelt	35	61	Iowa City
Shimek	4	81	Iowa City
Twain	57	43	Iowa City
Weber	24	66	Iowa City
Wickham	10	62	Iowa City
Wood	31	65	Iowa City
Black Hawk	35	46	Waterloo
Edison	62	41	Waterloo
Elk Run	56	48	Waterloo
Grant	81	36	Waterloo
Irving	45	52	Waterloo
Jewett	50	44	Waterloo
Kingsley	15	76	Waterloo
Kittrell	40	48	Waterloo
Lincoln	74	30	Waterloo
Longfellow	99	27	Waterloo
Lowell	82	28	Waterloo
McKinstry	81	20	Waterloo
Orange	38	56	Waterloo
Roosevelt	80	23	Waterloo
Arthur	13	75	Cedar Rapids
Cleveland	27	55	Cedar Rapids
Coolidge	10	72	Cedar Rapids
Erskine	25	67	Cedar Rapids
Garfield	39	46	Cedar Rapids
Grant Wood	44	55	Cedar Rapids
Harrison	55	35	Cedar Rapids
Hiawatha	27	56	Cedar Rapids
Hoover	30	66	Cedar Rapids
Jackson	7	69	Cedar Rapids
Johnson	59	51	Cedar Rapids
Kenwood	41	75	Cedar Rapids
Madison	16	70	Cedar Rapids
Nixon	21	62	Cedar Rapids
Pierce	3	75	Cedar Rapids
Polk	80	54	Cedar Rapids
Taylor	78	36	Cedar Rapids

(*Continued*)

TABLE 2.2 (Continued)

School	Poverty	Test Scores	City
Truman	10	57	Cedar Rapids
Van Buren	52	43	Cedar Rapids
Wilson	39	41	Cedar Rapids
Wright	27	53	Cedar Rapids
Adams	17	52	Davenport
Blue Grass	9	53	Davenport
Buchanan	57	37	Davenport
Buffalo	31	43	Davenport
Eisenhower	40	58	Davenport
Fillmore	57	39	Davenport
Garfield	49	43	Davenport
Grant	38	47	Davenport
Harrison	22	56	Davenport
Hayes	61	30	Davenport
Jackson	58	34	Davenport
Jefferson	89	21	Davenport
Johnson	53	40	Davenport
Lincoln	59	56	Davenport
Madison	87	29	Davenport
McKinley	50	49	Davenport
Monroe	73	36	Davenport
Perry	51	20	Davenport
Truman	40	48	Davenport
Walcott	23	59	Davenport
Washington	71	38	Davenport
Wilson	39	53	Davenport
Adams	50	58	Des Moines
Brooks/Lucas	79	32	Des Moines
Cattell	49	50	Des Moines
Douglas	37	54	Des Moines
Edmunds	77	28	Des Moines
Findley	61	51	Des Moines
Garton	55	27	Des Moines
Granger	47	49	Des Moines
Greenwood	32	67	Des Moines
Hanawalt	12	79	Des Moines
Hills	31	57	Des Moines
Howe	50	50	Des Moines
Hubbell	22	81	Des Moines
Jackson	40	45	Des Moines
Jefferson	3	74	Des Moines
Longfellow	80	50	Des Moines
Lovejoy	62	40	Des Moines
Madison	52	45	Des Moines
Mann	65	32	Des Moines
McKee	57	31	Des Moines
McKinley	78	36	Des Moines
Mitchell	54	46	Des Moines

TABLE 2.2 (Continued)

School	Poverty	Test Scores	City
Monroe	45	53	Des Moines
Moore	40	53	Des Moines
Moulton	83	30	Des Moines
Oak Park	52	49	Des Moines
Park Avenue	42	36	Des Moines
Perkins	65	51	Des Moines
Phillips	29	61	Des Moines
Pleasant Hill	17	68	Des Moines
Stowe	53	47	Des Moines
Strudebaker	25	53	Des Moines
Wallace	77	24	Des Moines
Watrous	39	47	Des Moines
Willard	84	42	Des Moines
Windsor	32	62	Des Moines
Woodlawn	35	59	Des Moines
Wright	28	60	Des Moines
Bryant	32	56	Sioux City
Clark	4	78	Sioux City
Crescent Park	49	65	Sioux City
Emerson	53	40	Sioux City
Everett	79	48	Sioux City
Grant	50	45	Sioux City
Hunt	72	43	Sioux City
Irving	86	27	Sioux City
Joy	33	65	Sioux City
Leeds	46	42	Sioux City
Lincoln	14	76	Sioux City
Longfellew	34	40	Sioux City
Lowell	54	57	Sioux City
McKinley	84	37	Sioux City
Nodland	10	74	Sioux City
Riverview	60	59	Sioux City
Roosevelt	48	50	Sioux City
Smith	72	39	Sioux City
Sunnyside	14	73	Sioux City
Washington	20	57	Sioux City
Whittier	39	48	Sioux City

In a subsequent chapter, we will use this data set to illustrate model checking. The question whether the model can be improved by adding a quadratic component of poverty will be addressed in Exercise 5.17. In addition, we will investigate whether "city" information adds explanatory power. It may be that irrespective of the proportion of children on subsidized lunch, students in college communities (e.g., Iowa City) score higher. If that were true, one would need to look for plausible explanations.

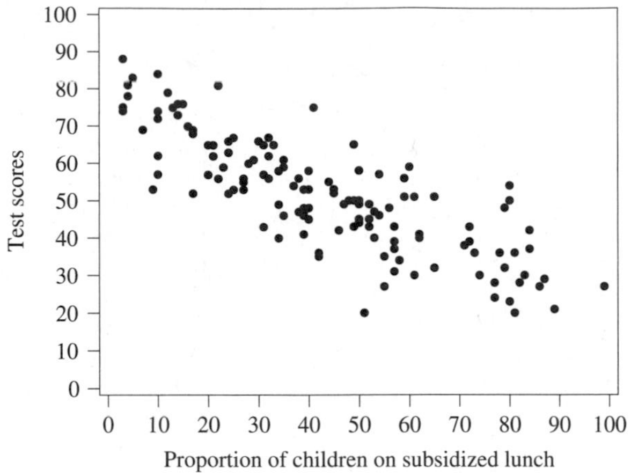

FIGURE 2.6
Scatterplot of Tests Scores Against Proportion of Children on Subsidized Lunch

TABLE 2.3 MINITAB OUTPUT OF TEST SCORES AGAINST PROPORTION OF CHILDREN ON SUBSIDIZED LUNCH

Test Scores = 74.6 −0.536 Poverty

Predictor	Coef	SE Coef	*T*	*P*
Constant	74.606	1.613	46.25	0.000
Poverty	−0.53578	0.03262	−16.43	0.000
$s = 8.766$	$R^2 = 67.3\%$			

Analysis of Variance

Source	DF	SS	MS	*F*	*P*
Regression	1	20731	20731	269.79	0.000
Residual Error	131	10066	77		
Total	132	30798			

The output from a standard regression program is listed below. Most computer packages, such as Minitab, SPSS, and SAS, provide very similar output. In Table 2.3, we show the output from Minitab, a popular statistics software.

The fitted regression equation,

$$\text{Test scores} = 74.6 - 0.536 \text{ poverty} \tag{2.37}$$

implies that with each additional unit (1%) increase in the proportion on subsidized lunch, average test scores decrease by 0.54 points.

The LSEs $\hat{\beta}_0 = 74.606$ and $\hat{\beta}_1 = -0.536$, their standard errors s.e.$(\hat{\beta}_0) = 1.613$ and s.e.$(\hat{\beta}_1) = 0.033$, and the t ratios $74.606/1.613 = 46.25$ and $-0.536/0.033 = -16.43$ are shown in the columns labeled Coef, SE Coef, and T. The last column labeled "P" contains the probability value of the regression coefficients. The t ratio for the slope β_1 is -16.43. It leads to a very small probability value. Under the null hypothesis of no relationship ($\beta_1 = 0$), there is almost no chance to get such an extreme value. Hence, we reject—very soundly—the null hypothesis

$\beta_1 = 0$. Yes, there is a very strong, and negative, association among test scores and the proportion of children on subsidized lunch.

The sum of squares, the degrees of freedom ($n - 1 = 133 - 1 = 132$ for total, $n - 2 = 133 - 2 = 131$ for error, and 1 for regression), and the mean squares are shown in the ANOVA table. The R^2 from the regression is 67.3%. We obtain this value by dividing the regression sum of squares SSR = 20,731 by the total sum of squares SST = 30,798. It says that 67.3% of the variation in average test scores can be explained through the linear association with poverty.

Another interpretation of "model fit" focuses on standard deviations. The standard deviation of the test scores, not keeping track of poverty, is given by $s_y = [\text{SST}/(n-1)]^{1/2} = [30{,}798/132]^{1/2} = 15.275$. After factoring in (or adjusting the analysis for) poverty, the standard deviation of the yet unexplained deviations is given by $s = [\text{SSE}/(n-2)]^{1/2} = [10{,}066/131]^{1/2} = 8.766$. This is the square root of the MSE. The reduction from $s_y = 15.275$ to $s = 8.766$ is 42.6%.

The last column of the ANOVA table contains the F ratio, $F = (\text{SSR}/1)/(\text{MSE}) = 269.79$. It serves as a test of the null hypothesis $\beta_1 = 0$. The probability value to the right of this number is the probability that an $F(1, 131)$ random variable exceeds this value. The probability is virtually zero, implying a solid rejection of the null hypothesis. Note that the F test in the simple linear regression model is equivalent to the test that looks at the t ratio. The square of the t ratio, $(-16.43)^2 = 269.79$, is identical to the F ratio.

2.9 RELATED MODELS

2.9.1 REGRESSION THROUGH THE ORIGIN

Theory or the patterns in the scatter plot may imply that the straight line should pass through the origin. Theory may suggest that the relationship between y and x is strictly proportional, implying that the line in the model $y_i = \beta x_i + \epsilon_i$ passes through the origin. The slope coefficient β can be estimated by minimizing the sum of squares

$$S(\beta) = \sum (y_i - \beta x_i)^2 \tag{2.38}$$

The minimization leads to

$$\hat{\beta} = \sum x_i y_i \Big/ \sum x_i^2 \tag{2.39}$$

and the residuals $y_i - \hat{\mu}_i = y_i - \hat{\beta} x_i$. The LSE of σ^2 is

$$s^2 = \sum (y_i - \hat{\beta} x_i)^2/(n-1) \tag{2.40}$$

Note that we divide by $(n - 1)$ degrees of freedom, because there is only one restriction among the residuals, $\sum e_i x_i = 0$.

One can show that $\hat{\beta}$ is unbiased [i.e., $E(\hat{\beta}) = \beta$] and that its variance is given by

$$V(\hat{\beta}) \equiv V\left[\frac{\sum x_i y_i}{\sum x_i^2}\right] = \frac{V(\sum x_i y_i)}{[\sum x_i^2]^2} = \frac{\sigma^2 \sum x_i^2}{[\sum x_i^2]^2} = \frac{\sigma^2}{\sum x_i^2} \tag{2.41}$$

Inference about β is similar to the one in the model with intercept, $y_i = \beta_0 + \beta_1 x_i + \epsilon_i$, except that now s^2 has $(n-1)$ rather than $(n-2)$ degrees of freedom.

A $100(1-\alpha)$ percent confidence interval for β is given by

$$\hat{\beta} \pm t\left(1 - \frac{\alpha}{2}; n-1\right) \text{s.e.}(\hat{\beta}) \tag{2.42}$$

where $\text{s.e.}(\hat{\beta}) = s/\sqrt{\sum x_i^2}$.

A $100(1-\alpha)$ percent confidence interval for the mean response at x_0, $\mu_0 = \beta x_0$, is

$$\hat{\mu}_0 \pm t\left(1 - \frac{\alpha}{2}; n-1\right) \text{ s.e. } (\hat{\mu}_0) \tag{2.43}$$

where

$$\hat{\mu}_0 = \hat{\beta} x_0 \text{ and s.e. } (\hat{\mu}_0) = s|x_0|/\sqrt{\sum x_i^2}$$

A $100(1-\alpha)$ percent prediction interval for a future observation y_p at x_p is

$$\hat{y}_p \pm t\left(1 - \frac{\alpha}{2}; n-1\right) s\sqrt{1 + \left(x_p^2/\sum x_i^2\right)} \tag{2.44}$$

where $\hat{y}_p = \hat{\mu}_p = \hat{\beta} x_p$.

2.9.2 THE CASE OF RANDOM x'S

In our discussions so far we have assumed that the x's are fixed constants. Thus, our inferences are based on repeated sampling, when the x's are kept the same from sample to sample.

Frequently, this assumption is not appropriate since fixed x's may not be possible. It may be preferable to consider both y and x as random variables having some joint distribution. Do the results of the previous sections still hold true in this situation?

We assume that y and x are jointly distributed as a bivariate normal distribution

$$f(y, x) = \frac{1}{2\pi\sigma_x\sigma_y\sqrt{1-\rho^2}} \exp\left\{-\frac{1}{(1-\rho^2)} Q\right\} \tag{2.45}$$

where

$$Q = \left(\frac{y-\mu_y}{\sigma_y}\right)^2 + \left(\frac{x-\mu_x}{\sigma_x}\right)^2 - 2\rho\left(\frac{y-\mu_y}{\sigma_y}\right)\left(\frac{x-\mu_x}{\sigma_x}\right)$$

Here, $\mu_y = E(y)$, $\mu_x = E(x)$ are the means; $\sigma_y^2 = V(y)$, $\sigma_x^2 = V(x)$ are the variances, and $\rho = E(y-\mu_y)(x-\mu_x)/\sigma_x\sigma_y$ is the correlation between y and x.

It can be shown that the conditional distribution of y given x is also normal with conditional mean $E(y|x) = \beta_0 + \beta_1 x$ and conditional variance $V(y|x) = \sigma_y^2(1-\rho^2)$. The regression coefficients β_0 and β_1 are related to the parameters of the bivariate normal distribution: $\beta_1 = (\sigma_y/\sigma_x)\rho$ and $\beta_0 = \mu_y - \rho(\sigma_y/\sigma_x)\mu_x$. Zero correlation ($\rho = 0$) implies $\beta_1 = 0$; then there is no linear association between y and x.

In this more general setup, one can also show that the maximum likelihood estimates of β_0 and β_1 are given by

$$\hat{\beta}_0 = \bar{y} - \hat{\beta}_1\bar{x}, \qquad \hat{\beta}_1 = s_{xy}/s_{xx}$$

which are exactly the previous estimates. Furthermore, $\hat{\rho} = s_{xy}/\sqrt{s_{xx}s_{yy}} = \hat{\beta}_1\sqrt{\dfrac{s_{xx}}{s_{yy}}}$. Hence, the regression model in which y and x are jointly normally distributed can be analyzed using the methods that treat x as fixed.

APPENDIX: UNIVARIATE DISTRIBUTIONS

1. THE NORMAL DISTRIBUTION

We say Y is a normal random variable if the density function of Y is given by

$$f(y) = \frac{1}{\sqrt{2\pi}\sigma}\exp\left\{-\frac{1}{2\sigma^2}(y-\mu)^2\right\}, \qquad -\infty < y < \infty$$

We use the notation $Y \sim N(\mu, \sigma^2)$. Note that $E(Y) = \mu$ and $V(Y) = \sigma^2$. The density function of the standard normal distribution with mean 0 and variance 1 is shown in Figure 2.7a.

To calculate probabilities for Y, we use the representation $Y = \mu + \sigma Z$, where Z is the standard normal distribution. Hence,

$$P(Y \le y) = P(\mu + \sigma Z \le y) = P\left(Z \le \frac{y-\mu}{\sigma}\right)$$

Table A at the end of the book gives the cumulative probabilities $P(Z \le z)$ for the standard normal distribution. For example, $P(Z \le 0) = 0.5$, $P(Z \le 1) = 0.8413$, $P(Z \le -0.85) = 0.1977$. The 97.5th percentile of the standard normal is 1.96; the 95th percentile is 1.645.

2. THE χ^2 DISTRIBUTION

We say that U follows a chi-square distribution with v degrees of freedom if the density function of U is given by

$$f(u) = cu^{(v/2)-1}e^{-u/2} \quad \text{for} \quad u \ge 0$$

c is a constant that makes the density integrate to 1. The parameter v is called the degrees of freedom. We write $U \sim \chi^2_v$. The density functions of three chi-square distributions ($v = 3, 6, 10$) are shown in Figure 2.7b. The mean of the chi-square distribution is given by $E(U) = v$, and the variance is given by $V(U) = 2v$. Table B (at the end of the book) gives selected percentiles. For example, in a χ^2_5 distribution, the 50, 95, and 99th percentiles are 4.3515, 11.0705, and 15.0863, respectively.

FIGURE 2.7
Densities of
(a) Standard Normal Distribution,
(b) Three Chi-Square Distributions,
(c) Standard Normal and Two *t* Distributions, and
(d) Four *F* Distributions

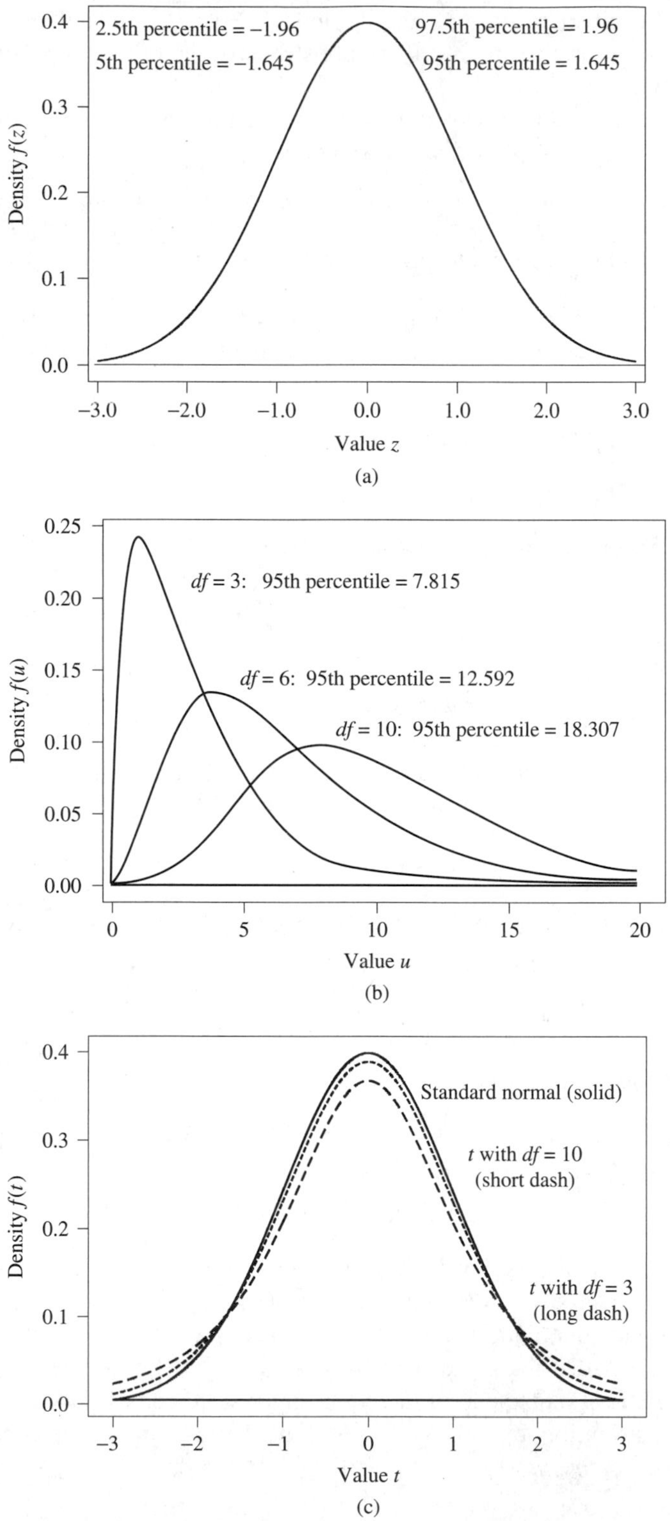

FIGURE 2.7
(Continued)

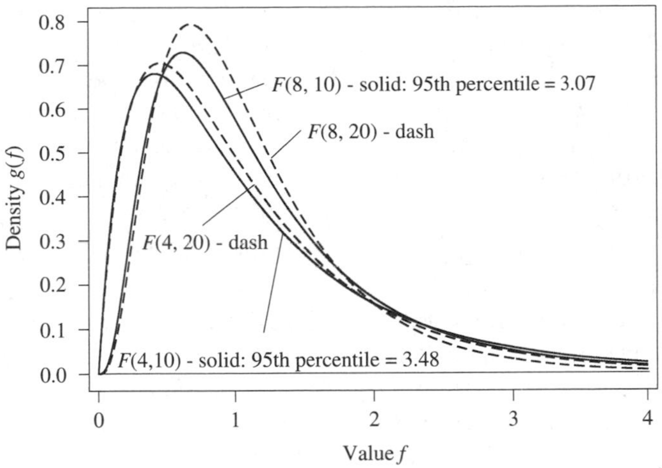

(d)

Suppose $Z_1, \ldots, Z_v$ are independent $N(0, 1)$ variables. Then Z_1^2 has a χ^2 distribution with 1 degree of freedom, and $U = Z_1^2 + Z_2^2 + \cdots + Z_v^2$ has a χ^2 distribution with v degrees of freedom.

3. THE STUDENT *t* DISTRIBUTION

We say that T follows a Student t distribution if the density function of T is given by

$$f(t) = \frac{c}{\left(1 + \frac{t^2}{v}\right)^{(v+1)/2}}, \qquad -\infty < t < \infty$$

c is a constant that makes the density integrate to 1. We write $T \sim t(v)$. The parameter v is called the degrees of freedom. The density functions of two t distributions with $v = 3$ and $v = 10$ degrees of freedom are shown in Figure 2.7c. The t distribution is symmetric with mean $E(T) = 0$ and variance $V(T) = v/(v-2)$ and is similar to the standard normal distribution, but with slightly heavier tails. As can be seen from Figure 2.7c, the t distribution is close to the standard normal distribution when the degrees of freedom (v) is large. Table C (at the end of the book) gives selected percentiles of Student t distributions. For example, for $v = 10$ d.f., the 90th percentile is 1.372, and the 5th percentile is -1.812.

Suppose $Z \sim N(0, 1)$ and $U \sim \chi_v^2$, with Z and U independent. Then $T = Z/\sqrt{U/v}$ has a Student t distribution with v degrees of freedom.

4. THE *F* DISTRIBUTION

We say that F follows an F distribution if the density function is given by

$$g(f) = c\frac{f^{(v/2)-1}}{\left(1 + \frac{v}{w}f\right)^{(v+w)/2}}, \qquad f \geq 0$$

c is a constant that makes the density integrate to 1. We write $F \sim F(v, w)$. The integer parameters v and w are called the degrees of freedom. The density functions of four F distributions are shown in Figure 2.7d. The mean of an F distribution is given by $E(F) = w/(w-2)$; it depends only on the second degrees of freedom and is slightly larger than 1. The variance depends on both v and w. Table D (at the end of the book) gives selected percentiles. For example, the 95 and 99th percentiles of an $F(5,8)$ distribution are given by 3.69 and 6.63, respectively.

Suppose $U \sim \chi_v^2$ and $V \sim \chi_w^2$, with U and V independent. Then $F = \frac{U/v}{V/w}$ has an F distribution with v and w degrees of freedom.

A comment on statistical tables: Most computer programs calculate cumulative probabilities and percentiles (the "inverse" of the cumulative probabilities) for a wide selection of distributions. For some programs (such as EXCEL) the calculation of percentiles requires the specification of the upper tail area.

EXERCISES

2.1. Determine the 95 and 99th percentiles of

a. The normal distribution with mean 10 and standard deviation 3;

b. The t distributions with 10 and 25 degrees of freedom;

c. The chi-square distributions with 1, 4, and 10 degrees of freedom;

d. The F distributions with 2 and 10, and 4 and 10 degrees of freedom.

2.2. It is a fact that two distributions are the same if (all) their percentiles are identical.

a. Convince yourself, by looking up several percentiles, that the square of a standard normal distribution is the same as a chi-square distribution with one degree of freedom. Determine the percentile of the χ_1^2 and the percentile of the square of a standard normal distribution, Z^2, and show that they are the same. Use the fact that $P(Z^2 \leq z) = P(-\sqrt{z} \leq Z \leq \sqrt{z})$. Hence, for example, the 95th percentile of Z^2 is the same as the 97.5th percentile of Z.

b. Convince yourself, by looking up several percentiles, that the square of a t distribution with v degrees of freedom is the same as the $F(1, v)$ distribution.

2.3. For each of the four sets of data given below (see Anscombe, 1973), plot y versus x. The data are given in the file **anscombe**. Fit a straight line model to each of the data sets giving least squares estimates, ANOVA table, and R^2. Compute the correlation coefficient between y and x for each data set. Comment on your results. Would a linear regression of y on x be appropriate in all cases? Discuss.

Set 1		Set 2		Set 3		Set 4	
x	y	x	y	x	y	x	y
4	4.26	4	3.10	4	5.39	8	6.58
5	5.68	5	4.74	5	5.73	8	5.76
6	7.24	6	6.13	6	6.08	8	7.71
7	4.82	7	7.26	7	6.42	8	8.84
8	6.95	8	8.14	8	6.77	8	8.47
9	8.81	9	8.77	9	7.11	8	7.04
10	8.04	10	9.14	10	7.46	8	5.25
11	8.33	11	9.26	11	7.81	8	5.56
12	10.84	12	9.13	12	8.15	8	7.91
13	7.58	13	8.74	13	12.74	8	6.89
14	9.96	14	8.10	14	8.84	19	12.50

2.4. A car dealer is interested in modeling the relationship between the weekly number of cars sold and the daily average number of salespeople who work on the showroom floor during that week. The dealer believes that the relationship between the two variables can be described by a straight line. The following data were supplied by the car dealer:

Week of	No. of Cars Sold y	Average. No. of Sales People on Duty x
January 30	20	6
June 29	18	6
March 2	10	4
October 26	6	2
February 7	11	3

a. Construct a scatter plot (y vs x) for the data.

b. Assuming that the relationship between the variables is described by a straight line, use the method of least squares to estimate the y intercept and the slope of the line.

c. Plot the least squares line on your scatter plot.

d. According to your least squares line, approximately how many cars should the dealer expect to sell in a week if an average of five salespeople are kept on the showroom floor each day?

e. Calculate the fitted value $\hat{\mu}$ for each observed x value. Use the fitted values to calculate the corresponding residuals. Plot the residuals against the fitted values. Are you satisfied with the fit?

f. Calculate an estimate of σ^2.

g. Construct a 95% confidence interval for β_1 and use it to assess the hypothesis that $\beta_1 = 0$.

h. Given the results of (a)–(g), what conclusions are you prepared to draw about the relationship between sales and number of salespeople on duty.

i. Would you be willing to use this model to help determine the number of salespeople to have on duty next year?

2.5. Use S-Plus or any other available statistics software for Exercise 2.4. Check your hand calculations with the results from these programs.

2.6. Dr. Joseph Hooker collected a set of 31 measurements on the boiling temperature of water (TEMP; in degrees Fahrenheit) and the atmospheric pressure (AP; in inches of mercury) at various locations in the Himalaya Mountains (see Weisberg, 1980). The data are given in the file **hooker**.

a. Plot TEMP vs AP. Does a linear model seem appropriate?

b. Repeat (a), plotting TEMP versus $x = 100 ln(AP)$.

c. Fit a linear model

$$\text{TEMP} = \beta_0 + \beta_1 x + \epsilon$$

and calculate the estimates of β_0, β_1, and σ^2. Draw the fitted line $\hat{y} = \hat{\beta}_0 + \hat{\beta}_1 x$ on the plot in (b). Does the model seem appropriate?

d. Find a 95% confidence interval for

i. β_1;

ii. the average temperature when the pressure is 25.

e. Suppose the temperature had been measured in °C instead of °F. Explain (think about it, but don't compute) how the estimates in (c) and the confidence intervals in (d) would change.

2.7. a. Consider the model

$$y_i = \beta + \epsilon_i,\ E(\epsilon_i) = 0,\ V(\epsilon_i) = \sigma^2,$$

$$\text{Cov}(\epsilon_i, \epsilon_j) = 0 \text{ for } i \neq j,\ i = 1, 2, \ldots, n$$

Find the LSEs of β and σ^2.

b. Discuss the following statements:

i. For the linear model

$$y_i = \beta_0 + \beta_1 x_i + \epsilon_i,\ i = 1, 2, \ldots, n$$

a 95% confidence interval for $\mu_k = \beta_0 + \beta_1 x_k$ is narrower than a 95% prediction interval for y_k.

ii. For the linear model
$y_i = \beta_0 + \beta_1 x_i + \epsilon_i, i = 1, 2, \ldots, n$
a 99% prediction interval for y_k is wider than a 95% prediction interval for y_k.

iii. For a certain regression situation it is reported that SST = 25, SSR = 30, and SSE = −5. These calculations are correct since SST = SSR + SSE.

2.8. Consider the annual number of cars sold and the revenues of the 10 largest car companies:

Company	Cars Sold (Millions)	Revenues (in Million Euros)
General Motors	8,149	1,996
Ford/Volvo	7,316	2,118
Renault/Nissan	4,778	1,174
Volkswagen	4,580	943
DaimlerChrysler	4,506	1,813
Toyota	4,454	1,175
Fiat	2,535	628
Honda	2,291	605
PSA	2,278	465
BMW	1,187	447

Consider the results of a simple linear regression model of y = revenues on x = sales:

a. Test whether the number of cars sold is an important predictor variable (use significance level 0.05).

b. Calculate a 95% confidence interval for the regression coefficient of number of cars sold.

c. Calculate a 90% confidence interval for the regression coefficient of number of cars sold.

d. Obtain the coefficient of determination.

e. Determine the standard deviation among revenues (y), after factoring in the explanatory variable sales (x). Compare this standard deviation to the standard deviation of y without considering the explanatory variable.

f. Estimate the revenues for BMW.

2.9. Grade point averages of 12 graduating MBA students, GPA, and their GMAT scores taken before entering the MBA program are given below. Use the GMAT scores as a predictor of GPA, and conduct a regression of GPA on GMAT scores.

x = GMAT	y = GPA
560	3.20
540	3.44
520	3.70
580	3.10
520	3.00
620	4.00
660	3.38
630	3.83
550	2.67
550	2.75
600	2.33
537	3.75

a. Obtain and interpret the coefficient of determination R^2.

b. Calculate the fitted value for the second person.

c. Test whether GMAT is an important predictor variable (use significance level 0.05).

2.10. The following are the results of a regression of fuel efficiency (gallons per 100 miles traveled) on the weight (in pounds) of the car. A total of 45 cars were considered.

The regression equation is

Gall/100 miles = 0.560 + 0.00102 Weight

Predictor	Coef	SE Coef	t	p
Constant	0.5598	0.1983	2.82	0.007
Weight	0.00102418	0.00007103	14.42	0.000

$R^2 = 82.9\%$ $R^2(\text{adj}) = 82.5\%$

Analysis of Variance

Source	DF	SS	MS	F	p
Regression	1	13.709	13.709	207.91	0.000
Error	43	2.835	0.066		
Total	44	16.544			

a. Determine an approximate 95% prediction interval for the fuel efficiency of an automobile weighing 2000 pounds. The computer output does not give you the information to construct exact prediction intervals. Approximate the prediction intervals, assuming that the sample size n is large enough to allow you to ignore the parameter estimation uncertainty.

b. Determine an approximate 95% prediction interval for the fuel efficiency of an automobile weighing 1500 pounds.

2.11. Discuss the functional relationship between the coefficient of determination R^2 and the F ratio.

2.12. Occasionally, a model is considered in which the intercept is known to be zero a priori. Such a model is given by

$$y_i = \beta_1 x_i + \epsilon_i, i = 1, 2, \ldots, n$$

where the errors ϵ_i follow the usual assumptions.

a. Obtain the LSEs $(\hat{\beta}_1, s^2)$ of (β_1, σ^2).

b. Define $e_i = y_i - \hat{\beta}_1 x_i$. Is it still true that $\sum_{i=1}^{n} e_i = 0$? Why or why not?

c. Show that $V(\hat{\beta}_1) = \sigma^2 / \sum_{i=1}^{n} x_i^2$.

2.13. The data listed in the file **sriver** include the water content of snow on April 1 (x) and the water yield from April to July (y) in the Snake River watershed in Wyoming. Information on $n = 17$ years (from 1919 to 1935) is listed (see Weisberg, 1980).

a. Fit a regression through the origin ($y = \beta_1 x + \epsilon$), and find $\hat{\beta}_1$ and s^2. Obtain a 95% confidence interval for β_1.

b. A more general model for the data includes an intercept,

$$y = \beta_0 + \beta_1 x + \epsilon.$$

Is there convincing evidence that suggests that the simpler model in (a) is an appropriate representation?

2.14. Often, researchers need to calibrate measurement processes. For that they use a set of known x's to obtain observed y's, then fit a model called the calibration model and use this model to convert future measured y's back into the corresponding x's.

The following is an example taken from analytical chemistry where the process is the assay of the element calcium. Determining calcium in the presence of other elements is quite tricky. The following table records the quantities of calcium in carefully prepared solutions (x) and the corresponding analytical results (y):

x	4	8	12.5	16	20	25	31	36	40	40
y	3.7	7.8	12.1	15.6	19.8	24.5	31.1	35.5	39.4	39.5

a. Fit a simple linear regression of y as a function of x. List the assumptions that you make.

b. Calculate a 95% confidence interval for the intercept of your model.

c. Calculate a 95% confidence interval for the slope of your model.

d. In this context two properties may be expected:

i. When $x = 0$, then $y = 0$; if there is no calcium present, your technique should not find any.

ii. If the empirical technique is any good at all, then the slope in the simple linear regression should be 1.

Is there evidence for (i)? For (ii)?

e. If you accept (i) as a condition to be imposed on the model a priori, then the model reduces to

$$y = \beta x + \epsilon$$

Redo part (c) and reexamine property (ii) for your new model.

f. Explain why the results in (d) and (e) are different.

2.15. The following data give the monthly machine maintenance cost (y) in hundreds of dollars and the number of machine hours per month (x), taken over the last 7 months.

Cost (y)	Hours (x)
26	110
25	98
20	121
18	116
30	90
40	88
30	84

a. Fit a linear regression. Construct the ANOVA table. Find R^2 and test the hypothesis that $\beta_1 = 0$ using the F ratio.

b. Obtain the standard errors of β_0 and β_1. Using the t distribution, test the hypothesis: (i) $\beta_0 = 0$; (ii) $\beta_1 = 0$. Construct a 99% confidence interval for β_1.

c. Find the fitted value $\hat{\mu}$ at $x = 100$ and estimate its standard error. Calculate the 95% confidence interval for $\beta_0 + \beta_1 100$.

d. Repeat (c), only this time take $x = 84$. Explain the change in the interval length.

2.16. A company builds custom electronic instruments and computer components. All jobs are manufactured to customer specifications. The firm wants to be able to estimate its overhead cost. As part of a preliminary investigation, the firm decides to focus on a particular department and investigates the relationship between total departmental overhead cost (y) and total direct labor hours (x). The data for the most recent 16 months are given below. They are also given in the file **overhead**.

Month Number	Total Departmental Overhead (y)	Total Direct Labor Hours (x)
1	25,835	878
2	27,451	1,088
3	28,611	1,281
4	32,361	1,340
5	28,967	1,090
6	24,817	1,067
7	29,795	1,188
8	26,135	928
9	31,361	1,319
10	26,006	790
11	27,812	934
12	28,612	871
13	22,992	781
14	31,836	1,236
15	26,252	902
16	26,977	1,140

The two objectives of this investigation are

a. Summarize for management the relationship between total departmental overhead and total direct labor hours.

b. Estimate the expected and predict the actual total departmental overhead from the total direct labor hours.

Analyze the overhead data and write a brief paragraph for your manager that summarizes the results that you have obtained about the two objectives. Include the computer output that you think is necessary for clarification of your discussion.

2.17. The following data, in the file **turtles**, are measurements on length and width (both in mm) of 10 painted female turtles (*Chrysemys picta marginta*):

Length (y)	Width (x)
100	81
103	86
109	88
123	94
133	102
134	100
137	98
141	105
150	107
155	115

a. Plot y against x on a scatter plot. Comment on this plot.

b. Assuming the model $y = \beta_0 + \beta_1 x + \epsilon$, obtain the LSEs of the coefficients and their corresponding 95% confidence limits.

c. Graph the fitted line on the plot in (a). Is this a good fit? Explain.

d. Predict the length of a female turtle if it is 100 mm wide, and obtain the 95% prediction limits.

e. Is the linear relationship a strong or a weak one? Explain.

2.18. The following data, in the file **bloodpressure**, are measurements of systolic blood pressure (SBP) and age for a sample of 15 individuals older than age 40 years:

SBP (y)	Age (x)
164	65
220	63
133	47
146	54
162	60
144	44
166	59
152	64
140	51
145	49
135	57
150	56
170	63
122	41
120	43

a. Plot systolic blood pressure against age. Comment on the plot.

b. Assuming the model $y = \beta_0 + \beta_1 x + \epsilon$, obtain the fitted equation.

c. Construct an ANOVA table for the simple linear model.

d. Use the results from the ANOVA table and the F ratio to test for a significant linear relationship at the 5% level.

e. Test the hypothesis $H_0 : \beta_1 = 0$ at the 5% level using a t test. Does your conclusion agree with the finding in part (d)?

f. Do you think that the individual with $x = 63$ and $y = 220$ is an unusual observation? Why? Check if this observation is influential. Remove it from the data set and redo steps (b)–(e). The observation is influential if there are substantial changes in the resulting fit. Do you think that there are substantial changes? Explain.

2.19. An experiment was conducted to determine the extent to which the growth rate of a certain fungus can be affected by filling test tubes containing the same medium at the same temperature with different inert gases. Three such experiments were performed for each of six different gases, and the average growth rate from these three tests was used as the response. The following table gives the molecular weight (x) of each gas used and the average growth rate (y) in milliliters per hour:

Gas	Average Growth Rate (y)	Molecular Weight (x)
A	3.85	4.0
B	3.48	20.2
C	3.27	28.2
D	3.08	39.9
E	2.56	83.8
F	2.21	131.3

a. Find the LSEs of the slope and the intercept for the linear model $y = \beta_0 + \beta_1 x + \epsilon$, and draw the fitted line on the scatter plot.

b. Is there a significant linear relationship between y and x at the 1% level? Comment on the fit of the line.

c. What information has not been used that may improve the sensitivity of the analysis?

d. Would it be appropriate to use this fitted line to estimate the growth rate of the fungus for a gas with a molecular weight of 200? Explain.

2.20. An investigation involving five factors has singled out temperature as having the greatest impact on the accelerated lifetime of a special type of heater. On the advice of the process engineer, temperatures 1,520, 1,620, 1,660, and 1,708° F were chosen.

Twenty-four heaters were selected at random from the current production and split randomly among the four temperatures. The life times of these heaters are given below.

Temperature T	Lifetime y (Hours)					
1,520	1,953	2,135	2,471	4,727	6,143	6,314
1,620	1,190	1,286	1,550	2,125	2,557	2,845
1,660	651	837	848	1,038	1,361	1,543
1,708	511	651	651	652	688	729

a. Plot the data and summarize the important features of the relationship.

b. Transform the y's to $\text{LY} = \ln y$ and replot the data. Comment on the functional relationship.

c. Fit the model

$$\text{LY} = \beta_0 + \beta_1 T + \epsilon$$

i. Assess the fit by adding the fitted line to the scatter plot.

ii. If you are not satisfied with the fit, state why. What other approach might you take to get a better fitting model?

2.21. The data are taken from Roberts, H. V., and Ling, R. F. *Conversational Statistics with IDA*. New York: Scientific Press/McGraw-Hill, 1982.

The iron content of crushed blast furnace slag needs to be determined. Two methods are available. One involves a chemical analysis in the laboratory, which is time-consuming and expensive. The other is a much cheaper and quicker magnetic test that can be carried out on-site. Measurements on 53 consecutive slags are listed below. The data are given in the file **ironcontent**.

Graph the results of the chemical test for iron (y) against the magnetic test (x). Fit a simple linear regression. Calculate and interpret the coefficient of determination R^2. Investigate the extent to which the results of the chemical tests of iron content can be predicted from a magnetic test of iron content.

y = Chemical	x = Magnetic
24	25
16	22
24	17
18	21
18	20
10	13
14	16
16	14
18	19
20	10
21	23
20	20
21	19
15	15
16	16
15	16
17	12
19	15
16	15
15	15
15	15
13	17
24	18
22	16
21	18
24	22
15	20
20	21
20	21
25	21
27	25
22	22
20	18
24	21
24	18
23	20
29	25
27	20
23	18
19	19
25	16
15	16
16	16
27	26

$y =$ Chemical	$x =$ Magnetic
27	28
30	28
29	30
26	32
25	28
25	36
32	40
28	33
25	33

2.22. The data are taken from Mosteller, F., Rourke, R. E. K. and Thomas, G. B.: *Probability with Statistical Applications*, (2nd ed.). Reading, MA: Addison-Wesley, 1970.

Average percentage memory retention was measured against passing time (in minutes). The measurements were taken five times during the first hour after the experimental subjects memorized a list of disconnected items and then at various times up to 1 week later. The data given in the file **memory**.

Graph memory retention (y) against time (x). Consider transformations such as the logarithm of y and/or the logarithm of x. Estimate and check the appropriate regression models.

A model such as $y = \alpha \exp(-\beta \text{Time})$ indicates geometric loss of memory. Discuss whether this is an appropriate model or whether there are other models that are equally (or better) suited to describe the data.

$x =$ Time (Minutes)	$y =$ Memory Retention (%)
1	0.84
5	0.71
15	0.61
30	0.56
60 (1 hour)	0.54
120	0.47
240	0.45
480	0.38
720	0.36
1,440	0.26
2,880	0.20
5,760	0.16
10,080	0.08

2.23. The data are taken from Gilchrist, W. *Statistical Modelling*. Chichester, UK: Wiley, 1984. These data give the distance by road and the straight line distance between 20 different pairs of points in Sheffield. The data are given in the file **distance**.

What is the relationship between the two variables? How well can you predict the road distance (y) from the linear distance (x)?

$x =$ Linear Distance	$y =$ Road Distance
9.5	10.7
5.0	6.5
23.0	29.4
15.2	17.2
11.4	18.4
11.8	19.7
12.1	16.6
22.0	29.0
28.2	40.5
12.1	14.2
9.8	11.7
19.0	25.6
14.6	16.3
8.3	9.5
21.6	28.8
26.5	31.2
4.8	6.5
21.7	25.7
18.0	26.5
28.0	33.1

2.24. The data are taken from Risebrough, R. W. Effects of environmental pollutants upon animals other than man. In *Proceedings of the 6th Berkeley Symposium on Mathematics and Statistics, VI*. Berkeley: University of California Press, 1972, pp. 443–463.

Polychlorinated biphenyl (PCB), an industrial pollutant, is thought to have harmful effects on the thickness of egg shells. The amount of PCB (in parts per million) and the thickness of the shell (in millimeters) of 65 Anacapa pelican eggs are given below. The data are also given in the file **pelicaneggs**.

Investigate the relationship between the thickness of the shell and the amount of PCB in pelican eggs. Construct a scatter plot and fit a linear regression model. Calculate a 95% confidence interval for the slope. Obtain the ANOVA table and the coefficient of determination R^2. Interpret the results and comment on the adequacy of the model.

x = Concentration of PCB	y = Thickness
452	0.14
139	0.21
166	0.23
175	0.24
260	0.26
204	0.28
138	0.29
316	0.29
396	0.30
46	0.31
218	0.34
173	0.36
220	0.37
147	0.39
216	0.42
216	0.46
206	0.49
184	0.19
177	0.22
246	0.23
296	0.25
188	0.26
89	0.28
198	0.29
122	0.30
250	0.30
256	0.31
261	0.34
132	0.36
212	0.37
171	0.40
164	0.42
199	0.46
115	0.20
214	0.22
177	0.23
205	0.25
208	0.26

x = Concentration of PCB	y = Thickness
320	0.28
191	0.29
305	0.30
230	0.30
204	0.32
143	0.35
175	0.36
119	0.39
216	0.41
185	0.42
236	0.47
315	0.20
356	0.22
289	0.23
324	0.26
109	0.27
265	0.29
193	0.29
203	0.30
214	0.30
150	0.34
229	0.35
236	0.37
144	0.39
232	0.41
87	0.44
237	0.49

2.25. The data are taken from Wallach, D., and Goffinet, B. Mean square error of prediction in models for studying ecological and agronomic systems. *Biometrics*, 43, 561–573, 1987.

The energy requirements (in Mcal/day) for a sample of 64 grazing merino sheep are given below, together with their body weights (kg). The data are given in the file **energyrequirement**. Construct a scatter plot and establish a model that explains the energy requirements as a linear function of body weight. Obtain a 95% confidence interval for the slope. Calculate and interpret the coefficient of determination R^2. Comment on the adequacy of the model. Discuss whether or not the variance of the measurements is constant across weight.

x = Weight	y = Energy Requirement
22.1	1.31
26.2	1.27
33.2	1.25
34.3	1.14
49.0	1.78
52.6	1.70
27.6	1.39
31.0	1.47
32.6	1.75
44.6	2.25
52.6	3.73
28.6	2.13
34.4	1.85
25.1	1.46
27.0	1.21
33.2	1.32
34.9	1.00
49.2	2.53
53.3	2.66
28.4	1.27
31.0	1.50
33.1	1.82
52.1	2.67
46.7	2.21
29.2	1.80
34.4	1.63
25.1	1.00
30.0	1.23
33.2	1.47
42.6	1.81
51.8	1.87
23.9	1.37
28.9	1.74
31.8	1.60
34.1	1.36
52.4	2.28
37.1	2.11
26.2	1.05
26.4	1.27
25.7	1.20
30.2	1.01
33.9	1.03
43.7	1.73
51.8	1.92

x = Weight	y = Energy Requirement
25.1	1.39
29.3	1.54
32.0	1.67
34.2	1.59
52.7	3.15
31.8	1.39
45.9	2.36
27.5	0.94
25.9	1.36
30.2	1.12
33.8	1.46
44.9	1.93
52.5	1.65
26.7	1.26
29.7	1.44
32.1	1.80
44.4	2.33
53.1	2.73
36.1	1.79
36.8	2.31

2.26. The data are taken from Atkinson, A. C. *Plots, Transformations, and Regression.* Oxford: Clarendon Press, 1985.

Here, we list 17 observations on the boiling point (°F) and the barometric pressure (in inches of mercury). The data are given in the file **boiling**. Relate boiling point to barometric pressure. Construct a scatter plot and establish a model that relates the boiling point to barometric pressure. Test the regression coefficients for their significance. Calculate and interpret the coefficient of determination R^2. Comment on the fit and the adequacy of the model.

Note that this exercise deals with the same problem as Exercise 2.6 but uses different data. Plot the data of the two exercises on the same graph, and add the two fitted regression lines that you found. Comment on the graph.

y = Boiling Point	x = Barometric Pressure
194.5	20.79
194.3	20.79
197.9	22.40
198.4	22.67

$y =$ Boiling Point	$x =$ Barometric Pressure
199.4	23.15
199.9	23.35
200.9	23.89
201.1	23.99
201.4	24.02
201.3	24.01
203.6	25.14
204.6	26.57
209.5	28.49
208.6	27.76
210.7	29.04
211.9	29.88
212.2	30.06

2.27. The data are taken from Bissell, A. F. Lines through the origin—IS NO INT the answer? *Journal of Applied Statistics*, 19, 193–210, 1992.

In a chemical process, batches of liquid are passed through a bed containing a certain ingredient. The ingredient gets absorbed by the liquid, and usually approximately 6–6.5% of the weight of the ingredient gets absorbed. In order to be sure that there is enough material, the bed is supplied with approximately 7.5% material. Excess material is costly and should be minimized because any excess cannot be recovered.

The interest is in the relationship between the material supplied (x) and the amount and/or the percentage of absorption. Develop appropriate regression models for both expressions of the response (in kg and as a percent) and comment on their fit and adequacy. The data are given in the file **absorption**.

$x =$ Liquid	$y =$ Take-up (kg)	$y =$ Take-up (%)
310	14.0	4.52
330	17.1	5.18
370	21.3	5.76
400	20.4	5.10
450	27.4	6.09
490	27.2	5.55
520	28.4	5.46
560	32.5	5.80
580	31.9	5.50
650	34.1	5.25
650	39.8	6.12
650	38.5	5.92
760	50.4	6.63
800	43.8	5.48
810	50.4	6.22
910	53.5	5.88
1,020	71.3	6.99
1,020	64.3	6.30
1,160	79.6	6.86
1,200	80.8	6.73
1,230	78.5	6.38
1,380	98.9	7.17
1,460	105.6	7.23
1,490	98.6	6.62

2.28. Search the Web for useful regression applets. Many such applets are available.

After entering points on a scatter plot, these applets calculate the least squares estimates, draw in the fitted regression line, and calculate summary statistics, such as the correlation coefficient or the coefficient of determination R^2. Applets allow you to change points and they illustrate the effect of such changes on the regression results.

Applets also illustrate the standard errors of the estimates. They take repeated samples of a certain size from a given population of points and for each sample they calculate an estimate of the regression slope. The results of repeated draws from the population are displayed in the form of histograms. This illustrates the sampling variability of the estimate.

Experiment with these applets and write a short discussion of what you can learn from them. Note that these applets are designed for the bivariate regression situation mostly because it is difficult to draw observations in higher dimensional space.

3 A Review of Matrix Algebra and Important Results on Random Vectors

In Chapter 2, we considered regression models that involve a single explanatory (regressor) variable. In the following chapters, we discuss regression models that contain several explanatory variables. Many of the results in Chapter 2 require tedious algebraic manipulations, and extending these results to more elaborate models would involve a considerable amount of complicated algebra. Much of the tedious work can be avoided if we perform the analysis with vectors and matrices. This chapter summarizes basic results on vectors and matrices that are needed in subsequent chapters, and we translate the regression results of Chapter 2 into the language of vectors and matrices. The chapter also includes a general discussion of random vectors and a summary of important results on the multivariate normal distribution.

3.1 REVIEW OF MATRIX ALGEBRA

Matrix A **matrix** A of dimension $p \times q$ is an array of pq elements, arranged in p rows and q columns. We write $A = (a_{ij})$, where a_{ij} is the entry in row i and column j of the matrix. A matrix is called **square**, if it has the same number of rows and columns ($p = q$). The **identity matrix** I is a square matrix with ones in the diagonal and zeros everyplace else. The zero matrix O is a matrix of all zeros.

Examples The 3×2 matrix $A = \begin{bmatrix} 2 & 5 \\ 1 & 3 \\ 1 & 2 \end{bmatrix}$; the 3×3 identity matrix $I = \begin{bmatrix} 1 & 0 & 0 \\ 0 & 1 & 0 \\ 0 & 0 & 1 \end{bmatrix}$; the 2×2 square matrix $B = \begin{bmatrix} 4 & 2 \\ 1 & 3 \end{bmatrix}$. ■

Vector A $p \times 1$ (column) **vector** $\boldsymbol{x} = (x_i)$ is a matrix consisting of a single column with elements $x_1, x_2, \ldots, x_p$. The unit vector, **1**, is the vector with all its elements equal to one. The zero vector, **0**, is the vector with all its elements equal to zero. A p-dimensional vector $\boldsymbol{x}$ with elements $x_1, x_2, \ldots, x_p$ can be represented geometrically in p-dimensional space as a directed line with component x_1 along the first axis, component x_2 along the second axis, ..., and component x_p along the pth axis.

Examples The 4×1 vector $\boldsymbol{x} = \begin{bmatrix} 3 \\ 2 \\ 5 \\ 1 \end{bmatrix}$; the 4×1 unit vector $\mathbf{1} = \begin{bmatrix} 1 \\ 1 \\ 1 \\ 1 \end{bmatrix}$; the 4×1 zero vector $\mathbf{0} = \begin{bmatrix} 0 \\ 0 \\ 0 \\ 0 \end{bmatrix}$. ■

Matrix/Vector Addition and Subtraction Let A and B be two matrices (or vectors) of the **same** dimension. Then $A + B = B + A = (a_{ij} + b_{ij})$, and $A - B = (a_{ij} - b_{ij})$. If c and d are two scalars, then $cA + dB = (ca_{ij} + db_{ij})$. Matrix addition/subtraction is defined elementwise.

Matrix/Vector Multiplication Let $A = (a_{ij})$ be a $p \times q$ matrix, and $B = (b_{ij})$ a $q \times t$ matrix. Then the matrix product $C = AB = (c_{ij})$ is a $(p \times t)$ matrix with elements $c_{ij} = \sum_{r=1}^{q} a_{ir} b_{rj}$.

Example The 3×2 matrix A and the 2×4 matrix B are given as

$$A = \begin{bmatrix} 2 & 5 \\ 1 & 3 \\ 1 & 2 \end{bmatrix}; B = \begin{bmatrix} 3 & 5 & -1 & 2 \\ -1 & 3 & 2 & 2 \end{bmatrix}.$$

The product is a 3×4 matrix given by

$$C = AB = \begin{bmatrix} 1 & 25 & 8 & 14 \\ 0 & 14 & 5 & 8 \\ 1 & 11 & 3 & 6 \end{bmatrix}$$ ■

Transpose of a Matrix Let $A = (a_{ij})$ be a $(p \times q)$ matrix. The transpose $A' = (a_{ji})$ is the $q \times p$ matrix obtained by interchanging rows and columns. Note that $(AB)' = B'A'$. A square matrix A is **symmetric** if $A = A'$.

The transpose of a $(p \times 1)$ column vector $\boldsymbol{x}$ is the $(1 \times p)$ row vector $\boldsymbol{x}'$. It is a matrix consisting of a single row.

Example The transpose of the 3×2 matrix A given previously is the 2×3 matrix $A' = \begin{bmatrix} 2 & 1 & 1 \\ 5 & 3 & 2 \end{bmatrix}$. The transpose of the 4×1 column vector $\boldsymbol{x}$ given previously is the 1×4 row vector $\boldsymbol{x}' = [3\ 2\ 5\ 1]$. ■

Product of Two Vectors The (inner) product of two $(p \times 1)$ vectors $\boldsymbol{x}$ and $\boldsymbol{y}$ is defined as the scalar $\boldsymbol{x}'\boldsymbol{y} = \sum_{i=1}^{p} x_i y_i$. Two vectors $\boldsymbol{x}$ and $\boldsymbol{y}$ are called **orthogonal** if their inner product $\boldsymbol{x}'\boldsymbol{y} = 0$. $\|\boldsymbol{x}\| = (\boldsymbol{x}'\boldsymbol{x})^{1/2} = [\sum_{i=1}^{p} (x_i)^2]^{1/2}$ is known as the Euclidean norm, or the length of the vector $\boldsymbol{x}$.

Example Consider the 4×1 vector $\boldsymbol{x}$ given previously and the 4×1 column vector $\boldsymbol{y} = \begin{bmatrix} 3 \\ 0 \\ -2 \\ 1 \end{bmatrix}$. The inner product $\boldsymbol{x}'\boldsymbol{y} = (3)(3) + (2)(0) + (5)(-2) + (1)(1) = 0$. The two vectors $\boldsymbol{x}$ and $\boldsymbol{y}$ are orthogonal. The length of the vector $\boldsymbol{x}$ is given by $\|\boldsymbol{x}\| = (\boldsymbol{x}'\boldsymbol{x})^{1/2} = (39)^{1/2} = 6.245$. The length of $\boldsymbol{y}$ is $\|\boldsymbol{y}\| = (14)^{1/2} = 3.742$. ■

Linearly Dependent and Linearly Independent Vectors A set of $n \times 1$ vectors $\boldsymbol{x}_1, \boldsymbol{x}_2, \ldots, \boldsymbol{x}_k$ is said to be **linearly dependent** if there exist scalars $c_1, c_2, \ldots, c_k$, not all simultaneously zero, such that $c_1\boldsymbol{x}_1 + c_2\boldsymbol{x}_2 + \cdots + c_k\boldsymbol{x}_k = \mathbf{0}$; otherwise, the set of vectors is called **linearly independent**. Linear dependence implies that at least one vector can be written as a linear combination of the remaining ones.

Example Consider the following four 5×1 vectors $\boldsymbol{x}_1, \boldsymbol{x}_2, \ldots, \boldsymbol{x}_4$, which we have represented as the four columns in a 5×4 matrix X:

$$X = [\boldsymbol{x}_1 \,|\, \boldsymbol{x}_2 \,|\, \boldsymbol{x}_3 \,|\, \boldsymbol{x}_4] = \begin{bmatrix} 1 & 2 & 3 & 7 \\ 1 & 3 & 4 & 6 \\ 1 & 5 & 6 & 4 \\ 1 & 2 & 3 & 7 \\ 1 & 1 & 2 & 8 \end{bmatrix}$$

Consider the first two vectors. These two vectors are linearly independent. Only the trivial solution $c_1 = c_2 = 0$ achieves $c_1\boldsymbol{x}_1 + c_2\boldsymbol{x}_2 = \mathbf{0}$. Now consider the first three vectors. One recognizes that $\boldsymbol{x}_1 + \boldsymbol{x}_2 = \boldsymbol{x}_3$. Hence, one can find coefficients $c_1 = 1$, $c_2 = 1$, and $c_3 = -1$ such that $c_1\boldsymbol{x}_1 + c_2\boldsymbol{x}_2 + c_3\boldsymbol{x}_3 = \mathbf{0}$. The first three vectors are linearly dependent. It is now obvious that the four vectors are also linearly dependent. Just set the coefficient $c_4 = 0$, and you have a nontrivial solution $c_1 = 1$, $c_2 = 1$, $c_3 = -1$, $c_4 = 0$ resulting in $c_1\boldsymbol{x}_1 + c_2\boldsymbol{x}_2 + c_3\boldsymbol{x}_3 + c_4\boldsymbol{x}_4 = \mathbf{0}$.

In this example, every possible set of three vectors from this matrix X results in three vectors that are linearly dependent. We already saw it for $\boldsymbol{x}_1, \boldsymbol{x}_2$, and $\boldsymbol{x}_3$. The same is true for $\boldsymbol{x}_1, \boldsymbol{x}_3$, and $\boldsymbol{x}_4$, as $\boldsymbol{x}_4 = 10\boldsymbol{x}_1 - \boldsymbol{x}_3$. Also for $\boldsymbol{x}_1, \boldsymbol{x}_2$, and $\boldsymbol{x}_4$, as $\boldsymbol{x}_4 = 10\boldsymbol{x}_1 - \boldsymbol{x}_3 = 10\boldsymbol{x}_1 - (\boldsymbol{x}_1 + \boldsymbol{x}_2) = 9\boldsymbol{x}_1 - \boldsymbol{x}_2$. Hence, we only need two columns to create the remaining columns of the matrix X. ■

Rank of a Matrix The rank of a $p \times q$ matrix A is the largest number of linearly independent columns (or equivalently, the largest number of linearly independent rows). The $p \times p$ square matrix AA' has the same rank as the $p \times q$ matrix A.

A $m \times m$ square matrix A is called **nonsingular** if its rank is m. We call the matrix **singular** if its rank is less than m.

Example The rank of the matrix X in the previous example is 2. Consider

$$A = \begin{bmatrix} 2 & 5 \\ 1 & 3 \\ 1 & 2 \end{bmatrix} \quad \text{and} \quad A' = \begin{bmatrix} 2 & 1 & 1 \\ 5 & 3 & 2 \end{bmatrix}$$

The rank of A, and also the rank of A', is 2. The 3×3 matrix

$$AA' = \begin{bmatrix} 29 & 17 & 12 \\ 17 & 10 & 7 \\ 12 & 7 & 5 \end{bmatrix}$$

Observe that the third column in this matrix is the difference of the first two. The matrix AA' has rank 2 (which is the rank of A). ■

Determinant of a Square Matrix The determinant of an $m \times m$ square matrix $A = (a_{ij})$ is the real value defined by $|A| = \sum \pm a_{1i}a_{2j} \dots a_{mp}$, where the summation is taken over all permutations $(i, j, \dots, p)$ of $(1, 2, \dots, m)$, with a plus sign if it is an even permutation and a minus sign if it is an odd permutation. A permutation is called even (odd) if its number of inversions is even (odd). An inversion in a permutation occurs whenever a larger integer precedes a smaller one. The determinant of A is different from zero if and only if the matrix A is nonsingular (i.e., has full rank m). For two $m \times m$ square matrices A and B, it holds that $|AB| = |A||B|$.

Example Consider the 2×2 and 3×3 matrices

$$A = \begin{bmatrix} 4 & 2 \\ 1 & 3 \end{bmatrix} \quad \text{and} \quad B = \begin{bmatrix} 2 & 1 & 3 \\ 1 & 2 & 3 \\ 4 & 6 & 10 \end{bmatrix}$$

The determinants are given by $|A| = (4)(3) - (2)(1) = 10$ and $|B| = (2)(2)(10) + (1)(6)(3) + (1)(3)(4) - (4)(2)(3) - (6)(3)(2) - (1)(1)(10) = 0$. The matrix B

is singular, and its rank is less than 3. We notice that the third column of B is the sum of the first two columns; hence, the matrix has rank at most 2. The matrix has rank 2 because we can find one pair of vectors that are linearly independent. ■

Inverse of a Nonsingular Matrix The inverse of a nonsingular matrix A is denoted by A^{-1}. The inverse A^{-1} satisfies $AA^{-1} = A^{-1}A = I$. If A and B are nonsingular $(m \times m)$ matrices, then $(AB)^{-1} = B^{-1}A^{-1}$. Various algorithms are available for numerically determining the inverse. Most algorithms become unstable if the matrix is close to a singular matrix.

Example Consider the 2×2 matrix A of the previous example. The inverse is given by

$$A^{-1} = \begin{bmatrix} 0.3 & -0.2 \\ -0.1 & 0.4 \end{bmatrix}$$

Convince yourself that $AA^{-1} = A^{-1}A = I$, where I is the 2×2 identity matrix. ■

Quadratic Forms Consider the $n \times 1$ vector $\mathbf{y} = (y_1, y_2, \ldots, y_n)'$. We call $\mathbf{y}'A\mathbf{y}$ a quadratic form in $\mathbf{y}$ provided that the $n \times n$ matrix A is symmetric (i.e., $A = A'$).

Positive Definite Matrix A symmetric matrix A is called positive definite if for all vectors $\mathbf{y} \neq \mathbf{0}$ the quadratic form $\mathbf{y}'A\mathbf{y} > 0$. We call the matrix positive semidefinite if the quadratic form $\mathbf{y}'A\mathbf{y} \geq 0$ for all $\mathbf{y}$ and $\mathbf{y}'A\mathbf{y} = 0$ for some vector $\mathbf{y} \neq \mathbf{0}$.

We call the symmetric matrix A negative definite if for all vectors $\mathbf{y} \neq \mathbf{0}$ the quadratic form $\mathbf{y}'A\mathbf{y} < 0$. We call the matrix negative semidefinite if the quadratic form $\mathbf{y}'A\mathbf{y} \leq 0$ for all $\mathbf{y}$ and $\mathbf{y}'A\mathbf{y} = 0$ for some vector $\mathbf{y} \neq \mathbf{0}$.

Orthogonal Matrix A square matrix A is called orthogonal if $AA' = I$. Since $A' = A^{-1}$ is the inverse of A, it follows that $A'A = I$. Hence, an orthogonal matrix satisfies $AA' = A'A = I$. The rows of an orthogonal matrix are mutually orthogonal and the length of the rows is one. The same can be said about the columns of A. Furthermore, the determinant $|A| = \pm 1$. This follows since $|AA'| = |A||A'| = |A|^2 = |I| = 1$.

Example Consider the 2×2 matrix $A = \begin{bmatrix} 1/\sqrt{2} & -1/\sqrt{2} \\ 1/\sqrt{2} & 1/\sqrt{2} \end{bmatrix}$. The inverse of this matrix is the same as the transpose, A'. The rows (columns) of the matrix are orthogonal, and the length of each row (column) vector is 1. A is an orthogonal matrix. The determinant of A is $+1$. ■

Trace of a Square Matrix The trace of a square $m \times m$ matrix A is defined as the sum of its diagonal elements; that is, $\text{tr}(A) = \sum_{i=1}^{m} a_{ii}$. The definition implies that $\text{tr}(A) = \text{tr}(A')$, $\text{tr}(A + B) = \text{tr}(A) + \text{tr}(B)$. Provided that the matrices C, D, and E are conformable, $\text{tr}(CDE) = \text{tr}(ECD) = \text{tr}(DEC)$. Conformable means that the dimensions of the matrices are such that all these products are defined.

Idempotent Matrix A square matrix A is idempotent if $AA = A$. The determinant of an idempotent matrix is either 0 or 1. The rank of an idempotent matrix is equal to the sum of its diagonal elements.

Example Consider the $n \times p$ matrix X, with $p \le n$, and assume that the matrix X has full column rank, p. Now consider the matrix $X'X$. This is a $p \times p$ matrix. The matrix is symmetric since $(X'X)' = X'(X')' = X'X$. Furthermore, it is nonsingular with rank p, and the inverse $(X'X)^{-1}$ exists.

Define the $n \times n$ matrix $H = X(X'X)^{-1}X'$. H is symmetric [since the transpose $(X(X'X)^{-1}X')' = X(X'X)^{-1}X'$]. H is idempotent since $HH = X(X'X)^{-1}X'X(X'X)^{-1}X' = X(X'X)^{-1}X' = H$. The rank of H is equal to the sum of its diagonal elements. The trace of H is given by $\text{tr}(H) = \text{tr}(X(X'X)^{-1}X') = \text{tr}(X'X(X'X)^{-1}) = \text{tr}(I_p) = p$ since $(X'X)$ is a $p \times p$ matrix and I_p is the $p \times p$ identity matrix. This shows that the rank of the $n \times n$ matrix H is p, the rank of X. The matrix H will become very useful in regression. ■

Vector Space A **vector space** V is the set of all vectors that is closed under addition and multiplication by a scalar and that contains the null vector $\mathbf{0}$. A set of linearly independent vectors $\mathbf{x}_1, \mathbf{x}_2, \ldots, \mathbf{x}_n$ is said to span a vector space of dimension n; any other member of this vector space can be generated through linear combinations of these vectors. We call this set a **basis** of the vector space. Basis vectors are not unique. Sometimes it is useful to work with **orthonormal** basis vectors. That is, vectors $\mathbf{u}_1, \mathbf{u}_2, \ldots, \mathbf{u}_n$ that are orthogonal ($\mathbf{u}_i'\mathbf{u}_j = 0$ for $i \ne j$) and have length one ($\mathbf{u}_i'\mathbf{u}_i = 1$). Such orthonormal basis vectors always exist, and the **Gram–Schmidt orthogonalization** shows how they can be constructed from a given set of linearly independent basis vectors $\mathbf{x}_1, \mathbf{x}_2, \ldots, \mathbf{x}_n$. The procedure works as follows:

$$
\begin{aligned}
\mathbf{u}_1 &= \mathbf{x}_1 \\
\mathbf{u}_2 &= \mathbf{x}_2 - a_{21}\mathbf{u}_1 \\
\mathbf{u}_3 &= \mathbf{x}_3 - a_{32}\mathbf{u}_2 - a_{31}\mathbf{u}_1 \\
&\ldots \\
\mathbf{u}_i &= \mathbf{x}_i - a_{i,i-1}\mathbf{u}_{i-1} - a_{i,i-2}\mathbf{u}_{i-2} - \cdots - a_{i,1}\mathbf{u}_1 \\
&\ldots
\end{aligned}
$$

The coefficients in the ith equation ($i = 2, \ldots, n$) are determined by setting the inner product of $\mathbf{u}_i$ with each of the previous vectors $\mathbf{u}_{i-1}, \ldots, \mathbf{u}_1$ equal to zero. We start with $\mathbf{u}_1 = \mathbf{x}_1$. We select a_{21} such that $\mathbf{u}_1'(\mathbf{x}_2 - a_{21}\mathbf{u}_1) = 0$, which implies that $a_{21} = (\boldsymbol{u}_1'\boldsymbol{x}_2)/(\boldsymbol{u}_1'\boldsymbol{u}_1)$. Next we determine a_{32} and a_{31} from $\boldsymbol{u}_1'(\boldsymbol{x}_3 - a_{32}\boldsymbol{u}_2 - a_{31}\boldsymbol{u}_1) = \boldsymbol{u}_1'(\boldsymbol{x}_3 - a_{31}\boldsymbol{u}_1) = 0$, and $\boldsymbol{u}_2'(\boldsymbol{x}_3 - a_{32}\boldsymbol{u}_2 - a_{31}\boldsymbol{u}_1) = \boldsymbol{u}_2'(\boldsymbol{x}_3 - a_{32}\boldsymbol{u}_2) = 0$. This implies $a_{31} = (\boldsymbol{u}_1'\boldsymbol{x}_3)/(\boldsymbol{u}_1'\boldsymbol{u}_1)$ and $a_{32} = (\boldsymbol{u}_2'\boldsymbol{x}_3)/(\boldsymbol{u}_2'\boldsymbol{u}_2)$, and so on. This process generates orthogonal vectors $\boldsymbol{u}_i$. The last step normalizes the vectors by dividing each vector $\boldsymbol{u}_i$ by $(\boldsymbol{u}_i'\boldsymbol{u}_i)^{0.5}$. The denominator is always different from zero, because the initial basis vectors are linearly independent.

The Euclidean space R^n is a vector space. The vectors $\boldsymbol{u}_i$, with all zero elements except for a one in the ith row, provide one natural orthonormal basis (but by no means the only one).

Subspace of R^n Consider $k < n$ linearly independent vectors in R^n, $\boldsymbol{x}_1, \boldsymbol{x}_2, \ldots, \boldsymbol{x}_k$. The linear combinations of $\boldsymbol{x}_1, \boldsymbol{x}_2, \ldots, \boldsymbol{x}_k$ span a certain k-dimensional subspace R^k within R^n. The Gram–Schmidt orthogonalization can be used to construct an orthonormal basis for this subspace.

Linear Transformations Matrices arise naturally in the study of linear transformations. A linear transformation of the $n \times 1$ vector $\boldsymbol{x}$ into the $n \times 1$ vector $\boldsymbol{y}$ is written as $\boldsymbol{y} = A\boldsymbol{x}$, where A is the $n \times n$ transformation matrix. For nonsingular matrices A the transformation is one-to-one.

Orthogonal Transformations A linear transformation $\boldsymbol{y} = A\boldsymbol{x}$ is called orthogonal if A is an orthogonal matrix satisfying $AA' = A'A = I$; that is, a matrix with mutually orthogonal rows (columns) of unit lengths. Orthogonal transformations correspond to rotations and reflections in n-dimensional space, and they preserve lengths (i.e., $\boldsymbol{y}'\boldsymbol{y} = \boldsymbol{x}'A'A\boldsymbol{x} = \boldsymbol{x}'\boldsymbol{x}$) and also volumes.

Projections Projection of a vector $\boldsymbol{x}$ in R^n on to a subspace S of R^n is the linear transformation $\boldsymbol{y} = P\boldsymbol{x}$, where the $n \times n$ matrix P is symmetric ($P = P'$) and idempotent (satisfying $PP = P$).

The subspace S onto which we project is spanned by the columns (or rows) of P, and the dimension of S is given by the rank of this matrix P. Since $\boldsymbol{y} = P\boldsymbol{x}$, $\boldsymbol{y}$ is a member of that subspace. The fact that $P\boldsymbol{y} = PP\boldsymbol{x} = P\boldsymbol{x}$ and hence $P(\boldsymbol{x} - \boldsymbol{y}) = P'(\boldsymbol{x} - \boldsymbol{y}) = \boldsymbol{0}$ implies that $\boldsymbol{x} - \boldsymbol{y}$ is orthogonal to the subspace S.

Eigenvalues and Eigenvectors of a Square Matrix The eigenvalues (or characteristic roots) $\lambda_1, \lambda_2, \ldots, \lambda_m$ of an $m \times m$ matrix A are the m solutions to the characteristic equation (determinant equation) $|A - \lambda I| = 0$. The eigenvectors $\boldsymbol{v}_1, \boldsymbol{v}_2, \ldots, \boldsymbol{v}_m$ of the matrix A are the (vector) solutions of the equations $(A - \lambda_i I)\boldsymbol{v}_i = \boldsymbol{0}$, for $i = 1, 2, \ldots, m$.

For symmetric matrices the eigenvalues are real numbers. For symmetric positive definite matrices all eigenvalues are positive. For symmetric positive semidefinite matrices one or more of the eigenvalues will be zero.

Spectral Representation (Canonical Reduction) of a Symmetric Matrix Let A be an $m \times m$ symmetric matrix (i.e., $A' = A$). Then there exist an orthogonal matrix P (i.e., satisfying $P'P = PP' = I$) and a diagonal matrix Λ with elements $\lambda_1 \geq \lambda_2 \geq \cdots \geq \lambda_m$ such that $P'AP = \Lambda$. Hence, $A = P\Lambda P'$.

The elements $\lambda_1 \geq \lambda_2 \geq \cdots \geq \lambda_m$ of the diagonal matrix Λ are the real-valued eigenvalues of the matrix A, and the m column vectors of the matrix $P = [\boldsymbol{p}_1, \boldsymbol{p}_2, \ldots, \boldsymbol{p}_m]$ are the corresponding eigenvectors, that satisfy $A\boldsymbol{p}_i = \lambda_i \boldsymbol{p}_i$ and have been normalized to lengths one.

3.2 MATRIX APPROACH TO SIMPLE LINEAR REGRESSION

In Chapters 1 and 2, we considered the hardness (y) and the temperature (x) of 14 coil springs. The linear model

$$y = \beta_0 + \beta_1 x + \epsilon$$

led us to the 14 equations in Eq. (2.2). These may be written as

$$\begin{bmatrix} 55.2 \\ 59.1 \\ \vdots \\ 16.9 \end{bmatrix} = \begin{bmatrix} \beta_0 + \beta_1 \cdot 30 \\ \beta_0 + \beta_1 \cdot 30 \\ \vdots \\ \beta_0 + \beta_1 \cdot 60 \end{bmatrix} + \begin{bmatrix} \epsilon_1 \\ \epsilon_2 \\ \vdots \\ \epsilon_{14} \end{bmatrix}$$

$$\boldsymbol{y} = X\boldsymbol{\beta} + \boldsymbol{\epsilon} \tag{3.1}$$

where $\boldsymbol{y}$ and $\boldsymbol{\epsilon}$ are (14×1) column vectors, $\boldsymbol{\beta}$ is a (2×1) column vector, and X is a (14×2) matrix. Specifically,

$$\boldsymbol{y} = \begin{bmatrix} y_1 \\ y_2 \\ \vdots \\ y_{14} \end{bmatrix} = \begin{bmatrix} 55.2 \\ 59.1 \\ \vdots \\ 16.9 \end{bmatrix}; \quad \boldsymbol{\epsilon} = \begin{bmatrix} \epsilon_1 \\ \epsilon_2 \\ \vdots \\ \epsilon_{14} \end{bmatrix}; \quad \boldsymbol{\beta} = \begin{bmatrix} \beta_0 \\ \beta_1 \end{bmatrix}; \quad X = \begin{bmatrix} 1 & x_1 \\ 1 & x_2 \\ \vdots & \vdots \\ 1 & x_{14} \end{bmatrix} = \begin{bmatrix} 1 & 30 \\ 1 & 30 \\ \vdots & \vdots \\ 1 & 60 \end{bmatrix}$$

With n cases, $\boldsymbol{y}$ and $\boldsymbol{\epsilon}$ are $(n \times 1)$ vectors, and X is a $(n \times 2)$ matrix.

3.2.1 LEAST SQUARES

The least squares estimates of $\boldsymbol{\beta}$ minimize the error sum of squares

$$S(\boldsymbol{\beta}) = S(\beta_0, \beta_1) = \sum_{i=1}^{n} \epsilon_i^2 = \boldsymbol{\epsilon}'\boldsymbol{\epsilon} = (\boldsymbol{y} - X\boldsymbol{\beta})'(\boldsymbol{y} - X\boldsymbol{\beta})$$

The minimization with respect to $\boldsymbol{\beta}$ leads to the normal equations in Eq. (2.6), which can be written in vector/matrix form as

$$\begin{bmatrix} n & \sum x_i \\ \sum x_i & \sum x_i^2 \end{bmatrix} \begin{bmatrix} \beta_0 \\ \beta_1 \end{bmatrix} = \begin{bmatrix} \sum y_i \\ \sum x_i y_i \end{bmatrix},$$

or

$$(X'X)\boldsymbol{\beta} = X'\mathbf{y} \tag{3.2}$$

The solution to this matrix equation, $\hat{\boldsymbol{\beta}}$, is given by

$$\hat{\boldsymbol{\beta}} = (X'X)^{-1}X'\mathbf{y} \tag{3.3}$$

provided that the matrix $X'X$ can be inverted (i. e., the matrix X has full column rank 2, which implies that $X'X$ is nonsingular).

The fitted values and residuals can also be expressed in vector form. The vector of **fitted values** is given by

$$\hat{\boldsymbol{\mu}} = \begin{bmatrix} \hat{\mu}_1 \\ \vdots \\ \hat{\mu}_n \end{bmatrix} = \begin{bmatrix} \hat{\beta}_0 + \hat{\beta}_1 x_1 \\ \vdots \\ \hat{\beta}_0 + \hat{\beta}_1 x_n \end{bmatrix} = \begin{bmatrix} 1 & x_1 \\ \vdots & \vdots \\ 1 & x_n \end{bmatrix} \begin{bmatrix} \hat{\beta}_0 \\ \hat{\beta}_1 \end{bmatrix} = X\hat{\boldsymbol{\beta}} \tag{3.4}$$

The vector of **residuals** is

$$\mathbf{e} = \begin{bmatrix} e_1 \\ \vdots \\ e_n \end{bmatrix} = \begin{bmatrix} y_1 - \hat{\mu}_1 \\ \vdots \\ y_n - \hat{\mu}_n \end{bmatrix} = \mathbf{y} - \hat{\boldsymbol{\mu}} = \mathbf{y} - X\hat{\boldsymbol{\beta}} \tag{3.5}$$

The least squares estimator of σ^2 is given as

$$s^2 = \frac{1}{n-2}\sum_{i=1}^{n} e_i^2 = \frac{1}{n-2}\mathbf{e}'\mathbf{e} \tag{3.6}$$

Since $\hat{\boldsymbol{\beta}}$ is the solution of Eq. (3.2), $X'X\hat{\boldsymbol{\beta}} = X'\mathbf{y}$, we find that

$$X'(\mathbf{y} - X\hat{\boldsymbol{\beta}}) = X'\mathbf{e} = \mathbf{0} \tag{3.7}$$

The equation

$$\begin{bmatrix} 1 & 1 & \cdots & 1 \\ x_1 & x_2 & \cdots & x_n \end{bmatrix} \begin{bmatrix} e_1 \\ \vdots \\ e_n \end{bmatrix} = \mathbf{0}$$

implies $\sum e_i = 0$ and $\sum x_i e_i = 0$, the two restrictions seen earlier.

Furthermore, through substitution of $\hat{\boldsymbol{\mu}} = X\hat{\boldsymbol{\beta}}$, we find that the scalar

$$\sum \hat{\mu}_i e_i = \hat{\boldsymbol{\mu}}'\mathbf{e} = (X\hat{\boldsymbol{\beta}})'\mathbf{e} = \hat{\boldsymbol{\beta}}'X'\mathbf{e} = 0 \tag{3.8}$$

which is another consequence of the least squares fit seen earlier.

This discussion shows that with vector/matrix algebra a number of results can be established quite easily. Without vector algebra, the analysis involves considerable algebraic manipulations. The results in Eqs. (3.7) and (3.8) also have interesting and useful geometric interpretations, which are discussed in Chapter 4.

3.2.2 HARDNESS EXAMPLE

Using the data in Chapter 2, we find

$$X'X = \begin{bmatrix} 14 & 630 \\ 630 & 30300 \end{bmatrix}, \quad (X'X)^{-1} = \begin{bmatrix} 1.1099 & -.0231 \\ -.0231 & .0005 \end{bmatrix}$$

$$X'\boldsymbol{y} = \begin{bmatrix} 2520.2 \\ 20940.0 \end{bmatrix} \quad \hat{\boldsymbol{\beta}} = (X'X)^{-1}X'\boldsymbol{y} = \begin{bmatrix} 94.134 \\ -1.266 \end{bmatrix}$$

$$\hat{\boldsymbol{\mu}} = \begin{bmatrix} \hat{\mu}_1 \\ \hat{\mu}_2 \\ \vdots \\ \hat{\mu}_{14} \end{bmatrix} = \begin{bmatrix} \hat{\beta}_0 + \hat{\beta}_1 \cdot 30 \\ \hat{\beta}_0 + \hat{\beta}_1 \cdot 30 \\ \vdots \\ \hat{\beta}_0 + \hat{\beta}_1 \cdot 60 \end{bmatrix} = \begin{bmatrix} 56.1495 \\ 56.1495 \\ \vdots \\ 18.1648 \end{bmatrix}$$

$$\boldsymbol{e} = \begin{bmatrix} e_1 \\ e_2 \\ \vdots \\ e_{14} \end{bmatrix} = \begin{bmatrix} 55.2 \\ 59.1 \\ \vdots \\ 16.9 \end{bmatrix} - \begin{bmatrix} 56.1495 \\ 56.1495 \\ \vdots \\ -18.1648 \end{bmatrix} = \begin{bmatrix} -0.3495 \\ 2.9505 \\ \vdots \\ -1.2648 \end{bmatrix}$$

3.3 VECTORS OF RANDOM VARIABLES

Equation (3.1) expresses the simple linear regression model in vector notation. The vectors $\boldsymbol{\epsilon} = (\epsilon_1, \ldots, \epsilon_n)'$ and $\boldsymbol{y} = (y_1, \ldots, y_n)'$ are called random vectors because their elements are random variables. In this section, we review random vectors and their properties. This will help us establish the general regression results. Note that in Chapter 2 we made a careful distinction between a random variable (denoted by capital letter Y) and a possible value of this random variable (denoted by lowercase letter y). From now on, we will not make this distinction and throughout this text we will use lowercase letters for random variables.

Definition The expected value of a random vector $\boldsymbol{y}$ is defined as

$$E(\boldsymbol{y}) = (E(y_1), E(y_2), \ldots, E(y_n))'$$

The expected value of a random vector is the vector of the respective expected values; we also call it the mean of the random vector. Similarly, if W is a matrix of random variables, then the expected value of W, $E(W)$, is the corresponding matrix of expected values.

For single random variables y_1 and y_2 and constants a_1 and a_2, we know that

$$\text{(i) } E(a_1 y_1 + a_2) = a_1 E(y_1) + a_2, \quad \text{(ii) } E(a_1 y_1 + a_2 y_2) = a_1 E(y_1) + a_2 E(y_2)$$

The second property is very important. It states that the expected value of a linear combination of random variables is the linear combination of the expected values. For vectors we have the corresponding properties

$$\text{(i) } E(a\boldsymbol{y} + \boldsymbol{b}) = aE(\boldsymbol{y}) + \boldsymbol{b}, \quad \text{(ii) } E(A_1\boldsymbol{y}) = A_1 E(\boldsymbol{y}); \quad E(\boldsymbol{y}'A_2) = E(\boldsymbol{y})'A_2$$

where a is a scalar constant, $\boldsymbol{b}$ is a vector of constants, and A_1 and A_2 are matrices of constants. These results follow easily by writing out the expressions after the expectation sign and using the definition of the expectation of a random vector.

Recall that for scalar random variables y_1 and y_2 with means $E(y_1) = \mu_1$ and $E(y_2) = \mu_2$, the variances of y_1 and y_2 are

$$V(y_1) = E(y_1 - \mu_1)^2 \quad \text{and} \quad V(y_2) = E(y_2 - \mu_2)^2$$

and the covariance between y_1 and y_2 is

$$\text{Cov}(y_1, y_2) = E(y_1 - \mu_1)(y_2 - \mu_2)$$

In the context of a $(n \times 1)$ random vector $\boldsymbol{y}$ we collect both variances and covariances in a single matrix called the **covariance matrix** $V(\boldsymbol{y})$.

Definition

Let $\boldsymbol{y} = (y_1, \ldots, y_n)'$ be a vector of random variables with mean vector $E(\boldsymbol{y}) = \boldsymbol{\mu} = (\mu_1, \ldots, \mu_n)'$. The covariance matrix $V(\boldsymbol{y})$ is the matrix with diagonal elements $V(y_i)$ and (i, j)th off-diagonal elements $\text{Cov}(y_i, y_j)$. That is,

$$V(\boldsymbol{y}) = \begin{bmatrix} V(y_1) & \text{Cov}(y_1, y_2) & \cdots & \text{Cov}(y_1, y_n) \\ & V(y_2) & \cdots & \text{Cov}(y_2, y_n) \\ \vdots & & \vdots & \\ \text{Cov}(y_n, y_1) & & \cdots & V(y_n) \end{bmatrix}$$ ■

Here we list several important properties of vectors of random variables:

i. The covariance matrix $\Sigma = V(\boldsymbol{y})$ is symmetric $(\Sigma = \Sigma')$ since $\text{Cov}(y_i, y_j) = \text{Cov}(y_j, y_i)$.

ii. For uncorrelated random variables $y_1, \ldots, y_n$ with $\text{Cov}(y_i, y_j) = 0$, $i \neq j$, the covariance matrix $V(\boldsymbol{y})$ is diagonal.

iii. We can express the covariance matrix as

$$V(\boldsymbol{y}) = E(\boldsymbol{y} - \boldsymbol{\mu})(\boldsymbol{y} - \boldsymbol{\mu})'$$

One can see this by writing

$$E(\boldsymbol{y} - \boldsymbol{\mu})(\boldsymbol{y} - \boldsymbol{\mu})' = E\left\{\begin{bmatrix} y_1 - \mu_1 \\ \cdots \\ y_n - \mu_n \end{bmatrix} [y_1 - \mu_1, \ldots, y_n - \mu_n]\right\}$$

$$= \begin{bmatrix} E(y_1 - \mu_1)^2 & E(y_1 - \mu_1)(y_2 - \mu_2) & \cdots & E(y_1 - \mu_1)(y_n - \mu_n) \\ & E(y_2 - \mu_2)^2 & \cdots & E(y_2 - \mu_2)(y_n - \mu_n) \\ \text{Symmetric} & & & \vdots \\ & & & E(y_n - \mu_n)^2 \end{bmatrix}$$

$$= V(\boldsymbol{y})$$

iv. Consider a system of p linear transformations $\boldsymbol{u} = A\boldsymbol{y}$, where A is a $p \times n$ matrix of constants. The $p \times 1$ vector $\boldsymbol{u}$ is a random vector. Its mean vector is given by the $p \times 1$ vector $E(\boldsymbol{u}) = E(A\boldsymbol{y}) = AE(\boldsymbol{y}) = A\boldsymbol{\mu}$. The $p \times p$ covariance matrix of $\boldsymbol{u}$ is given by $V(\boldsymbol{u}) = V(A\boldsymbol{y}) = AV(\boldsymbol{y})A'$. The result for the mean vector follows from the definition. The result on the covariance matrix can be proved as follows:

$$\begin{aligned} V(\boldsymbol{u}) &= E[\boldsymbol{u} - E(\boldsymbol{u})][\boldsymbol{u} - E(\boldsymbol{u})]' \\ &= E[(A\boldsymbol{y} - A\boldsymbol{\mu})(A\boldsymbol{y} - A\boldsymbol{\mu})'] \\ &= E[A(\boldsymbol{y} - \boldsymbol{\mu})(\boldsymbol{y} - \boldsymbol{\mu})'A'] \\ &= AE[(\boldsymbol{y} - \boldsymbol{\mu})(\boldsymbol{y} - \boldsymbol{\mu})']A' \\ &= AV(\boldsymbol{y})A' \end{aligned}$$

If the linear transformation includes a constant vector $\boldsymbol{b}$ (i.e., $\boldsymbol{u} = \boldsymbol{b} + A\boldsymbol{y}$), then the mean vector changes to $E(\boldsymbol{u}) = \boldsymbol{b} + A\boldsymbol{\mu}$. However, since $\boldsymbol{b}$ is nonstochastic, the covariance matrix is as given previously.

Example Consider the random vector $\boldsymbol{y} = (y_1, y_2, y_3)'$ such that $E(\boldsymbol{y}) = \boldsymbol{\mu} = (1, 2, 3)'$ and

$$V(\boldsymbol{y}) = \begin{bmatrix} 1 & 0 & 1 \\ 0 & 2 & -1 \\ 1 & -1 & 3 \end{bmatrix}$$

Consider the system of linear combinations

$$\begin{aligned} u_1 &= y_1 - y_2 + 2y_3 \\ u_2 &= y_1 + y_3 \end{aligned}$$

That is, $\boldsymbol{u} = A\boldsymbol{y}$, where $A = \begin{bmatrix} 1 & -1 & 2 \\ 1 & 0 & 1 \end{bmatrix}$. Then

$$E(\boldsymbol{u}) = A\boldsymbol{\mu} = \begin{bmatrix} 1 & -1 & 2 \\ 1 & 0 & 1 \end{bmatrix} \begin{bmatrix} 1 \\ 2 \\ 3 \end{bmatrix} = \begin{bmatrix} 5 \\ 4 \end{bmatrix}$$

and

$$V(\boldsymbol{u}) = \begin{bmatrix} 1 & -1 & 2 \\ 1 & 0 & 1 \end{bmatrix} \begin{bmatrix} 1 & 0 & 1 \\ 0 & 2 & -1 \\ 1 & -1 & 3 \end{bmatrix} \begin{bmatrix} 1 & 1 \\ -1 & 0 \\ 2 & 1 \end{bmatrix} = \begin{bmatrix} 23 & 11 \\ 11 & 6 \end{bmatrix}$$

This implies that $V(u_1) = 23$, $V(u_2) = 6$, $\text{Cov}(u_1, u_2) = 11$, and

$$\text{Corr}(u_1, u_2) = \frac{\text{Cov}(u_1, u_2)}{\sqrt{V(u_1)V(u_2)}} = \frac{11}{\sqrt{23}\sqrt{6}} = 0.936$$ ■

v. Consider $p = 1$, and the single linear combination $u = \boldsymbol{a}'\boldsymbol{y}$, where $\boldsymbol{a}' = (a_1, \ldots, a_n)$ is a $1 \times n$ row vector. The mean of the scalar random variable u is $E(u) = \boldsymbol{a}'\boldsymbol{\mu}$ and its variance is $V(u) = \boldsymbol{a}'V(\boldsymbol{y})\boldsymbol{a}$.

vi. The covariance matrix $\Sigma = V(\boldsymbol{y})$ must be positive semidefinite. Variances of linear combinations $\boldsymbol{a}'\boldsymbol{y}$ can never be negative; hence, we require that $V(\boldsymbol{a}'\boldsymbol{y}) = \boldsymbol{a}'V(\boldsymbol{y})\boldsymbol{a} \geq 0$. Note that the requirement of being positive **semi**definite allows for nontrivial linear combinations of $\boldsymbol{y}$ with zero variance. In such cases, we have deterministic linear relationship(s) among the n random variables, and the distribution of $\boldsymbol{y}$ can be defined on a lower (lower than n) dimensional space. In this case, the covariance matrix $V(\boldsymbol{y})$ is singular.

vii. Eigenvalues of a positive semidefinite symmetric matrix are nonnegative. Hence, the covariance matrix Σ of a random vector $\boldsymbol{y}$ can always be written as $\Sigma = P\Lambda^{1/2}\Lambda^{1/2}P' = P^*P^{*'}$, where $P^* = P\Lambda^{1/2}$, and where $\Lambda^{1/2}$ is the diagonal matrix with elements $(\lambda_1)^{1/2} \geq (\lambda_2)^{1/2} \geq \cdots \geq (\lambda_m)^{1/2}$.

This is an important result. For any random vector $\boldsymbol{y}$ with general nonsingular covariance matrix $\Sigma = P\Lambda P' = P^*P^{*'}$, we can find a linear transformation $\boldsymbol{z} = (\Lambda^{-1/2}P')\boldsymbol{y}$ that transforms the correlated random variables $\boldsymbol{y}$ into uncorrelated random variables with unit variances. Here, $\Lambda^{-1/2}$ is a diagonal matrix with elements $(\lambda_1)^{-1/2} \leq (\lambda_2)^{-1/2} \leq \cdots \leq (\lambda_m)^{-1/2}$, and P is the matrix of normalized eigenvectors of Σ.

You can see this result from our general result on covariance matrices of linear transformations. The covariance matrix of $\boldsymbol{z}$ is given by $(\Lambda^{-1/2}P')\Sigma(P\Lambda^{-1/2}) = \Lambda^{-1/2}\Lambda\Lambda^{-1/2} = I$. Note that we have assumed that the covariance matrix Σ is nonsingular. In this case, all eigenvalues $(\lambda_1, \lambda_2, \ldots, \lambda_m)$ are strictly larger than zero, and it is possible to take the reciprocal of the eigenvalues.

Example

Consider the 2×2 matrix $\Sigma = \begin{bmatrix} 1 & 0.5 \\ 0.5 & 1 \end{bmatrix}$. This is a symmetric and positive definite matrix. It is a nonsingular covariance matrix of a bivariate (nondegenerate) random vector. The eigenvalues are the solution to the equation $(1-\lambda)(1-\lambda) - 0.25 = 0$, and they are given by 0.5 and 1.5. The eigenvector corresponding to the first eigenvalue satisfies the equation $\begin{bmatrix} 1 & 0.5 \\ 0.5 & 1 \end{bmatrix}\begin{bmatrix} x_1 \\ x_2 \end{bmatrix} - (0.5)\begin{bmatrix} x_1 \\ x_2 \end{bmatrix} = \begin{bmatrix} 0 \\ 0 \end{bmatrix}$. This leads to the equation $x_1 + x_2 = 0$ (note that the second equation is identical). Solving this equation and imposing the restriction that the length of the resulting solution is 1 [i.e., $(x_1)^2 + (x_2)^2 = 1$] leads to the first eigenvector $\boldsymbol{p}_1 = \begin{bmatrix} 1/\sqrt{2} \\ -1/\sqrt{2} \end{bmatrix}$. The second eigenvector satisfies the equation $\begin{bmatrix} 1 & 0.5 \\ 0.5 & 1 \end{bmatrix}\begin{bmatrix} x_1 \\ x_2 \end{bmatrix} - (1.5)\begin{bmatrix} x_1 \\ x_2 \end{bmatrix} = \begin{bmatrix} 0 \\ 0 \end{bmatrix}$, which leads to the equation $x_1 - x_2 = 0$ and the second eigenvector $\boldsymbol{p}_2 = \begin{bmatrix} 1/\sqrt{2} \\ 1/\sqrt{2} \end{bmatrix}$. Hence, the orthogonal matrix $P = \begin{bmatrix} 1/\sqrt{2} & 1/\sqrt{2} \\ -1/\sqrt{2} & 1/\sqrt{2} \end{bmatrix}$ and $\Lambda = \begin{bmatrix} 0.5 & 0 \\ 0 & 1.5 \end{bmatrix}$. You can convince yourself that $\Sigma = P \Lambda P'$. ■

So far, we have discussed the mean $E(\boldsymbol{y})$ and the covariance matrix $V(\boldsymbol{y})$ of a random vector $\boldsymbol{y}$. We can also consider the distribution of random vectors. Such distributions are referred to as **multivariate distributions**. The most important multivariate distribution is the multivariate normal distribution.

3.4 THE MULTIVARIATE NORMAL DISTRIBUTION

In the Appendix in Chapter 2, we discussed the **univariate normal distribution** with mean μ and standard deviation $\sigma > 0$. The density function of the univariate normal distribution is given by

$$f(y) = \frac{1}{(2\pi)^{1/2}\sigma} \exp\left[-\frac{1}{2\sigma^2}(y-\mu)^2 \right] \quad \text{for} \quad -\infty < y < \infty$$

The **standard normal distribution** is the normal distribution with mean 0 and variance 1. Percentiles of the standard normal distribution are given in Table A at the end of this book.

The ***n* variate normal distribution** for an $n \times 1$ random vector $\boldsymbol{y} = (y_1, y_2, \ldots, y_n)'$ with mean vector $\boldsymbol{\mu} = (\mu_1, \mu_2, \ldots, \mu_n)'$ and $n \times n$ covariance matrix $\Sigma = (\sigma_{ij})$ is given by

$$f(\boldsymbol{y}) = \left[\frac{1}{2\pi} \right]^{n/2} |\Sigma|^{-1/2} \exp\left[-\frac{1}{2}(\boldsymbol{y}-\boldsymbol{\mu})'\Sigma^{-1}(\boldsymbol{y}-\boldsymbol{\mu}) \right] \quad \text{for} \quad -\infty < y_i < \infty$$

The distribution is completely specified by a mean vector $\boldsymbol{\mu}$ and a symmetric positive definite covariance matrix Σ. This assumption ensures that Σ is nonsingular and has an inverse. Note that a positive **semi**definite covariance matrix Σ implies at least one nontrivial linear combination of $\boldsymbol{y}$ with zero variance. In this case, there exists a deterministic linear relationship among the n random variables, and the distribution of $\boldsymbol{y}$ can be defined on a lower (lower than n) dimensional space. The resulting covariance matrix Σ is singular, and one needs to consider a special type of inverse called generalized inverse of Σ.

Important properties of the multivariate normal distribution are listed as follows:

i. **The reproductive property: Linear transforms of multivariate normals are again normal.**
Consider the (vector) linear transformation of $\boldsymbol{y}$, $\boldsymbol{u} = A\boldsymbol{y}$, where $\boldsymbol{u}$ is a $p \times 1$ vector with $p \le n$, and where A is a $p \times n$ matrix of rank p. It follows that $\boldsymbol{u}$ has a p **variate normal distribution** with mean vector $E(\boldsymbol{u}) = A\boldsymbol{\mu}$ and covariance matrix $A\Sigma A'$.

The result about the mean vector and covariance matrix of a (vector) linear transformation was shown in Section 3.3. The importance of the present result is that the distribution of linear combinations of a multivariate normal vector is again multivariate normal.

As a special case, consider a **single** linear transformation $u = \boldsymbol{a}'\boldsymbol{y}$, where $\boldsymbol{a} = (a_1, a_2, \ldots, a_n)'$ is a known $n \times 1$ vector of constants. The (scalar) random variable u follows a **univariate** normal distribution with mean $\boldsymbol{a}'\boldsymbol{\mu}$ and variance $\boldsymbol{a}'\Sigma\boldsymbol{a}$.

ii. **Marginal distributions of multivariate normals are again normal.**
Consider $\boldsymbol{y}_1$, a $p \times 1$ subvector of the $n \times 1$ vector $\boldsymbol{y} = (y_1, y_2, \ldots, y_n)'$. Assume without loss of generality that $\boldsymbol{y}_1$ consists of the first p components of $\boldsymbol{y}$. Partition the vector $\boldsymbol{y}$ accordingly, and let $\boldsymbol{y} = \begin{bmatrix} \boldsymbol{y}_1 \\ \boldsymbol{y}_2 \end{bmatrix}$, with mean vector $\boldsymbol{\mu} = \begin{bmatrix} \boldsymbol{\mu}_1 \\ \boldsymbol{\mu}_2 \end{bmatrix}$ and covariance matrix $\Sigma = \begin{bmatrix} \Sigma_{11} & \Sigma_{12} \\ \Sigma_{12}' & \Sigma_{22} \end{bmatrix}$. We can write $\boldsymbol{y}_1$ as a (vector) linear combination of $\boldsymbol{y}$. That is,

$$\boldsymbol{y}_1 = \begin{bmatrix} I_{p \times p} & O_{p \times (n-p)} \end{bmatrix} \boldsymbol{y}$$

where $I_{p \times p}$ is the $p \times p$ identity matrix, and $O_{p \times (n-p)}$ is the $p \times (n-p)$ matrix of zeros.

Applying property i, we find that $\boldsymbol{y}_1$ has a p variate normal distribution with mean vector $\boldsymbol{\mu}_1$ and covariance matrix

$$\begin{bmatrix} I_{p \times p} & O_{p \times (n-p)} \end{bmatrix} \begin{bmatrix} \Sigma_{11} & \Sigma_{12} \\ \Sigma_{12}' & \Sigma_{22} \end{bmatrix} \begin{bmatrix} I_{p \times p} \\ O_{(n-p) \times p} \end{bmatrix} = \Sigma_{11}$$

Let us consider $p = 1$ and an individual component y_i (where i is any index between 1 and n). It follows that the distribution of y_i is univariate normal with mean μ_i and variance σ_{ii}.

iii. **Conditional distributions of multivariate normal distributions are again normal.**
Partition the multivariate normal vector $\boldsymbol{y}$ into nonoverlapping subvectors $\boldsymbol{y}_1$ and $\boldsymbol{y}_2$ as shown in (ii). The conditional distribution of $\boldsymbol{y}_1$ given certain specified values for $\boldsymbol{y}_2$ is a p variate **normal distribution** with the $p \times 1$ mean vector

$$\boldsymbol{\mu}_{1 \cdot 2} = \boldsymbol{\mu}_1 + \Sigma_{12}(\Sigma_{22})^{-1}(\boldsymbol{y}_2 - \boldsymbol{\mu}_2)$$

and the $p \times p$ covariance matrix

$$\Sigma_{1 \cdot 2} = \Sigma_{11} - \Sigma_{12}(\Sigma_{22})^{-1}\Sigma_{12}'$$

iv. **Equivalence of zero covariance and independence.**
We know that for any distribution, the independence of (two) random variables implies zero covariance between the random variables. Under a multivariate normal distribution, the converse also holds, and zero covariance implies statistical independence.

You can see that with $\sigma_{12} = \text{Cov}(y_1, y_2) = 0$, the bivariate normal density of y_1 and y_2 factors into

$$f(y_1, y_2) = \frac{1}{(2\pi)(\sigma_{11}\sigma_{22})^{1/2}} \exp\left[-\frac{(y_1-\mu_1)^2}{2\sigma_{11}} - \frac{(y_2-\mu_2)^2}{2\sigma_{22}}\right]$$

$$= \left\{\frac{1}{(2\pi)^{1/2}\sigma_{11}^{1/2}} \exp\left[-\frac{(y_1-\mu_1)^2}{2\sigma_{11}}\right]\right\}\left\{\frac{1}{(2\pi)^{1/2}\sigma_{22}^{1/2}} \exp\left[-\frac{(y_2-\mu_2)^2}{2\sigma_{22}}\right]\right\},$$

the product of two marginal normal densities.

The assumption of joint normality is essential for the equivalence of uncorrelatedness and independence. The following example shows that, in general, uncorrelated random variables need not be independent. Take the random variable y with zero odd moments; that is, $E(y) = E(y^3) = E(y^5) = \cdots = 0$. Take as a second random variable $x = y^2$. Then $\text{Cov}(x, y) = E[(y^2 - E(y^2))(y - E(y))] = E(y^3) - E(y)E(y^2) = 0$, but x and y are functionally related and hence not independent.

Consider two (vector) linear transforms of a multivariate normal random vector $\mathbf{y}$ with covariance matrix Σ. That is, $z_1 = A_1\mathbf{y}$ and $z_2 = A_2\mathbf{y}$, where A_1 and A_2 are matrices of dimensions $p_1 \times n$ and $p_2 \times n$. The transforms are independent if and only if $A_1\Sigma A_2' = O$, a matrix of $p_1 \times p_2$ zeros.

Result i implies that the vector $z = \begin{bmatrix} z_1 \\ z_2 \end{bmatrix} = \begin{bmatrix} A_1 \\ A_2 \end{bmatrix} \mathbf{y}$ has a multivariate normal distribution with covariance matrix

$\begin{bmatrix} A_1 \\ A_2 \end{bmatrix} \Sigma \begin{bmatrix} A_1' & A_2' \end{bmatrix} = \begin{bmatrix} A_1\Sigma A_1' & A_1\Sigma A_2' \\ A_2\Sigma A_1' & A_2\Sigma A_2' \end{bmatrix}$. The covariance is zero if and only if $A_1\Sigma A_2' = O$.

3.5 IMPORTANT RESULTS ON QUADRATIC FORMS

We call $Q = \mathbf{y}'A\mathbf{y}$ a quadratic form in $\mathbf{y} = (y_1, y_2, \ldots, y_n)'$ provided that the $n \times n$ matrix A is symmetric (i.e., $A = A'$).

i. **The distribution of quadratic forms**
Assume that the $n \times 1$ vector $\mathbf{y} = (y_1, y_2, \ldots, y_n)'$ follows a multivariate normal distribution with mean vector zero and nonsingular covariance matrix Σ. Assume that the symmetric $n \times n$ matrix A has rank $r \le n$. A necessary and sufficient condition that $Q = \mathbf{y}'A\mathbf{y}$ follows a chi-square distribution with r degrees of freedom is provided by the matrix equality $A\Sigma A = A$.

Consider the special case in which $y_1, y_2, \ldots, y_n$ are independent univariate normal random variables with means zero and a constant variance σ^2. This implies that the $n \times 1$ vector z with elements

$z_1 = y_1/\sigma, z_2 = y_2/\sigma, \ldots, z_n = y_n/\sigma$ follows a multivariate normal distribution with mean vector zero and covariance matrix I_n. Hence, the quadratic form $Q = \mathbf{y}'A\mathbf{y}/\sigma^2$ follows a chi-square distribution with r degrees of freedom if and only if the matrix A is idempotent ($AA = A$).

ii. **Independence of two quadratic forms**
Assume that the $n \times 1$ vector $\mathbf{y} = (y_1, y_2, \ldots, y_n)'$ follows a multivariate normal distribution with mean vector zero and nonsingular covariance matrix Σ. Consider the two quadratic forms $Q_1 = \mathbf{y}'A\mathbf{y}$ and $Q_2 = \mathbf{y}'B\mathbf{y}$, where A and B are symmetric $n \times n$ matrices. The random variables $Q_1 = \mathbf{y}'A\mathbf{y}$ and $Q_2 = \mathbf{y}'B\mathbf{y}$ are independent if and only if $A\Sigma B = O$, an $n \times n$ matrix of zeros. If $\Sigma = \sigma^2 I$, the condition simplifies to $AB = O$.

iii. **Results on the distribution of a sum of quadratic forms**
Assume that the $n \times 1$ vector $\mathbf{y} = (y_1, y_2, \ldots, y_n)'$ follows a multivariate normal distribution with mean vector zero and nonsingular covariance matrix $\Sigma = \sigma^2 I$. Let $Q = Q_1 + \cdots + Q_{k-1} + Q_k$, where $Q, Q_1, \ldots, Q_{k-1}, Q_k$ are $k + 1$ quadratic forms in $\mathbf{y}$. Let Q/σ^2 be chi-square with r degrees of freedom, let Q_i/σ^2 be chi-square with r_i degrees of freedom ($i = 1, 2, \ldots, k - 1$), and let Q_k be nonnegative. Then the random variables $Q_1, \ldots, Q_{k-1}, Q_k$ are mutually independent and, hence, Q_k/σ^2 is chi-square with $r_k = r - (r_1 + r_2 + \cdots + r_{k-1})$ degrees of freedom.

EXERCISES

3.1. Consider the matrix

$$A = \begin{bmatrix} 2 & 0 & 1 \\ 3 & 2 & 2 \\ 2 & 1 & 4 \end{bmatrix}$$

a. Obtain the transpose A' of A.

b. Calculate $A'A$.

c. Obtain the trace $\text{tr}(A)$ of A; obtain $\text{tr}(A'A)$.

d. Calculate the determinant $\det(A)$ of A; obtain $\det(A'A)$.

3.2. Consider the matrix

$$X = \begin{bmatrix} 1 & -1 & -1 \\ 1 & +1 & -1 \\ 1 & -1 & +1 \\ 1 & +1 & +1 \end{bmatrix} \quad \text{and the vector} \quad \mathbf{y} = \begin{bmatrix} 4 \\ 3 \\ 5 \\ 7 \end{bmatrix}$$

a. Find $X'X$, $(X'X)^{-1}$, $X'\mathbf{y}$, and $(X'X)^{-1} X'\mathbf{y}$.

b. Describe the structure of $X'X$ and $(X'X)^{-1}$.

3.3. Consider the matrix

$$X = \begin{bmatrix} 1 & x_{11} & x_{12} \\ 1 & x_{21} & x_{22} \\ 1 & x_{31} & x_{32} \\ 1 & x_{41} & x_{42} \\ 1 & x_{51} & x_{52} \end{bmatrix}$$

a. Find $X'X$.

b. Obtain

$$\sum_{i=1}^{5} x_{i1}, \quad \sum_{i=1}^{5} x_{i2}, \quad \sum_{i=1}^{5} x_{i1}^2, \quad \sum_{i=1}^{5} x_{i2}^2,$$

and

$$\sum_{i=1}^{5} x_{i1}x_{i2}$$

How are these quantities related to the elements of the $X'X$ matrix?

3.4. Consider the matrix

$$A = \begin{bmatrix} 2 & -1 \\ -1 & 2 \end{bmatrix}$$

a. Calculate the determinant and the inverse of the matrix A.

b. Obtain the eigenvalues and the eigenvectors of the matrix A.

c. Determine the spectral representation of the matrix A.

d. Determine whether the matrix A can be a covariance matrix. If so, determine the corresponding correlation matrix.

3.5. Consider the matrix

$$A = \begin{bmatrix} 3 & 1 & 1 \\ 1 & 4 & 2 \\ 1 & 2 & 2 \end{bmatrix}$$

a. Calculate the determinant and the inverse of the matrix A.

b. Obtain the eigenvalues and the eigenvectors of the matrix A.

c. Determine the spectral representation of the matrix A.

d. Determine whether the matrix A can be a covariance matrix. If so, determine the corresponding correlation matrix.

3.6. Consider the matrix

$$A = \begin{bmatrix} 2 & 1 & 1 \\ 1 & 4 & 0 \\ 1 & 0 & 1 \end{bmatrix}$$

a. Calculate the determinant and the inverse of the matrix A.

b. Obtain the eigenvalues and the eigenvectors of the matrix A.

c. Determine whether or not the matrix A can be a covariance matrix. If so, determine the corresponding correlation matrix.

3.7. Consider the matrix

$$A = \begin{bmatrix} 2 & 1 & 3 \\ 1 & 2 & 3 \\ 3 & 3 & 6 \end{bmatrix}$$

a. Calculate the determinant and the inverse of the matrix A.

b. Obtain the eigenvalues and the eigenvectors of the matrix A.

c. Determine whether the matrix A can be a covariance matrix. If so, determine the corresponding correlation matrix. Identify the linear combination that has variance zero.

3.8. Consider the matrices

$$A = \begin{bmatrix} 1 & 4 & 2 \\ 3 & 1 & 2 \end{bmatrix} \quad \text{and} \quad B = \begin{bmatrix} 4 & 1 \\ 2 & 2 \\ 2 & 4 \end{bmatrix}$$

a. Determine the matrix product AB and obtain its rank.

b. Determine the matrix product BA and obtain its rank.

3.9. Obtain a 3×3 orthogonal matrix other than the trivial case of the identity matrix.

3.10. Specify a linear regression model for the hardness data in Table 1.1. Specify the 14×2 matrix X and the 14×1 vector of responses $\mathbf{y}$. Determine the 2×2 matrix $X'X$ and its inverse $(X'X)^{-1}$. Using matrix algebra, write down the expression for the least squares estimates in $\hat{\boldsymbol{\beta}} = (X'X)^{-1}X'\mathbf{y}$.

3.11. Consider a trivariate normal distribution $\mathbf{y} = (y_1, y_2, y_3)'$ with mean vector $E(\mathbf{y}) = (2, 6, 4)'$ and covariance matrix

$$V(\mathbf{y}) = \begin{bmatrix} 1 & 0 & 1 \\ 0 & 2 & -1 \\ 1 & -1 & 3 \end{bmatrix}$$

a. Determine the marginal bivariate distribution of $(y_1, y_2)'$.

b. Determine the conditional bivariate distribution of $(y_1, y_2)'$, given that $y_3 = 5$.

3.12. Let z_1, z_2, z_3 be random variables with mean vector and covariance matrix

$$\boldsymbol{\mu} = \begin{bmatrix} 1 \\ 2 \\ 3 \end{bmatrix}; \quad V = \begin{bmatrix} 3 & 2 & 1 \\ 2 & 2 & 1 \\ 1 & 1 & 1 \end{bmatrix}$$

Define the new variables

$$y_1 = z_1 + 2z_3; \quad y_2 = z_1 + z_2 - z_3;$$
$$y_3 = 2z_1 + z_2 + z_3 - 7$$

a. Find the mean vector and the covariance matrix of (y_1, y_2, y_3).

b. Find the mean and variance of $y = \frac{1}{3}(y_1 + y_2 + y_3)$.

3.13. Let X be an $n \times p$ matrix. Assume that the inverse $(X'X)^{-1}$ exists, and define

$A = (X'X)^{-1}X'$ and $H = XA$.

a. Show that (i) $HH = H$;
(ii) $(I - H)(I - H) = (I - H)$; and
(iii) $HX = X$.

b. Find (i) $A(I - H)$; (ii) $(I - H)A'$;
(iii) $H(I - H)$; and (iv) $(I - H)'H'$.

3.14. Suppose that the covariance matrix of a vector $\mathbf{y}$ is $\sigma^2 I$, where I is an $n \times n$ identity matrix. Using the matrices A and H in Exercise 3.13, find the covariance matrix of

a. $A\mathbf{y}$

b. $H\mathbf{y}$

c. $(I - H)\mathbf{y}$

d. $\begin{bmatrix} A \\ -- \\ I - H \end{bmatrix}\mathbf{y}$

3.15. Consider the bivariate random vector with covariance matrix

$$A = \begin{bmatrix} 1 & \rho \\ \rho & 1 \end{bmatrix};$$

$|\rho| < 1$ is the correlation coefficient

a. Show that the eigenvalues of the matrix A are given by $1 + \rho$ and $1 - \rho$.

b. Show that the normalized eigenvectors that correspond to these two eigenvalues are given by

$$\mathbf{p}_1 = \begin{bmatrix} 1/\sqrt{2} \\ 1/\sqrt{2} \end{bmatrix} \quad \text{and} \quad \mathbf{p}_2 = \begin{bmatrix} -1/\sqrt{2} \\ 1/\sqrt{2} \end{bmatrix}$$

c. Confirm the spectral representation of the covariance matrix A. That is, show that

$$P\Lambda P' = \begin{bmatrix} 1/\sqrt{2} & -1/\sqrt{2} \\ 1/\sqrt{2} & 1/\sqrt{2} \end{bmatrix} \begin{bmatrix} 1+\rho & 0 \\ 0 & 1-\rho \end{bmatrix}$$

$$\begin{bmatrix} 1/\sqrt{2} & 1/\sqrt{2} \\ -1/\sqrt{2} & 1/\sqrt{2} \end{bmatrix} = A = \begin{bmatrix} 1 & \rho \\ \rho & 1 \end{bmatrix}$$

d. Generate $n = 20$ independent random vectors from a bivariate normal distribution with mean vector zero and covariance matrix A.

The result in (c) helps you with the generation (simulation) of correlated random variables. Assume that you want to generate bivariate normal random variables (y_1, y_2) with covariance matrix A given previously. You can achieve this by generating independent random variables (x_1, x_2) with variances $1 + \rho$ and $1 - \rho$ and applying the transformation

$$\begin{bmatrix} y_1 \\ y_2 \end{bmatrix} = \begin{bmatrix} 1/\sqrt{2} & -1/\sqrt{2} \\ 1/\sqrt{2} & 1/\sqrt{2} \end{bmatrix} \begin{bmatrix} x_1 \\ x_2 \end{bmatrix}$$

The resulting random vector (y_1, y_2) has a bivariate normal distribution with covariance matrix A.

Most computer programs make it easy to generate univariate normal random variables, but they lack routines for simulating correlated random vectors. For generating multivariate normal random variables with a certain specified covariance matrix, one can use the spectral representation of the covariance matrix to determine the matrix that transforms independent random variables into correlated random variables.

3.16. The linear regression model in Chapter 2, $y = \beta_0 + \beta_1 x + \varepsilon$, assumes that the settings of the regressor variable are fixed (nonstochastic). In Section 2.9.2, we showed that the standard regression results still apply in the random x case, provided that the random components on the right-hand side of the model (the regressor x and the error ε) are independent. As this exercise now shows, difficulties arise if the error and the regressor are dependent.

Assume that the vector (ε, x) in the linear regression model $y = \beta_0 + \beta_1 x + \varepsilon$ follows a bivariate normal distribution with mean vector $(0, \mu_x)$ and covariance matrix

$$V(\varepsilon, x) = \begin{bmatrix} \sigma_\varepsilon^2 & \rho_{\varepsilon x}\sigma_\varepsilon\sigma_x \\ \rho_{\varepsilon x}\sigma_\varepsilon\sigma_x & \sigma_x^2 \end{bmatrix}$$

$\rho_{\varepsilon x}$ is the correlation between the error and the random regressor.

a. Use the result on linear transformations of (jointly) normal random variables in Section 3.4 and show that the distribution of

$$\begin{bmatrix} y \\ x \end{bmatrix} = \begin{bmatrix} \beta_0 \\ 0 \end{bmatrix} + \begin{bmatrix} \beta_1 & 1 \\ 1 & 0 \end{bmatrix} \begin{bmatrix} x \\ \varepsilon \end{bmatrix}$$

is bivariate normal with mean vector $(\beta_0 + \beta_1 \mu_x, \mu_x)'$ and covariance matrix

$$V(y, x) = \begin{bmatrix} \sigma_y^2 & \rho_{yx}\sigma_y\sigma_x \\ \rho_{yx}\sigma_y\sigma_x & \sigma_x^2 \end{bmatrix}$$

$$= \begin{bmatrix} (\beta_1)^2\sigma_x^2 + 2\beta_1\rho_{\varepsilon x}\sigma_\varepsilon\sigma_x + \sigma_\varepsilon^2 & \beta_1\sigma_x^2 + \rho_{\varepsilon x}\sigma_\varepsilon\sigma_x \\ \beta_1\sigma_x^2 + \rho_{\varepsilon x}\sigma_\varepsilon\sigma_x & \sigma_x^2 \end{bmatrix}$$

b. Use the result in Section 3.4 on conditional distributions and show that the conditional distribution of y given x is (univariate) normal with mean

$$\mu_{y.x} = \beta_0 + \beta_1\mu_x + \frac{\beta_1\sigma_x^2 + \rho_{\varepsilon x}\sigma_\varepsilon\sigma_x}{\sigma_x^2}(x - \mu_x)$$
$$= [\beta_0 - (\sigma_\varepsilon/\sigma_x)\rho_{\varepsilon x}\mu_x] + [\beta_1 + (\sigma_\varepsilon/\sigma_x)\rho_{\varepsilon x}]x$$
$$= \beta_0^* + \beta_1^* x$$

and variance

$$\sigma_{y.x}^2 = \sigma_y^2 - \frac{[\rho_{yx}\sigma_y\sigma_x]^2}{\sigma_x^2} = \sigma_y^2(1 - \rho_{yx}^2)$$

c. Interpret the result in (b).

The least squares estimate $\hat{\beta}_1 = \sum (x_i - \bar{x})y_i / \sum (x_i - \bar{x})^2$ estimates the slope in the conditional expectation $\mu_{y.x} = \beta_0^* + \beta_1^* x$. The slope $\beta_1^* = \beta_1 + (\sigma_\varepsilon/\sigma_x)\rho_{\varepsilon x}$ equals β_1 only if the error and the covariate are independent. This result shows that with correlation between the error and the regressor, the least squares estimate is no longer an unbiased estimate of β_1. Discuss why this has important implications. Can you think of situations in which the error in the regression relationship depends on the value of the regressor variable?

In Section 2.9.2, we assumed that the conditional expectation $\mu_{y.x}$ is β_1, tacitly implying that $\rho_{\varepsilon x} = 0$. In this situation, the least squares estimate is an unbiased estimate of β_1.

d. Consider the following simulation experiment. Generate n = 20 independent random vectors $(\varepsilon, x)$with mean vector $(0, 1)$and covariance matrix $V(\varepsilon, x) =$

$$\begin{bmatrix} 1 & \rho_{\varepsilon x} \\ \rho_{\varepsilon x} & 1 \end{bmatrix}; \rho_{\varepsilon x} = 0.5.$$ Use the approach in Exercise 3.15(d) if your computer software does not allow you to simulate multivariate normal random variables. Generate the 20 × 1 response vector $\mathbf{y}$ from $\mathbf{y} = 2\mathbf{x} + \varepsilon$ (i.e., $\beta_0 = 0$ and $\beta_1 = 2$). Obtain the least squares estimate $\hat{\beta}_1$. Repeat this exercise for 1,000 independent samples, and obtain the sampling distribution of $\hat{\beta}_1$. Confirm that the mean of the sampling distribution is given by $\beta_1 + \rho_{\varepsilon x} = 2 + \rho_{\varepsilon x}$. Repeat the simulation experiment for $\rho_{\varepsilon x} = -0.5$ and $\rho_{\varepsilon x} = 0$. Demonstrate that the ordinary least squares estimate is not an unbiased estimate of β_1 if the independence between the regressor and the error is violated.

3.17. Consider a trivariate normal distribution $\mathbf{y} = (y_1, y_2, y_3)'$ with mean vector zero and covariance matrix $V(\mathbf{y}) = \sigma^2 I$. Determine the distribution of the quadratic form $y_1^2 + 0.5y_2^2 + 0.5y_3^2 + y_2 y_3$.

3.18. Consider a bivariate normal distribution $\mathbf{y} = (y_1, y_2)'$ with mean vector zero and covariance matrix $V(\mathbf{y}) = \sigma^2 I$. Show the independence of the two quadratic forms $(y_1 - y_2)^2$and $(y_1 + y_2)^2$.

4 Multiple Linear Regression Model

4.1 INTRODUCTION

In this chapter we consider the general linear model introduced in Eq. (1.10),

$$y = \beta_0 + \beta_1 x_1 + \cdots + \beta_p x_p + \epsilon \tag{4.1}$$

which links a response variable y to several independent (also called explanatory or predictor) variables $x_1, x_2, \ldots, x_p$. We discuss how to estimate the model parameters $\boldsymbol{\beta} = (\beta_0, \beta_1, \ldots, \beta_p)'$ and how to test various hypotheses about them. You may find the subsequent discussion interesting from a theoretical standpoint because it uses linear algebra to establish general results. It also maps out an elegant geometric approach to least squares regression. Be prepared for subspaces, basis vectors, and orthogonal projections.

4.1.1 TWO EXAMPLES

In order to motivate the general notation, we start our discussion with two examples: the urea formaldehyde foam insulation (UFFI) example and the gas consumption data of Chapter 1.

UFFI Example

In Example 1.2.5 of Chapter 1, we considered 12 homes without UFFI ($x_1 = 0$) and 12 homes with insulation ($x_1 = 1$). For each home we obtained an air-tightness measure (x_2) and a reading of its ambient formaldehyde concentration (y). The model in Eq. (1.6) relates the ambient formaldehyde concentration (y) of the ith home to its air tightness (x_2) and the presence of UFFI (x_1):

$$y_i = \beta_0 + \beta_1 x_{i1} + \beta_2 x_{i2} + \epsilon_i, \quad i = 1, 2, \ldots, 24 \tag{4.2}$$

Table 1.2 lists the information on the 12 houses without UFFI ($x_1 = 0$) first; the remaining 12 homes with UFFI ($x_1 = 1$) are listed second. Note that Chapter 1 uses z and x for the predictors x_1 and x_2. The 24 equations resulting from model (4.2),

$$\begin{aligned}
31.33 &= \beta_0 + \beta_1 0 + \beta_2 0 + \epsilon_1 \\
28.57 &= \beta_0 + \beta_1 0 + \beta_2 1 + \epsilon_2 \\
\vdots & \qquad \vdots \\
56.67 &= \beta_0 + \beta_1 0 + \beta_2 9 + \epsilon_{12} \\
43.58 &= \beta_0 + \beta_1 1 + \beta_2 1 + \epsilon_{13} \\
\vdots & \qquad \vdots \\
70.34 &= \beta_0 + \beta_1 1 + \beta_2 10 + \epsilon_{24}
\end{aligned}$$

can be written in vector form,

$$\begin{bmatrix} 31.33 \\ 28.57 \\ \vdots \\ 56.67 \\ 43.58 \\ \vdots \\ 70.34 \end{bmatrix} = \begin{bmatrix} 1 & 0 & 0 \\ 1 & 0 & 1 \\ \vdots & \vdots & \vdots \\ 1 & 0 & 9 \\ 1 & 1 & 1 \\ \vdots & \vdots & \vdots \\ 1 & 1 & 10 \end{bmatrix} \begin{bmatrix} \beta_0 \\ \beta_1 \\ \beta_2 \end{bmatrix} + \begin{bmatrix} \epsilon_1 \\ \epsilon_2 \\ \vdots \\ \epsilon_{12} \\ \epsilon_{13} \\ \vdots \\ \epsilon_{24} \end{bmatrix}$$

In short,

$$\boldsymbol{y} = X\boldsymbol{\beta} + \boldsymbol{\epsilon} \tag{4.3}$$

where

$$\boldsymbol{y} = \begin{bmatrix} 31.33 \\ 28.57 \\ \vdots \\ 70.34 \end{bmatrix}; \quad X = \begin{bmatrix} 1 & 0 & 0 \\ 1 & 0 & 1 \\ \vdots & \vdots & \vdots \\ 1 & 0 & 9 \\ 1 & 1 & 1 \\ \vdots & \vdots & \vdots \\ 1 & 1 & 10 \end{bmatrix}; \quad \boldsymbol{\beta} = \begin{bmatrix} \beta_0 \\ \beta_1 \\ \beta_2 \end{bmatrix}; \quad \text{and} \quad \boldsymbol{\epsilon} = \begin{bmatrix} \epsilon_1 \\ \epsilon_2 \\ \vdots \\ \epsilon_{24} \end{bmatrix}$$

Gas Consumption Data

In Example 1.2.7 of Chapter 1 we relate the fuel efficiency on each of 38 cars to their weight, engine displacement, and number of cylinders. Consider the model

$$y_i = \beta_0 + \beta_1 x_{i1} + \beta_2 x_{i2} + \beta_3 x_{i3} + \epsilon_i, \quad i = 1, 2, \ldots, 38 \tag{4.4}$$

where y_i = gas consumption (miles per gallon) for the ith car

x_{i1} = weight of the ith car

x_{i2} = engine displacement for the ith car

x_{i3} = number of cylinders for the ith car

The resulting 38 equations

$$\begin{aligned}
16.9 &= \beta_0 + \beta_1 4.360 + \beta_2 350 + \beta_3 8 + \epsilon_1 \\
15.5 &= \beta_0 + \beta_1 4.054 + \beta_2 351 + \beta_3 8 + \epsilon_2 \\
\vdots \quad & \quad \vdots \qquad\qquad\qquad \vdots \\
31.9 &= \beta_0 + \beta_1 1.925 + \beta_2 89 + \beta_3 4 + \epsilon_{38}
\end{aligned}$$

can be written in vector form as

$$\begin{bmatrix} 16.9 \\ 15.5 \\ \vdots \\ 31.9 \end{bmatrix} = \begin{bmatrix} 1 & 4.360 & 350 & 8 \\ 1 & 4.054 & 351 & 8 \\ \vdots & \vdots & \vdots & \vdots \\ 1 & 1.925 & 89 & 4 \end{bmatrix} \begin{bmatrix} \beta_0 \\ \beta_1 \\ \beta_2 \\ \beta_3 \end{bmatrix} + \begin{bmatrix} \epsilon_1 \\ \epsilon_2 \\ \vdots \\ \epsilon_{38} \end{bmatrix}$$

In short,

$$\boldsymbol{y} = X\boldsymbol{\beta} + \boldsymbol{\epsilon}$$

where

$$\boldsymbol{y} = \begin{bmatrix} 16.9 \\ 15.5 \\ \vdots \\ 31.9 \end{bmatrix}; \quad X = \begin{bmatrix} 1 & 4.360 & 350 & 8 \\ 1 & 4.054 & 351 & 8 \\ \vdots & \vdots & \vdots & \vdots \\ 1 & 1.925 & 89 & 4 \end{bmatrix}; \quad \boldsymbol{\beta} = \begin{bmatrix} \beta_0 \\ \beta_1 \\ \beta_2 \\ \beta_3 \end{bmatrix}; \quad \text{and} \quad \boldsymbol{\epsilon} = \begin{bmatrix} \epsilon_1 \\ \epsilon_2 \\ \vdots \\ \epsilon_{38} \end{bmatrix} \tag{4.5}$$

4.1.2 THE GENERAL LINEAR MODEL

These two examples show us how we can write the general linear model (4.1) in vector form. Suppose that we have information on n cases, or subjects $i = 1, 2, \ldots, n$. Let y_i be the observed value on the response variable and let x_{i1}, $x_{i2}, \ldots, x_{ip}$ be the values on the independent or predictor variables of the ith case. The values of the p predictor variables are treated as fixed constants; however, the responses are subject to variability. The model for the response of case i is written as

$$\begin{aligned}
y_i &= \beta_0 + \beta_1 x_{i1} + \cdots + \beta_p x_{ip} + \epsilon_i \\
&= \mu_i + \epsilon_i
\end{aligned} \tag{4.6}$$

where $\mu_i = \beta_0 + \beta_1 x_{i1} + \cdots + \beta_p x_{ip}$ is a deterministic component that is affected by the regressor variables and ϵ_i is a term that captures the effect of all other variables that are not included in the model.

We assume that ϵ_i is a random variable with mean $E(\epsilon_i) = 0$ and variance $V(\epsilon_i) = \sigma^2$, and we suppose that the ϵ_i are normally distributed. Furthermore, we assume that the errors from different cases, $\epsilon_1, \ldots, \epsilon_n$, are independent random variables. These assumptions imply that the responses $y_1, \ldots, y_n$ are independent normal random variables with mean $E(y_i) = \mu_i = \beta_0 + \beta_1 x_{i1} + \cdots + \beta_p x_{ip}$ and variance $V(y_i) = \sigma^2$.

We assume that the variance $V(y_i)$ is the same for each case. Note that this is an assumption that needs to be checked because one needs to check all other model assumptions, such as the form of the deterministic relationship and the normal distribution of the errors.

The n equations in (4.6) can be rewritten in vector form,

$$\begin{bmatrix} y_1 \\ \vdots \\ y_n \end{bmatrix} = \begin{bmatrix} \beta_0 + \beta_1 x_{11} + \cdots + \beta_p x_{1p} \\ \vdots \\ \beta_0 + \beta_1 x_{n1} + \cdots + \beta_p x_{np} \end{bmatrix} + \begin{bmatrix} \epsilon_1 \\ \vdots \\ \epsilon_n \end{bmatrix}$$

In short,

$$\mathbf{y} = X\boldsymbol{\beta} + \boldsymbol{\epsilon} \tag{4.7}$$

where

$$\mathbf{y} = \begin{bmatrix} y_1 \\ y_2 \\ \vdots \\ y_n \end{bmatrix}; \quad X = \begin{bmatrix} 1 & x_{11} & x_{12} & \cdots & x_{1p} \\ 1 & x_{21} & x_{22} & \cdots & x_{2p} \\ \vdots & \vdots & \vdots & & \vdots \\ 1 & x_{n1} & x_{n2} & \cdots & x_{np} \end{bmatrix}; \quad \boldsymbol{\beta} = \begin{bmatrix} \beta_0 \\ \beta_1 \\ \vdots \\ \beta_p \end{bmatrix}; \quad \text{and} \quad \boldsymbol{\epsilon} = \begin{bmatrix} \epsilon_1 \\ \epsilon_2 \\ \vdots \\ \epsilon_n \end{bmatrix}$$

You should convince yourself that this representation is correct by multiplying out the first few elements of $X\boldsymbol{\beta}$.

The assumptions on the errors in this model can also be written in vector form. We write $\boldsymbol{\epsilon} \sim N(\mathbf{0}, \sigma^2 I)$, a multivariate normal distribution with mean vector $E(\boldsymbol{\epsilon}) = \mathbf{0}$ and covariance matrix $V(\boldsymbol{\epsilon}) = \sigma^2 I$. Similarly, we write $\mathbf{y} \sim N(X\boldsymbol{\beta}, \sigma^2 I)$, a multivariate normal distribution with mean vector $E(\mathbf{y}) = X\boldsymbol{\beta}$ and covariance matrix $V(\mathbf{y}) = \sigma^2 I$.

4.2 ESTIMATION OF THE MODEL

We now consider the estimation of the unknown parameters: the $(p+1)$ regression parameters $\boldsymbol{\beta}$, and the variance of the errors σ^2. Since $y_i \sim N(\mu_i, \sigma^2)$ with $\mu_i = \beta_0 + \beta_1 x_{i1} + \cdots + \beta_p x_{ip}$ are independent, it is straightforward to write down the joint probability density $p(y_1, \ldots, y_n \mid \boldsymbol{\beta}, \sigma^2)$. Treating this, for given data $\mathbf{y}$, as a function of the parameters leads to the likelihood function

$$L(\boldsymbol{\beta}, \sigma^2 \mid y_1, \ldots, y_n) = (1/\sqrt{2\pi}\sigma)^n \exp\left[-\sum_{i=1}^{n} (y_i - \mu_i)^2 / 2\sigma^2 \right] \tag{4.8}$$

Maximizing the likelihood function L with respect to $\boldsymbol{\beta}$ is equivalent to minimizing $S(\boldsymbol{\beta}) = \sum_{i=1}^{n} (y_i - \mu_i)^2$ with respect to $\boldsymbol{\beta}$. This is because the exponent in Eq. (4.8) is the only term containing $\boldsymbol{\beta}$. The sum of squares $S(\boldsymbol{\beta})$ can be written in vector notation,

$$S(\boldsymbol{\beta}) = (\mathbf{y} - \boldsymbol{\mu})'(\mathbf{y} - \boldsymbol{\mu}) = (\mathbf{y} - X\boldsymbol{\beta})'(\mathbf{y} - X\boldsymbol{\beta}), \quad \text{since } \boldsymbol{\mu} = X\boldsymbol{\beta} \tag{4.9}$$

The minimization of $S(\boldsymbol{\beta})$ with respect to $\boldsymbol{\beta}$ is known as **least squares estimation**, and for normal errors it is equivalent to maximum likelihood estimation. We determine the least squares estimates by obtaining the first derivatives of $S(\boldsymbol{\beta})$ with respect to the parameters $\beta_0, \beta_1, \ldots, \beta_p$, and by setting these $(p+1)$ derivatives equal to zero.

The appendix shows that this leads to the $(p+1)$ equations

$$X'X\hat{\boldsymbol{\beta}} = X'\boldsymbol{y} \tag{4.10}$$

These equations are referred to as the **normal equations**. The matrix X is assumed to have full column rank $p+1$. Hence, the $(p+1) \times (p+1)$ matrix $X'X$ is nonsingular and the solution of Eq. (4.10) is given by

$$\hat{\boldsymbol{\beta}} = (X'X)^{-1}X'\boldsymbol{y} \tag{4.11}$$

The estimate $\hat{\boldsymbol{\beta}}$ in Eq. (4.11) minimizes $S(\boldsymbol{\beta})$, and is known as the **least squares estimate** (LSE) of $\boldsymbol{\beta}$.

4.2.1 A GEOMETRIC INTERPRETATION OF LEAST SQUARES

The model in Eq. (4.7) can be written as

$$\begin{aligned} \boldsymbol{y} &= \beta_0\mathbf{1} + \beta_1\boldsymbol{x}_1 + \cdots + \beta_p\boldsymbol{x}_p + \boldsymbol{\epsilon} \\ &= \boldsymbol{\mu} + \boldsymbol{\epsilon} \end{aligned} \tag{4.12}$$

where the $(n \times 1)$ vectors $\boldsymbol{y}$ and $\boldsymbol{\epsilon}$ are as defined before, and the $(n \times 1)$ vectors $\mathbf{1} = (1, 1, \ldots, 1)'$ and $\boldsymbol{x}_j = (x_{1j}, x_{2j}, \ldots, x_{nj})'$, for $j = 1, 2, \ldots, p$, represent the columns of the matrix X. Thus, $X = (\mathbf{1}, \boldsymbol{x}_1, \ldots, \boldsymbol{x}_p)$ and $\boldsymbol{\mu} = X\boldsymbol{\beta} = \beta_0\mathbf{1} + \beta_1\boldsymbol{x}_1 + \cdots + \beta_p\boldsymbol{x}_p$.

The representation in Eq. (4.12) shows that the deterministic component $\boldsymbol{\mu}$ is a linear combination of the vectors $\mathbf{1}, \boldsymbol{x}_1, \ldots, \boldsymbol{x}_p$. Let $L(\mathbf{1}, \boldsymbol{x}_1, \ldots, \boldsymbol{x}_p)$ be the set of all linear combinations of these vectors. If we assume that these vectors are not linearly dependent, $L(X) = L(\mathbf{1}, \boldsymbol{x}_1, \ldots, \boldsymbol{x}_p)$ is a subspace of R^n of dimension $p+1$. Note that the assumption that $\mathbf{1}, \boldsymbol{x}_1, \ldots, \boldsymbol{x}_p$ are not linearly dependent is the same as saying that X has rank $p+1$.

We want to explain these concepts slowly because they are essential for understanding the geometric interpretation that follows. First, note that the dimension of the regressor vectors $\mathbf{1}, \boldsymbol{x}_1, \ldots, \boldsymbol{x}_p$ is n, the number of cases. When we display the $(p+1)$ regressor vectors, we do that in n-dimensional Euclidean space R^n. The coordinates on each regressor vector correspond to the regressor's values on the n cases. For example, the regressor vector $\boldsymbol{x}$ may represent the air tightness of a home, and the dimension of this vector is 24, if measurements on 24 homes are taken. Note that for models with an intercept, one of the regressor columns is always the vector of ones, $\mathbf{1}$.

Obviously, it is impossible to graph vectors in 24-dimensional space, but you can get a good idea of this by considering lower dimensional situations. Consider the case in which $n = 3$, and use two regressor columns: the unit vector

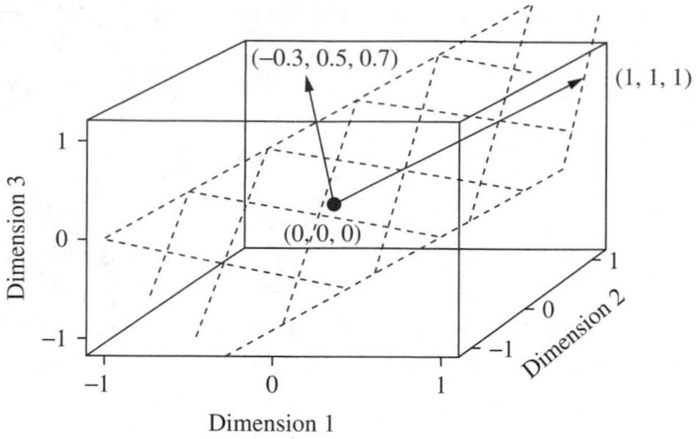

FIGURE 4.1 Two Vectors in Three-Dimensional Space, and the Two-Dimensional Space Spanned by These Two Vectors

$\mathbf{1} = (1, 1, 1)'$ and $\boldsymbol{x} = (-0.3, 0.5, 0.7)'$. These two vectors are graphed in three-dimensional space in Figure 4.1. Any linear combination of these two vectors results in a vector that lies in the two-dimensional space that is spanned by the vectors $\mathbf{1}$ and $\boldsymbol{x}$. We highlight this by shading the plane that contains all linear combinations. We see that $L(\mathbf{1}, \boldsymbol{x})$ is a subspace of R^3, and its dimension is 2.

Observe that we have selected two vectors $\mathbf{1}$ and $\boldsymbol{x}$ that are **not** linearly dependent. This means that one of the two vectors cannot be written as a multiple of the other. This is the case in our example. Note that the matrix

$$X = [\mathbf{1}, \boldsymbol{x}] = \begin{bmatrix} 1 & -0.3 \\ 1 & 0.5 \\ 1 & 0.7 \end{bmatrix}$$

has full column rank, 2.

What would happen if two regressor vectors were linearly dependent; for example, if $\mathbf{1} = (1, 1, 1)'$ and $\boldsymbol{x} = (0.5, 0.5, 0.5)'$? Here, every linear combination of $\mathbf{1}$ and $\boldsymbol{x}$, $\alpha_1 \mathbf{1} + \alpha_2 \boldsymbol{x} = \alpha_1 \mathbf{1} + \alpha_2 (0.5) \mathbf{1} = (\alpha_1 + 0.5\alpha_2)\mathbf{1}$, is a multiple of $\mathbf{1}$. Hence, the set of all linear combinations are points along the unit vector, and $L(\mathbf{1}, \boldsymbol{x})$ defines a subspace of dimension 1. You can also see this from the rank of the matrix X: The rank of

$$X = [\mathbf{1}, \boldsymbol{x}] = \begin{bmatrix} 1 & 0.5 \\ 1 & 0.5 \\ 1 & 0.5 \end{bmatrix}$$

is one; X does not have full column rank.

If we contemplate a model with two regressor columns, $\mathbf{1}$ and $\boldsymbol{x}$, then we suppose that $\mathbf{1}$ and $\boldsymbol{x}$ are not linearly dependent. If they were linearly dependent, we would encounter difficulties because an infinite number of linear combinations could be used to represent each point in the subspace spanned by $\mathbf{1}$ and $\boldsymbol{x}$. You can see this from our example. There is an infinite number of values for α_1 and α_2 that result in a given value $\alpha_1 + 0.5\alpha_2 = c$.

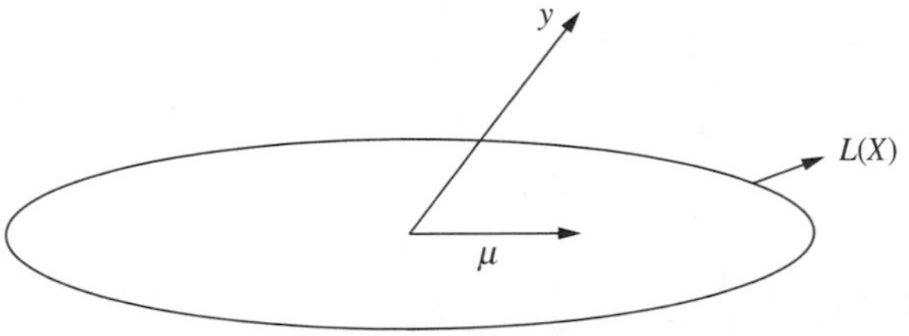

FIGURE 4.2 Geometric Representation of the Response Vector y and the Subspace $L(X)$

Now we are ready to go to the more general case with a large number of cases, n. Suppose that there are two regressors ($p = 2$) and three regressor columns $\mathbf{1}$, $\boldsymbol{x}_1$, and $\boldsymbol{x}_2$. We assume that these three columns are not linearly dependent and that the matrix $X = [\mathbf{1}, \boldsymbol{x}_1, \boldsymbol{x}_2]$ has full column rank, rank 3. The regressor vectors are elements in R^n, and the set of all linear combinations of $\mathbf{1}, \boldsymbol{x}_1, \boldsymbol{x}_2$, $L(\mathbf{1}, \boldsymbol{x}_1, \boldsymbol{x}_2)$, defines a three-dimensional subspace of R^n. If $\mathbf{1}, \boldsymbol{x}_1, \boldsymbol{x}_2$ were linearly dependent, then the subspace would be of lower dimension (either 2 or 1).

Now we consider the case with p regressors shown in Figure 4.2. The oval represents the subspace $L(X)$. The vector $\boldsymbol{\mu} = \beta_0\mathbf{1} + \beta_1\boldsymbol{x}_1 + \cdots + \beta_p\boldsymbol{x}_p$ is a linear combination of $\mathbf{1}, \boldsymbol{x}_1, \ldots, \boldsymbol{x}_p$, and is part of the subspace $L(X)$. This picture is simplified as it tries to illustrate a higher dimensional space. You need to use your imagination.

Until now, we have talked about the subspace of R^n that is spanned by the $p + 1$ regressor vectors $\mathbf{1}, \boldsymbol{x}_1, \ldots, \boldsymbol{x}_p$. Next, let us add the $(n \times 1)$ response vector $\boldsymbol{y}$ to the picture (see Figure 4.2). The response vector $\boldsymbol{y}$ is not part of the subspace $L(X)$. For a given value of $\boldsymbol{\beta}$, $X\boldsymbol{\beta}$ is a vector in the subspace; $\boldsymbol{y} - X\boldsymbol{\beta}$ is the difference between the response vector $\boldsymbol{y}$ and the vector in the subspace, and $S(\boldsymbol{\beta}) = (\boldsymbol{y} - X\boldsymbol{\beta})'(\boldsymbol{y} - X\boldsymbol{\beta})$ represents the squared length of this difference. Minimizing $S(\boldsymbol{\beta})$ with respect to $\boldsymbol{\beta}$ corresponds to finding $\hat{\boldsymbol{\beta}}$ so that $\boldsymbol{y} - X\hat{\boldsymbol{\beta}}$ has minimum length.

In other words, we must find a vector $X\hat{\boldsymbol{\beta}}$ in the subspace $L(X)$ that is "closest" to $\boldsymbol{y}$. The vector in the subspace $L(X)$ that is closest to $\boldsymbol{y}$ is obtained by making the difference $\boldsymbol{y} - X\hat{\boldsymbol{\beta}}$ perpendicular to the subspace $L(X)$; see Figure 4.3. Since $\mathbf{1}, \boldsymbol{x}_1, \ldots, \boldsymbol{x}_p$ are in the subspace, we require that $\boldsymbol{y} - X\hat{\boldsymbol{\beta}}$ is perpendicular to $\mathbf{1}$, $\boldsymbol{x}_1, \ldots,$ and $\boldsymbol{x}_p$.

This implies the equations

$$\mathbf{1}'(\boldsymbol{y} - X\hat{\boldsymbol{\beta}}) = 0$$
$$\boldsymbol{x}_1'(\boldsymbol{y} - X\hat{\boldsymbol{\beta}}) = 0$$
$$\cdots$$
$$\boldsymbol{x}_p'(\boldsymbol{y} - X\hat{\boldsymbol{\beta}}) = 0$$

Combining these $p + 1$ equations leads to

$$X'(\boldsymbol{y} - X\hat{\boldsymbol{\beta}}) = \mathbf{0}$$

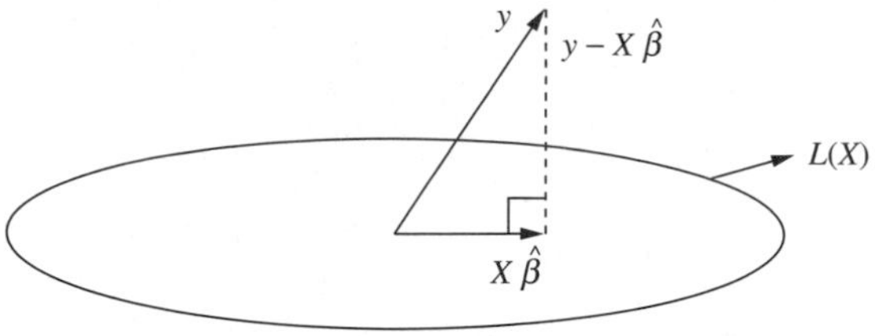

FIGURE 4.3 A Geometric View of Least Squares

and

$$X'X\hat{\beta} = X'y$$

the normal equations in Eq. (4.10) that we previously derived algebraically.

We assume that X has full column rank, $p+1$. Hence, $X'X$ has rank $(p+1)$, the inverse $(X'X)^{-1}$ exists, and the least squares estimate is given by $\hat{\beta} = (X'X)^{-1}X'y$. Notice that we have obtained the LSE solely through a geometric argument; no algebraic derivation was involved.

The vector of fitted values is given by $\hat{\mu} = X\hat{\beta}$, and the vector of residuals is $e = y - \hat{\mu} = y - X\hat{\beta}$. The geometric interpretation of least squares is quite simple. Least squares estimation amounts to finding the vector $\hat{\mu} = X\hat{\beta}$ in the subspace $L(X)$ that is closest to the observation vector y. This requires that the difference (i.e., the residual vector) is perpendicular (or othogonal) to the subspace $L(X)$. Hence, the vector of fitted values $\hat{\mu} = X\hat{\beta}$ is the **orthogonal projection** of y onto the subspace $L(X)$. In algebraic terms,

$$\hat{\mu} = X\hat{\beta} = X(X'X)^{-1}X'y = Hy$$

where $H = X(X'X)^{-1}X'$ is an $n \times n$ symmetric and idempotent matrix. It is easy to confirm that H is idempotent as

$$HH = X(X'X)^{-1}X'X(X'X)^{-1}X' = H$$

The matrix H is an important matrix because it represents the orthogonal projection of y onto $L(X)$. It is referred to as the "hat" matrix.

The vector of residuals $e = y - \hat{\mu} = y - X(X'X)^{-1}X'y = (I - H)y$ is also a projection of y, this time on the subspace of R^n that is perpendicular to $L(X)$.

The vector of fitted values $\hat{\mu} = X\hat{\beta}$ and the vector of residuals e are orthogonal, which means algebraically that

$$X'e = X'(y - X\hat{\beta}) = \mathbf{0}$$

See the normal equations in Eq. (4.10). Hence, least squares decomposes the response vector

$$y = \hat{\mu} + e = X\hat{\beta} + (y - X\hat{\beta})$$

into two orthogonal pieces. The vector of fitted values $X\hat{\beta}$ is in $L(X)$, whereas the vector of residuals $y - X\hat{\beta}$ is in the space orthogonal to $L(X)$.

It may help you to look at this in the very simplest special case in which we have $n = 2$ cases and just a single regressor column, $\mathbf{1} = (1, 1)'$. This represents the "mean" regression model, $y_i = \beta_0 + \epsilon_i$, with $i = 1, 2$. How does this look geometrically? Since the number of cases is 2, we are looking at the two-dimensional Euclidean space. Draw in the unit vector $\mathbf{1} = (1, 1)'$ and the response vector $\boldsymbol{y} = (y_1, y_2)'$. For illustration, take $\boldsymbol{y} = (0, 1)'$. We project $\boldsymbol{y} = (y_1, y_2)' = (0, 1)'$ onto the subspace $L(\mathbf{1})$, which is the 45-degree line in the two-dimensional Euclidean space. The projection leads to the vector of fitted values $\hat{\boldsymbol{\mu}} = 0.5\mathbf{1} = (0.5, 0.5)'$ and the LSE $\hat{\beta}_0 = 0.5$. The estimate is the average of the two observations, 0 and 1. The residual vector $\boldsymbol{e} = \boldsymbol{y} - \hat{\boldsymbol{\mu}} = (0 - 0.5, 1 - 0.5)' = (-0.5, 0.5)'$ and the vector of fitted values $\hat{\boldsymbol{\mu}} = (0.5, 0.5)'$ are orthogonal; that is, $\boldsymbol{e}'\hat{\boldsymbol{\mu}} = -(0.5)^2 + (0.5)^2 = 0$.

4.2.2 USEFUL PROPERTIES OF ESTIMATES AND OTHER RELATED VECTORS

Recall our model

$$\boldsymbol{y} = X\boldsymbol{\beta} + \boldsymbol{\epsilon}$$

where X is a fixed (nonrandom) matrix with full rank, and the random error $\boldsymbol{\epsilon}$ follows a distribution with mean $E(\boldsymbol{\epsilon}) = \mathbf{0}$ and covariance matrix $V(\boldsymbol{\epsilon}) = \sigma^2 I$. Usually, we also assume a normal distribution. The model implies that $E(\boldsymbol{y}) = X\boldsymbol{\beta}$ and $V(\boldsymbol{y}) = \sigma^2 I$. The LSE of the parameter vector $\boldsymbol{\beta}$ is $\hat{\boldsymbol{\beta}} = (X'X)^{-1}X'\boldsymbol{y}$. The vector of fitted values is $\hat{\boldsymbol{\mu}} = X\hat{\boldsymbol{\beta}} = H\boldsymbol{y}$ and the residual vector is $\boldsymbol{e} = \boldsymbol{y} - \hat{\boldsymbol{\mu}} = (I - H)\boldsymbol{y}$. We now study properties of these vectors and other related quantities, always assuming that the model is true.

i. Estimate $\hat{\boldsymbol{\beta}}$:

$$\begin{aligned} E(\hat{\boldsymbol{\beta}}) &= E(X'X)^{-1}X'\boldsymbol{y} \\ &= (X'X)^{-1}X'E(\boldsymbol{y}) = (X'X)^{-1}X'X\boldsymbol{\beta} = \boldsymbol{\beta} \end{aligned} \tag{4.13}$$

showing that $\hat{\boldsymbol{\beta}}$ is an unbiased estimator of $\boldsymbol{\beta}$.

$$\begin{aligned} V(\hat{\boldsymbol{\beta}}) &= V[(X'X)^{-1}X'\boldsymbol{y}] \\ &= (X'X)^{-1}X'V(\boldsymbol{y})X(X'X)^{-1} \\ &= (X'X)^{-1}X'(\sigma^2 I)(X'X)^{-1} \\ &= (X'X)^{-1}X'X(X'X)^{-1}\sigma^2 = (X'X)^{-1}\sigma^2 \end{aligned} \tag{4.14}$$

The matrix in Eq. (4.14) contains the variances of the estimates in the diagonal and the covariances in the off-diagonal elements. Let v_{ij} denote the elements of the matrix $(X'X)^{-1}$. Then $V(\hat{\beta}_i) = \sigma^2 v_{ii}$, $\text{Cov}(\hat{\beta}_i, \hat{\beta}_j) = \sigma^2 v_{ij}$, and $\text{Corr}(\hat{\beta}_i, \hat{\beta}_j) = \dfrac{v_{ij}}{(v_{ii}v_{jj})^{1/2}}$.

ii. Linear combination of estimates, $\boldsymbol{a}'\hat{\boldsymbol{\beta}}$:
The linear combination $\boldsymbol{a}'\boldsymbol{\beta}$, where $\boldsymbol{a}$ is a vector of constants of appropriate dimension, can be estimated by $\boldsymbol{a}'\hat{\boldsymbol{\beta}}$. We find

$$E(\boldsymbol{a}'\hat{\boldsymbol{\beta}}) = \boldsymbol{a}'E(\hat{\boldsymbol{\beta}}) = \boldsymbol{a}'\boldsymbol{\beta}$$

and

$$V(\boldsymbol{a}'\hat{\boldsymbol{\beta}}) = \boldsymbol{a}'V(\hat{\boldsymbol{\beta}})\boldsymbol{a} = \boldsymbol{a}'(X'X)^{-1}\boldsymbol{a}\sigma^2 \tag{4.15}$$

iii. Fitted values $\hat{\mu} = X\hat{\beta}$:

$$E(\hat{\mu}) = E(X\hat{\beta}) = XE(\hat{\beta}) = X\beta = \mu$$

and

$$\begin{aligned} V(\hat{\mu}) &= V(X\hat{\beta}) = XV(\hat{\beta})X' = X(X'X)^{-1}X'\sigma^2 \\ &= H\sigma^2 \end{aligned} \tag{4.16}$$

where $H = X(X'X)^{-1}X'$ is the idempotent projection matrix defined earlier.

iv. Residual vector $\boldsymbol{e} = \boldsymbol{y} - X\hat{\beta}$:

$$E(\boldsymbol{e}) = E(\boldsymbol{y} - X\hat{\beta}) = E(\boldsymbol{y}) - XE(\hat{\beta}) = X\beta - X\beta = \boldsymbol{0} \tag{4.17}$$

$$\begin{aligned} V(\boldsymbol{e}) &= V[(I - H)\boldsymbol{y}] = (I - H)V(\boldsymbol{y})(I - H)' \\ &= (I - H)(I - H)\sigma^2 = (I - H)\sigma^2 \end{aligned} \tag{4.18}$$

v. Statistical independence between $\hat{\beta}$ and $\boldsymbol{e}$: We stack the $(p + 1)$ vector $\hat{\beta}$ and the $(n \times 1)$ vector of residuals $\boldsymbol{e}$ to obtain the $(p + 1 + n) \times 1$ vector

$$\begin{bmatrix} \hat{\beta} \\ --- \\ \boldsymbol{e} \end{bmatrix} = \begin{bmatrix} A \\ --- \\ I - H \end{bmatrix} \boldsymbol{y} = P\boldsymbol{y}$$

with $A = (X'X)^{-1}X'$ and $H = X(X'X)^{-1}X'$. The stacked vector is a linear transformation of $\boldsymbol{y}$. Our assumption of independent normal random variables for y_i implies a multivariate normal distribution for the vector $\boldsymbol{y}$. Hence, the linear transform $P\boldsymbol{y}$ follows a multivariate normal distribution with mean

$$E\begin{bmatrix} \hat{\beta} \\ --- \\ \boldsymbol{e} \end{bmatrix} = PE(\boldsymbol{y}) = \begin{bmatrix} A \\ --- \\ I - H \end{bmatrix} X\beta = \begin{bmatrix} \beta \\ --- \\ \boldsymbol{0} \end{bmatrix}$$

and covariance matrix

$$\begin{aligned} V\begin{bmatrix} \hat{\beta} \\ --- \\ \boldsymbol{e} \end{bmatrix} &= PV(\boldsymbol{y})P' = \sigma^2 \begin{bmatrix} A \\ --- \\ I - H \end{bmatrix} [A' \mid (I - H)] \\ &= \sigma^2 \begin{bmatrix} AA' & | & A(I - H) \\ --- & & --- \\ (I - H)A' & | & (I - H)(I - H) \end{bmatrix} \\ &= \sigma^2 \begin{bmatrix} (X'X)^{-1} & | & O \\ --- & & --- \\ O' & | & (I - H) \end{bmatrix} \end{aligned}$$

Hence,

$$\begin{pmatrix} \hat{\beta} \\ \boldsymbol{e} \end{pmatrix} \sim N\left\{ \begin{pmatrix} \beta \\ --- \\ \boldsymbol{0} \end{pmatrix}, \begin{bmatrix} (X'X)^{-1}\sigma^2 & | & O \\ --- & & --- \\ O' & | & (I - H)\sigma^2 \end{bmatrix} \right\} \tag{4.19}$$

Marginal distributions of multivariate normal distributions are themselves normal. Hence, it follows that

$$\hat{\boldsymbol{\beta}} \sim N(\boldsymbol{\beta}, \sigma^2(X'X)^{-1}) \quad \text{and} \quad \boldsymbol{e} \sim N(0, \sigma^2(I - H))$$

Equation (4.19) confirms our earlier results on the means and variances of $\hat{\boldsymbol{\beta}}$ and $\boldsymbol{e}$ in Eqs. (4.13), (4.14) and (4.17), (4.18). Also note that the matrices $(X'X)^{-1}$, H, and $(I - H)$ are very useful quantities, and they will appear repeatedly.

Let $\text{Cov}(\boldsymbol{e}, \hat{\boldsymbol{\beta}})$ represents the $n \times (p+1)$ matrix of covariances between the residuals $\hat{\boldsymbol{e}}$ and the parameter estimates $\hat{\boldsymbol{\beta}}$. That is,

$$\text{Cov}(\boldsymbol{e}, \hat{\boldsymbol{\beta}}) = \begin{bmatrix} \text{Cov}(e_1, \hat{\beta}_0) & \text{Cov}(e_1, \hat{\beta}_1) & \cdots & \text{Cov}(e_1, \hat{\beta}_p) \\ \vdots & \vdots & & \vdots \\ \text{Cov}(e_n, \hat{\beta}_0) & \text{Cov}(e_n, \hat{\beta}_1) & \cdots & \text{Cov}(e_n, \hat{\beta}_p) \end{bmatrix} \tag{4.20}$$

Equation (4.19) shows that $\boldsymbol{e}$ and $\hat{\boldsymbol{\beta}}$ are uncorrelated:

$$\text{Cov}(\boldsymbol{e}, \hat{\boldsymbol{\beta}}) = O'$$

which is an $n \times (p+1)$ matrix of zeros. Since they are jointly normal, they are statistically independent and not just uncorrelated. This also implies that $S(\hat{\boldsymbol{\beta}}) = \boldsymbol{e}'\boldsymbol{e}$, which is a function of just $\boldsymbol{e}$, and $\hat{\boldsymbol{\beta}}$ are statistically independent (an alternate proof of this result is given in the appendix). It should be noted that any linear combination $\boldsymbol{a}'\hat{\boldsymbol{\beta}}$ of $\hat{\boldsymbol{\beta}}$ is a linear combination of normal random variables and hence itself normally distributed. That is

$$\boldsymbol{a}'\hat{\boldsymbol{\beta}} \sim N(\boldsymbol{a}'\boldsymbol{\beta}, \sigma^2\boldsymbol{a}'(X'X)^{-1}\boldsymbol{a})$$

vi. $S(\hat{\boldsymbol{\beta}})/\sigma^2 = \boldsymbol{e}'\boldsymbol{e}/\sigma^2 \sim \chi^2_{n-p-1}$, a chi-square distribution with $n - p - 1$ degrees of freedom. This result is shown in the appendix. The degrees of freedom are easy to remember: $n - p - 1$ is the difference between the number of observations and the number of estimated regression parameters.

The appendix to Chapter 2 mentions that the mean of a chi-square random variable is equal to its degrees of freedom. Hence,

$$E\left(\frac{\boldsymbol{e}'\boldsymbol{e}}{\sigma^2}\right) = \frac{E(\boldsymbol{e}'\boldsymbol{e})}{\sigma^2} = n - p - 1$$

and

$$s^2 = \frac{\boldsymbol{e}'\boldsymbol{e}}{n - p - 1} = \frac{S(\hat{\boldsymbol{\beta}})}{n - p - 1} \tag{4.21}$$

is an unbiased estimate of σ^2. You can see this from

$$E(s^2) = \frac{E(\boldsymbol{e}'\boldsymbol{e})}{n - p - 1} = \frac{\sigma^2(n - p - 1)}{n - p - 1} = \sigma^2.$$

vii. The residuals $\boldsymbol{e}$ and the fitted values $\hat{\boldsymbol{\mu}}$ are statistically independent.
We have already shown that $\boldsymbol{e}$ and $\hat{\boldsymbol{\beta}}$ are independent. X is a fixed (nonrandom) matrix and hence $\boldsymbol{e}$ and $X\hat{\boldsymbol{\beta}} = \hat{\boldsymbol{\mu}}$ are separate nonoverlapping functions of independent random variables. Hence, they are independent. This can be proved directly as well; see Exercise 4.18.

viii. **Gauss–Markov Theorem**
Assume that the usual regression assumptions are satisfied and that the $n \times 1$ response vector $\boldsymbol{y}$ has mean $E(\boldsymbol{y}) = \boldsymbol{\mu} = X\boldsymbol{\beta}$ and covariance matrix $V(\boldsymbol{y}) = \sigma^2 I$. The Gauss–Markov theorem says that among all **linear unbiased** estimators, the LSE $\hat{\boldsymbol{\beta}} = (X'X)^{-1}X'\boldsymbol{y}$ has the smallest variance. "Smallest" variance means that the covariance matrix of any other linear unbiased estimator exceeds the covariance matrix of $\hat{\boldsymbol{\beta}}$ by a positive semidefinite matrix.

Proof: The LSE $\hat{\boldsymbol{\beta}} = (X'X)^{-1}X'\boldsymbol{y}$ is a linear combination of the random response vector $\boldsymbol{y}$. Consider any other linear transformation, for example, $\hat{\boldsymbol{b}} = M^*\boldsymbol{y}$, where M^* is a $(p+1) \times n$ matrix of fixed coefficients. Define $M = M^* - (X'X)^{-1}X'$, and write the new estimator as

$$\begin{aligned}\hat{\boldsymbol{b}} &= [M + (X'X)^{-1}X']\boldsymbol{y} = [M + (X'X)^{-1}X'][X\boldsymbol{\beta} + \boldsymbol{\epsilon}] \\ &= [MX\boldsymbol{\beta} + \boldsymbol{\beta}] + [M + (X'X)^{-1}X']\boldsymbol{\epsilon}\end{aligned}$$

The requirement of unbiasedness for $\hat{\boldsymbol{b}}$ implies that $MX = O$, a $(p+1) \times (p+1)$ matrix of zeros. With this condition imposed, the covariance matrix for $\hat{\boldsymbol{b}}$ becomes

$$\begin{aligned}V(\hat{\boldsymbol{b}}) &= E[(\hat{\boldsymbol{b}} - \boldsymbol{\beta})(\hat{\boldsymbol{b}} - \boldsymbol{\beta})'] = E\{[M + (X'X)^{-1}X']\boldsymbol{\epsilon}\boldsymbol{\epsilon}'[M + (X'X)^{-1}X']'\} \\ &= [M + (X'X)^{-1}X']E(\boldsymbol{\epsilon}\boldsymbol{\epsilon}')[M + (X'X)^{-1}X']' \\ &= [M + (X'X)^{-1}X']\sigma^2 I[M + (X'X)^{-1}X']' \\ &= \sigma^2[M + (X'X)^{-1}X'][M + (X'X)^{-1}X']' \\ &= \sigma^2[MM' + (X'X)^{-1}] = \sigma^2(X'X)^{-1} + \sigma^2 MM' \\ &= V(\hat{\boldsymbol{\beta}}) + \sigma^2 MM'\end{aligned}$$

Here we have used the fact that $MX = O$, and hence $MX(X'X)^{-1} = O$ and $(X'X)^{-1}X'M' = O$.

This result shows that the variance of the new linear estimator $\hat{\boldsymbol{b}}$ exceeds the variance of the LSE $\hat{\boldsymbol{\beta}}$ by the matrix $\sigma^2 MM'$. However, this matrix is positive semidefinite because for any vector $\boldsymbol{a}$ the quadratic form $\boldsymbol{a}'MM'\boldsymbol{a} = \tilde{\boldsymbol{a}}'\tilde{\boldsymbol{a}} = \sum(\tilde{a}_i)^2 \geq 0$. ■

The Gauss–Markov result also holds when estimating an arbitrary linear combination of the regression parameters. Consider the linear combination $\boldsymbol{a}'\boldsymbol{\beta}$ and the two estimators $\boldsymbol{a}'\hat{\boldsymbol{\beta}}$ and $\boldsymbol{a}'\hat{\boldsymbol{b}}$. The first estimator uses the LSE, whereas the second uses the linear unbiased estimator studied previously. The variances of these estimators are given by

$$V(\boldsymbol{a}'\hat{\boldsymbol{\beta}}) = \sigma^2 \boldsymbol{a}'(X'X)^{-1}\boldsymbol{a}$$

and

$$V(\boldsymbol{a}'\hat{\boldsymbol{b}}) = \sigma^2 \boldsymbol{a}'(X'X)^{-1}\boldsymbol{a} + \sigma^2 \boldsymbol{a}'MM'\boldsymbol{a}$$

Since $\boldsymbol{a}'MM'\boldsymbol{a} \geq 0$, the estimator using the LSE has the smaller variance. As a special case, consider the vector $\boldsymbol{a}$ with a one in the ith position and zeros everywhere else. Then the Gauss–Markov result implies that the LSE of the (individual) coefficient β_i has the smallest variance among all other linear unbiased estimators.

Note that it is not necessary to make any assumption about the **form** of the error distribution in order to get the Gauss–Markov property. However, it must be emphasized that the result only proves that the LSE is best within the class of **linear** estimators. For certain nonnormal error distributions, it is possible to find a nonlinear estimator that has smaller variance than the LSE. For normal errors, however, this cannot be done, and in this case the LSE is the best estimator among all estimators—linear as well as nonlinear.

4.2.3 PRELIMINARY DISCUSSION OF RESIDUALS

The residual vector is $\boldsymbol{e} = \boldsymbol{y} - \hat{\boldsymbol{\mu}}$. The ith component $e_i = y_i - \hat{\mu}_i$ is the residual associated with the ith individual or case in the experiment. The residual represents the deviation between the response and the fitted value and hence estimates the random component ϵ in the model. Any misspecification or departure from the underlying assumptions in the model will show up as patterns in the residuals. Hence, the analysis of the residuals is an effective way of discovering model inadequacies. Let us examine some important properties of the residuals.

i. The vector of residuals $\boldsymbol{e}$ is orthogonal to $L(X)$, and hence $\boldsymbol{e}'\mathbf{1} = 0$ if β_0 is in the model. This means that $\sum_{i=1}^{n} e_i = 0$ and $\bar{e} = 0$.

ii. $\boldsymbol{e}$ is orthogonal to $\hat{\boldsymbol{\mu}}$.

These two properties are direct consequences of the least squares fitting procedure. They always hold, whether or not the model is adequate.

Next, let us summarize the properties of $\boldsymbol{e}$ that only hold if the model is correct. We assume that the funtional form is correct; that is, $E(\boldsymbol{y})$ is in $L(X)$. In addition, we suppose that the errors ϵ are multivariate normal with covariance matrix $\sigma^2 I$.

i. $E(\boldsymbol{e}) = \mathbf{0}$. If $E(\boldsymbol{y})$ is not in the subspace $L(X)$ and the assumed functional form of the model is incorrect, then this property does not hold. We will discuss this more fully in Chapter 6.

ii. $\boldsymbol{e}$ and $\hat{\boldsymbol{\mu}}$ are independent.

iii. $\boldsymbol{e} \sim N(\mathbf{0}, \sigma^2(I - H))$.

If the errors ϵ do not have a normal distribution with constant variance σ^2, then the residuals $\boldsymbol{e}$ will not satisfy properties (ii) and (iii).

We construct several graphical residual checks that investigate whether the residuals exhibit the properties in (i)–(iii). These graphs can tell us whether the fitted model is an adequate representation. If the model is adequate, we do not expect systematic patterns in the residuals, and hence a plot of the residuals e_i versus the order i should exhibit the noninformative pattern depicted in Figure 4.4(a); that is, the e_i's should fall within an approximate horizontal band around $\bar{e} = 0$. A similar plot should result if e_i is plotted against the values of the jth predictor x_{ij}, $(j = 1, 2, \ldots, p)$. Also, a plot of the residuals e_i against the fitted values $\hat{\mu}_i$ should show no systematic patterns and should look like Figure 4.4(b). Departures from these patterns indicate model inadequacies, and we will discuss those more fully in Chapter 6.

FIGURE 4.4 **Adequate Residual Plots**

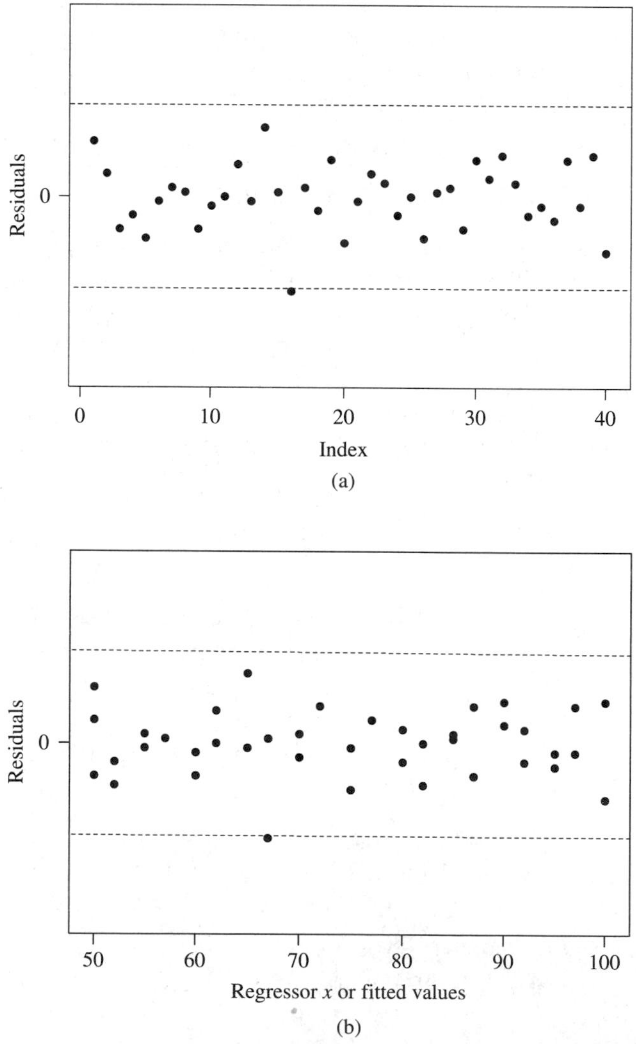

UFFI Example Continued

Consider Example 1.2.5 in Chapter 1 and the data in Table 1.2, where we relate the formaldehyde concentrations y_i to the presence or absence of UFFI ($x_{i1}=1$ or $x_{i1}=0$) and the airtightness (TIGHT, x_{i2}) of the home. The model in Eq. (4.2) specifies $y_i=\beta_0+\beta_1 x_{i1}+\beta_2 x_{i2}+\epsilon_i$, with the usual assumptions on ϵ_i. The $\mathbf{y}$ vector and the X matrix are given in Eq. (4.3). One can compute the LSE $\hat{\beta}$ and s^2, the unbiased estimator of $V(\epsilon_i)=\sigma^2$, from Eqs. (4.11) and (4.21).

$$\hat{\beta}=(X'X)^{-1}X'\mathbf{y}$$

and

$$s^2=\frac{1}{n-p-1}(\mathbf{y}-X\hat{\beta})'(\mathbf{y}-X\hat{\beta})$$

These computations are usually performed by a statistical software package (such as S-Plus, R, Minitab, SAS, SPSS, or even EXCEL). Computational details (commands and outputs from the well-known software S-Plus) are shown in Table 4.1.

TABLE 4.1 S-PLUS INPUT AND OUTPUT

```
> ch2o<-matrix(scan('uffi.dat',multi.line = T),byrow = T,ncol = 3,nrow = 24)
> uffi<-ch2o[,1]
> tight<-ch2o[,2]
> form<-ch2o[,3]
> ch2fit<-lm(form ~ uffi+tight)
> summary(ch2fit)
Call: lm(formula = form ~ uffi + tight)
Residuals:
Min          1Q          Median        3Q          Max
-9.546       -3.131      -0.1389       3.578       8.362
Coefficients:
                Value       Std. Error    t value     Pr(> |t|)
(Intercept)     31.3734     2.4607        12.7500     0.0000
       uffi      9.3120     2.1325         4.3666     0.0003
      tight      2.8545     0.3764         7.5843     0.0000
Residual standard error: 5.223 on 21 degrees of freedom
Multiple R-Squared: 0.7827
F-statistic: 37.82 on 2 and 21 degrees of freedom, the p-value is 1.095e-07
Correlation of Coefficients:
             (Intercept)     uffi
uffi         -0.4449
tight        -0.7903         0.0147
> X11()
> par(mfrow = c(2,2))
> obsno< - 1:24
> plot(obsno,ch2fit$res,xlab = 'Observation Number',
+ylab = 'Residuals',main = 'Residuals vs Obs. No.')
> plot(ch2fit$ fit,ch2fit $ res,xlab = 'Fitted Values',
+ylab = 'Residuals',main = 'Residuals vs Fitted Values')
>plot(tight,ch2fit $ res, xlab = 'Airtightness',ylab = 'Residuals',
+ main = 'Residuals Vs Airtightness')
```

You may want to consider other packages and convince yourself that the output from other programs will be similar. For the time being, we ignore much of the output and concentrate on the vector of LSEs $\hat{\beta}' = (31.37, 9.31, 2.85)$, the sum of squared errors $S(\hat{\beta}) = 572.72$, and the estimate of σ^2, $s^2 = 572.72/21 = 27.27$. The square root of s^2 is listed in the output. S-Plus calls it the residual standard error.

In addition, the software can calculate and store the vector of fitted values and the vector of residuals. This is useful for generating residual plots that help us check the model assumptions. Figure 4.5(a) shows plots of the residuals e_i against the order i, Figure 4.5(b) residuals e_i against fitted values $\hat{\mu}_i$, and Figure 4.5(c) residuals e_i against the explanatory variable airtightness. These plots do not show any systematic patterns in the residuals. Hence, we conclude, at least for now, that the model assumptions are reasonable. We will revisit this topic in a later chapter.

The estimate $\hat{\beta}_1 = 9.31$ implies that, on average, there is a difference of 9.31 parts per billion (ppb) in the ambient formaldehyde concentration in two homes having identical airtightness but different insulations—one with UFFI present and the other without it.

4.3 STATISTICAL INFERENCE

For the following discussion we assume that the errors in Eq. (4.6) are normally distributed. We discuss how to construct confidence intervals and how to test hypotheses.

4.3.1 CONFIDENCE INTERVALS AND TESTS OF HYPOTHESES FOR A SINGLE PARAMETER

Usually, one is interested in making inferences about a single regression parameter β_i or about a single linear combination of the coefficients $\theta = \boldsymbol{a}'\boldsymbol{\beta}$. We have studied the distribution of $\hat{\boldsymbol{\beta}}$ previously and have found that

$$\hat{\boldsymbol{\beta}} \sim N(\boldsymbol{\beta}, \sigma^2(X'X)^{-1})$$

Suppose we are interested in making inferences about one of these coefficients, β_i. The estimate of β_i is given by $\hat{\beta}_i$, and its variance is given by $\sigma^2 v_{ii}$, where v_{ii} is the corresponding diagonal element in the matrix $(X'X)^{-1}$. The sampling distribution of $\hat{\beta}_i$ is

$$\hat{\beta}_i \sim N(\beta_i, \sigma^2 v_{ii}) \tag{4.22}$$

The variance of the errors, σ^2, is unknown and must be estimated. As estimate we use the unbiased estimator of σ^2,

$$s^2 = \frac{1}{n-p-1} S(\hat{\boldsymbol{\beta}}) \tag{4.23}$$

FIGURE 4.5
Residual Plots: UFFI Data

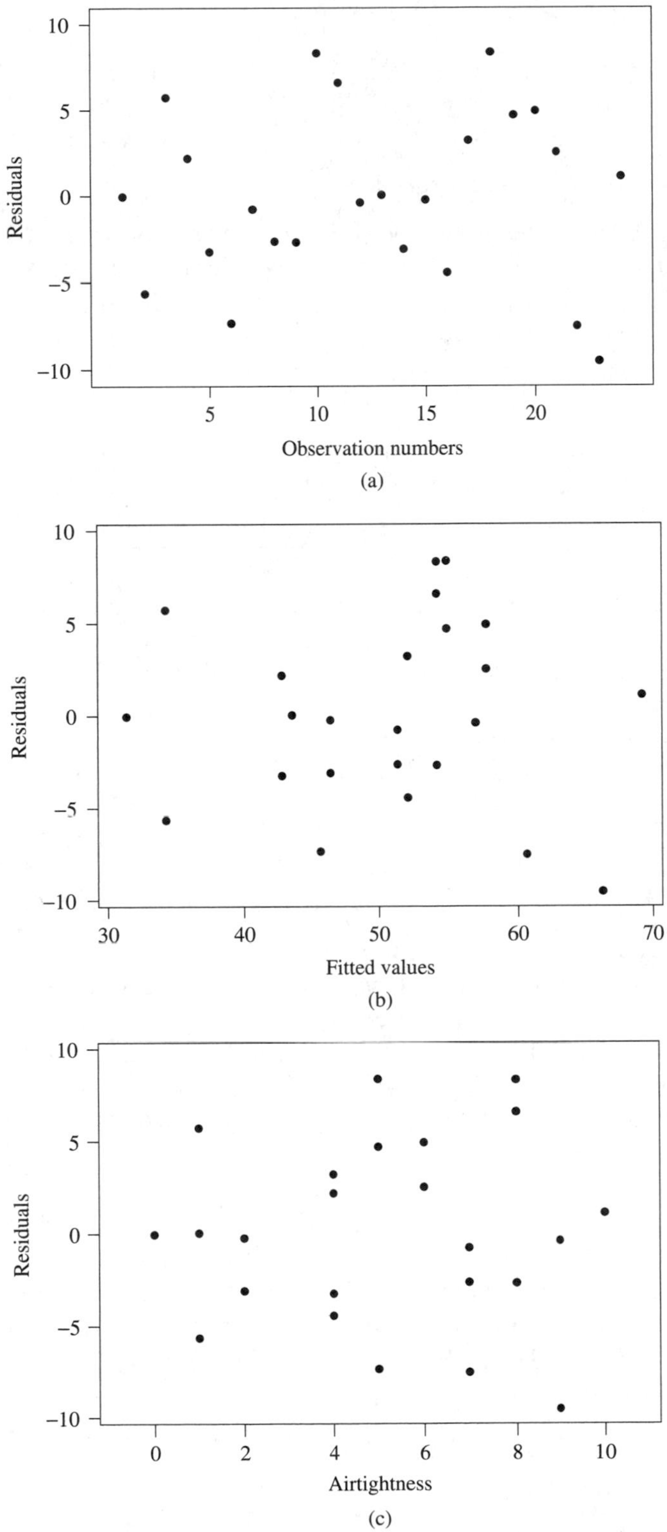

We know that

i. $(\hat{\beta}_i - \beta_i)/\sigma\sqrt{v_{ii}} \sim N(0, 1)$. This follows from Eq. (4.22).

ii. $\dfrac{(n-p-1)s^2}{\sigma^2} \sim \chi^2_{n-p-1}$.

iii. s^2 and $\hat{\beta}$ are independent .

The results in (ii) and (iii) were shown in Section 4.2.2 and are also shown in the appendix. It follows from properties of the t distribution (see appendix to Chapter 2) that

$$T = \frac{\hat{\beta}_i - \beta_i}{s\sqrt{v_{ii}}} = \frac{(\hat{\beta}_i - \beta_i)/\sigma\sqrt{v_{ii}}}{\sqrt{\frac{(n-p-1)s^2}{\sigma^2}/n-p-1}} \sim t(n-p-1) \tag{4.24}$$

This quantity is used to construct confidence intervals and to test hypotheses about β_i. The ratio T is easy to remember:

$$T = \frac{\text{estimate} - \text{parameter}}{\text{s.e.(estimate)}}$$

relates the difference between the estimate and the true value to the standard error of the estimate, s.e.$(\hat{\beta}_i) = s\sqrt{v_{ii}}$. The standard error estimates the overall variability of the estimate in repeated random samples.

UFFI Example Continued

In this example, $\hat{\beta}' = (31.37, 9.31, 2.85)$, $s^2 = 27.2$, and

$$(X'X)^{-1} = \begin{bmatrix} 0.2219 & -0.0856 & -0.0268 \\ -0.0856 & 0.1667 & 0.0004 \\ -0.0268 & 0.0004 & 0.0052 \end{bmatrix}$$

Let us study β_1, the effect of formaldehyde insulation on the ambient formaldehyde concentration. Is there a difference in the average concentration between homes of equal airtightness but different UFFI insulation? If insulation does not matter, then $\beta_1 = 0$. To answer this question, we test the hypothesis $\beta_1 = 0$. We know that $\hat{\beta}_1 = 9.31$, $V(\hat{\beta}_1) = 0.1667\sigma^2$, and s.e.$(\hat{\beta}_1) = s\sqrt{0.1667} = 5.22\sqrt{0.1667} = 2.13$. The t statistic for the coefficient $\hat{\beta}_1$ is

$$t_0(\hat{\beta}_1) = (\hat{\beta}_1 - 0)/\text{s.e.}(\hat{\beta}_1) = 9.31/2.13 = 4.37$$

The subscript zero indicates that we test the hypothesis $\beta_1 = 0$; the argument $\hat{\beta}_1$ in parentheses indicates that the statistic refers to the estimate $\hat{\beta}_1$. If there is no danger of confusion, we just write $t(\hat{\beta}_1)$. Since there are 24 observations and three parameters in the model, the residual sum of squares has 21 degrees of freedom. The probability value of this test statistic for a two-sided alternative $(\beta_1 \neq 0)$ is given by

$$P(|T| > 4.37) = 2P(T > 4.37) \approx 0.0003$$

Here, we use the t distribution with 21 degrees of freedom. The probability is very small—smaller than any reasonable significance level. Thus, there is very strong evidence that β_1 differs from 0. There is very strong evidence that homes with UFFI insulation have higher formaldehyde concentration levels.

A 95% confidence interval for β_1 is given by

$$\hat{\beta}_1 \pm t(0.975; 21)\text{s.e.}(\hat{\beta}_1),$$
$$9.31 \pm (2.08)(2.13), \quad 9.31 \pm 4.43, \quad \text{or} \quad (4.88, 13.74)$$

Note that $t(0.975; 21) = 2.08$ and $t(0.025; 21) = -2.08$ are the 97.5th and the 2.5th percentiles of the t distribution with 21 degrees of freedom, respectively. We are 95% confident that the interval (4.88, 13.74) covers the true, but unknown, difference in the average ambient formaldehyde concentration of homes with and without UFFI insulation.

One can repeat these calculations for the other parameter β_2, which represents the effect of airtightness on the average ambient formaldehyde concentration. The relevant diagonal element of $(X'X)^{-1}$ is 0.0052, and the standard error is

$$\text{s.e.}(\hat{\beta}_2) = s\sqrt{0.0052} = (5.22)\sqrt{0.0052} = 0.37$$

The t ratio, $t(\hat{\beta}_2) = (2.85 - 0)/0.37 = 7.58$ is very large, and the probability of obtaining such an extreme value from a t distribution with 21 degrees of freedom is negligible; the probability value for a two-sided alternative, $2P(T > 7.58)$, is essentially zero. Hence, there is little doubt that airtightness of a home increases the formaldehyde concentration in the home. A 99% confidence interval for β_2 is given by

$$2.85 \pm t(0.995; 21)(0.37)$$
$$2.85 \pm (2.83)(0.37), \quad \text{or from } 1.80 \text{ to } 3.90$$

We could repeat the calculations for the intercept β_0, which mathematically is the average concentration for homes without UFFI and with airtightness zero. Here (and also in many other applications) the intercept does not have much physical meaning, and we skip the calculation.

Note that estimates, standard errors, t ratios, and probability values for the coefficients are standard output of all statistical software packages.

Linear Combination of Coefficients

Suppose that we are interested in a linear combination of the regression coefficients. For instance, suppose we are interested in estimating the average formaldehyde concentration in homes with UFFI and with airtightness 5. That is,we are interested in

$$\theta = \beta_0 + \beta_1 + 5\beta_2 = \boldsymbol{a}'\boldsymbol{\beta}$$

where $\boldsymbol{a}' = (1,\ 1,\ 5)$ is a vector of known coefficients. The estimate of θ is given by

$$\hat{\theta} = \boldsymbol{a}'\hat{\boldsymbol{\beta}} = (1,\ 1,\ 5)\begin{bmatrix} 31.37 \\ 9.31 \\ 2.85 \end{bmatrix} = 54.96$$

Before we can construct a confidence interval for θ, we need to study the sampling distribution of $\hat{\theta}$. From properties of linear combinations of normal random variables, we know that

$$\hat{\theta} = \boldsymbol{a}'\hat{\boldsymbol{\beta}} \sim N(\theta, \sigma^2\boldsymbol{a}'(X'X)^{-1}\boldsymbol{a})$$

Replacing σ^2 by the estimate s^2, and after going through similar steps as those in Eq. (4.24), we find that

$$T = \frac{\hat{\theta} - \theta}{s\sqrt{\boldsymbol{a}'(X'X)^{-1}\boldsymbol{a}}} \sim t(21) \tag{4.25}$$

Hence, a 95% confidence interval is given by

$$\hat{\theta} \pm t(0.975;\ 21)s\sqrt{\boldsymbol{a}'(X'X)^{-1}\boldsymbol{a}} \tag{4.26}$$

With $s = 5.22$ and

$$\boldsymbol{a}'(X'X)^{-1}\boldsymbol{a} = (1,\ 1,\ 5)(X'X)^{-1}\begin{pmatrix} 1 \\ 1 \\ 5 \end{pmatrix} = 0.0833$$

the 95% confidence interval for θ is

$$54.96 \pm 2.08(5.22\sqrt{0.0833}),$$
$$54.96 \pm (2.08)(1.51),\quad 54.96 \pm 3.13,\quad \text{or } (51.83, 58.09)$$

We are 95% confident that the interval (51.83, 58.09) will cover the true average concentration of homes with UFFI and airtightness 5. Most statistics software packages allow you to ask for this information.

Gas Consumption Example

In this example, we are interested in predicting the gas consumption of an automobile from its size and engine characteristics. The data are given in Table 1.4. There are $n = 38$ cars and measurements on fuel efficiency in miles per gallon (y), weight (x_1), engine displacement (x_2), number of cylinders (x_3), horsepower (x_4), acceleration (x_5), and engine type (x_6). Part of the data and variables x_1, x_2, and x_3 were discussed in Chapter 1 and also at the beginning of this chapter. Now let us consider all six regressor variables. Initial exploration with the data indicates that it is preferable to consider $z = 100/y$, the gas consumption per 100 traveled miles, as the response. A thorough discussion of this point will be given in Section 6.5 when we discuss transformations. In the following, we consider

$$z = \beta_0 + \beta_1 x_1 + \beta_2 x_2 + \beta_3 x_3 + \beta_4 x_4 + \beta_5 x_5 + \beta_6 x_6 + \epsilon \tag{4.27}$$

TABLE 4.2 SAS OUTPUT OF THE FUEL CONSUMPTION EXAMPLE

The SAS System
Model: MODEL1
Dependent Variable: Z
Analysis of Variance

Source	DF	Sum of Squares	Mean Square	F Value	Prob > F
Model	6	46.41156	7.73526	79.015	0.0001
Error	31	3.03477	0.09790		
C Total	37	49.44632			

Root MSE 0.31288 R-square 0.9386
Dep Mean 4.33061 Adj R-sq 0.9267
C.V. 7.22491

Parameter Estimates

Variable	DF	Parameter Estimate	Standard Error	T for H_0: Parameter = 0	Prob > $\|T\|$
INTERCEP	1	−2.599749	0.66312133	−3.920	0.0005
X_1	1	0.787706	0.45173293	1.744	0.0911
X_2	1	−0.004892	0.00269495	−1.815	0.0792
X_3	1	0.444251	0.12263114	3.623	0.0010
X_4	1	0.023605	0.00673885	3.503	0.0014
X_5	1	0.068804	0.04419393	1.557	0.1297
X_6	1	−0.959720	0.26667148	−3.599	0.0011

Least squares estimates of the parameters, their standard errors, t ratios, and probability values are given in Table 4.2, the output from SAS, another popular software package. Different software packages will use slightly different formats, but all of them will supply most of the information in Table 4.2.

Furthermore, in this example, $s^2 = 0.0979$, and the inverse of the matrix $X'X$, is given by

$$(X'X)^{-1} = \begin{bmatrix} 4.4918 & 0.6045 & 0.0019 & -0.1734 & -0.0210 & -0.2361 & 0.1872 \\ & 2.0845 & -0.0092 & -0.2052 & -0.0239 & -0.1081 & 0.7099 \\ & & 0.0001 & -0.0005 & 0.0001 & 0.0005 & -0.0030 \\ & & & 0.1536 & 0.0009 & -0.0011 & -0.2001 \\ & \textit{Symmetric} & & & 0.0005 & 0.0017 & -0.0073 \\ & & & & & 0.0200 & -0.0051 \\ & & & & & & 0.7264 \end{bmatrix}$$

Consider the parameter β_5, which measures the effect of x_5 (acceleration) on the average fuel consumption. The estimate is $\hat{\beta}_5 = 0.0688$. From the relevant diagonal element in $(X'X)^{-1}$, we find that $V(\hat{\beta}_5) = \sigma^2(0.0200)$, and s.e.$(\hat{\beta}_5) = \sqrt{0.0979}\,\sqrt{0.0200} = 0.0442$.

For a test of the hypothesis $\beta_5 = 0$, we consider the t statistic

$$t(\hat{\beta}_5) = \hat{\beta}_5/\text{s.e.}(\hat{\beta}_5) = 0.0688/0.0442 = 1.56$$

and its corresponding p value

$$P(|T| > 1.56) = 2P(T > 1.56) = 0.1297 \tag{4.28}$$

Note that the appropriate degrees of freedom are $n - 7 = 38 - 7 = 31$.

The probability value indicates that at the 5% significance level one cannot reject the hypothesis that $\beta_5 = 0$, given that the other variables x_1, x_2, x_3, x_4, x_6 have been included in the model. The t ratio $t(\hat{\beta}_5)$ assesses the potential effect of x_5, having adjusted the analysis for all other variables in the model. The result implies that on top of x_1, x_2, x_3, x_4, x_6 in the model, x_5 is not an important predictor of gas consumption. On the other hand, the probability value for $\hat{\beta}_6$ indicates that there is evidence for rejecting the hypothesis that $\beta_6 = 0$. It implies that x_6 is important in predicting gas consumption, even if x_1, x_2, x_3, x_4, x_5 are already in the model.

We need to remember that any inference procedure depends on the validity of the assumptions that we make about the errors ϵ. We must always check the residuals for any violations of the assumptions. The residual plots in Figure 4.6 [(a) residuals against observation number, (b) residuals against fitted values, (c) residuals against weight, and (d) residuals against displacement] indicate no systematic unusual patterns, and we conclude that the model assumptions are justified.

4.3.2 PREDICTION OF A NEW OBSERVATION

Suppose that we are interested in predicting the response of a new case, for example, the formaldehyde concentration in a new home with UFFI insulation and airtightness 5. Let y_p represent this unknown concentration,

$$y_p = \beta_0 + \beta_1 1 + \beta_2 5 + \epsilon_p = \mu_p + \epsilon_p$$

In other words, $y_p \sim N(\mu_p, \sigma^2)$. The mean $\mu_p = (1, 1, 5)\boldsymbol{\beta} = \boldsymbol{a}'\boldsymbol{\beta}$ depends on the specified (fixed) levels of the regressor variables and the parameter $\boldsymbol{\beta}$. If the parameter $\boldsymbol{\beta}$ were known exactly, then we could use μ_p as our prediction. Any other choice would have a larger expected squared error. You can see this by using any other prediction f and considering the expected squared future error,

$$\begin{aligned} E(y_p - f)^2 &= E[(y_p - \mu_p) + (\mu_p - f)]^2 \\ &= E(y_p - \mu_p)^2 + (\mu_p - f)^2 + (\mu_p - f)E(y_p - \mu_p) \\ &= \sigma^2 + (\mu_p - f)^2 \geq \sigma^2 \end{aligned}$$

However, since the parameter $\boldsymbol{\beta}$ and μ_p are unknown, we need to replace them with their LSEs. Our point prediction for the response at the new case is given by $\hat{\mu}_p = (1, 1, 5)\hat{\boldsymbol{\beta}} = 54.96$. We had calculated this earlier, and had denoted it by $\hat{\theta}$. To assess the precision of this prediction, we need to take account of two sources of variability. First, we have only an estimate $\hat{\mu}_p$ of μ_p, and there is uncertainty from the estimation. Second, there is variability of a single observation y_p around its mean μ_p. Consider the prediction error,

$$y_p - \hat{\mu}_p = \mu_p - \hat{\mu}_p + \epsilon_p$$

FIGURE 4.6
Residual Plots: Gas Consumption Data

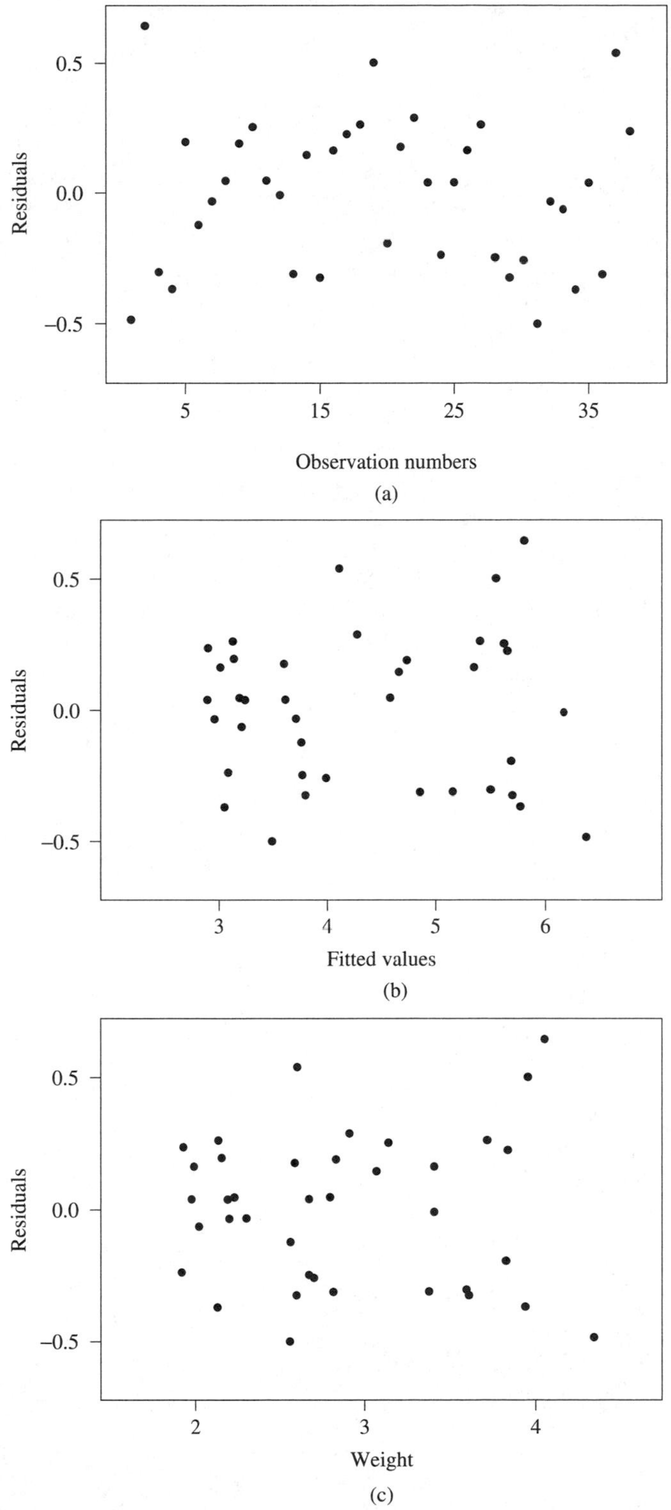

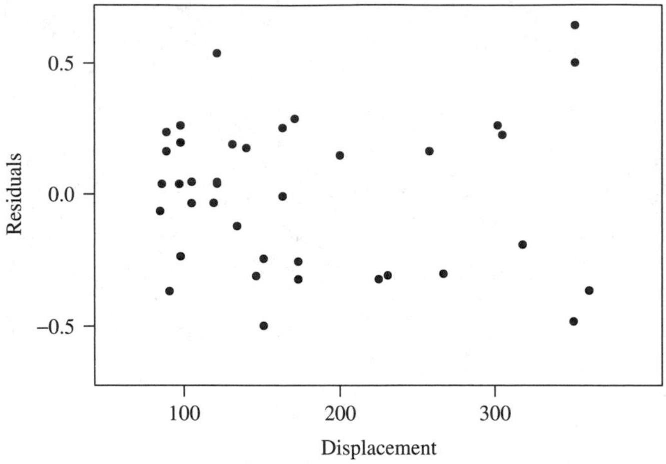

FIGURE 4.6 (Continued)

(d)

Since we consider a new case, and since the error for a new observation is independent of the errors in the observations that we used for estimating $\boldsymbol{\beta}$, the two errors, $(\mu_p - \hat{\mu}_p)$ and ϵ_p, are independent. Hence, the variance is given by

$$V(y_p - \hat{\mu}_p) = V(\hat{\mu}_p) + V(\epsilon_p) = \sigma^2 \boldsymbol{a}'(X'X)^{-1}\boldsymbol{a} + \sigma^2$$
$$= (1 + \boldsymbol{a}'(X'X)^{-1}\boldsymbol{a})\sigma^2$$

where in our special case $\boldsymbol{a}' = (1, \ 1, \ 5)$. Hence,

$$y_p - \hat{\mu}_p \sim N(0, \sigma^2(1 + \boldsymbol{a}'(X'X)^{-1}\boldsymbol{a}))$$

and

$$T = \frac{y_p - \hat{\mu}_p}{s\sqrt{1 + \boldsymbol{a}'(X'X)^{-1}\boldsymbol{a}}} \sim t(n - p - 1)$$

The denominator in this ratio is the standard error of the prediction error

$$\text{s.e.}(y_p - \hat{\mu}_p) = s\sqrt{1 + \boldsymbol{a}'(X'X)^{-1}\boldsymbol{a}}$$

Here we have used the same argument as in the derivation of the t ratios for individual coefficients; see Eq. (4.24).

This result implies that

$$P\left(-t\left(1 - \frac{\alpha}{2}; n - p - 1\right) \le \frac{y_p - \hat{\mu}_p}{\text{s.e.}(y_p - \hat{\mu}_p)} \le t\left(1 - \frac{\alpha}{2}; n - p - 1\right)\right) = 1 - \alpha$$

$$P\left(\hat{\mu}_p - t\left(1 - \frac{\alpha}{2}; n - p - 1\right)\text{s.e.}(y_p - \hat{\mu}_p)\right.$$
$$\left. < y_p < \hat{\mu}_p + t\left(1 - \frac{\alpha}{2}; n - p - 1\right)\text{s.e.}(y_p - \hat{\mu}_p)\right) = 1 - \alpha$$

Hence, a $100(1-\alpha)\%$ prediction interval for y_p is given as

$$\hat{\mu}_p \pm t\left(1-\frac{\alpha}{2}; n-p-1\right)s\sqrt{1+\boldsymbol{a}'(X'X)^{-1}\boldsymbol{a}} \tag{4.29}$$

For our example with $n-p-1=24-3=21$, $\boldsymbol{a}'=(1,1,5)$, and the estimates in Table 4.1, we obtain the 95% prediction interval

$$54.96 \pm (2.08)(5.22)\sqrt{1+0.0833}$$
$$54.96 \pm (2.08)(5.43),\ 54.96 \pm 11.30, \quad \text{or} \quad (43.66, 66.26)$$

Our best prediction is 54.96; our uncertainty for the new value ranges from 43.7 to 66.3. Note that the prediction interval is much wider than the confidence interval for the mean response $\mu_p(=\theta)$. This is because a prediction interval is concerned with a single new observation and not the average for a fixed setting on the regressor variables.

A Caution on Predictions One should be cautious when predicting values of y for sets of x's that are very different from those that are used to fit the model. Extrapolation beyond the experimental region may lead to unreasonable results because the model is descriptive of the relationship between y and $x_1, \ldots, x_p$ only in the region of the observed x's. We are always unsure about the form of the model in a region of the x's for which we have no data. For illustration, consider Figure 4.7, which displays the relationship between a dependent variable y and a single regressor variable x. For values of x in the range from 50 to 120, we entertain a quadratic model and the fitted curve is shown in the figure. Now suppose that we had x only over the range from 50 to 80. Then a straight line model will fit the data quite well. However, Figure 4.7 shows that the prediction from the linear model of y for $x=120$ would be very misleading.

Good predictions require a valid regression model—that is, a model in which the predictor variables are significant. A model in which the influence of regressor varibles is established poorly will not do much for prediction.

FIGURE 4.7
Predicting Outside the Study Region

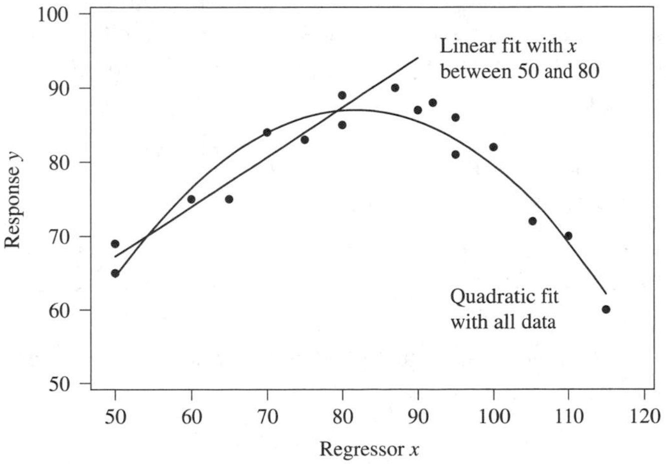

4.4 THE ADDITIONAL SUM OF SQUARES PRINCIPLE

4.4.1 INTRODUCTION

In this section, we describe a procedure for testing simultaneous statements about several parameters. For illustration, consider the model

$$y = \beta_0 + \beta_1 x_1 + \beta_2 x_2 + \beta_3 x_3 + \epsilon \tag{4.30}$$

Suppose that previous studies suggest

$$\beta_1 = 2\beta_2 \quad \text{and} \quad \beta_3 = 0$$

How can we simulteneously test these two restrictions? The restrictions specify values for two linear combinations of the parameters. Our restrictions can be restated in vector form as

$$\begin{bmatrix} 0 & 1 & -2 & 0 \\ 0 & 0 & 0 & 1 \end{bmatrix} \begin{bmatrix} \beta_0 \\ \beta_1 \\ \beta_2 \\ \beta_3 \end{bmatrix} = \begin{bmatrix} 0 \\ 0 \end{bmatrix}$$

or as $A\boldsymbol{\beta} = \mathbf{0}$, where the 2×4 matrix $A = \begin{bmatrix} 0 & 1 & -2 & 0 \\ 0 & 0 & 0 & 1 \end{bmatrix}$, $\boldsymbol{\beta}' = (\beta_0, \beta_1, \beta_2, \beta_3)$, and $\mathbf{0} = \begin{bmatrix} 0 \\ 0 \end{bmatrix}$.

Note that the two linear restrictions are **not** linearly dependent, and hence the matrix A has rank 2. A situation in which this would not be the case is $\beta_1 = 2\beta_2$ and $4\beta_1 = 8\beta_2$. In this case, the second condition is superfluous and can be ignored. The rank of the implied A matrix would be one.

Under the null hypothesis $H_0 : A\boldsymbol{\beta} = \mathbf{0}$, the **full model** in Eq. (4.30) simplifies to

$$y = \beta_0 + \beta_2(2x_1 + x_2) + \epsilon \tag{4.31}$$

We call this the **restricted model** because its form is constrained by H_0. For a test of $A\boldsymbol{\beta} = \mathbf{0}$ we compare two models: the full model in Eq. (4.30) and the restricted model in Eq. (4.31). We illustrate the general approach with the following two examples.

Gas Consumption Example Continued

Previously, we considered the model

$$z = \beta_0 + \beta_1 x_1 + \beta_2 x_2 + \beta_3 x_3 + \beta_4 x_4 + \beta_5 x_5 + \beta_6 x_6 + \epsilon \tag{4.32}$$

The estimates, t ratios, and probability values of the estimates were listed in Table 4.2.

Consider a test of the hypothesis that the last three regressor variables can be omitted from the model. The hypothesis

$$\beta_4 = \beta_5 = \beta_6 = 0$$

can be written in matrix form

$$H_0 : A\boldsymbol{\beta} = \mathbf{0}$$

where the 3×7 matrix

$$A = \begin{bmatrix} 0 & 0 & 0 & 0 & 1 & 0 & 0 \\ 0 & 0 & 0 & 0 & 0 & 1 & 0 \\ 0 & 0 & 0 & 0 & 0 & 0 & 1 \end{bmatrix}; \quad \boldsymbol{\beta} = \begin{bmatrix} \beta_0 \\ \beta_1 \\ \vdots \\ \beta_6 \end{bmatrix}; \quad \text{and} \quad \mathbf{0} = \begin{bmatrix} 0 \\ 0 \\ 0 \end{bmatrix}$$

Under H_0, the model reduces to the restricted model

$$z = \beta_0 + \beta_1 x_1 + \beta_2 x_2 + \beta_3 x_3 + \epsilon \tag{4.33}$$

Note that we have reduced the number of model parameters from seven to four. The matrix A has rank 3, and our hypothesis involves three independent restrictions. Failing to reject the null hypothesis implies that the associated variables x_4, x_5, x_6 are not important, given that the rest of the variables are already in the model. On the other hand, a rejection of the null hypothesis indicates that at least one of the variables x_4, x_5, x_6 is important, in addition to the regressor variables x_1, x_2, and x_3.

OC Example

Our model in Chapter 1 relates the HDL (cholesterol) at the end of the study (y) to the initial HDL (z) and indicators for five different drug regimes,

$$y = \alpha z + \beta_1 x_1 + \cdots + \beta_5 x_5 + \epsilon \tag{4.34}$$

Here $x_1, \ldots, x_5$ are indicator variables denoting the five oral contraceptive groups. The most interesting question is whether there are differences among the five groups. In terms of our parameters, we ask whether there is any evidence that $\beta_1, \ldots, \beta_5$ differ. To examine this question, we consider the hypothesis

$$H_0 : \beta_1 = \beta_2 = \beta_3 = \beta_4 = \beta_5$$

The model (4.34), written in vector notation, is

$$\boldsymbol{y} = \alpha \boldsymbol{z} + \beta_1 \boldsymbol{x}_1 + \cdots + \beta_5 \boldsymbol{x}_5 + \boldsymbol{\epsilon} \tag{4.35}$$

Under the null hypothesis that the five β's are equal,

$$\beta_1 = \beta_2 = \beta_3 = \beta_4 = \beta_5 = \gamma \tag{4.36}$$

the model becomes

$$\begin{aligned} \boldsymbol{y} &= \alpha \boldsymbol{z} + \gamma \boldsymbol{x}_1 + \gamma \boldsymbol{x}_2 + \gamma \boldsymbol{x}_3 + \gamma \boldsymbol{x}_4 + \gamma \boldsymbol{x}_5 + \boldsymbol{\epsilon} \\ &= \gamma \mathbf{1} + \alpha \boldsymbol{z} + \boldsymbol{\epsilon} \end{aligned} \tag{4.37}$$

since the indicator structure of $\boldsymbol{x}_i$ implies that $\boldsymbol{x_1} + \boldsymbol{x_2} + \cdots + \boldsymbol{x_5} = \mathbf{1}$, a $(n \times 1)$ vector of ones. The full model contains six coefficients and the restricted model only two. Hence, we reduced the dimension of the model from six parameters $(\alpha, \beta_1, \ldots, \beta_5)$ in Eq. (4.35) to just two parameters (α, γ) in Eq. (4.37).

Geometrically, one can visualize the restriction as follows. In the original model, $E(\boldsymbol{y})$ is an element of $L(\boldsymbol{z}, \boldsymbol{x}_1, \ldots, \boldsymbol{x}_5)$, a six-dimensional subspace of R^{50}. Under the null hypothesis, $E(\boldsymbol{y})$ is an element of $L(\boldsymbol{z}, \mathbf{1})$, a two-dimensional subspace of the subspace $L(\boldsymbol{z}, \boldsymbol{x}_1, \ldots, \boldsymbol{x}_5)$.

The restrictions in Eq. (4.36) are equivalent to

$$\begin{aligned} \beta_1 - \beta_2 &= 0 \\ \beta_1 - \beta_3 &= 0 \\ \beta_1 - \beta_4 &= 0 \\ \beta_1 - \beta_5 &= 0 \end{aligned} \tag{4.38}$$

and can also be written as

$$A\boldsymbol{\beta} = \mathbf{0} \tag{4.39}$$

where

$$A = \begin{bmatrix} 0 & 1 & -1 & 0 & 0 & 0 \\ 0 & 1 & 0 & -1 & 0 & 0 \\ 0 & 1 & 0 & 0 & -1 & 0 \\ 0 & 1 & 0 & 0 & 0 & -1 \end{bmatrix}; \quad \boldsymbol{\beta}' = [\alpha, \beta_1, \beta_2, \beta_3, \beta_4, \beta_5]$$

The rank of the matrix A is 4; we are testing four linearly independent restrictions among the six parameters. Note that there are many other ways of parameterizing the restrictions. One could write $\beta_1 = \beta_2$, $\beta_2 = \beta_3$, $\beta_3 = \beta_4$, $\beta_4 = \beta_5$, and select

$$A = \begin{bmatrix} 0 & 1 & -1 & 0 & 0 & 0 \\ 0 & 0 & 1 & -1 & 0 & 0 \\ 0 & 0 & 0 & 1 & -1 & 0 \\ 0 & 0 & 0 & 0 & 1 & -1 \end{bmatrix}$$

which is another matrix of rank 4. It turns out that the particular parameterization does not matter. Testing the hypothesis in Eq. (4.39) is usually referred to as testing a general linear hypothesis since the test involves linear functions of the parameters.

4.4.2 TEST OF A SET OF LINEAR HYPOTHESES

Suppose we have the model

$$\boldsymbol{y} = \beta_0 \mathbf{1} + \beta_1 \boldsymbol{x}_1 + \cdots + \beta_p \boldsymbol{x}_p + \boldsymbol{\epsilon} \tag{4.40}$$

and want to test the hypothesis that a certain set of linear combinations of $\beta_0, \beta_1, \ldots, \beta_p$ are zero. That is,

$$A\boldsymbol{\beta} = \mathbf{0} \tag{4.41}$$

where A is an $l \times (p+1)$ matrix of rank l.

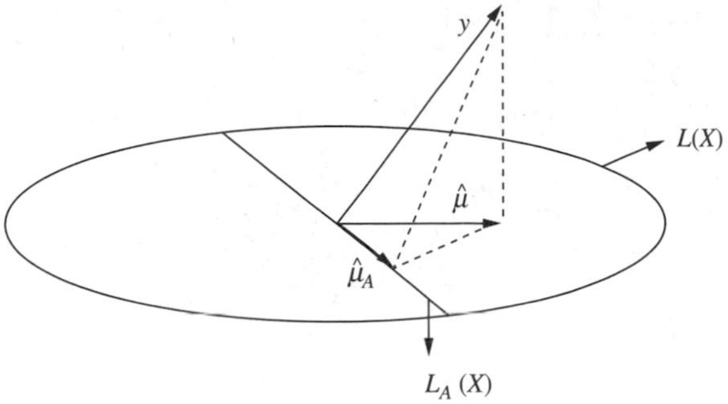

FIGURE 4.8 Geometric Representation of $L(X)$ and $L_A(X)$

Our test procedure relies on the **additional sum of squares** principle. First, we look at the problem geometrically. As usual, we write the model (4.40) in vector notation,

$$\boldsymbol{y} = \boldsymbol{\mu} + \boldsymbol{\epsilon} = X\boldsymbol{\beta} + \boldsymbol{\epsilon}$$

The model component $\boldsymbol{\mu} = X\boldsymbol{\beta}$ is in the $p+1$ dimensional subspace $L(X)$ that is spanned by the regressor vectors in $X = [\mathbf{1}, \boldsymbol{x}_1, \ldots, \boldsymbol{x}_p]$.
Let

$$L_A(X) = \{\beta_0\mathbf{1} + \beta_1\boldsymbol{x}_1 + \cdots + \beta_p\boldsymbol{x}_p \mid A\boldsymbol{\beta} = \mathbf{0}\}$$

be the subspace spanned by all linear combinations of the regressor vectors $\mathbf{1}, \boldsymbol{x}_1, \ldots, \boldsymbol{x}_p$ with coefficients satisfying the restriction $A\boldsymbol{\beta} = \mathbf{0}$. $L_A(X)$ is a subspace of $L(X)$, with dimension $p + 1 - l$. This is easy to see because every linear combination in $L_A(X)$ is also an element in $L(X)$ (see Figure 4.8). If $E(\boldsymbol{y})$ is an element of $L_A(X)$, then the hypothesis $A\boldsymbol{\beta} = \mathbf{0}$ is true exactly.

Let $\hat{\boldsymbol{\mu}}$ be the orthogonal projection of $\boldsymbol{y}$ onto $L(X)$, and let $\hat{\boldsymbol{\mu}}_A$ be the orthogonal projection of $\boldsymbol{y}$ onto the subspace $L_A(X)$. If the null hypothesis is true, then $\hat{\boldsymbol{\mu}}$ should be close to $L_A(X)$, and the difference $\hat{\boldsymbol{\mu}} - \hat{\boldsymbol{\mu}}_A$ and its squared length $(\hat{\boldsymbol{\mu}} - \hat{\boldsymbol{\mu}}_A)'(\hat{\boldsymbol{\mu}} - \hat{\boldsymbol{\mu}}_A) = \|\hat{\boldsymbol{\mu}} - \hat{\boldsymbol{\mu}}_A\|^2$ should be small. We would be surprised if this quantity was exactly 0 because there is random variation in the model. The formal procedure takes this variability into account. In the results that follow, we calculate the distribution of the random variable $\|\hat{\boldsymbol{\mu}} - \hat{\boldsymbol{\mu}}_A\|^2$ under the hypothesis $A\boldsymbol{\beta} = \mathbf{0}$.

Once again, we use a technique similar to the one we used in showing that $\hat{\boldsymbol{\beta}}$ and the residual sum of squares $S(\hat{\boldsymbol{\beta}})$ are independent.

Theorem Suppose that $\boldsymbol{y} = X\boldsymbol{\beta} + \boldsymbol{\epsilon}$, $\boldsymbol{\epsilon} \sim N(\mathbf{0}, \sigma^2 I)$. Let $\hat{\boldsymbol{\mu}}$ be the orthogonal projection of $\boldsymbol{y}$ onto $L(X)$ and $\hat{\boldsymbol{\mu}}_A$ the orthogonal projection of $\boldsymbol{y}$ onto $L_A(X)$. Assume that $A\boldsymbol{\beta} = \mathbf{0}$. We can show that (i) $\|\hat{\boldsymbol{\mu}} - \hat{\boldsymbol{\mu}}_A\|^2/\sigma^2 \sim \chi^2_l$, and (ii) $\|\hat{\boldsymbol{\mu}} - \hat{\boldsymbol{\mu}}_A\|^2$ is independent of $S(\hat{\boldsymbol{\beta}}) = (\boldsymbol{y} - \hat{\boldsymbol{\mu}})'(\boldsymbol{y} - \hat{\boldsymbol{\mu}})$.

Proof: See the appendix. ■

The Theorem implies the following

Corollary: Under the hypothesis $A\boldsymbol{\beta} = \mathbf{0}$, the ratio

$$F = \frac{\|\hat{\boldsymbol{\mu}} - \hat{\boldsymbol{\mu}}_A\|^2 / l}{S(\hat{\boldsymbol{\beta}})/(n-p-1)} \tag{4.42}$$

has an F distribution with l and $n-p-1$ degrees of freedom.

Proof: The Theorem states that

$$\|\hat{\boldsymbol{\mu}} - \hat{\boldsymbol{\mu}}_A\|^2 / \sigma^2 \sim \chi_l^2$$

and that it is independent of $S(\hat{\boldsymbol{\beta}})$. Earlier, we showed that

$$S(\hat{\boldsymbol{\beta}})/\sigma^2 \sim \chi_{n-p-1}^2$$

In the appendix to Chapter 2, we stated that the ratio of two independent normalized chi-square distributions leads to an F distribution. Hence,

$$\frac{\|\hat{\boldsymbol{\mu}} - \hat{\boldsymbol{\mu}}_A\|^2 / \sigma^2 l}{S(\hat{\boldsymbol{\beta}})/\sigma^2(n-p-1)} = F \sim F(l, n-p-1)$$

The quantity F in Eq. (4.42) helps us test the hypothesis $A\boldsymbol{\beta} = \mathbf{0}$. Large values of F provide evidence against the hypothesis. ■

Comments

It is easy to see why F in Eq. (4.42) is a sensible test statistic. The denominator in F is the unbiased estimator of σ^2, that we have used previously. It gives an estimate of σ^2, irrespective of whether or not H_0 is true. If $A\boldsymbol{\beta} = \mathbf{0}$, the numerator has distribution $\sigma^2 \chi_l^2 / l$ and expected value σ^2. The numerator of F also estimates σ^2, but only if $A\boldsymbol{\beta} = \mathbf{0}$. If the hypothesis is not true, we expect $\hat{\boldsymbol{\mu}}$ to differ substantially from $\hat{\boldsymbol{\mu}}_A$, and as a consequence $E(\|\hat{\boldsymbol{\mu}} - \hat{\boldsymbol{\mu}}_A\|^2 / l)$ will exceed σ^2. Hence, the ratio F in Eq. (4.42) will tend to be larger than 1 if the hypothesis is false.

The term $\|\hat{\boldsymbol{\mu}} - \hat{\boldsymbol{\mu}}_A\|^2$ can be interpreted as the "**additional sum of squares**," hence the title of this section. We can see this from the geometry. Look at the diagram in Figure 4.9, now slightly relabeled. Consider the right-angled triangle with end points ABC. Denote the squared distances between the points by AB^2, AC^2, and BC^2. You notice that

$$BC^2 = S(\hat{\boldsymbol{\beta}}),\ AB^2 = \|\hat{\boldsymbol{\mu}} - \hat{\boldsymbol{\mu}}_A\|^2,\ AC^2 = S(\hat{\boldsymbol{\beta}}_A)$$

where $S(\hat{\boldsymbol{\beta}}_A)$ is the minimum value of $S(\boldsymbol{\beta})$ when $\boldsymbol{\beta}$ is restricted so that $A\boldsymbol{\beta} = \mathbf{0}$. Pythagoras theorem tells us that

$$\|\hat{\boldsymbol{\mu}} - \hat{\boldsymbol{\mu}}_A\|^2 = AC^2 - BC^2 = S(\hat{\boldsymbol{\beta}}_A) - S(\hat{\boldsymbol{\beta}}) \tag{4.43}$$

Hence, the numerator of our test statistic in Eq. (4.42) is the difference of two error sum of squares: $S(\hat{\boldsymbol{\beta}}) = (\mathbf{y} - X\hat{\boldsymbol{\beta}})'(\mathbf{y} - X\hat{\boldsymbol{\beta}})$ is the error sum of squares in

FIGURE 4.9 Geometric Interpretation of Additional Sum of Squares

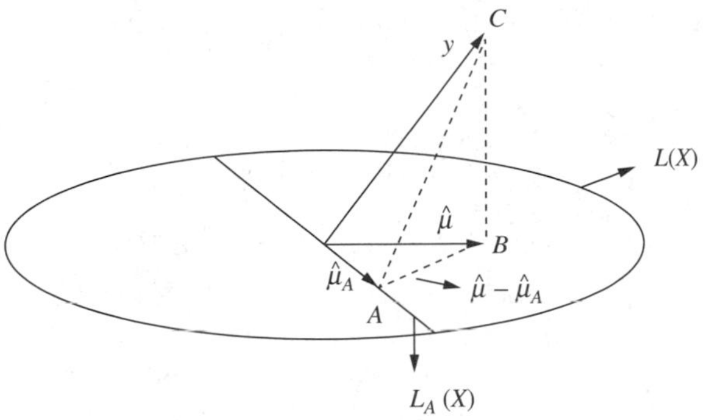

the full model and $S(\hat{\beta}_A) = (y - X\hat{\beta}_A)'(y - X\hat{\beta}_A)$ is the error sum of squares in the restricted model—that is, the model under $A\beta = \mathbf{0}$. This cannot be smaller than $S(\hat{\beta})$ because we have restricted the minimization. The difference $S(\hat{\beta}_A) - S(\hat{\beta}) = \|\hat{\mu} - \hat{\mu}_A\|^2$ is the **additional or extra sum of squares** that our restricted model has failed to pick up. We can also think of it as the extra sum of squares that is picked up when omitting the constraints $A\beta = \mathbf{0}$.

Note

i. Here we have considered the hypothesis $A\beta = \mathbf{0}$. The vector $\mathbf{0}$ on the right-hand side can be replaced by any known vector, for example, $\boldsymbol{\delta}$. The results will remain the same.

   Consider the following illustration. Take the full model $y = \beta_0 + \beta_1 x_1 + \beta_2 x_2 + \epsilon$ and the restriction $2\beta_1 + 3\beta_2 = 5$. We can write this restriction as $\beta_2 = (5/3) - (2/3)\beta_1$ and obtain the restricted model as

$$\begin{aligned} y &= \beta_0 + \beta_1 x_1 + [(5/3) - (2/3)\beta_1]x_2 + \epsilon \\ &= (5/3)x_2 + \beta_0 + \beta_1[x_1 - (2/3)x_2] + \epsilon \end{aligned}$$

   The restricted estimates β_A can be obtained by regressing the transformed response $y - (5/3)x_2$ on the new regressor $x_1 - (2/3)x_2$. From the estimates $\hat{\beta}_{0,A}$ and $\hat{\beta}_{1,A}$ we can obtain $\hat{\beta}_{2,A} = (5/3) - (2/3)\hat{\beta}_{1,A}$. From these estimates we can obtain the residual vector $y - [\hat{\beta}_{0,A} + \hat{\beta}_{1,A}x_1 + \hat{\beta}_{2,A}x_2]$ and compute $S(\hat{\beta}_A)$.

ii. The test for the hypothesis $A\beta = \delta$ can be implemented in a slightly different way as well. Consider the statistic

$$F^* = \frac{(A\hat{\beta} - \delta)'[A'(X'X)^{-1}A]^{-1}(A\hat{\beta} - \delta)/l}{S(\hat{\beta})/(n - p - 1)}$$

   where l is the rank of the constraint matrix A as defined earlier. It can be shown that F^* has an F distribution with l and $(n - p - 1)$ degrees of freedom, and that the test based on F^* is identical to our earlier approach in Eq. (4.42). Some software packages provide the F^* statistic and its associated probability value automatically if you supply the A matrix and the δ vector. However, we prefer our approach in Eq. (4.42) because we believe it to be more intuitive.

Gas Consumption Example Continued

Let us return to the restriction that we had specified,

$$H_0: \beta_4 = \beta_5 = \beta_6 = 0$$

Not rejecting this hypothesis implies that the variables x_4, x_5, and x_6 are not important in predicting the fuel consumption z, given that the variables x_1, x_2, and x_3 are already in the model. On the other hand, rejecting the null hypothesis means that one or more of the regressor variables x_4, x_5, or x_6 contribute explanatory power beyond that provided by the variables x_1, x_2, and x_3.

The restricted model under the null hypothesis is

$$z = \beta_0 + \beta_1 x_1 + \beta_2 x_2 + \beta_3 x_3 + \epsilon \tag{4.44}$$

The LSEs can be obtained, and it turns out that the residual sum of squares from this restricted model is $S(\hat{\beta}_A) = 4.8036$.

Previously, we obtained the residual sum of squares, $S(\hat{\beta}) = 3.0348$, with $n - p - 1 = 38 - 7 = 31$ degrees of freedom. Hence, the additional sum of squares is $S(\hat{\beta}_A) - S(\hat{\beta}) = 4.8036 - 3.0348 = 1.7688$. The full model reduces the error sum of squares by 1.7688; in other words, the constraints in the parameters have cost us an extra sum of squares of 1.7688. This sum of squares has $l = 3$ degrees of freedom since we have constrained three parameters, or equivalently since there are three independent rows in A.

Thus,

$$F = \frac{\text{additional sum of squares}/3}{S(\hat{\beta})/31} = \frac{1.7688/3}{3.0348/31} = 6.02$$

The sampling distribution of the F ratio under the null hypothesis is F with 3 and 31 degrees of freedom. The probability value is given by

$$P(F(3, 31) > 6.02) \simeq 0.01$$

The probability value expresses the likelihood of obtaining the observed F ratio 6.02 under the null hypothesis. It is small, which makes the null hypothesis unlikely. Hence, one can rule out the null hypothesis and reject $\beta_4 = \beta_5 = \beta_6 = 0$. This implies that at least one of the variables x_4, x_5, and x_6 is important, even if variables x_1, x_2, and x_3 are already in the model.

Note that the t tests on individual parameters that we discussed earlier can also be carried out within the additional sums of squares framework. For example, consider the test $\beta_5 = 0$. This test can be formulated as testing the hypothesis

$$H_0: A\beta = 0$$

where $A = (0, 0, 0, 0, 0, 1, 0)$ is a row vector, and β is the (7×1) vector of parameters in the full model. Under this hypothesis, the restricted model becomes

$$z = \beta_0 + \beta_1 x_1 + \beta_2 x_2 + \beta_3 x_3 + \beta_4 x_4 + \beta_6 x_6 + \epsilon \tag{4.45}$$

It is the model without x_5. Estimates in the restricted model, $\hat{\boldsymbol{\beta}}_A$, can be obtained. The residual sum of squares from this restricted model is $S(\hat{\boldsymbol{\beta}}_A) = 3.2720$. Comparing this with the residual sum of squares from the original (full) model, $S(\hat{\boldsymbol{\beta}}) = 3.0348$, gives us the additional sum of squares $S(\hat{\boldsymbol{\beta}}_A) - S(\hat{\boldsymbol{\beta}}) = 3.2720 - 3.0348 = 0.2372$. It has $l = 1$ degree of freedom. Hence,

$$F = \frac{\text{additional sum of squares}/1}{S(\hat{\boldsymbol{\beta}})/(n-p-1)} = \frac{0.2372}{3.0348/(38-7)} = 2.42 \tag{4.46}$$

with probability value

$$P(F(1,31) > 2.42) = 0.13$$

Note that the computed F in Eq. (4.46) is the square of the t ratio for $\hat{\beta}_5$ in the full model: $2.42 = (1.56)^2$. This equality can be shown in general. Furthermore, note that the previous probability value is exactly the same as the one we found for the t statistic for testing $\beta_5 = 0$. Hence, the conclusions from both tests, the F test for one extra parameter and the t test for an individual coefficient, are identical. We know in general that the square of a t distributed random variable follows a certain F distribution ($t_{df}^2 = F(1, df)$; see Chapter 2, Exercise 2.2). Such F tests for one extra parameter are referred to as **partial F tests**, and the **additional sum of squares** is referred to as the **partial sum of squares** due to this extra variable (in our case, x_5), given that all other variables are already in the model.

OC Example Continued

The full model is

$$\mathbf{y} = \alpha\mathbf{z} + \beta_1\mathbf{x}_1 + \cdots + \beta_5\mathbf{x}_5 + \boldsymbol{\epsilon} \tag{4.47}$$

A test of the hypothesis that there are no differences among the five drugs restricts the parameters as follows:

$$\beta_1 = \beta_2 = \beta_3 = \beta_4 = \beta_5 = \gamma \text{ (say)},$$

or

$$A\boldsymbol{\beta} = \mathbf{0}$$

where A and $\boldsymbol{\beta}$ are given in Eq. (4.39). The restricted model

$$\mathbf{y} = \gamma\mathbf{1} + \alpha\mathbf{z} + \boldsymbol{\epsilon} \tag{4.48}$$

can be estimated, and the residual sum of squares can be calculated. We find $S(\hat{\boldsymbol{\beta}}_A) = 2{,}932.0$. The residual sum of squares of the full model in Eq. (4.47) is given by $S(\hat{\boldsymbol{\beta}}) = 2{,}505.0$. Hence, the additional sum of squares is

$$\|\hat{\boldsymbol{\mu}} - \hat{\boldsymbol{\mu}}_A\|^2 = S(\hat{\boldsymbol{\beta}}_A) - S(\hat{\boldsymbol{\beta}}) = 2{,}932.0 - 2{,}505.0 = 427.0$$

We have placed $l = 4$ linear restrictions on the six coefficients $\alpha, \beta_1, \ldots, \beta_5$, and there are $n - p - 1 = 50 - 6 = 44$ degrees of freedom for the error sum of squares in the full model. Hence,

$$F = \frac{427/4}{2{,}505/44} = 1.87$$

We use the $F(4, 44)$ distribution to obtain the probability value,

$$p \text{ value} = P(F(4, 44) > 1.87) \simeq 0.13$$

Since this is quite large (certainly larger than the commonly used significance level 0.05), we believe in our null hypothesis. Hence, we find no real evidence that the drugs differ in their effect on the final HDLC.

Next, suppose we are also interested in knowing whether or not the drugs have an effect at all. This hypothesis specifies that $\beta_1 = \cdots = \beta_5 = 0$. Under this hypothesis, the model (4.47) becomes

$$y = \alpha z + \epsilon \tag{4.49}$$

The LSE of α can be found $\left(\text{it is } \hat{\alpha}_A = \frac{\sum z_i y_i}{\sum z_i^2}\right)$, leading to the residual sum of squares $S(\hat{\beta}_A) = \sum (y_i - \hat{\alpha}_A z_i)^2 = 3,410.68$. Hence, the additional sum of squares is

$$\|\hat{\mu} - \hat{\mu}_A\|^2 = S(\hat{\beta}_A) - S(\hat{\beta}) = 3,410.68 - 2,505.0 = 905.68$$

Since there are five restrictions, 5 degrees of freedom are associated with this extra sum of squares. The test statistic for the previous hypothesis is

$$F = \frac{905.68/5}{2505/44} = 3.18$$

The probability value is given by

$$P(F(5, 44) > 3.18) \simeq 0.02$$

The probability value is small—smaller than the usual significance level 0.05. $F = 3.18$ is an extreme value under the null hypothesis. We can reject H_0 and conclude that there is evidence that the drugs affect the final HDLC.

4.4.3 JOINT CONFIDENCE REGIONS FOR SEVERAL PARAMETERS

In the UFFI example, we constructed confidence intervals for individual parameters. For instance, the 95% confidence interval for β_1 has the form

$$P(L_1 \le \beta_1 \le U_1) = 0.95$$

where $L_1 = \hat{\beta}_1 - t(0.975; n-p-1)\text{s.e.}(\hat{\beta}_1)$, $U_1 = \hat{\beta}_1 + t(0.975; n-p-1)$ s.e.$(\hat{\beta}_1)$, and $t(0.975; n-p-1)$ is the 97.5% percentile of a t distribution with degrees of freedom $n-p-1$ (see Section 4.3.1). A 95% confidence interval for β_2 has a similar form with lower and upper limits L_2 and U_2. For our data, $L_1 = 4.88$, $U_1 = 13.74$ and $L_2 = 2.08$, $U_2 = 3.62$.

In some contexts, it may be necessary to make joint confidence statements about β_1 and β_2. For example, we may want to construct a confidence region CR such that $P((\beta_1, \beta_2) \text{ is in CR}) = 0.95$. It is known that

$$P(L_1 \le \beta_1 \le U_1, L_2 \le \beta_2 \le U_2) \le P(L_1 \le \beta_1 \le U_1)P(L_2 \le \beta_2 \le U_2)$$
$$= 0.95^2 = 0.9025$$

The confidence level associated with the rectangular region obtained by taking the two marginal intervals as shown above is less than 0.95.

There is a procedure, however, to construct a joint confidence region for a set of parameters that has the required coverage. In the linear model with parameter vector $\boldsymbol{\beta}$ it can be shown that

$$\frac{(\hat{\boldsymbol{\beta}} - \boldsymbol{\beta})' X' X (\hat{\boldsymbol{\beta}} - \boldsymbol{\beta})}{(p+1)s^2} \sim F(p+1, n-p-1)$$

where $F(p+1, n-p-1)$ denotes an F distribution with degrees of freedom $(p+1)$ and $(n-p-1)$. This result can be shown as follows.

For the linear model, $\mathbf{y} = X\boldsymbol{\beta} + \boldsymbol{\epsilon}$ with the standard assumptions, $\hat{\boldsymbol{\beta}} \sim N(\boldsymbol{\beta}, \sigma^2(X'X)^{-1})$, and hence $(\hat{\boldsymbol{\beta}} - \boldsymbol{\beta}) \sim N(\mathbf{0}, \sigma^2(X'X)^{-1})$. The results on the distribution of quadratic forms in Section 3.4 imply that

$$(\hat{\boldsymbol{\beta}} - \boldsymbol{\beta})'[\sigma^2(X'X)^{-1}]^{-1}(\hat{\boldsymbol{\beta}} - \boldsymbol{\beta}) = (\hat{\boldsymbol{\beta}} - \boldsymbol{\beta})' X' X (\hat{\boldsymbol{\beta}} - \boldsymbol{\beta})/\sigma^2 \sim \chi^2_{p+1}$$

We also know from previous results in this chapter that $(n-p-1)s^2/\sigma^2 \sim \chi^2_{n-p-1}$, and that $\hat{\boldsymbol{\beta}}$ and s^2 are statistically independent. The ratio of two independent chi-square random variables, standardized by their degrees of freedom, has an F distribution; see the appendix in Chapter 2. Hence,

$$\frac{(\hat{\boldsymbol{\beta}} - \boldsymbol{\beta})' X' X (\hat{\boldsymbol{\beta}} - \boldsymbol{\beta})/(p+1)\sigma^2}{(n-p-1)s^2/(n-p-1)\sigma^2} = \frac{(\hat{\boldsymbol{\beta}} - \boldsymbol{\beta})' X' X (\hat{\boldsymbol{\beta}} - \boldsymbol{\beta})}{(p+1)s^2} \sim F(p+1, n-p-1)$$

This result implies that a $100(1-\alpha)\%$ joint confidence region for all parameters in $\boldsymbol{\beta}$ is given by

$$\frac{(\hat{\boldsymbol{\beta}} - \boldsymbol{\beta})' X' X (\hat{\boldsymbol{\beta}} - \boldsymbol{\beta})}{(p+1)s^2} \leq F(1-\alpha; p+1, n-p-1)$$

where $F(1-\alpha; p+1, n-p-1)$ is the $100(1-\alpha)$ percentile of an F distribution with degrees of freedom $(p+1)$ and $(n-p-1)$. Algebraically, and this is somewhat cumbersome, one needs to find the values $\boldsymbol{\beta}$ such that the previous equality is satisfied. Choosing submatrices of $(X'X)^{-1}$ appropriately, confidence regions for subsets of parameters in $\boldsymbol{\beta}$ can also be obtained. For example, a joint confidence region for β_1, β_2 uses a submatrix of $(X'X)^{-1}$ that corresponds to these coefficients (see Exercise 4.24).

For a pair of parameters, the joint confidence region is an ellipse on a two-dimensional plot. For more than two parameters it is an ellipsoid. Since joint confidence regions are rarely used in practice, we will not pursue the topic further.

4.5 THE ANALYSIS OF VARIANCE AND THE COEFFICIENT OF DETERMINATION, R^2

Let us consider the general linear model

$$y = \beta_0 + \beta_1 x_1 + \cdots + \beta_p x_p + \epsilon \tag{4.50}$$

and the hypothesis

$$H_0 : \beta_1 = \beta_2 = \cdots = \beta_p = 0$$

Under this hypothesis, the model reduces to

$$y = \beta_0 + \epsilon \tag{4.51}$$

This model implies that the response y has a mean $E(y) = \beta_0$ that is not affected by any of the explanatory variables. The hypothesis expresses the fact that $x_1, \ldots, x_p$ do not influence the response.

We can use the **additional sum of squares principle** to test this hypothesis. The residual sum of squares of the full model in (Eq. 4.50) is given by $S(\hat{\boldsymbol{\beta}})$. In the restricted model (Eq. 4.51), the estimate of β_0 is given by $\bar{y}$ and the "residual sum of squares" by

$$\sum_{i=1}^{n} (y_i - \bar{y})^2 \tag{4.52}$$

This is called the **total sum of squares**, corrected for the mean. It is a measure of how the observations fluctuate around their mean. The **additional sum of squares** from the p regressor variables is given by $\sum (y_i - \bar{y})^2 - S(\hat{\boldsymbol{\beta}})$. It has $l = p$ degrees of freedom since our null hypothesis specifies p independent constraints. This quantity is usually called the **regression**, or **the model sum of squares**. It tells us how much variability is explained by the full model, over and above the simple mean model. It can be shown that the regression sum of squares (SSR) is given by

$$\begin{aligned} \text{SSR} &= \sum (y_i - \bar{y})^2 - S(\hat{\boldsymbol{\beta}}) = \mathbf{y}'\mathbf{y} - n\bar{y}^2 - (\mathbf{y} - X\hat{\boldsymbol{\beta}})'(\mathbf{y} - X\hat{\boldsymbol{\beta}}) \\ &= \hat{\boldsymbol{\beta}}' X'\mathbf{y} - n\bar{y}^2 = \hat{\boldsymbol{\beta}}' X'X\hat{\boldsymbol{\beta}} - n\bar{y}^2 \end{aligned} \tag{4.53}$$

The three sums of squares—the regression sum of squares, the residual sum of squares, and the total sum of squares—are usually displayed in a table called the analysis of variance (ANOVA) table (Table 4.3).

The degrees of freedom column in the table contains the relevant degrees of freedom. The regression sum of squares, SSR, has p degrees of freedom, because there are p regressor variables that make up the model. The error sum of squares has $n - p - 1$ degrees of freedom; the number of observations n minus the number of parameters in the model, $p + 1$. The total sum of squares has

TABLE 4.3 ANALYSIS OF VARIANCE (ANOVA) TABLE

Source	df	Sum of Squares	Mean Squares	F
Model (Regression)	p	$\text{SSR} = \hat{\boldsymbol{\beta}}' X'\mathbf{y} - n\bar{y}^2$	$\text{MSR} = \text{SSR}/p$	$\dfrac{\text{MSR}}{\text{MSE}}$
Residual (Error)	$n - p - 1$	$\text{SSE} = S(\hat{\boldsymbol{\beta}}) = (\mathbf{y} - X\hat{\boldsymbol{\beta}})'(\mathbf{y} - X\hat{\boldsymbol{\beta}})$	$\text{MSE} = \dfrac{\text{SSE}}{(n-p-1)}$	
Corrected total	$n - 1$	$\text{SST} = \sum (y_i - \bar{y})^2$		

$n-1$ degrees of freedom, because there are n deviations from the mean, but the sum of these deviations is zero.

The fourth column contains the **mean squares**, the sums of squares divided by their respective degrees of freedom. The mean square error, $\text{MSE}=\text{SSE}/(n-p-1)=S(\hat{\beta})/(n-p-1)$, was seen earlier. It is the unbiased estimator of σ^2. The fifth column contains the F ratio for testing the hypothesis $H_0:\beta_1=\beta_2=\cdots=\beta_p=0$,

$$F=\frac{\text{additional sum of squares}/p}{S(\hat{\beta})/(n-p-1)}=\frac{\text{SSR}/p}{\text{SSE}/(n-p-1)} \tag{4.54}$$

Observe that, by construction, the regression sum of squares and the error sum of squares must add up to the total sum of squares. Hence, the ANOVA table partitions the variability (the total sum of squares) into two interpretable components: a sum of squares that is explained by the model and a sum of squares that has been left unexplained.

Gas Consumption Example Continued

The basic model is

$$z=100/y=\beta_0+\beta_1x_1+\beta_2x_2+\beta_3x_3+\beta_4x_4+\beta_5x_5+\beta_6x_6+\epsilon \tag{4.55}$$

The hypothesis

$$H_0:\beta_1=\beta_2=\beta_3=\beta_4=\beta_5=\beta_6=0$$

can be written as $A\boldsymbol{\beta}=\mathbf{0}$, where A is the 6×7 matrix of rank 6,

$$A=\begin{bmatrix}0&1&0&0&0&0&0\\0&0&1&0&0&0&0\\0&0&0&1&0&0&0\\0&0&0&0&1&0&0\\0&0&0&0&0&1&0\\0&0&0&0&0&0&1\end{bmatrix}$$

$\boldsymbol{\beta}$ is the (7×1) vector of parameters, and $\mathbf{0}$ a (6×1) vector of zeros. Failure to reject this hypothesis implies that none of the variables are important in predicting the gas consumption of the vehicle. Rejection of the hypothesis implies that at least one of the variables is important in predicting gas consumption. Under the null hypothesis, the model reduces to

$$y=\beta_0+\epsilon \tag{4.56}$$

Estimation of the full model in Eq. (4.55) gives $\text{SSE}=S(\hat{\beta})=3.0348$. The total sum of squares (corrected for the mean) is easy to calculate; $\text{SST}=\sum(y_i-\bar{y})^2=\sum y_i^2-n\bar{y}^2=49.4463$. Hence, by subtraction, we find the regression sum of squares, $\text{SSR}=\text{SST}-\text{SSE}=46.4115$. These are the sum of squares entries in Table 4.4.

The degrees of freedom are 6 (as there are six regressor variables), 31 (because we estimate seven coefficients from $n=38$ cases), and 37 $(=n-1)$. The F ratio,

TABLE 4.4 ANOVA TABLE FOR GAS CONSUMPTION DATA

Source	df	Sum of Squares	Mean Squares	F	Prob > F
Model (Regression)	6	46.4115	7.7352	79.015	0.0001
Residual (Error)	31	3.0348	0.0979		
Corrected total	37	49.4463			

Note that this table was part of the SAS output; see Section 4.3.1.

$F = 79.015$, is large; its probability value 0.0001 is tiny. It indicates that there is strong evidence against the claim that none of the regressor variables have an influence ($\beta_1 = \beta_2 = \beta_3 = \beta_4 = \beta_5 = \beta_6 = 0$). In other words, we cannot discard x_1, x_2, x_3, x_4, x_5, and x_6 simultaneously; at least one of the variables is important in predicting z.

The F test in the ANOVA table is also known as a test for the overall significance of the regression. If we reject H_0, some regression relations exist. Which ones, we do not know at this point.

4.5.1 COEFFICIENT OF DETERMINATION, R^2

The ANOVA table partitions the total response variation into two components: SST = SSR + SSE, the variation that is explained by the regression model (SSR), and the variation that is left unexplained (SSE). The coefficient of determination R^2 is defined as the proportion of the total response variation that is explained by the model,

$$R^2 = \frac{\text{SSR}}{\text{SST}} = \frac{\text{SST} - \text{SSE}}{\text{SST}} = 1 - \frac{\text{SSE}}{\text{SST}} \tag{4.57}$$

For the gas consumption example and model (4.55), the total sum of squares is SST = 49.4463, the residual sum of squares (the unexplained response variation) is SSE = 3.0348, and the response variation that is explained by the model (regression sum of squares) is SSR = 49.4463 − 3.0348 = 46.4115. Hence, $R^2 = 46.4115/49.4463 = 0.9386$. This means that 94% of the variation in the response is explained by the linear model with the regressor variables $x_1, x_2, x_3, x_4, x_5, x_6$.

R^2 is a useful summary measure. It provides an overall measure of how well the model fits. It can also give feedback on the importance of adding a variable to (or deleting a variable from) a model. For instance, if we delete x_5, x_6 from the model in Eq. (4.55), the regression sum of squares reduces to 44.9505 and $R^2 = 0.9091$. This is slightly smaller, but it appears that a model without x_5 and x_6 is not much worse than the full model. This casts doubt on the inclusion of these two variables in the model. Note that adding a variable to a model increases the regression sum of squares, and hence the R^2. (In the worst case, it can stay the same.) R^2 can be made 1 by adding increasingly more explanatory variables. If we fit a model with $(n - 1)$ explanatory variables to n cases, the fit is perfect; the residual sum of squares will be zero, and $R^2 = 1$. One certainly does not want to do this because one would "overfit" the data, trying to find an explanation for every random perturbation. Hence, the use of large numbers of explanatory variables, especially when n is small, is not a good idea.

FIGURE 4.10
Different Data Plots Yielding Identical R^2

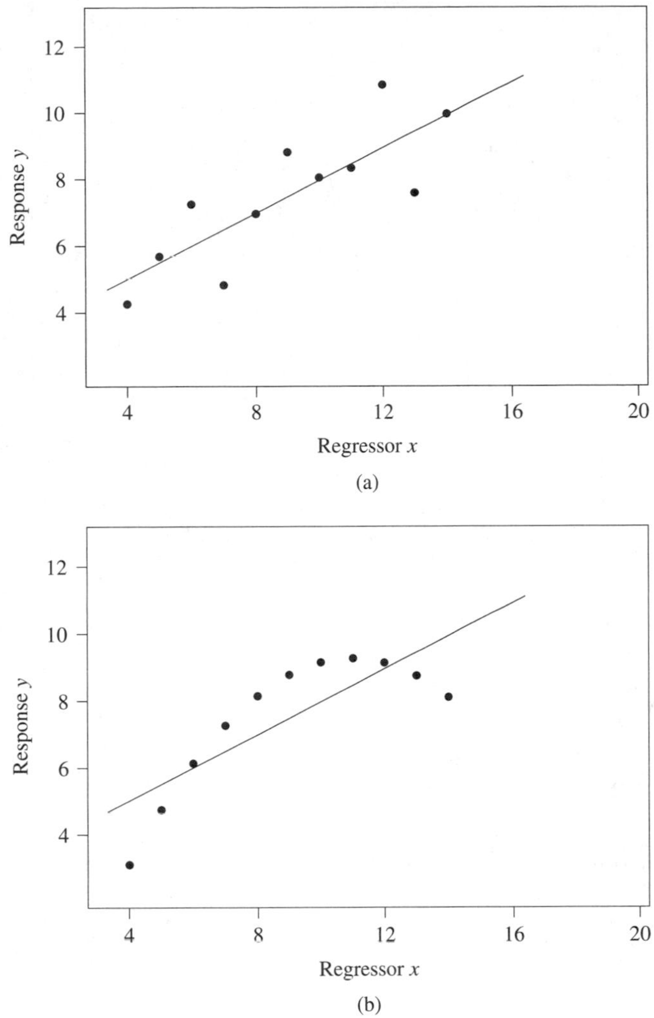

R^2 is just one summary measure of the regression fit. It alone does not tell us whether the fitted model is appropriate. Look at the data that are listed in Exercise 2.3 and plotted in Figure 4.10. It turns out (you should check this) that all four data sets lead to the same least squares estimates, the same ANOVA table, and identical R^2. However, there is only one situation (case a) in which one would say that a simple linear regression describes the data. One needs to be careful when interpreting the R^2.

4.6 GENERALIZED LEAST SQUARES

4.6.1 INTRODUCTION

The standard regression $\mathbf{y} = X\boldsymbol{\beta} + \boldsymbol{\epsilon}$ assumes that the vector of errors $\boldsymbol{\epsilon}$ has zero mean and covariance matrix $V(\boldsymbol{\epsilon}) = \sigma^2 I$, which implies that all errors have the

FIGURE 4.10 (Continued)

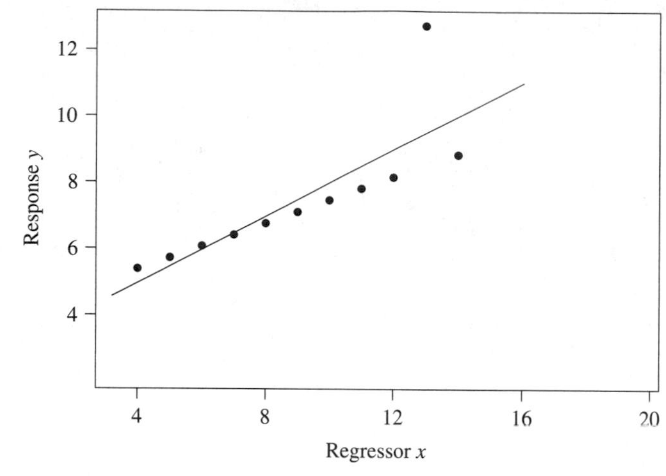

(c)

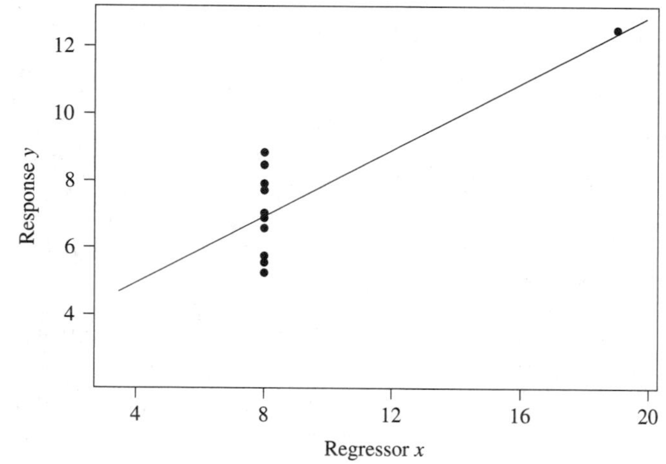

(d)

same precision and that they are uncorrelated. In some situations, these assumptions will not be reasonable.

In certain applications, some errors have more variability than others. Consider the situation in which the response for the ith case is obtained as an average of several measurements and the number of measurements that go into that average changes from case to case. In this situation, $V(y_i) = V(\epsilon_i) = \sigma^2/n_i$, where n_i represents the number of measurements in the average y_i. The assumption of equal variance is clearly violated.

Consider the case in which responses are taken over time. For example, consider modeling the relationship between the sales of your product, its price, as well as the prices of major competitors, and the amount your company spends on advertisement. Suppose that monthly observations (e.g., the past 5 years) are available to estimate the coefficients in the regression model. You expect that a regression of sales on price and advertising will be useful, and that these regressor variables will "explain" sales. However, even after controlling for prices and advertising, deviations of the sales from their implied expected levels tend to exhibit "runs." If sales in a certain month are unusually high, then there is a good chance that they will also be high in adjacent months. This is because the economic "driving forces" (which are not in your model) are persistent, moving only slowly over time; if the economy is poor today, then it tends to be poor also in preceeding and following months. You could try to specify an additional variable, "state of the economy," and use this as an additional regressor variable. This may help, but most likely there will be some other unknown and slowly changing variables that affect your sales, and measurement errors from different periods will tend to be correlated. We refer to this as **autocorrelation** or **serial correlation** because the errors are correlated among themselves at different lags. Very often, the amount of (auto)correlation diminishes with the lag. For example, adjacent observations are the most strongly correlated, whereas errors several steps apart are less correlated. Usually, one assumes a certain structure for the auto- (or serial) correlation. Often, one assumes that $\text{Cov}(\epsilon_i, \epsilon_{i-1}) = \text{Cov}(\epsilon_i, \epsilon_{i+1}) = \sigma^2\phi$, $\text{Cov}(\epsilon_i, \epsilon_{i-2}) = \text{Cov}(\epsilon_i, \epsilon_{i+2}) = \sigma^2\phi^2, \ldots, \text{Cov}(\epsilon_i, \epsilon_{i-k}) = \text{Cov}(\epsilon_i, \epsilon_{i+k}) = \sigma^2\phi^k$, for lags $k = 1, 2, \ldots$. In this case, the $n \times n$ covariance matrix of the errors is given by

$$V(\epsilon) = \sigma^2 \begin{bmatrix} 1 & \phi & \phi^2 & \ldots & \phi^{n-1} \\ \phi & 1 & \phi & \ldots & \phi^{n-2} \\ \phi^2 & \phi & 1 & \ldots & \phi^{n-3} \\ \phi^3 & \ldots & \ldots & \ldots & \ldots \\ \ldots & \ldots & \ldots & \ldots & \ldots \\ \phi^{n-1} & \phi^{n-2} & \ldots & \ldots & 1 \end{bmatrix}$$

The model that corresponds to this covariance matrix is known as the **first-order autoregressive model**. It is a particularly simple and useful parameterization because it requires only one additional parameter, the autoregressive parameter ϕ. However, many other models are available for representing autocorrelations among observations. In Chapter 10 on regression time series models, we discuss these models in detail.

A third example in which independence of the errors is violated arises when spatial observations are involved. Consider measurements on a certain groundwater pollutant that are taken at the same time but at different locations. In this situation, it is likely that errors for measurements taken in close spatial proximity are correlated. Many different models have been developed to characterize **spatial correlation**, and most express the spatial correlation as a function of the (Euclidean) distance between the measurement locations. A common assumption is that the correlation decreases with distance among measurement sites.

Expand on this example, and consider the situation when spatial observations are involved but when observations are also taken at several time periods. Here, one faces the situation in which observations (or errors) exhibit a spatial as well as a temporal correlation structure. A common approach is to model the covariance matrix of the errors with several (hopefully few) additional parameters that characterize the spatial and temporal correlations and then estimate the parameters in the regression model $y = X\beta + \epsilon$ under this more general error model.

4.6.2 GENERALIZED LEAST SQUARES ESTIMATION

Assume that the vector of errors ϵ in the regression model $y = X\beta + \epsilon$ has zero mean and general covariance matrix $V(\epsilon) = E(\epsilon\epsilon') = \sigma^2 V$. Now V is no longer the identity matrix. The proportionality coefficient, σ^2, is unknown, but we assume—at least initially—that all elements in the matrix V are known.

We will try to find a linear transformation, $L\epsilon$ of ϵ, which satisfies the assumptions of the standard model. The matrix V is symmetric and positive definite, and we can apply our results in Chapter 3 on the spectral decomposition of a symmetric matrix. We can write the matrix as $V = P\Lambda P' = P\Lambda^{1/2}\Lambda^{1/2}P'$, where the matrix Λ is diagonal. Its elements $\lambda_1 \geq \lambda_2 \geq \cdots \geq \lambda_m > 0$ are the eigenvalues of the positive definite matrix V; the column vectors of the matrix P are the corresponding normalized eigenvectors. Note that $V^{-1} = (P\Lambda^{1/2}\Lambda^{1/2}P')^{-1} = P\Lambda^{-1/2}\Lambda^{-1/2}P' = L'L$.

Premultiplying the regression model by the matrix $L = \Lambda^{-1/2}P'$ results in the model

$$Ly = LX\beta + L\epsilon = LX\beta + \tilde{\epsilon}$$

where the vector Ly represents the transformed response, and the columns in the matrix LX represent the transformed regressor variables. Then $E(\tilde{\epsilon}) = E(L\epsilon) = LE(\epsilon) = \mathbf{0}$ and

$$V(\tilde{\epsilon}) = LV(\epsilon)L' = LVL'\sigma^2 = \Lambda^{-1/2}P'P\Lambda^{1/2}\Lambda^{1/2}P'P\Lambda^{-1/2}\sigma^2 = I\sigma^2$$

The new disturbance vector $\tilde{\epsilon}$ satisfies the standard regression assumptions. According to the Gauss–Markov theorem, least squares—applied to the transformed variables—will yield the best linear unbiased estimator of β. Replacing y and X in the standard least squares estimator in Eq. (4.11) by Ly and LX, respectively, leads to the **generalized least squares (GLS) estimator**

$$\hat{\beta}^{GLS} = (X'L'LX)^{-1}X'L'Ly = (X'V^{-1}X)^{-1}X'V^{-1}y \tag{4.58}$$

and its covariance matrix

$$\begin{aligned} V(\hat{\beta}^{GLS}) &= (X'L'LX)^{-1}X'L'V(Ly)LX(X'L'LX)^{-1} \\ &= \sigma^2(X'L'LX)^{-1}X'L'LX(X'L'LX)^{-1} \\ &= \sigma^2(X'L'LX)^{-1} = \sigma^2(X'V^{-1}X)^{-1} \end{aligned} \tag{4.59}$$

The GLS estimator can be used to compute the error sum of squares in the transformed model,

$$\begin{aligned} S(\hat{\beta}^{\text{GLS}}) &= (Ly - LX\hat{\beta}^{\text{GLS}})'(Ly - LX\hat{\beta}^{\text{GLS}}) \\ &= (y - X\hat{\beta}^{\text{GLS}})'L'L(y - X\hat{\beta}^{\text{GLS}}) \\ &= (y - X\hat{\beta}^{\text{GLS}})'V^{-1}(y - X\hat{\beta}^{\text{GLS}}) \end{aligned} \tag{4.60}$$

The model in the transformed variables satisfies the standard regression assumptions. Hence, $S(\hat{\beta}^{\text{GLS}})/\sigma^2$ follows a chi-square distribution with $n - (p + 1)$ degrees of freedom, where n represents the number of cases and $p + 1$ the number of regression coefficients. Hence,

$$s^2_{\text{GLS}} = S(\hat{\beta}^{\text{GLS}})/(n - p - 1) \tag{4.61}$$

is an unbiased estimator of σ^2. This can be used in Eq. (4.59) to obtain an estimate of $V(\hat{\beta}^{\text{GLS}})$.

What are the properties of the standard least squares estimator $\hat{\beta} = (X'X)^{-1} X'y$ that has been derived under the wrong assumption of independent and equally precise errors? It also is unbiased, but it is no longer "best" among all linear unbiased estimators. The Gauss–Markov result has already shown us that it is the GLS estimator $\hat{\beta}^{\text{GLS}}$ that has the smallest covariance matrix. The covariance matrix of the standard least squares estimator

$$\begin{aligned} V(\hat{\beta}) &= V[(X'X)^{-1}X'y] = (X'X)^{-1}X'V(y)X(X'X)^{-1} \\ &= \sigma^2(X'X)^{-1}X'VX(X'X)^{-1} \end{aligned}$$

exceeds the covariance matrix in Eq. (4.59) by a positive semidefinite matrix.

Remark

So far, our analysis has assumed that all elements in the matrix V are specified. For this reason, we call the estimator in Eq. (4.58) the **feasible** generalized least squares estimator. In the first example of our introduction, the precision $V(y_i) = V(\epsilon_i) = \sigma^2/n_i$ depends on the **known** number of measurements that go into the observation y_i. Here, V is specified, and the generalized least squares estimator can be calculated. In the second illustration, the matrix V contains the autoregressive parameter ϕ. In practice, this parameter is unknown, and one must estimate the regression coefficients β and ϕ jointly. This issue will be addressed in Chapter 10, when we discuss regression models with time series errors.

4.6.3 WEIGHTED LEAST SQUARES

Weighted least squares is a special case of generalized least squares. The weighted least squares estimator minimizes the weighted error sum of squares

$$S(\beta) = \sum_{i=1}^{n} w_i(y_i - x_i'\beta)^2$$

where $w_i > 0$ are known specified weights. This criterion is equivalent to the one for generalized least squares, with V^{-1} a diagonal matrix having diagonal elements w_i.

Equations (4.58) and (4.59) imply that the weighted least squares (WLS) estimator is given by

$$\hat{\boldsymbol{\beta}}^{\text{WLS}} = \left[\sum_{i=1}^{n} w_i \boldsymbol{x}_i \boldsymbol{x}_i'\right]^{-1} \left[\sum_{i=1}^{n} w_i \boldsymbol{x}_i y_i\right] \tag{4.62}$$

with variance

$$V(\hat{\boldsymbol{\beta}}^{\text{WLS}}) = \sigma^2 \left[\sum_{i=1}^{n} w_i \boldsymbol{x}_i \boldsymbol{x}_i'\right]^{-1} \tag{4.63}$$

APPENDIX: PROOFS OF RESULTS

1. MINIMIZATION OF $S(\boldsymbol{\beta})$ IN EQ. (4.9)

We wish to minimize

$$S(\boldsymbol{\beta}) = \sum_{i=1}^{n} (y_i - \mu_i)^2 = \sum_{i=1}^{n} \epsilon_i^2$$

subject to the restriction that $\boldsymbol{\mu} = X\boldsymbol{\beta}$. The elements of the vector $\boldsymbol{\mu}$ are given by

$$\mu_i = \beta_0 + \beta_1 x_{i1} + \cdots + \beta_p x_{ip} = \beta_0 + \sum_{j=1}^{p} x_{ij}\beta_j$$

for $i = 1, 2, \ldots, n$. The partial derivatives of $S(\boldsymbol{\beta})$ with respect to the parameters $\beta_0, \beta_1, \ldots, \beta_p$ are

$$\frac{\partial S(\boldsymbol{\beta})}{\partial \beta_0} = 2\sum_{i=1}^{n} \epsilon_i \frac{\partial \epsilon_i}{\partial \beta_0} = -2\sum_{i=1}^{n} \epsilon_i$$

and

$$\frac{\partial S(\boldsymbol{\beta})}{\partial \beta_j} = 2\sum_{i=1}^{n} \epsilon_i \frac{\partial \epsilon_i}{\partial \beta_j} = -2\sum_{i=1}^{n} x_{ij}\epsilon_i, \quad j = 1, 2, \ldots, p$$

At the minimum of $S(\boldsymbol{\beta})$ these derivatives are zero. Hence,

$$\sum_{i=1}^{n} \epsilon_i = \sum (y_i - \mu_i) = 0$$

$$\sum_{i=1}^{n} x_{ij}(y_i - \mu_i) = 0, \quad j = 1, 2, \ldots, p$$

In vector form,

$$\mathbf{1}'(\boldsymbol{y} - \boldsymbol{\mu}) = 0$$

$$\boldsymbol{x}_j'(\boldsymbol{y} - \boldsymbol{\mu}) = 0, \quad j = 1, 2, \ldots, p$$

where the $n \times 1$ vector $\mathbf{1}$, a vector of ones, and $\boldsymbol{x}_j$, the vector with elements $x_{1j}, x_{2j}, \ldots, x_{nj}$, are columns of the matrix $X = [\mathbf{1}, \boldsymbol{x}_1, \ldots, \boldsymbol{x}_p]$. Combining these $p+1$ equations, we obtain

$$X'(\boldsymbol{y} - X\boldsymbol{\beta}) = \mathbf{0}$$

or

$$X'X\boldsymbol{\beta} = X'\boldsymbol{y}$$

Let $\hat{\boldsymbol{\beta}}$ denote a solution of this equation. Solving the normal equations $(X'X)\hat{\boldsymbol{\beta}} = X'\boldsymbol{y}$ leads to $\hat{\boldsymbol{\beta}} = (X'X)^{-1}X'\boldsymbol{y}$; the inverse $(X'X)^{-1}$ exists since we assume that X has full column rank. In order to prove that $\hat{\boldsymbol{\beta}}$ actually minimizes $S(\boldsymbol{\beta})$ we show that any other estimate will lead to a larger value of $S(\boldsymbol{\beta})$:

$$\begin{aligned} S(\boldsymbol{\beta}) &= (\boldsymbol{y} - X\boldsymbol{\beta})'(\boldsymbol{y} - X\boldsymbol{\beta}) \\ &= (\boldsymbol{y} - X\hat{\boldsymbol{\beta}} + X\hat{\boldsymbol{\beta}} - X\boldsymbol{\beta})'(\boldsymbol{y} - X\hat{\boldsymbol{\beta}} + X\hat{\boldsymbol{\beta}} - X\boldsymbol{\beta}) \\ &= (\boldsymbol{y} - X\hat{\boldsymbol{\beta}})'(\boldsymbol{y} - X\hat{\boldsymbol{\beta}}) + (\hat{\boldsymbol{\beta}} - \boldsymbol{\beta})'X'X(\hat{\boldsymbol{\beta}} - \boldsymbol{\beta}) \end{aligned}$$

since the normal equations imply that the cross-product term $(\hat{\boldsymbol{\beta}} - \boldsymbol{\beta})'X'(\boldsymbol{y} - X\hat{\boldsymbol{\beta}}) = (\hat{\boldsymbol{\beta}} - \boldsymbol{\beta})'(X'\boldsymbol{y} - X'X\hat{\boldsymbol{\beta}}) = 0$. Thus,

$$S(\boldsymbol{\beta}) = S(\hat{\boldsymbol{\beta}}) + \boldsymbol{c}'\boldsymbol{c}$$

where $\boldsymbol{c} = X(\hat{\boldsymbol{\beta}} - \boldsymbol{\beta})$. Since $\boldsymbol{c}'\boldsymbol{c} = \sum_{i=1}^{n} c_i^2 \geq 0$, $S(\boldsymbol{\beta}) \geq S(\hat{\boldsymbol{\beta}})$; the equality is true if and only if $\boldsymbol{\beta} = \hat{\boldsymbol{\beta}}$.

2. ANOTHER PROOF OF THE UNBIASEDNESS OF s^2 AS AN ESTIMATE OF σ^2: $E(\sum_{i=1}^{n} e_i^2) = (n-p-1)\sigma^2$

Consider

$$\begin{aligned} E\left(\sum_{i=1}^{n} e_i^2\right) &= E(\boldsymbol{e}'\boldsymbol{e}) = E[\boldsymbol{y}'(I-H)(I-H)\boldsymbol{y}] && \text{since } \boldsymbol{e} = (I-H)\boldsymbol{y} \\ &= E[\boldsymbol{y}'(I-H)\boldsymbol{y}] && \text{since } (I-H) \text{ is idempotent} \\ &= E[\text{tr}(\boldsymbol{y}'(I-H)\boldsymbol{y})] && \text{since } \boldsymbol{y}'(I-H)\boldsymbol{y} \text{ is a scalar} \\ &= E[\text{tr}(I-H)\boldsymbol{y}\boldsymbol{y}'] && \text{since tr } AB = \text{tr } BA \\ &= \text{tr}[(I-H)E(\boldsymbol{y}\boldsymbol{y}')] \end{aligned}$$

Now

$$\begin{aligned} E(\boldsymbol{y}\boldsymbol{y}') &= E[(X\boldsymbol{\beta} + \boldsymbol{\epsilon})(X\boldsymbol{\beta} + \boldsymbol{\epsilon})'] \\ &= X\boldsymbol{\beta}\boldsymbol{\beta}'X' + E(\boldsymbol{\epsilon}\boldsymbol{\epsilon}') = X\boldsymbol{\beta}\boldsymbol{\beta}'X' + \sigma^2 I \end{aligned}$$

Here we have used the fact that $E(\boldsymbol{\epsilon}) = \mathbf{0}$ and $V(\boldsymbol{\epsilon}) = E(\boldsymbol{\epsilon}\boldsymbol{\epsilon}') = \sigma^2 I$. Hence,

$$\begin{aligned} E\left(\sum_{i=1}^{n} e_i^2\right) &= \text{tr}[(I-H)(\sigma^2 I + X\boldsymbol{\beta}\boldsymbol{\beta}'X')] \\ &= \text{tr}[I-H]\sigma^2, \text{ since } \text{tr}(A+B) = \text{tr}(A) + \text{tr}(B), \text{ and } (I-H)X = O \\ &= \sigma^2[n - \text{tr}X(X'X)^{-1}X'] \\ &= \sigma^2[n - (p+1)] = (n-p-1)\sigma^2 \end{aligned}$$

since $\text{tr}[X(X'X)^{-1}X'] = \text{tr}[(X'X)^{-1}X'X] = \text{tr}(I_{p+1})$, where I_{p+1} is the $(p+1) \times (p+1)$ identity matrix.

3. DIRECT PROOF THAT $\hat{\beta}$ AND $S(\hat{\beta})$ ARE STATISTICALLY INDEPENDENT AND THAT $S(\hat{\beta})/\sigma^2$ FOLLOWS A χ^2_{n-p-1} DISTRIBUTION

The set of all linear functions of $L(X) = L(\mathbf{1}, \boldsymbol{x}_1, \ldots, \boldsymbol{x}_p)$ forms a $(p+1)$-dimensional subspace of R^n. We can always find $p+1$ orthonormal vectors $\boldsymbol{c}_1, \ldots, \boldsymbol{c}_{p+1}$ (i.e., $\boldsymbol{c}_i'\boldsymbol{c}_i = 1, \boldsymbol{c}_i'\boldsymbol{c}_j = 0, i \neq j$) that form a basis of $L(X)$. These orthonormal vectors are linearly related to the regressor columns. The Gram–Schmidt orthogonalization procedure (see Chapter 3) shows us how to obtain these vectors.

$L(X)$ is a subset of R^n. Hence, we need to add $(n-p-1)$ additional orthonormal vectors $\boldsymbol{c}_{p+2}, \ldots, \boldsymbol{c}_n$ such that $(\boldsymbol{c}_1, \ldots, \boldsymbol{c}_n)$ forms an orthonormal basis of the larger space R^n. You can visualize the construction as follows:

$$\underbrace{\underbrace{\boldsymbol{c}_1, \ldots, \boldsymbol{c}_{p+1}}_{L(X)}, \boldsymbol{c}_{p+2}, \ldots, \boldsymbol{c}_n}_{R^n}$$

The vectors in the matrix

$$P = (\boldsymbol{c}_1, \ldots, \boldsymbol{c}_n) = (P_1, P_2)$$

where $P_1 = (\boldsymbol{c}_1, \ldots, \boldsymbol{c}_{p+1})$ and $P_2 = (\boldsymbol{c}_{p+2}, \ldots, \boldsymbol{c}_n)$ are $n \times (p+1)$ and $n \times (n-p-1)$ matrices, provide an orthonormal basis. By construction, P is an orthogonal matrix. That is, $P'P = PP' = I$.

Our model specifies $\boldsymbol{y} \sim N(\boldsymbol{\mu}, \sigma^2 I)$, where $\boldsymbol{\mu} = X\boldsymbol{\beta}$ is in $L(X)$. Consider the orthogonal transformation

$$\boldsymbol{z} = P'\boldsymbol{y} = \begin{pmatrix} P_1' \\ P_2' \end{pmatrix} \boldsymbol{y}$$

Then $\boldsymbol{z} \sim N(P'\boldsymbol{\mu}, \sigma^2 I)$ since $P'P = I$. This says that the z_i's are independent and have the same variance σ^2. Furthermore,

$$P'\boldsymbol{\mu} = \begin{pmatrix} P_1'\boldsymbol{\mu} \\ P_2'\boldsymbol{\mu} \end{pmatrix} = \begin{pmatrix} P_1'\boldsymbol{\mu} \\ \mathbf{0} \end{pmatrix}$$

since $P_2'\boldsymbol{\mu} = \mathbf{0}$. This is because $\boldsymbol{\mu}$ is in $L(X)$ and the columns of P_2 are perpendicular to $L(X)$.

Turning the transformation around results in

$$\begin{aligned} \boldsymbol{y} = (P')^{-1}\boldsymbol{z} = P\boldsymbol{z} &= \sum_{i=1}^{n} \boldsymbol{c}_i z_i \\ &= \sum_{i=1}^{p+1} \boldsymbol{c}_i z_i + \sum_{i=p+2}^{n} \boldsymbol{c}_i z_i \\ &= \hat{\boldsymbol{\mu}} + (\boldsymbol{y} - \hat{\boldsymbol{\mu}}) \end{aligned}$$

Here we have used the fact that P is orthogonal, and $P^{-1} = P'$. Now,

$$S(\hat{\boldsymbol{\beta}}) = (\boldsymbol{y} - X\hat{\boldsymbol{\beta}})'(\boldsymbol{y} - X\hat{\boldsymbol{\beta}}) = \|\boldsymbol{y} - \hat{\boldsymbol{\mu}}\|^2 = \left(\sum_{i=p+2}^{n} \boldsymbol{c}_i z_i\right)' \left(\sum_{i=p+2}^{n} \boldsymbol{c}_i z_i\right)$$

$$= \sum_{i=p+2}^{n} \sum_{j=p+2}^{n} z_i z_j \boldsymbol{c}_i' \boldsymbol{c}_j$$

$$= \sum_{i=p+2}^{n} z_i^2 \text{ since } \boldsymbol{c}_i'\boldsymbol{c}_i = 1 \text{ and } \boldsymbol{c}_i'\boldsymbol{c}_j = 0, i \neq j$$

Since $z_{p+2}, \ldots, z_n$ are i.i.d. $N(0, \sigma^2)$, it follows that

$$\frac{S(\hat{\boldsymbol{\beta}})}{\sigma^2} = \sum_{i=p+2}^{n} z_i^2/\sigma^2$$

is the sum of $(n - p - 1)$ independent χ_1^2 random variables. It has a χ_{n-p-1}^2 distribution. Furthermore, this is independent of $z_1, \ldots, z_{p+1}$. Now

$$\hat{\boldsymbol{\beta}} = (X'X)^{-1}X'\boldsymbol{y} = (X'X)^{-1}X'P\boldsymbol{z}$$

$$= (X'X)^{-1}X'(P_1, P_2)\boldsymbol{z}$$

However, $X'(P_1, P_2) = (X'P_1, O)$ since the columns of P_2 are perpendicular to $L(X)$ while rows of X' are in $L(X)$. Hence, $\hat{\boldsymbol{\beta}} = (X'X)^{-1}X'P_1\boldsymbol{z}_{(1)}$, where $\boldsymbol{z}_{(1)} = (z_1, \ldots, z_{p+1})'$. The least squares estimator $\hat{\boldsymbol{\beta}}$ depends on $z_1, \ldots, z_{p+1}$, whereas $S(\hat{\boldsymbol{\beta}})$ depends on $z_{p+2}, \ldots, z_n$. Thus, $\hat{\boldsymbol{\beta}}$ is independent of $S(\hat{\boldsymbol{\beta}})$.

4. PROOF OF THEOREM

Proof: $L_A(X)$ is a subset of $L(X)$, and $L(X)$ is a subset of R^n. $L(X)$ is of dimension $p + 1$. $L_A(X)$ imposes l independent restrictions on the subset $L(X)$. Hence, the dimension of $L_A(X)$ is $p + 1 - l$. Choose an orthonormal basis $(\boldsymbol{c}_1, \ldots, \boldsymbol{c}_{p+1-l})$ for $L_A(X)$ and extend it successively to form orthonormal bases for $L(X)$ and R^n. Visualize the process as follows:

$$\underbrace{\underbrace{\underbrace{\boldsymbol{c}_1, \ldots, \boldsymbol{c}_{p+1-l}}_{L_A(X)}, \boldsymbol{c}_{p+2-l}, \ldots, \boldsymbol{c}_{p+1}}_{L(X)}, \boldsymbol{c}_{p+2}, \ldots, \boldsymbol{c}_n}_{R^n}$$

The vectors are collected in the $n \times n$ matrix P,

$$P = (\boldsymbol{c}_1, \ldots, \boldsymbol{c}_n) = (P_1, P_2, P_3)$$

where $P_1 = (\boldsymbol{c}_1, \ldots, \boldsymbol{c}_{p+1-\ell})$, $P_2 = (\boldsymbol{c}_{p+2-\ell}, \ldots, \boldsymbol{c}_{p+1})$, and $P_3 = (\boldsymbol{c}_{p+2}, \ldots, \boldsymbol{c}_n)$ are $n \times (p + 1 - l)$, $n \times l$, and $n \times (n - p - 1)$ matrices. The matrix P is orthogonal: $PP' = P'P = I$.

Consider the orthogonal transformation $z = P'y$ and its inverse

$$y = Pz = \sum_{i=1}^{n} c_i z_i$$

$$= \sum_{i=1}^{p+1-l} c_i z_i + \sum_{i=p+2-l}^{p+1} c_i z_i + \sum_{i=p+2}^{n} c_i z_i$$

$$= \hat{\mu}_A + (\hat{\mu} - \hat{\mu}_A) + (y - \hat{\mu})$$

where $\hat{\mu}_A$ is the projection of y on $L_A(X)$, and $\hat{\mu}$ is the projection of y on $L(X)$; $\hat{\mu} - \hat{\mu}_A$ is in $L(X)$ and perpendicular to $L_A(X)$; $\|y - \hat{\mu}\|^2 = \sum_{i=p+2}^{n} z_i^2 = S(\hat{\beta})$ and $\|\hat{\mu} - \hat{\mu}_A\|^2 = \sum_{i=p+2-l}^{p+1} z_i^2$.

Since $y \sim N(\mu, \sigma^2 I)$, it follows that $z = P'y \sim N(P'\mu, \sigma^2 I)$. Under the null hypothesis $A\beta = \mathbf{0}$, the mean vector

$$P'\mu = \begin{bmatrix} P_1'\mu \\ P_2'\mu \\ P_3'\mu \end{bmatrix} = \begin{bmatrix} P_1'\mu \\ \mathbf{0} \\ \mathbf{0} \end{bmatrix}$$

This is because under the null hypothesis $\hat{\mu} = \hat{\mu}_A$ is in $L_A(X)$ and the columns of P_2 are perpendicular to $L_A(X)$. In addition, $P_3'\mu = \mathbf{0}$ since the columns of P_3 are perpendicular to $L(X)$. Hence,

i. $z_1, z_2, \ldots, z_n$ are independent normal random variables with variance σ^2.
ii. $z_{p+2-l}, \ldots, z_{p+1}$ have zero means under the null hypothesis $A\beta = \mathbf{0}$.
iii. $z_{p+2}, \ldots, z_n$ have zero means under the original model, even if the null hypothesis is false.

Thus,

i. $\|\hat{\mu} - \hat{\mu}_A\|^2 / \sigma^2 = \sum_{p+2-l}^{p+1} z_i^2$ is the sum of l independent χ_1^2 random variables. It has a χ_l^2 distribution.
ii. $S(\hat{\beta})$ is a function of $z_{p+2}, \ldots, z_n$, whereas $\|\hat{\mu} - \hat{\mu}_A\|^2$ is a function of $z_{p+2-l}, \ldots, z_{p+1}$. Furthermore, $z_1, z_2, \ldots, z_n$ are independent. This shows that $S(\hat{\beta})$ and $\|\hat{\mu} - \hat{\mu}_A\|^2$ are independent.

■

EXERCISES

4.1. Consider the regression on time, $y_t = \beta_0 + \beta_1 t + \epsilon_t$, with $t = 1, 2, \ldots, n$. Here, the regressor vector is $x' = (1, 2, \ldots, n)$. Take $n = 10$. Write down the matrices $X'X$, $(X'X)^{-1}$, $V(\hat{\beta})$, and the variances of $\hat{\beta}_0$ and $\hat{\beta}_1$.

4.2. For the regression model $y_t = \beta_0 + \epsilon_t$ with $n = 2$ and $y' = (2, 4)$, draw the data in two-dimensional space. Identify the orthogonal projection of y onto $L(X) = L(\mathbf{1})$. Explain geometrically $\hat{\beta}_0$, $\hat{\mu}$, and e.

4.3. Consider the regression model $y_i = \beta_0 + \beta_1 x_i + \epsilon_i$, $i = 1, 2, 3$. With

$$\boldsymbol{x} = \begin{bmatrix} 1 \\ 3 \\ 2 \end{bmatrix} \quad \boldsymbol{y} = \begin{bmatrix} 2.2 \\ 3.9 \\ 3.1 \end{bmatrix}$$

draw the data in three-dimensional space and identify the orthogonal projection of $\boldsymbol{y}$ onto $L(X) = L(\mathbf{1}, \boldsymbol{x})$. Explain geometrically $\hat{\boldsymbol{\beta}}$, $\hat{\boldsymbol{\mu}}$, and $\boldsymbol{e}$.

4.4. Consider the regression model $y_i = \beta_0 + \beta_1 x_i + \epsilon_i$, $i = 1, 2, 3$. With

$$\boldsymbol{x} = \begin{bmatrix} 1 \\ 3 \\ 2 \end{bmatrix} \quad \boldsymbol{y} = \begin{bmatrix} 2 \\ 4 \\ 6 \end{bmatrix}$$

draw the data in three-dimensional space and identify the orthogonal projection of $\boldsymbol{y}$ onto $L(X) = L(\mathbf{1}, \boldsymbol{x})$. Explain geometrically $\hat{\boldsymbol{\beta}}$, $\hat{\boldsymbol{\mu}}$, and $\boldsymbol{e}$.

4.5. After fitting the regression model,

$$y = \beta_0 + \beta_1 x_1 + \beta_2 x_2 + \beta_3 x_3 + \epsilon$$

on 15 cases, it is found that the mean square error $s^2 = 3$ and

$$(X'X)^{-1} = \begin{bmatrix} 0.5 & 0.3 & 0.2 & 0.6 \\ 0.3 & 6.0 & 0.5 & 0.4 \\ 0.2 & 0.5 & 0.2 & 0.7 \\ 0.6 & 0.4 & 0.7 & 3.0 \end{bmatrix}$$

Find

a. The estimate of $V(\hat{\beta}_1)$.

b. The estimate of $\text{Cov}(\hat{\beta}_1, \hat{\beta}_3)$.

c. The estimate of $\text{Corr}(\hat{\beta}_1, \hat{\beta}_3)$.

d. The estimate of $V(\hat{\beta}_1 - \hat{\beta}_3)$.

4.6. When fitting the model

$$E(y) = \beta_0 + \beta_1 x_1 + \beta_2 x_2$$

to a set of $n = 15$ cases, we obtained the least squares estimates $\hat{\beta}_0 = 10$, $\hat{\beta}_1 = 12$, $\hat{\beta}_2 = 15$, and $s^2 = 2$. It is also known that

$$(X'X)^{-1} = \begin{bmatrix} 1 & 0.25 & 0.25 \\ 0.25 & 0.5 & -0.25 \\ 0.25 & -0.25 & 2 \end{bmatrix}$$

a. Estimate $V(\hat{\beta}_2)$.

b. Test the hypothesis that $\beta_2 = 0$.

c. Estimate the covariance between $\hat{\beta}_1$ and $\hat{\beta}_2$.

d. Test the hypothesis that $\beta_1 = \beta_2$, using both the t ratio and the 95% confidence interval.

e. The corrected total sum of squares, SST = 120. Construct the ANOVA table and test the hypothesis that $\beta_1 = \beta_2 = 0$. Obtain the percentage of variation in y that is explained by the model.

4.7. Consider a multiple regression model of the price of houses (y) on three explanatory variables: taxes paid (x_1), number of bathrooms (x_2), and square feet (x_3). The incomplete (Minitab) output from a regression on $n = 28$ houses is given as follows:

The regression equation is price = −10.7 + 0.190 taxes + 81.9 baths + 0.101 sqft

Predictor	Coef	SE Coef	t	p
Constant	−10.65	24.02		
taxes	0.18966	0.05623		
baths	81.87	47.82		
sqft	0.10063	0.03125		

Analysis of variance

Source	DF	SS	MS	F	p
Regression	3	504541			
Residual Error					
Total	27	541119			

a. Calculate the coefficient of determination R^2.

b. Test the null hypothesis that all three regression coefficients are zero (H_0: $\beta_1 = \beta_2 = \beta_3 = 0$). Use significance level 0.05.

c. Obtain a 95% confidence interval of the regression coefficient for "taxes." Can you simplify the model by dropping "taxes"? Obtain a 95% confidence interval of the regression coefficient for "baths." Can you simplify the model by dropping "baths"?

4.8. Continuation of Exercise 4.7. The incomplete (Minitab) output from a multiple regression of the price of houses on the two explanatory variables, taxes paid and square feet, is given as follows:

The regression equation is price = 4.9 + 0.242 taxes + 0.134 sqft

Predictor	Coef	SE Coef	t	p
Constant	4.89	23.08		
taxes	0.24237	0.04884		
sqft	0.13397	0.02537		

Analysis of variance

Source	DF	SS	MS	F	p
Regression	2	500074	250037		
Residual Error					
Total		541119			

a. Calculate the coefficient of determination R^2.

b. Test the null hypothesis that both regression coefficients are zero (H_0: $\beta_1 = \beta_2 = 0$). Use significance level 0.05.

c. Test whether you can omit the variable "taxes" from the regression model. Use significance level 0.05.

d. Comment on the fact that the regression coefficients for taxes and square feet are different than those shown in Exercise 4.7.

4.9. Fitting the regression $y_i = \beta_0 + \beta_1 x_{i1} + \beta_2 x_{i2} + \varepsilon_i$ on $n = 30$ cases leads to the following results:

$$X'X = \begin{bmatrix} 30 & 2{,}108 & 5{,}414 \\ 2{,}108 & 152{,}422 & 376{,}562 \\ 5{,}414 & 376{,}562 & 1{,}015{,}780 \end{bmatrix}$$

$$X'\mathbf{y} = \begin{bmatrix} 5{,}263 \\ 346{,}867 \\ 921{,}939 \end{bmatrix} \quad \text{and} \quad \mathbf{y}'\mathbf{y} = 1{,}148{,}317$$

a. Use computer software to find $(X'X)^{-1}$. Obtain the least squares estimates and their standard errors.

b. Compute the t statistics to test the simple hypotheses that each regression coefficient is zero.

c. Determine the coefficient of variation R^2. (The complete data are given in the file **abrasion**.)

4.10. The following matrices were computed for a certain regression problem:

$$X'X = \begin{bmatrix} 15 & 3{,}626 & 44{,}428 \\ 3{,}626 & 1{,}067{,}614 & 11{,}419{,}181 \\ 44{,}428 & 11{,}419{,}181 & 139{,}063{,}428 \end{bmatrix},$$

$$X'\mathbf{y} = \begin{bmatrix} 2{,}259 \\ 647{,}107 \\ 7{,}096{,}619 \end{bmatrix}$$

$$(X'X)^{-1} = \begin{bmatrix} 1.2463484 & 2.1296642 \times 10^{-4} & -4.1567125 \times 10^{-4} \\ & 7.7329030 \times 10^{-6} & -7.0302518 \times 10^{-7} \\ & & 1.9771851 \times 10^{-7} \end{bmatrix},$$

$$\hat{\beta} = \begin{bmatrix} 3.452613 \\ 0.496005 \\ 0.009191 \end{bmatrix}$$

$$\mathbf{y}'\mathbf{y} = 394{,}107$$

a. Write down the estimated regression equation. Obtain the standard errors of the regression coefficients.

b. Compute the t statistics to test the simple hypotheses that each regression coefficient is equal to zero. Carry out these tests. State your conclusions.

4.11. A study was conducted to investigate the determinants of survival size of nonprofit U.S. hospitals. Survival size, y, was defined to be the largest U.S. hospital (in terms of the number of beds) exhibiting growth in market share. For the investigation, 10 states were selected at random, and the survival size for nonprofit hospitals in each of the selected states was determined for two time periods t: 1981–1982 and 1984–1985.

Furthermore, the following characteristics were collected on each selected state for each of the two time periods:

x_1 = Percentage of beds that are in for-profit hospitals.

x_2 = Number of people enrolled in health maintenance organizations as a fraction of the number of people covered by hospital insurance.

x_3 = State population in thousands.

x_4 = Percentage of state that is urban.

The data are given in the file **hospital.**

a. Fit the model

$$y = \beta_0 + \beta_1 x_1 + \beta_2 x_2 + \beta_3 x_3 + \beta_4 x_4 + \epsilon$$

b. The influence of the percentage of beds in for-profit hospitals was of particular interest to the investigators. What does the analysis tell us?

c. What further investigation might you do with this data set. Give reasons?

d. Rather than selecting 10 states at random, how else might you collect the data on survival size? Would your approach be an improvement over the random selection?

4.12. The amount of water used by the production facilities of a plant varies. Observations on water usage and other, possibly related, variables were collected for 17 months. The data are given in the file **water**. The explanatory variables are

TEMP = average monthly temperature(°F)

PROD = amount of production

DAYS = number of operating days in the month

PAYR = number of people on the monthly plant payroll

HOUR = number of hours shut down for maintenance

The response variable is USAGE = monthly water usage (gallons/100).

a. Fit the model containing all five independent variables,

$$y = \beta_0 + \beta_1 \text{ TEMP} + \beta_2 \text{ PROD} + \beta_3 \text{ DAYS} + \beta_4 \text{ PAYR} + \beta_5 \text{ HOUR} + \epsilon$$

Plot residuals against fitted values and residuals against the case index, and comment about model adequacy.

b. Test the hypothesis that $\beta_1 = \beta_3 = \beta_5 = 0$.

c. Which model or set of models would you suggest for predictive purposes? Briefly justify.

d. Which independent variable seems to be the most important one in determining the amount of water used?

e. Write a **nontechnical** paragraph that summarizes your conclusions about plant water usage that is supported by the data.

4.13. Data on last year's sales (y, in 100,000s of dollars) in 15 sales districts are given in the file **sales**. This file also contains promotional expenditures (x_1, in thousands of dollars), the number of active accounts (x_2), the number of competing brands (x_3), and the district potential (x_4, coded) for each of the districts.

a. A model with all four regressors is proposed:

$$y = \beta_0 + \beta_1 x_1 + \beta_2 x_2 + \beta_3 x_3 + \beta_4 x_4 + \epsilon, \quad \epsilon \sim N(0, \sigma^2)$$

Interpret the parameters β_0, β_1, and β_4.

b. Fit the proposed model in (a) and calculate estimates of $\beta_i, i = 0, 1, \ldots, 4$, and σ^2.

c. Test the following hypotheses:

(i) $\beta_4 = 0$; (ii) $\beta_3 = \beta_4 = 0$; (iii) $\beta_2 = \beta_3$; (iv) $\beta_1 = \beta_2 = \beta_3 = \beta_4 = 0$

d. Consider the reduced (restricted) model with $\beta_4 = 0$. Estimate its coefficients and give an expression for the expected sales.

e. Using the model in (d), obtain a prediction for the sales in a district where $x_1 = 3.0$, $x_2 = 45$, and $x_3 = 10$. Obtain the corresponding 95% prediction interval.

4.14. The survival rate (in percentage) of bull semen after storage is measured at various combinations of concentrations of three materials (additives) that are thought to increase the chance of survival. The data listed below are given in the file **bsemen**.

% Survival (y)	% Weight 1 (x_1)	% Weight 2 (x_2)	% Weight 3 (x_3)
25.5	1.74	5.30	10.80
31.2	6.32	5.42	9.40
25.9	6.22	8.41	7.20
38.4	10.52	4.63	8.50
18.4	1.19	11.60	9.40
26.7	1.22	5.85	9.90

% Survival (y)	% Weight 1 (x_1)	% Weight 2 (x_2)	% Weight 3 (x_3)
26.4	4.10	6.62	8.00
25.9	6.32	8.72	9.10
32.0	4.08	4.42	8.70
25.2	4.15	7.60	9.20
39.7	10.15	4.83	9.40
35.9	1.72	3.12	7.60
26.5	1.70	5.30	8.20

Assume the model $y = \beta_0 + \beta_1 x_1 + \beta_2 x_2 + \beta_3 x_3 + \epsilon$.

a. Compute $X'X$, $(X'X)^{-1}$, and $X'\mathbf{y}$.

b. Plot the response y versus each predictor variable. Comment on these plots.

c. Obtain the least squares estimates of $\boldsymbol{\beta}$ and give the fitted equation.

d. Construct a 90% confidence interval for

  i. the predicted mean value of y when $x_1 = 3$, $x_2 = 8$, and $x_3 = 9$;

  ii. the predicted individual value of y when $x_1 = 3$, $x_2 = 8$, and $x_3 = 9$.

e. Construct the ANOVA table and test for a significant linear relationship between y and the three predictor variables.

4.15. An experiment was conducted to study the toxic action of a certain chemical on silkworm larvae. The relationship of $\log_{10}$(survival time) to $\log_{10}$(dose) and $\log_{10}$(larvae weight) was investigated. The data, obtained by feeding each larvae a precisely measured dose of the chemical in an aqueous solution and recording the survival time until death, are given in the following table. The data are stored in the file **silkw**.

$\log_{10}$ Survival Time (y)	$\log_{10}$ Dose (x_1)	$\log_{10}$ Weight (x_2)
2.836	0.150	0.425
2.966	0.214	0.439
2.687	0.487	0.301
2.679	0.509	0.325
2.827	0.570	0.371
2.442	0.590	0.093
2.421	0.640	0.140
2.602	0.781	0.406
2.556	0.739	0.364
2.441	0.832	0.156
2.420	0.865	0.247
2.439	0.904	0.278
2.385	0.942	0.141
2.452	1.090	0.289
2.351	1.194	0.193

Assume the model $y = \beta_0 + \beta_1 x_1 + \beta_2 x_2 + \epsilon$.

a. Plot the response y versus each predictor variable. Comment on these plots.

b. Obtain the least squares estimates for $\boldsymbol{\beta}$ and give the fitted equation.

c. Construct the ANOVA table and test for a significant linear relationship between y and the two predictor variables.

d. Which independent variable do you consider to be the better predictor of log(survival time)? What are your reasons?

e. Of the models involving one or both of the independent variables, which do you prefer, and why?

4.16. You are given the following matrices computed for a regression analysis:

$$X'X = \begin{bmatrix} 9 & 136 & 269 & 260 \\ 136 & 2{,}114 & 4{,}176 & 3{,}583 \\ 269 & 4{,}176 & 8{,}257 & 7{,}104 \\ 260 & 3{,}583 & 7{,}104 & 12{,}276 \end{bmatrix}$$

$$X'\mathbf{y} = \begin{bmatrix} 45 \\ 648 \\ 1{,}283 \\ 1{,}821 \end{bmatrix}$$

$$(X'X)^{-1} = \begin{bmatrix} 9.610 & 0.008 & -0.279 & -0.044 \\ 0.008 & 0.509 & -0.258 & 0.001 \\ -0.279 & -0.258 & 0.139 & 0.001 \\ -0.044 & 0.001 & 0.001 & 0.0003 \end{bmatrix}$$

$$\hat{\beta} = (X'X)^{-1}(X'y) = \begin{bmatrix} -1.163461 \\ 0.135270 \\ 0.019950 \\ 0.121954 \end{bmatrix}$$

$y'y = 285$

a. Use these results to construct the analysis of variance table.

b. Give the computed regression equation and the standard errors of the regression coefficients.

c. Compare each estimated regression coefficient to its standard error and use the t test to test the simple hypotheses that each individual regression coefficient is equal to zero. State your conclusions about β_1, β_2, and β_3.

4.17. Consider the following two models:

$$\text{Model A}: \quad y_i = \beta_0 + \beta_1 x_i + \epsilon_i$$
$$\text{Model B}: \quad y_i = \beta_1 x_i + \epsilon_i$$

Suppose that model A is fitted to 22 data points (x_i, y_i) with the following results:

$$\hat{\beta}' = (\hat{\beta}_0, \hat{\beta}_1) = (4.0, -4.5), \quad V(\hat{\beta}_0) = 4.0,$$
$$V(\hat{\beta}_1) = 9.0, \quad \text{and} \quad \text{Cov}(\hat{\beta}_0, \hat{\beta}_1) = 0.0$$

a. Construct individual 95% confidence intervals for β_0 and for β_1. What conclusions can you draw?

b. Construct a joint 95% confidence region for (β_0, β_1). Draw this confidence region on the plane of possible values for (β_0, β_1). On the basis of this region, what conclusions can you draw about the relative merits of models A and B?

c. Do the results of (a) and (b) conflict? Carefully explain your reasoning.

4.18. Consider the model

$$\boldsymbol{y} = X\boldsymbol{\beta} + \boldsymbol{\epsilon}, \quad \boldsymbol{\epsilon} \sim N(\boldsymbol{0}, \sigma^2 I)$$

Let $\hat{\boldsymbol{\beta}} = (X'X)^{-1}X'\boldsymbol{y}$, $\hat{\boldsymbol{\mu}} = H\boldsymbol{y}$, and $\boldsymbol{e} = (I - H)\boldsymbol{y}$, where $H = X(X'X)^{-1}X'$. Show that $\hat{\boldsymbol{\mu}}$ and $\boldsymbol{e}$ are statistically independent.

4.19. Consider a regression through the origin,

$$y_i = \beta x_i + \epsilon_i, \quad \text{with } E(\epsilon_i) = 0,$$
$$V(\epsilon_i) = \sigma^2 x_i^2, \quad i = 1, 2, \ldots, 12$$

a. Derive the generalized least squares estmate of β in Eq. (4.58) and obtain its variance. Note that the covariance matrix V and its inverse V^{-1} are diagonal matrices. The generalized least squares estimate minimizes a weighted sum of squares with weights given by the diagonal elements in V^{-1}. Hence, one refers to it as the **weighted least squares** estimate.

b. Suppose that $z_i = y_i/x_i$ and $\sum_{i=1}^{12} z_i = 30$. Find the numerical value for the weighted least squares estimate in (a) and express its variance as a function of σ^2.

4.20. Consider a regression through the origin,

$$y_i = \beta x_i + \epsilon_i, \quad \text{with } E(\epsilon_i) = 0,$$
$$V(\epsilon_i) = \sigma^2 x_i, \quad x_i > 0, \quad i = 1, 2, \ldots, 10$$

a. Derive the generalized (weighted) least squares estmator of β and obtain its variance.

b. Assume that the experimenter recorded only the sample means $\bar{x} = 15$ and $\bar{y} = 30$. If possible, obtain a numerical value for the weighted least squares estimate in (a) and express its variance as a function of σ^2.

4.21. The data are taken from Davies, O. L., and Goldsmith, P. L. (Eds.). *Statistical Methods in Research and Production* (4th ed.). Edinburgh, UK: Oliver & Boyd, 1972. The data are given in the file **abrasion**.

The hardness and the tensile strength of rubber affect its resistance to abrasion. Thirty samples of rubber are tested for hardness (in degrees Shore; the larger the number, the harder the rubber) and tensile strength (in kilograms per square centimeter). Each sample was subjected to steady abrasion for a certain fixed period of time, and the loss of rubber (in grams per hour of testing) was measured.

Develop a model that relates the abrasion loss to hardness and tensile strength.

Construct scatter plots of abrasion loss against hardness and tensile strength. Fit appropriate regression models, obtain and interpret the estimates of the coefficients, calculate the ANOVA table, and discuss the adequacy of the model fit. Use your model(s) to obtain a 95% confidence interval for the mean abrasion loss for rubber with hardness 70 and tensile strength 200.

$y =$ Abrasion Loss (g/hr)	$x_1 =$ Hardness (degree Shore)	$x_2 =$ Tensile Strength (kg/cm^2)
372	45	162
206	55	233
175	61	232
154	66	231
136	71	231
112	71	237
55	81	224
45	86	219
221	53	203
166	60	189
164	64	210
113	68	210
82	79	196
32	81	180
228	56	200
196	68	173
128	75	188
97	83	161
64	88	119
249	59	161
219	71	151
186	80	165
155	82	151
114	89	128
341	51	161
340	59	146
283	65	148
267	74	144
215	81	134
148	86	127

4.22. The data are taken from Joglekar, G., Schuenemeyer, J. H., and LaRiccia, V. Lack-of-fit testing when replicates are not available. *American Statistician,* 43, 135–143, 1989. The data are given in the file **woodstrength**.

The tensile strength of Kraft paper (in pounds per square inch) is measured against the percentage of hardwood in the batch of pulp from which the paper was produced. Data for 19 observations are given here.

Develop a model that relates tensile strength to the percentage of hardwood in the paper. Construct scatter plots of tensile strength against the percentage of hardwood.

a. Fit a linear model and comment on your findings.

b. Consider a model that also includes the square of the percentage of hardwood. Fit the quadratic model, obtain and interpret the estimates of the coefficients, calculate the ANOVA table, and discuss the adequacy of the model fit. Add the fitted line to your scatter plot. Discuss whether the quadratic component is needed. Use your model to obtain a 95% confidence interval for the mean tensile strength of paper with 6% hardwood content. How is this interval different from a corresponding prediction interval? Discuss whether it is reasonable to obtain a confidence interval for the mean tensile strength of paper with 20% hardwood content.

$x =$ Hardwood Concentration	$y =$ Tensile Strength
1.0	6.3
1.5	11.1
2.0	20.0
3.0	24.0
4.0	26.1
4.5	30.0
5.0	33.8
5.5	34.0
6.0	38.1
6.5	39.9
7.0	42.0
8.0	46.1
9.0	53.1
10.0	52.0

x = Hardwood Concentration	y = Tensile Strength
11.0	52.5
12.0	48.0
13.0	42.8
14.0	27.8
15.0	21.9

4.23. The data are taken from Humphreys, R. M. Studies of luminous stars in nearby galaxies. I. Supergiants and O stars in the Milky Way. *Astrophysics Journal, Supplementary Series,* 38, 309–350, 1978. The data are given in the file **lightintensity**.

Light intensity and surface temperature were determined for 47 stars taken from the Hertzsprung–Russel diagram of Star Cluster CYG OB1. The objective is to find a relationship between light intensity and surface temperature.

Construct a scatter plot of light intensity against surface temperature. Fit a quadratic regression model, obtain and interpret the estimates of the coefficients, calculate the ANOVA table, and discuss the adequacy of the model fit. Add the fitted line to your scatter plot.

What other interpretations of the scatter plot are possible? For example, could it be that four stars are different in the sense that they do not follow the linear pattern established by the other stars? What questions would you ask the astrophysicist?

Index	x = Log Surface Temp	y = Log Light Intensity
1	4.37	5.23
2	4.56	5.74
3	4.26	4.93
4	4.56	5.74
5	4.30	5.19
6	4.46	5.46
7	3.84	4.65
8	4.57	5.27
9	4.26	5.57
10	4.37	5.12
11	3.49	5.73
12	4.43	5.45
13	4.48	5.42
14	4.01	4.05
15	4.29	4.26
16	4.42	4.58
17	4.23	3.94
18	4.42	4.18
19	4.23	4.18
20	3.49	5.89
21	4.29	4.38
22	4.29	4.22
23	4.42	4.42
24	4.49	4.85
25	4.38	5.02
26	4.42	4.66
27	4.29	4.66
28	4.38	4.90
29	4.22	4.39
30	3.48	6.05
31	4.38	4.42
32	4.56	5.10
33	4.45	5.22
34	3.49	6.29
35	4.23	4.34
36	4.62	5.62
37	4.53	5.10
38	4.45	5.22
39	4.53	5.18
40	4.43	5.57
41	4.38	4.62
42	4.45	5.06
43	4.50	5.34
44	4.45	5.34
45	4.55	5.54
46	4.45	4.98
47	4.42	4.50

4.24. Consider the UFFI data set in Table 1.2 ($n = 24$ observations). Estimate the model with three regression coefficients, $y = \beta_0 + \beta_1 x_1(\text{UFFI}) + \beta_2 x_2(\text{TIGHT}) + \varepsilon$.

a. Use the statistical software of your choice and confirm the regression results in Table 4.1.

b. Determine the 3×3 matrix $X'X$ and its inverse $(X'X)^{-1}$. Determine the standard errors of the three estimates and the pairwise correlations among the estimates (there are three correlations).

c. Determine a 95% confidence region (ellipse) for the two slopes $\beta = (\beta_1, \beta_2)'$. We know that the marginal distribution of $\hat{\beta} = (\hat{\beta}_1, \hat{\beta}_2)'$ is a bivariate normal distribution with covariance matrix $\sigma^2 A^{-1}$, where A^{-1} is the appropriate 2×2 submatrix of $(X'X)^{-1}$ found in (b). Hence, the contours of the confidence ellipse can be traced out by solving $(\hat{\beta} - \beta)' A (\hat{\beta} - \beta) = 2s^2 F(0.95; 2, n-3)$. Here, $F(0.95; 2, n-3=21)$ is the 95th percentile of the F distribution, and s^2 is the mean square error.

4.25. Confidence intervals for regression coefficients and the mean response and prediction intervals for future observations in Section 4.3 make use of the t distribution. The t distribution as the resulting sampling distribution of the coefficient estimates in Eq. (4.24) depends critically on the model assumptions, in particular the assumption that the independent errors are normally distributed. The distribution in Eq. (4.24) is not a t distribution and it is no longer known if the distribution of the errors is nonnormal.

Bootstrapping (or resampling) methods are commonly used to overcome problems of unknown sampling distributions. The bootstrap, originally proposed by Efron (1979), approximates the unknown theoretical sampling distribution of the coefficient estimates by an empirical distribution that is obtained through a resampling process.

Several versions of the bootstrap are proposed for the regression situation, and the references listed at the end of this exercise will give you more details. Here, we discuss the "bootstrap in pairs" method, which resamples directly from the original data $(y_i, \boldsymbol{x}_i)$, $i = 1, 2, \ldots, n$. This method repeats the following steps B times. Sample with replacement n pairs from the original n observations $(y_i, \boldsymbol{x}_i)$. From these n sampled pairs, calculate the least squares estimates and denote the jth coefficient estimate by $\hat{\beta}_j^{*(b)}$. The superscript asterisk denotes the fact that the estimate is obtained from data generated by the bootstrap procedure, the superscript b denotes the bth replication, and the subscript j refers to a particular scalar coefficient. The B independent replications supply the empirical bootstrap distribution function.

Percentile bootstrap intervals are proposed as confidence intervals for the regression coefficients. One approach determines the $100(\alpha/2)$ and $100(1-(\alpha/2))$ percentiles of the empirical bootstrap distribution function, $\hat{\beta}_j^*(\alpha/2)$ and $\hat{\beta}_j^*(1-(\alpha/2))$, and computes a $100(1-\alpha)\%$ bootstrap confidence interval for the parameter β_j as

$$\hat{\beta}_j^*(\alpha/2), \quad \hat{\beta}_j^*(1-(\alpha/2))$$

Here, we have given the very simplest bootstrap method for the regression situation. Modifications that improve on this simple procedure have been proposed and are discussed in the references. The modifications involve sampling residuals (compared to the resampling of cases discussed here) and refinements for improving the coverage properties of percentile bootstrap intervals [one modification calculates the lower and upper limits as $\hat{\beta}_j - [\hat{\beta}_j^*(1-(\alpha/2)) - \hat{\beta}_j]$ and $\hat{\beta}_j - [\hat{\beta}_j^*(\alpha/2) - \hat{\beta}_j]$, where $\hat{\beta}_j$ is the estimate from the original sample].

a. Select one or more of the listed references and write a brief summary that explains the bootstrap methods in regression and discusses their importance.

b. Consider the simple linear regression model. Use the fuel efficiency data in Table 1.3 and regress fuel efficiency (gallons per 100 traveled miles) on the weight of the car. Obtain a 95% bootstrap confidence interval for the slope. Use $B = 1{,}000$ and 2,000 replications. Relate the results to the standard confidence interval based on the t distribution.

Literature on the Bootstrap and Its Applications to Regression

Davison, A. C., and Hinkley, D. V. *Bootstrap Methods and Their Applications*. New York: Cambridge University Press, 1997.

Efron, B. Bootstrap methods: Another look at the jackknife. *Annals of Statistics,* 7, 1–26, 1979.

Efron, B., and Tibshirani, R. J. *An Introduction to the Bootstrap*. New York: Chapman & Hall, 1993.

Horowitz, J. L. The Bootstrap. In *Handbook of Econometrics* (Vol. 6). Amsterdam: North Holland, 1999.

5 Specification Issues in Regression Models

In Chapter 4, we considered the general linear model

$$y = X\beta + \epsilon \qquad (5.1)$$

where $y' = (y_1, \ldots, y_n)$, $\epsilon' = (\epsilon_1, \ldots, \epsilon_n)$,

$$X = \begin{bmatrix} 1 & x_{11} & x_{12} & \cdots & x_{1p} \\ 1 & x_{21} & x_{22} & \cdots & x_{2p} \\ \vdots & \vdots & \vdots & & \vdots \\ 1 & x_{n1} & x_{n2} & \cdots & x_{np} \end{bmatrix}, \qquad \beta = \begin{bmatrix} \beta_0 \\ \beta_1 \\ \vdots \\ \beta_p \end{bmatrix},$$

and $\epsilon \sim N(0, \sigma^2 I)$. The model can also be writen as $y \sim N(\mu, \sigma^2 I)$, with mean vector

$$\mu = E(y) = X\beta \qquad (5.2)$$

It is the mean vector $\mu = X\beta$ that is at the center of our interest. By properly defining the X matrix and the β vector, we can adapt the mean vector to represent various models of interest. Consider the following special cases.

5.1 ELEMENTARY SPECIAL CASES

5.1.1 ONE-SAMPLE PROBLEM

Suppose that $y_1, \ldots, y_n$ are observations taken under uniform conditions from a stable process with mean level β_0. We can write the data-generating model as

$$y_i = \beta_0 + \epsilon_i \quad \text{with mean} \quad E(y_i) = \beta_0, \qquad i = 1, 2, \ldots, n \qquad (5.3)$$

In this case

$$E(y) = X\beta$$

where

$$\mathbf{y} = \begin{bmatrix} y_1 \\ y_2 \\ \vdots \\ y_n \end{bmatrix}; \quad X = \begin{bmatrix} 1 \\ 1 \\ \vdots \\ 1 \end{bmatrix}; \quad \text{and} \quad \boldsymbol{\beta} = \beta_0$$

5.1.2 TWO-SAMPLE PROBLEM

Suppose that the first m observations $y_1, \ldots, y_m$ are taken under one set of conditions (e.g., the standard process), whereas the remaining $n - m$ observations $y_{m+1}, y_{m+2}, \ldots, y_n$ are taken under a different set of conditions (the new process). Let β_1 denote the mean of the standard process and β_2 that of the new process. Then

$$y_i = \begin{cases} \beta_1 + \epsilon_i & i = 1, 2, \ldots, m \\ \beta_2 + \epsilon_i & i = m + 1, \ldots, n \end{cases}$$

This can also be written as

$$\begin{aligned} y_i &= \beta_1 x_{i1} + \beta_2 x_{i2} + \epsilon_i \\ \text{or} \quad E(y_i) &= \beta_1 x_{i1} + \beta_2 x_{i2} \end{aligned} \tag{5.4}$$

where x_{i1} and x_{i2} are indicator variables such that $x_{i1} = 1$ if $i = 1, 2, \ldots, m$ and zero for $i = m + 1, \ldots, n$, and $x_{i2} = 0$ if $i = 1, 2, \ldots, m$ and one for $i = m + 1, \ldots, n$.

The n equations can be combined as

$$E \begin{bmatrix} y_1 \\ \vdots \\ y_m \\ -- \\ y_{m+1} \\ \vdots \\ y_n \end{bmatrix} = \begin{bmatrix} 1 \\ \vdots \\ 1 \\ -- \\ 0 \\ \vdots \\ 0 \end{bmatrix} \beta_1 + \begin{bmatrix} 0 \\ \vdots \\ 0 \\ -- \\ 1 \\ \vdots \\ 1 \end{bmatrix} \beta_2 \tag{5.5}$$

In matrix form

$$E(\mathbf{y}) = X\boldsymbol{\beta}$$

where

$$X = \begin{bmatrix} 1 & 0 \\ \vdots & \vdots \\ 1 & 0 \\ -- & -- \\ 0 & 1 \\ \vdots & \vdots \\ 0 & 1 \end{bmatrix}$$

and $\boldsymbol{\beta} = \begin{bmatrix} \beta_1 \\ \beta_2 \end{bmatrix}$

Our interest is in examining whether the two processes have the same mean. We wish to test the hypothesis $\beta_1 = \beta_2$.

An Equivalent Formulation

Let us write

$$\beta_2 = \beta_1 + \delta$$

where $\delta = \beta_2 - \beta_1$ represents the difference of the process means. Corresponding to Eq. (5.4), the mean in our model becomes

$$E(y_i) = \beta_1 + \delta x_{i2} \tag{5.6}$$

where x_{i2} is the indicator defined earlier,

$$x_{i2} = \begin{cases} 0 & \text{if } i = 1, 2, \ldots, m \\ 1 & \text{if } i = m+1, \ldots, n \end{cases}$$

Hence,

$$E\begin{bmatrix} y_1 \\ \vdots \\ y_m \\ -- \\ y_{m+1} \\ \vdots \\ y_n \end{bmatrix} = \begin{bmatrix} 1 \\ \vdots \\ 1 \\ -- \\ 1 \\ \vdots \\ 1 \end{bmatrix} \beta_1 + \begin{bmatrix} 0 \\ \vdots \\ 0 \\ -- \\ 1 \\ \vdots \\ 1 \end{bmatrix} \delta \tag{5.7}$$

or

$$E(\mathbf{y}) = X\boldsymbol{\beta}$$

where

$$X = \begin{bmatrix} 1 & 0 \\ \vdots & \vdots \\ 1 & 0 \\ -- & -- \\ 1 & 1 \\ \vdots & \vdots \\ 1 & 1 \end{bmatrix}; \quad \boldsymbol{\beta} = \begin{bmatrix} \beta_1 \\ \delta \end{bmatrix}$$

Our hypothesis of interest $\beta_1 = \beta_2$ is now expressed as $\delta = \beta_2 - \beta_1 = 0$.

5.1.3 POLYNOMIAL MODELS

Let $y_i, i = 1, 2, \ldots, n$ represent the yields of a chemical process at operating temperatures $t_1, t_2, \ldots, t_n$. Suppose that the expected yield changes linearly with temperature, suggesting a model of the form

$$y_i = \beta_0 + \beta_1 t_i + \epsilon_i, \qquad i = 1, 2, \ldots, n$$

In this case,

$$\mu_i = E(y_i) = \beta_0 + \beta_1 t_i, \qquad i = 1, 2, \ldots, n \tag{5.8}$$

and $\boldsymbol{\mu} = E(\mathbf{y})$ can be written as

$$E(\mathbf{y}) = X\boldsymbol{\beta}$$

where

$$X = \begin{bmatrix} 1 & t_1 \\ 1 & t_2 \\ \vdots & \vdots \\ 1 & t_n \end{bmatrix}; \quad \text{and} \quad \boldsymbol{\beta} = \begin{bmatrix} \beta_0 \\ \beta_1 \end{bmatrix}$$

Next, let us assume that the expected yield is quadratic in time. That is,

$$\mu_i = \beta_0 + \beta_1 t_i + \beta_2 t_i^2, \qquad i = 1, 2, \ldots, n \tag{5.9}$$

Then

$$E(\mathbf{y}) = X\boldsymbol{\beta}$$

where

$$X = \begin{bmatrix} 1 & t_1 & t_1^2 \\ 1 & t_2 & t_2^2 \\ \vdots & \vdots & \vdots \\ 1 & t_n & t_n^2 \end{bmatrix}; \quad \text{and} \quad \boldsymbol{\beta} = \begin{bmatrix} \beta_0 \\ \beta_1 \\ \beta_2 \end{bmatrix}$$

This model is quadratic in time but still linear in the parameters $\boldsymbol{\beta}$.

5.2 SYSTEMS OF STRAIGHT LINES

Suppose $y_1, \ldots, y_m$ represent the yields of a chemical process at temperatures $t_1, \ldots, t_m$ in the absence of a catalyst ($x_i = 0$), and $y_{m+1}, \ldots, y_{2m}$ represent yields at the very same temperatures in the presence of the catalyst ($x_i = 1$). Suppose that the expected yield changes linearly with temperature. Several possibilities exist.

Case a The catalyst has an effect, and this effect is the same at all temperatures:

$$\mu_i = \begin{cases} \beta_0 + \beta_1 t_i & i = 1, 2, \ldots, m \\ \beta_0 + \beta_2 + \beta_1 t_{i-m} & i = m+1, \ldots, 2m \end{cases}$$

The parameter β_2 expresses the effect of the catalyst. Using an indicator variable for the presence of the catalyst, we can write the model as

$$E(y_i) = \beta_0 + \beta_1 t_i + \beta_2 x_i, \quad i = 1, 2, \ldots, 2m \tag{5.10}$$

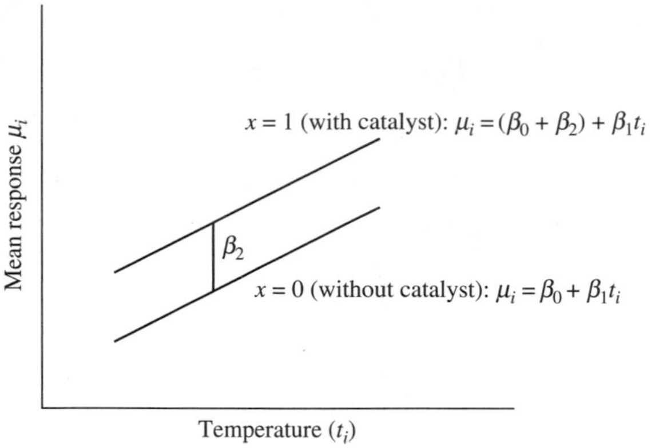

FIGURE 5.1 Same Catalyst Effect for All Temperatures

where $x_i = 0$ if $i = 1, 2, \ldots, m$ and 1 if $i = m+1, \ldots, 2m$, and $t_{i+m} = t_i$, $i = 1, 2, \ldots, m$.

In matrix form,

$$E(\mathbf{y}) = X\boldsymbol{\beta}$$

where

$$\mathbf{y} = \begin{bmatrix} y_1 \\ y_2 \\ \vdots \\ y_m \\ -- \\ y_{m+1} \\ \vdots \\ y_{2m} \end{bmatrix}, \quad X = \begin{bmatrix} 1 & t_1 & 0 \\ 1 & t_2 & 0 \\ \vdots & \vdots & \vdots \\ 1 & t_m & 0 \\ -- & -- & -- \\ 1 & t_1 & 1 \\ \vdots & \vdots & \vdots \\ 1 & t_m & 1 \end{bmatrix}; \quad \text{and} \quad \boldsymbol{\beta} = \begin{bmatrix} \beta_0 \\ \beta_1 \\ \beta_2 \end{bmatrix}$$

Figure 5.1 illustrates this model graphically. This model represents two parallel straight lines (identical slopes); β_2 represents the change due to the catalyst. The effect of the catalyst is the same for all temperatures $t_1, \ldots, t_m$. The hypothesis $\beta_2 = 0$ implies that the catalyst has no effect.

Case b The catalyst has an effect, but its effect changes with temperature. This situation can be expressed with the model

$$\mu_i = E(y_i) = \beta_0 + \beta_1 t_i + \beta_2 x_i + \beta_3 t_i x_i, \quad i = 1, 2, \ldots, 2m \tag{5.11}$$

where x_i is the indicator defined earlier.

If the catalyst is absent ($x_i = 0$),

$$\mu_i = \beta_0 + \beta_1 t_i, \quad i = 1, 2, \ldots, m$$

If the catalyst is present ($x_i = 1$),

$$\begin{aligned} \mu_i &= \beta_0 + \beta_1 t_{i-m} + \beta_2 + \beta_3 t_{i-m}, \quad i = m+1, \ldots, 2m \\ &= (\beta_0 + \beta_2) + (\beta_1 + \beta_3) t_{i-m} \end{aligned}$$

FIGURE 5.2
Catalyst Effect Depends on Temperature

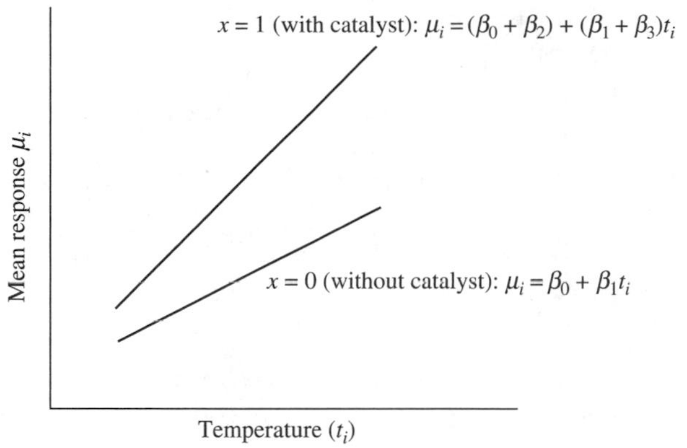

In matrix form we can write this model as

$$\boldsymbol{\mu} = E(\boldsymbol{y}) = X\boldsymbol{\beta}$$

where

$$\boldsymbol{y} = \begin{bmatrix} y_1 \\ y_2 \\ \vdots \\ y_m \\ \hline y_{m+1} \\ \vdots \\ y_{2m} \end{bmatrix}, \quad X = \begin{bmatrix} 1 & t_1 & 0 & 0 \\ 1 & t_2 & 0 & 0 \\ \vdots & \vdots & \vdots & \vdots \\ 1 & t_m & 0 & 0 \\ \hline 1 & t_1 & 1 & t_1 \\ \vdots & \vdots & \vdots & \vdots \\ 1 & t_m & 1 & t_m \end{bmatrix}; \quad \text{and} \quad \boldsymbol{\beta} = \begin{bmatrix} \beta_0 \\ \beta_1 \\ \beta_2 \\ \beta_3 \end{bmatrix}$$

Graphically, this model represents a pair of straight lines with different intercepts and different slopes; see Figure 5.2.

To test whether there is any catalyst effect, we test the hypothesis $\beta_2 = \beta_3 = 0$. If this hypothesis cannot be rejected, then the catalyst has no effect. A test of just $\beta_3 = 0$ in the model (5.11) tests whether the catalyst effect depends on temperature.

UFFI Example Revisited

For these data (see Figure 1.2 and Table 1.2), we consider the model

$$\mu_i = E(y_i) = \beta_0 + \beta_1 x_{i1} + \beta_2 x_{i2} \tag{5.12}$$

where y_i is the ambient formaldehyde concentration for house i,

$$x_{i1} = \begin{cases} 1 & \text{if house } i \text{ has UFFI} \\ 0 & \text{otherwise} \end{cases}$$

$$x_{i2} = \text{Airtightness of house } i$$

TABLE 5.1 UFFI DATA: ESTIMATES, STANDARD ERRORS, *t* RATIOS, AND *p* VALUES FOR MODEL (5.13)

	Estimate	Standard Error	*t* Ratio	*p* Value
Intercept	29.9976	3.0107	9.9635	0.0000
x_1 = UFFI	12.4781	4.4746	2.7887	0.0113
x_2 = TIGHT	3.1208	0.5030	6.2049	0.0000
x_1x_2 = UFFI * TIGHT	−0.6185	0.7665	−0.8069	0.4292

This model represents a pair of parallel lines when graphing μ against x_2. In Chapter 4, we found that $\hat{\beta}_1$ was significant, which indicates that houses with UFFI have increased levels of formaldehyde concentration. The model assumes that this increase is the same for all levels of airtightness.

Next, consider the following more general model that allows the possibility that the effect of UFFI depends on the level of airtightness:

$$E(y_i) = \beta_0 + \beta_1 x_{i1} + \beta_2 x_{i2} + \beta_3 x_{i1} x_{i2}, \quad i = 1, 2, \ldots, 24 \tag{5.13}$$

where x_{i1} and x_{i2} are as defined before. For houses without UFFI ($x_{i1} = 0$),

$$E(y_i) = \beta_0 + \beta_2 x_{i2} \tag{5.14}$$

whereas for houses with UFFI ($x_{i1} = 1$),

$$E(y_i) = (\beta_0 + \beta_1) + (\beta_2 + \beta_3) x_{i2} \tag{5.15}$$

This model represents two lines with different slopes and intercepts. A test of $\beta_3 = 0$ examines whether the effect of UFFI depends on airtightness. A test of $\beta_1 = \beta_3 = 0$ indicates whether or not UFFI has any effect at all. The estimation results for model (5.13) are given in Table 5.1.

The estimate $\hat{\beta}_3 = -0.6185$ and its standard error s.e.$(\hat{\beta}_3) = 0.7665$ can be used to test $\beta_3 = 0$. The t ratio is $t(\hat{\beta}_3) = -0.6185/0.7665 = -0.81$, and its probability value for a two-sided alternative is given by $2P(T \geq 0.81) = 0.4292$. Note that the degrees of freedom are $n - p - 1 = 24 - 4 = 20$. The probability value is quite large (certainly larger than commonly used significance levels), indicating that such a t ratio could have easily resulted if the null hypothesis $\beta_3 = 0$ actually were true. This probability value indicates that $\beta_3 = 0$ is a plausible hypothesis. The effect of UFFI does not depend on the airtightness of the house.

Next, we test whether UFFI has an effect at all. This means testing $\beta_1 = \beta_3 = 0$. This test cannot be performed by looking at the t ratios in Table 5.1. We must use the additional sum of squares principle to perform this test. Estimates of the full model (5.13) are given in Table 5.1. The residual sum of squares is given by $S(\hat{\beta}) = 554.834$, with 20 degrees of freedom. The null hypothesis $\beta_1 = \beta_3 = 0$ constrains the model to

$$E(y_i) = \beta_0 + \beta_2 x_{i2} \tag{5.16}$$

Fitting this model results in new estimates for β_0 and β_2, and the residual sum of squares $S(\hat{\beta}_A) = 1{,}093.067$ with $24 - 2 = 22$ degrees of freedom. The additional sum of squares is given by

$$S(\hat{\beta}_A) - S(\hat{\beta}) = 1{,}093.067 - 554.834 = 538.233$$

with 2 degrees of freedom.

The relevant test statistic for testing $\beta_1 = \beta_3 = 0$ is given by

$$F = \frac{538.233/2}{554.834/20} = 9.70$$

and the probability value using the F distribution with 2 and 20 degrees of freedom is $P(F \geq 9.7) \approx 0.001$. The probability value is very small, providing strong evidence against the hypothesis $\beta_1 = \beta_3 = 0$. This states that UFFI has an effect on the ambient formaldehyde.

5.3 COMPARISON OF SEVERAL "TREATMENTS"

This is also known as one-way classification, or the k-sample problem. We generalize the previously considered two-sample problem in Section 5.1.2 to $k > 2$ groups. This situation arises, for example, if we compare (i) the output from several machines, (ii) the reliability of several suppliers, or (iii) the effectiveness of several catalysts.

Suppose we are concerned with the effects of k catalysts on the yield of a chemical process. Assume that we take n_i observations with the ith catalyst, resulting in a total of $n = n_1 + n_2 + \cdots + n_k$ observations. The data can be organized as in the following table:

Catalyst	Observations			
1	y_{11}	y_{12}	$\cdots$	y_{1n_1}
2	y_{21}	y_{22}	$\cdots$	y_{2n_2}
$\vdots$	$\vdots$	$\vdots$	$\vdots$	$\vdots$
k	y_{k1}	y_{k2}	$\cdots$	y_{kn_k}

As an example, consider $k = 4$ groups and an equal number of observations in each group, $n_1 = n_2 = n_3 = n_4 = 5$. The observations for a special example are listed here, as are the averages for the four catalysts:

Catalyst	Observations					$\bar{y}_i$
1	91.5	92.1	93.9	91.0	94.5	92.60
2	94.1	91.7	93.5	89.9	92.0	92.24
3	84.4	85.7	86.5	88.5	87.4	86.50
4	86.0	87.3	85.5	84.8	83.2	85.36

Our model assumes different means for the k (catalyst) groups. Let y_{ij} be the jth observation from the ith catalyst group. We assume that $E(y_{ij}) = \beta_i$ for all $j = 1, 2, \dots, n_i$. As in all our previous regression models, we assume that the observations are independent and normally distributed with constant variance σ^2.

In matrix notation, the mean vector of our response becomes

$$E(\mathbf{y}) = X\boldsymbol{\beta} = \beta_1\mathbf{x_1} + \beta_2\mathbf{x_2} + \cdots + \beta_k\mathbf{x_k} \tag{5.17}$$

The regressor vectors $\mathbf{x}_i$ are strings of zeros and ones, indicating the group membership of the observations. That is, $x_{ji} = 1$ if y_{ij} is from group i, and 0 otherwise.

For $k = 4$ groups,

$$\mathbf{y} = \begin{bmatrix} y_{11} \\ \vdots \\ y_{1n_1} \\ -- \\ y_{21} \\ \vdots \\ y_{2n_2} \\ -- \\ y_{31} \\ \vdots \\ y_{3n_3} \\ -- \\ y_{41} \\ \vdots \\ y_{4n_4} \end{bmatrix}; \quad X = [\mathbf{x_1}, \mathbf{x_2}, \mathbf{x_3}, \mathbf{x_4}] = \begin{bmatrix} 1 & 0 & 0 & 0 \\ \vdots & \vdots & \vdots & \vdots \\ 1 & 0 & 0 & 0 \\ -- & -- & -- & -- \\ 0 & 1 & 0 & 0 \\ \vdots & \vdots & \vdots & \vdots \\ 0 & 1 & 0 & 0 \\ -- & -- & -- & -- \\ 0 & 0 & 1 & 0 \\ \vdots & \vdots & \vdots & \vdots \\ 0 & 0 & 1 & 0 \\ -- & -- & -- & -- \\ 0 & 0 & 0 & 1 \\ \vdots & \vdots & \vdots & \vdots \\ 0 & 0 & 0 & 1 \end{bmatrix}; \quad \text{and} \quad \boldsymbol{\beta} = \begin{bmatrix} \beta_1 \\ \beta_2 \\ \beta_3 \\ \beta_4 \end{bmatrix}$$

The least squares estimator of $\boldsymbol{\beta} = (\beta_1, \beta_2, \dots, \beta_k)'$, $\hat{\boldsymbol{\beta}} = (X'X)^{-1}X'\mathbf{y}$, is easy to obtain. Its elements are the respective group means, $\hat{\beta}_1 = \bar{y}_1$, $\hat{\beta}_2 = \bar{y}_2, \dots, \hat{\beta}_k = \bar{y}_k$.

The hypothesis of interest is the equality of the k means, $\beta_1 = \beta_2 = \ldots = \beta_k$. An equivalent, but for the following discussion somewhat more convenient, representation relates the group means to the mean of a reference group; in our case, the mean of the first group, β_1.

Let $\beta_i = \beta_1 + \delta_i$, $i = 2, 3, \dots, k$. Then we can write the model as

$$E(y_{ij}) = \begin{cases} \beta_1 & \text{for catalyst } i = 1 \\ \beta_1 + \delta_i & \text{for catalyst } i, \quad i = 2, \dots, k \end{cases} \tag{5.18}$$

The mean vector of the response is

$$E(\mathbf{y}) = X\boldsymbol{\beta} = [\mathbf{1}, \mathbf{x_2}, \mathbf{x_3}, \dots, \mathbf{x_k}]\boldsymbol{\beta} \tag{5.19}$$

For $k = 4$ groups, the matrix X and the vector of parameters $\boldsymbol{\beta}$ are given as

$$X = \begin{bmatrix} 1 & 0 & 0 & 0 \\ \vdots & \vdots & \vdots & \vdots \\ 1 & 0 & 0 & 0 \\ -- & -- & -- & -- \\ 1 & 1 & 0 & 0 \\ \vdots & \vdots & \vdots & \vdots \\ 1 & 1 & 0 & 0 \\ -- & -- & -- & -- \\ 1 & 0 & 1 & 0 \\ \vdots & \vdots & \vdots & \vdots \\ 1 & 0 & 1 & 0 \\ -- & -- & -- & -- \\ 1 & 0 & 0 & 1 \\ \vdots & \vdots & \vdots & \vdots \\ 1 & 0 & 0 & 1 \end{bmatrix}; \quad \text{and} \quad \boldsymbol{\beta} = \begin{bmatrix} \beta_1 \\ \delta_2 \\ \delta_3 \\ \delta_4 \end{bmatrix}$$

The null hypothesis is now expressed as $\delta_2 = \delta_3 = \delta_4 = 0$. If this hypothesis cannot be rejected, then the process means can be considered the same. If the hypothesis is rejected, then at least one of the means differs from the others.

Let $\bar{y}_i = \left(\sum_{j=1}^{n_i} y_{ij}\right)/n_i$ denote the average of the ith group, and let $\bar{y} = \left(\sum_{i=1}^{k}\sum_{j=1}^{n_i} y_{ij}\right)/\left(\sum_{i=1}^{k} n_i\right)$ be the overall average. The least squares estimator of $\boldsymbol{\beta}$ in model (5.19) is given by

$$\hat{\boldsymbol{\beta}} = (X'X)^{-1}X'\mathbf{y} = \begin{bmatrix} \bar{y}_1 \\ \bar{y}_2 - \bar{y}_1 \\ \vdots \\ \bar{y}_k - \bar{y}_1 \end{bmatrix} \tag{5.20}$$

The regression sum of squares is given by

$$\hat{\boldsymbol{\beta}}'X'\mathbf{y} - n\bar{y}^2 = \sum_{i=1}^{k} n_i(\bar{y}_i - \bar{y})^2 \tag{5.21}$$

Since there are $k - 1$ "regressor" variables (in addition to the intercept), the degrees of freedom for the regression sum of squares are $k - 1$. You can convince yourself of the sum of squares result by first working out the inverse of $(X'X)$; this is somewhat cumbersome because $(X'X)$ is not diagonal. Even simpler, you can work from the estimates $\hat{\beta}_1 = \bar{y}_1$, and $\hat{\delta}_i = \bar{y}_i - \bar{y}_1$, for $i = 2, \ldots, k$.

As an illustration consider the data set given earlier. The corresponding analysis of variance (ANOVA) table is given in Table 5.2.

The residual sum of squares is given by

$$S(\hat{\boldsymbol{\beta}}) = \mathbf{y}'\mathbf{y} - \hat{\boldsymbol{\beta}}'X'\mathbf{y} = \sum_{i=1}^{k}\sum_{j=1}^{n_i}(y_{ij} - \bar{y}_i)^2 \tag{5.22}$$

TABLE 5.2 ANOVA TABLE FOR THE CATALYST DATA

Source	df	Sum of Squares	MS	F	p Value
Regression (Treatment)	$k - 1 = 3$	$\sum_{i=1}^{k} n_i(\bar{y}_i - \bar{y})^2 = 214.17$	71.39	29.12	< 0.0001
Residual (Error)	$\sum_{i=1}^{k}(n_i - 1) = 16$	$\sum_{i=1}^{k}\sum_{j=1}^{n_i}(y_{ij} - \bar{y}_i)^2 = 39.22$	2.45		
Total	$\sum_{i=1}^{k} n_i - 1 = 19$	$\sum_{i=1}^{k}\sum_{j=1}^{n_i}(y_{ij} - \bar{y})^2 = 253.40$			

The associated degrees of freedom are $n - k$, where $n = \sum_{i=1}^{k} n_i$. Note that in this example $k = 4$, $n_1 = n_2 = n_3 = n_4 = 5$, $n = 20$, and $n - k = 16$.

We are interested in testing the null hypothesis $\beta_1 = \beta_2 = \beta_3 = \beta_4$, or equivalently, $\delta_2 = \delta_3 = \delta_4 = 0$. The F statistic

$$F = \frac{\text{regression SS}/(k - 1)}{\text{residual SS}/(n - k)} \tag{5.23}$$

in the fifth column of the ANOVA table can be used. We find that F is very large, and its probability value $P(F(3, 16) \geq 29.12)$ is very small. Hence, there is ample evidence to reject the hypothesis $\delta_2 = \delta_3 = \delta_4 = 0$. This implies that at least one of the means is different from the others.

When testing the equality of group means, we often attach different labels to the sums of squares in the ANOVA table. We refer to the regression sum of squares as the treatment sum of squares or the between group sum of squares. This is because this sum of squares picks up the variablity between the groups. The residual sum of squares is sometimes called the within group sum of squares.

5.4 *X* MATRICES WITH NEARLY LINEAR-DEPENDENT COLUMNS

In the general linear model we assumed that the columns $(\mathbf{1}, \boldsymbol{x}_1, \ldots, \boldsymbol{x}_p)$ of the X matrix are not linearly dependent. In some contexts, especially when working with observational data, these columns are close to being linearly dependent. What are the consequences of such a situation? The following illustration will show us what can happen.

EXAMPLE: PIZZA SALES DATA

A manager of a pizza outlet has collected monthly sales data over a 16-month period. During this time span, the outlet has been running a series of different advertisements. The manager has kept track of the cost of these advertisements (in hundreds of dollars) as well as the number of advertisements that have appeared. The data are shown in Table 5.3, where

$$\begin{aligned} y &= \text{Sales (in thousands of dollars)} \\ x_1 &= \text{Number of advertisements} \\ x_2 &= \text{Cost of advertisements (in hundreds of dollars)} \end{aligned}$$

TABLE 5.3 PIZZA SALES DATA

Month	Number of Ads	Cost of Ads (Hundred $)	Sales (Thousand $)
Jan	11	14.0	49.4
Feb	8	11.8	47.5
Mar	11	15.7	52.6
Apr	14	15.5	49.3
May	17	19.5	61.1
Jun	15	16.8	53.2
Jul	12	12.8	47.4
Aug	10	13.6	49.4
Sep	17	18.2	62.0
Oct	11	16.0	47.9
Nov	8	13.0	47.3
Dec	18	20.0	61.5
Jan	12	15.1	54.2
Feb	10	14.2	44.7
Mar	13	17.3	53.6
Apr	12	15.9	55.4

Figure 5.3 shows the scatter plots of sales against the number of advertisements and sales against the cost of advertisements. The graphs show that sales increase as the number of advertisements increases and also as the amount spent on advertising increases. The manager fits the model

$$y_i = \beta_0 + \beta_1 x_{i1} + \beta_2 x_{i2} + \epsilon_i \tag{5.24}$$

The results are given in Table 5.4.

The R^2 from this regression, $R^2 = 348.1428/446.9644 = 0.78$, indicates that the regressors x_1 and x_2 explain a large part of the variability in sales. The F ratio in the ANOVA in Table 5.5 ($F = 22.90$, with probability value 0.0001) indicates that there is strong evidence to reject the hypothesis $\beta_1 = \beta_2 = 0$. This means that at least one of the coefficients β_1 and β_2 is nonzero. In other words, at least one of the x variables is important in explaining the variation in y.

However, an examination of the individual p values (0.24 and 0.10 for β_1 and β_2) in Table 5.4 indicates that we cannot reject the hypothesis $\beta_1 = 0$ if x_2 is already included in the model. Similarly, we cannot reject $\beta_2 = 0$ if x_1 is in the model. In other words, if one of the variables is in the model, then the extra contribution of the other variable toward the regression is not important. Keep in mind the correct interpretation of the individual probability values. The individual t test results in Table 5.4 state that you do not need one variable if you already have included the other. This is certainly **not** an indication that you can omit from the model both x_1 and x_2 at the same time.

What would happen if you consider a model with just one of the two x variables? The results of fitting the regressions of sales on each variable separately, and the results of fitting sales on both x_1 and x_2, are shown in Table 5.6.

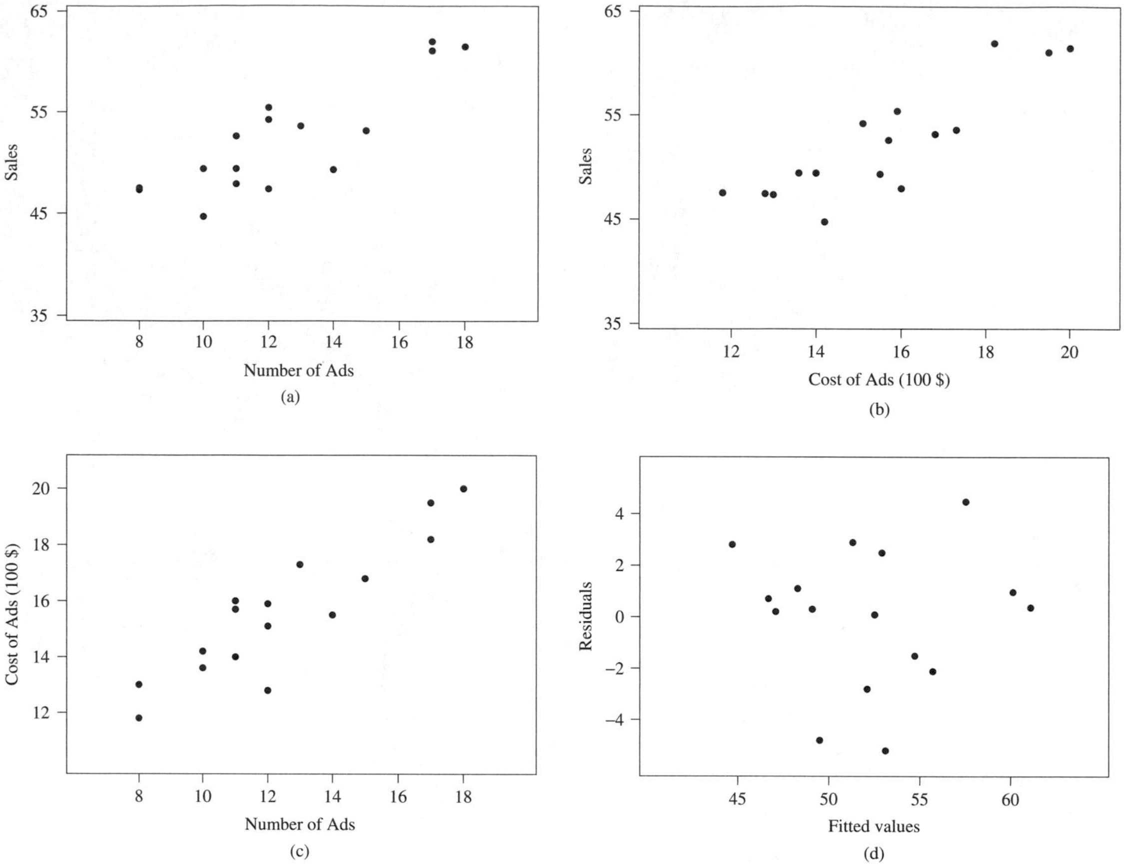

FIGURE 5.3 Plots for Pizza Sales

TABLE 5.4 LEAST SQUARES ESTIMATES FOR THE PIZZA SALES DATA

	Estimate	Standard Error	t Value	p Value
Intercept	24.8231	5.6611	4.3848	0.0007
Number of ads, x_1	0.6626	0.5386	1.2303	0.2404
Cost of ads, x_2	1.2329	0.6962	1.7709	0.1000

TABLE 5.5 ANOVA TABLE FOR THE PIZZA SALES DATA

Source	df	SS	MS	F	p Value
Regression	2	348.1428	174.0714	22.8990	0.0001
Residual	13	98.8216	7.6017		
Total	15	446.9644			

TABLE 5.6 SUMMARY OF THE REGRESSION RESULTS FOR PIZZA SALES DATA[a]

	Coefficients				
Variables in the Model	$\hat{\beta}_0$	$\hat{\beta}_1$	$\hat{\beta}_2$	Regression SS	R^2
x_1 only	33.3473 (0.0000)	1.5223 (0.0000)		324.3042	0.7256
x_2 only	21.0278 (0.0007)		2.0050 (0.0000)	336.6371	0.7532
x_1, x_2	24.8231 (0.0007)	0.6626 (0.2404)	1.2329 (0.1000)	348.1428	0.7789

[a] Numbers in parentheses are the probability values

We find:

i. A regression on x_1 alone explains 72.56% of the variability in sales. A regression on x_2 alone explains 75.32%. Both together (x_1 and x_2 in the model) explain 77.89%.

ii. In the single-variable model y on x_1, the regression coefficient $\hat{\beta}_1 = 1.5223$ is highly significant. In the two-variable model, $\hat{\beta}_1 = 0.6626$ is not significant, given that x_2 is in the model. Also notice that the estimate of β_1 changes considerably. The same comments apply to β_2. In the single-variable model, $\hat{\beta}_2 = 2.0050$ is highly significant. In the two variable model, the estimate of β_2 is not significant, and the estimate of β_2 changes considerably.

iii. If x_2 is in the model, then it is not important to include x_1 (and vice versa). In the presence of one variable, the other is not important enough to have it included. This is because variables x_1 and x_2 are highly correlated. The two variables express the same information, so there is no point to include both. A graph of x_1 against x_2 in Figure 5.3 shows that x_1 is strongly linearly related to x_2. This phenomenon is known as **multicollinearity**.

In the general linear model we assumed that the columns $(\mathbf{1}, \boldsymbol{x}_1, \ldots, \boldsymbol{x}_p)$ of the X matrix are not linearly dependent. When the regressor columns are close to being linearly dependent, then we can approximate one of the columns in the X matrix as a linear combination of the others. This states that one of the regressor variables is strongly influenced by some or all of the other explanatory variables. Hence, the fitting results of one variable are strongly affected by the presence or absence of other variables in the model. This is a consequence of multicollinearity, and this is exactly what we see in the pizza sales data. When x_2 is in the model, x_1 has little to contribute because it is highly "correlated" with x_2. A model with x_2 alone is sufficient. Note that the increase in R^2 from 0.75 in the model with just x_2 to $R^2 = 0.78$ with both x_1 and x_2 included is rather small. Also, the plot of the residuals vs fitted values from the simple model with just x_2 in Figure 5.3 indicates no systematic patterns.

What happens if the multicollinearity is perfect? What happens if the correlation between x_1 and x_2 is $+1$ or -1, exactly? Algebraically, the $n \times 3$ matrix X has rank 2, and not $3 = p + 1$ as usually assumed. As a consequence, the 3×3 matrix $X'X$ has rank 2, and it is not possible to obtain the inverse $(X'X)^{-1}$. The computer program would crash, or at least complain about the multicollinearity.

5.4.1 DETECTION OF MULTICOLLINEARITY

Correlations Among Regressor Variables

Suppose we have p regressors and we calculate the sample correlations, r_{ij}, between pairs of regressors x_i and x_j,

$$r_{ij} = \frac{\sum_{\ell=1}^{n}(x_{i\ell} - \bar{x}_i)(x_{j\ell} - \bar{x}_j)}{\sqrt{\sum_{\ell=1}^{n}(x_{i\ell} - \bar{x}_i)^2 \sum_{\ell=1}^{n}(x_{j\ell} - \bar{x}_j)^2}}, \quad i, j = 1, 2, \ldots, p \tag{5.25}$$

where $\bar{x}_i$ denotes the average of the measurements on the variable x_i. The sample correlation r_{ij} measures the linear association between x_i and x_j. A matrix of the correlations

$$C = \begin{bmatrix} 1 & r_{12} & \cdots & r_{1p} \\ r_{12} & 1 & \cdots & r_{2p} \\ \vdots & \vdots & & \vdots \\ r_{1p} & r_{2p} & \cdots & 1 \end{bmatrix} \tag{5.26}$$

provides an indication of the pairwise associations among the explanatory variables. If the off-diagonal elements of C are large in absolute value (close to ± 1), then there is strong pairwise linear association among the corresponding variables. For instance, if r_{12} is large, then x_1 and x_2 are linearly associated and multicollinearity exists. For the pizza sales data in Table 5.3,

$$C = \begin{bmatrix} 1 & r_{12} \\ r_{12} & 1 \end{bmatrix} = \begin{bmatrix} 1 & 0.8518 \\ 0.8518 & 1 \end{bmatrix}$$

Hence, the correlation between x_1 and x_2 is high and the two regressors are strongly linearly related. There is no need to have both variables together in the model. It should be noted that if the r_{ij}'s in the correlation matrix C are zero, then the regressor variables are **orthogonal** to each other. We discuss this situation later.

Variance Inflation Factors

Consider the regression model in Eq. (5.1) with an intercept and p regressors. Suppose we standardize the y and x variables,

$$y_i^* = \frac{y_i - \bar{y}}{s_y} \quad \text{and} \quad z_{ij} = \frac{x_{ij} - \bar{x}_j}{s_j}, \quad j = 1, 2, \ldots, p$$

where $\bar{y}$ and $\bar{x}_j$ are the corresponding sample means, and s_y and s_j are the appropriate sample standard deviations. Hence, the linear model can be expressed as

$$y^* = \alpha_1 z_1 + \alpha_2 z_2 + \cdots + \alpha_p z_p + \epsilon^* \tag{5.27}$$

Note that there are only p regression coefficients and that there is no intercept in this model. The covariance matrix of the least squares estimates of the parameters in the linear model Eq. (5.1) is given by $V(\hat{\beta}) = (X'X)^{-1}\sigma^2$. In the standardized model (Eq. 5.27), the matrix that corresponds to $X'X$ reduces to the correlation matrix C. Hence, $V(\hat{\alpha}) = C^{-1}\sigma^2$, where $\alpha = (\alpha_1, \ldots, \alpha_p)'$. The diagonal elements of C^{-1} are the scaled variances of the least squares estimates, $V(\hat{\alpha}_i)/\sigma^2$. For illustration, consider the special case, $p = 2$. Then the model is

$$y^* = \alpha_1 z_1 + \alpha_2 z_2 + \epsilon^* \tag{5.28}$$

and

$$C = \begin{bmatrix} 1 & r_{12} \\ r_{12} & 1 \end{bmatrix}, \quad \text{with } C^{-1} = \left(1 - r_{12}^2\right)^{-1} \begin{bmatrix} 1 & -r_{12} \\ -r_{12} & 1 \end{bmatrix}$$

If r_{12} were zero, then C^{-1} has ones in its diagonal, and $\dfrac{V(\hat{\alpha}_1)}{\sigma^2} = \dfrac{V(\hat{\alpha}_2)}{\sigma^2} = 1$. If r_{12} is large, then the diagonal elements of C^{-1} are larger than one, and $\dfrac{V(\hat{\alpha}_1)}{\sigma^2} = \dfrac{V(\hat{\alpha}_2)}{\sigma^2} > 1$. The values $\dfrac{V(\hat{\alpha}_i)}{\sigma^2}$, $i = 1, 2$ are called **variance inflation factors** (VIF) because they measure how the correlation among the regressor variables inflates the variance of the estimates. If these factors are much larger than one then there is multicollinearity. For the pizza sales data,

$$C = \begin{bmatrix} 1 & r_{12} \\ r_{12} & 1 \end{bmatrix} = \begin{bmatrix} 1 & 0.8518 \\ 0.8518 & 1 \end{bmatrix}, \quad C^{-1} = \begin{bmatrix} 3.6439 & -3.1038 \\ -3.1038 & 3.6439 \end{bmatrix}$$

Thus, $\text{VIF}_1 = \text{VIF}_2 = 3.6439$. The variance is inflated 3.64-fold. This is considerably larger than one; hence, there is evidence of multicollinearity.

In the general case of p regressors, it can be shown that the VIF of the coefficient estimate corresponding to the jth regressor x_j is

$$\text{VIF}_j = 1/\left(1 - R_j^2\right) \tag{5.29}$$

where R_j^2 is the coefficient of determination (see Section 4.5.1) from the regression of x_j on all other regressors. If x_j is linearly dependent on the other regressors, then R_j^2 will be large (close to one), and VIF_j will be large as well. Values of VIF larger than 10 are taken as solid evidence of multicollinearity.

5.4.2 GUARDING AGAINST MULTICOLLINEARITY

How can you guard against multicollinearity and its associated problems? A careful model specification is the key. You should avoid adding regressor variables more than once. For example, in a model for a car's fuel efficiency, you would not want to include both weight in kilograms and weight in pounds. This example is trivial because no reasonable person would make the mistake of including both. In many observational studies, however, the decision is not as clear-cut. Consider describing the state of the economy with such variables as interest rates, gross national product, employment, unemployment, etc. Although these variables are not perfectly related, some fairly strong relations are certain to exist.

5.5 *X* MATRICES WITH ORTHOGONAL COLUMNS

In many experimental situations, experimenters can set the values of the explanatory variables in such a way that the columns of the X matrix (also called the design matrix) are orthogonal. Orthogonality is an attractive property and there are advantages to choosing the regressor vectors as orthogonal. We illustrate this by considering an example.

Example: Excess Shrinkage Data

In an investigation to find the causes of excess shrinkage of parts produced by an injection molding operation, the team considered the following design factors (regressor variables): $x_1 =$ mold temperature (T), $x_2 =$ holding pressure (P), and $x_3 =$ screw speed (S). It was decided to study these variables at two levels each, with coding -1 (low) and $+1$ (high). A total of eight runs were taken. The values of the design variables and the corresponding results, shrinkage (y in percent), are given in Table 5.7. The first run describes the experiment in which all three variables are at their low values. The second run has P and S at the low levels, whereas T is set at its high level. Table 5.7 lists the eight runs in **standard order** in which "−" and "+" signs alternate in groups of one, two, and four. This order makes it easy to write down the settings for the eight runs. However, note that the order in which the runs are carried out should be randomized.

Consider a linear model of the form

$$y = \beta_0 + \beta_1 x_1 + \beta_2 x_2 + \beta_3 x_3 + \epsilon \tag{5.30}$$

TABLE 5.7 SHRINKAGE DATA

Run	T	P	S	Shrinkage
1	−1	−1	−1	19.7
2	+1	−1	−1	19.1
3	−1	+1	−1	20.0
4	+1	+1	−1	19.5
5	−1	−1	+1	15.9
6	+1	−1	+1	15.3
7	−1	+1	+1	25.5
8	+1	+1	+1	24.9

In matrix form this model can be written as

$$y = X\beta + \epsilon \tag{5.31}$$

where

$$y = \begin{bmatrix} 19.7 \\ 19.1 \\ 20.0 \\ 19.5 \\ 15.9 \\ 15.3 \\ 25.5 \\ 24.9 \end{bmatrix}; X = [\mathbf{1}, \mathbf{x}_1, \mathbf{x}_2, \mathbf{x}_3] = \begin{bmatrix} 1 & -1 & -1 & -1 \\ 1 & 1 & -1 & -1 \\ 1 & -1 & 1 & -1 \\ 1 & 1 & 1 & -1 \\ 1 & -1 & -1 & 1 \\ 1 & 1 & -1 & 1 \\ 1 & -1 & 1 & 1 \\ 1 & 1 & 1 & 1 \end{bmatrix}; \beta = \begin{bmatrix} \beta_0 \\ \beta_1 \\ \beta_2 \\ \beta_3 \end{bmatrix}; \text{ and } \epsilon = \begin{bmatrix} \epsilon_1 \\ \epsilon_2 \\ \vdots \\ \epsilon_8 \end{bmatrix}$$

Convince yourself that the columns in the matrix X are orthogonal. It is easy to see that $\mathbf{1}'\mathbf{x}_1 = \mathbf{1}'\mathbf{x}_2 = \mathbf{1}'\mathbf{x}_3 = 0$. Furthermore, check that $\mathbf{x}_1'\mathbf{x}_2 = \mathbf{x}_1'\mathbf{x}_3 = \mathbf{x}_2'\mathbf{x}_3 = 0$. The matrix $X'X$ is diagonal with diagonal element 8.

Fitting the model in Eq. (5.30) with least squares leads to the estimates

$$\hat{\beta} = (X'X)^{-1}X'y = \begin{bmatrix} 8 & 0 & 0 & 0 \\ 0 & 8 & 0 & 0 \\ 0 & 0 & 8 & 0 \\ 0 & 0 & 0 & 8 \end{bmatrix}^{-1} \begin{bmatrix} \sum y_i \\ \sum x_{i1}y_i \\ \sum x_{i2}y_i \\ \sum x_{i3}y_i \end{bmatrix} = \begin{bmatrix} 19.9875 \\ -0.2875 \\ 2.4875 \\ 0.4125 \end{bmatrix} \tag{5.32}$$

Changing x_1 (temperature) by one unit reduces shrinkage by 0.2875%. A change in temperature from the low to the high level reduces shrinkage by $(2)(0.2875) = 0.575\%$. A similar interpretation applies to the other coefficients. The regression sum of squares due to x_1, x_2, x_3 is given by $\text{SSR}(x_1, x_2, x_3) = \hat{\beta}'X'y - n\bar{y}^2 = 51.5238$, with 3 degrees of freedom. The residual sum of squares is SSE $= 42.7850$, with $n - 4 = 8 - 4 = 4$ degrees of freedom. The F statistic for the overall significance and t ratios for each coefficient can be readily obtained.

Suppose that we consider the regression of y on x_1 alone, $y = \beta_0 + \beta_1 x_1 + \epsilon$. The least squares estimates are

$$\hat{\beta}_0 = 19.9875 \quad \text{and} \quad \hat{\beta}_1 = -0.2875$$

and the regression sum of squares due to x_1 is $\text{SSR}(x_1) = 0.6613$. We note that the estimate of β_1, $\hat{\beta}_1 = -0.2875$, is the same whether x_1 is the only variable in the model or all three (x_1, x_2, x_3) are included. Table 5.8 shows the results of fitting all possible models with the three variables. There are three models with just one x variable, three models with two x variables, and one model with all three variables. The results show that $\hat{\beta}_1$ is the same in all models. It does not matter whether the estimate comes from a one-variable model (x_1 alone), a two-variable model $[(x_1, x_2)$ or $(x_1, x_3)]$, or the three-variable model (x_1, x_2, x_3). This is true for the other parameter estimates as well.

TABLE 5.8 REGRESSION RESULTS FOR SHRINKAGE DATA

Variables in the Model	Coefficients $\hat{\beta}_0$	$\hat{\beta}_1$	$\hat{\beta}_2$	$\hat{\beta}_3$	Regression SS (SSR)
x_1	19.9875	−0.2875			0.6613
x_2	19.9875		2.4875		49.5012
x_1, x_2	19.9875	−0.2875	2.4875		50.1625
x_3	19.9875			0.4125	1.36125
x_1, x_3	19.9875	−0.2875		0.4125	2.0225
x_2, x_3	19.9875		2.4875	0.4125	50.8625
x_1, x_2, x_3	19.9875	−0.2875	2.4875	0.4125	51.5238

Another point to note from Table 5.8 concerns the regression sums of squares. One notices that they are additive:

$$\text{SSR}(x_1, x_2) = \text{SSR}(x_1) + \text{SSR}(x_2), \quad \text{or} \quad 50.1625 = 0.6613 + 49.5012$$

$$\text{SSR}(x_1, x_2, x_3) = \text{SSR}(x_1) + \text{SSR}(x_2) + \text{SSR}(x_3) \quad \text{or}$$

$$51.5238 = 0.6613 + 49.5012 + 1.3613$$

etc.

The nonchanging estimates and the additivity of the regression sums of squares are special consequences of orthogonality. The special orthogonal structure of the design matrix X implies a diagonal $X'X$ matrix, with diagonal elements $\boldsymbol{x}_i'\boldsymbol{x}_i$. Consequently, the inverse $(X'X)^{-1}$ is also diagonal with diagonal elements $(\boldsymbol{x}_i'\boldsymbol{x}_i)^{-1}$. In our case,

$$X'X = \begin{bmatrix} 8 & 0 & 0 & 0 \\ 0 & 8 & 0 & 0 \\ 0 & 0 & 8 & 0 \\ 0 & 0 & 0 & 8 \end{bmatrix}, \quad \text{and} \quad (X'X)^{-1} = \begin{bmatrix} \frac{1}{8} & 0 & 0 & 0 \\ 0 & \frac{1}{8} & 0 & 0 \\ 0 & 0 & \frac{1}{8} & 0 \\ 0 & 0 & 0 & \frac{1}{8} \end{bmatrix} \tag{5.33}$$

With orthogonality, the least squares estimates $\hat{\boldsymbol{\beta}} = (X'X)^{-1}X'\boldsymbol{y}$ in the general regression model $\boldsymbol{y} = \beta_0\boldsymbol{x_0} + \beta_1\boldsymbol{x_1} + \cdots + \beta_p\boldsymbol{x_p} + \boldsymbol{\epsilon}$ (where, $\boldsymbol{x_0} = \boldsymbol{1}$) are given by $\hat{\beta}_i = (\boldsymbol{x}_i'\boldsymbol{y})/(\boldsymbol{x}_i'\boldsymbol{x}_i)$. The estimate of β_i in the model with just $\boldsymbol{x}_i$, $\boldsymbol{y} = \beta_i\boldsymbol{x}_i + \boldsymbol{\epsilon}$, is given by $\hat{\beta}_i = (\boldsymbol{x}_i'\boldsymbol{y})/(\boldsymbol{x}_i'\boldsymbol{x}_i)$, and we can see that the two estimates are the same.

The regression sum of squares of the full model is $\text{SSR} = \hat{\boldsymbol{\beta}}'X'\boldsymbol{y} = \sum_{i=0}^{p} \hat{\beta}_i(\boldsymbol{x}_i'\boldsymbol{y})$. Note that here we are not correcting this sum of squares for the constant. Correcting for the constant would require the subtraction of $\hat{\beta}_0(\boldsymbol{x_0}'\boldsymbol{y}) = [(\boldsymbol{1}'\boldsymbol{y})/\boldsymbol{1}'\boldsymbol{1}](\boldsymbol{1}'\boldsymbol{y}) = n\bar{y}^2$. The regression sum of squares of the model $\boldsymbol{y} = \beta_i\boldsymbol{x}_i + \boldsymbol{\epsilon}$ is given by $\text{SSR}(\boldsymbol{x}_i) = \hat{\beta}_i(\boldsymbol{x}_i'\boldsymbol{y})$. This shows that the regression sums of squares are additive.

We also note that $V(\hat{\boldsymbol{\beta}}) = (X'X)^{-1}\sigma^2$. With orthogonality, this is a diagonal matrix and the covariances between the elements of $\hat{\boldsymbol{\beta}}' = (\hat{\beta}_0, \hat{\beta}_1, \hat{\beta}_2, \hat{\beta}_3)$ are zero. The additional assumption of normal errors implies that the least squares estimators are statistically independent.

EXERCISES

5.1. Consider the following regression model:

$$\text{Salary (in \$1,000)} = 20 + 2x + 5z + 0.7xz$$

where x is the number of years of experience, and z is an indicator variable that is 1 if you have obtained an MBA degree and 0 otherwise; xz is the product between years of experience and the indicator variable z.

Graph salary (y) against years of experience (x). Do this for both groups (without MBA and with MBA) on the same graph, and comment on the degree of interaction.

5.2. You are interested in the starting salaries of accounting, management information systems, and economics majors. You consider a model that factors in the GPA of students, obtaining the following regression model:

$$\text{Salary (in \$1,000)} = -15 + (18)\text{GPA} + (3)\text{IND}_{\text{acc}} + (2.1)\text{IND}_{\text{mis}}$$

IND_{acc} is an indicator variable that is 1 if the student is in accounting and 0 otherwise. IND_{mis} is an indicator variable that is 1 if the student is an MIS student and 0 otherwise.

a. Calculate the expected salary difference between an accounting and an economics student with the same GPA.

b. Calculate the expected salary difference between an accounting and an MIS student with the same GPA.

5.3. The data are taken from Mazess, R. B., Peppler, W. W., and Gibbons, M. Total body composition by dualphoton (^{153}Gd) absorptiometry. *American Journal of Clinical Nutrition,* 40, 834–839, 1983. The data are given in the file **bodyfat**.

A new method of measuring the body fat percentage is investigated. The body fat, age (between 23 and 61 years), and gender (4 males and 14 females) of 18 normal adults are listed below.

Graph body fat against age and gender (you may want to overlay these two on the same graph). Consider a regression model with age and gender as the explanatory variables. Interpret the results, and discuss the effects of age and gender. Is it useful to include an interaction term for age and gender?

y = % Fat	x_1 = Age	x_2 = Gender
9.5	23	1
27.9	23	0
7.8	27	1
17.8	27	1
31.4	39	0
25.9	41	0
27.4	45	1
25.2	49	0
31.1	50	0
34.7	53	0
42.0	53	0
29.1	54	0
32.5	56	0
30.3	57	0
33.0	58	0
33.8	58	0
41.1	60	0
34.5	61	0

5.4. You are regressing fuel efficiency (y) on three predictor variables, x_1, x_2, and x_3, and you obtain the following fitted regression model:

$$\hat{\mu} = \hat{\beta}_0 + \hat{\beta}_1 x_1 + \hat{\beta}_2 x_2 + \hat{\beta}_3 x_3$$

The coefficient of determination for this regression model is R^2 =90%.

A regression of x_1 on x_2, x_3 gives you an R^2 of 60%;

A regression of x_2 on x_1, x_3 gives you an R^2 of 80%; and

A regression of x_3 on x_1, x_2 gives you an R^2 of 90%.

Calculate and interpret the variance inflation factors for the regression coefficients $\hat{\beta}_1$, $\hat{\beta}_2$, and $\hat{\beta}_3$.

5.5. Which one of the following statements suggests the presence of a multicollinearity problem:

a. High R^2 and high t ratios

b. High correlation between explanatory variables and dependent variable

c. Low pairwise correlation among independent variables

d. Low R^2 and low t ratios

e. High $R^2$and mostly insignificant t ratios

5.6. The data are taken from Latter, H. O. The cuckoo's egg. *Biometrika,* 1, 164–176, 1901. The data are given in the file **cuckoo**.

The female cuckoo lays her eggs into the nest of foster parents. The foster parents are usually deceived, probably because of the similarity in the sizes of the eggs. Latter investigated this possible explanation and measured the lengths of cuckoo eggs (in millimeters) that were found in the nests of the following three species:

Hedge Sparrow:					
22.0	23.9	20.9	23.8	25.0	24.0
21.7	23.8	22.8	23.1	23.1	23.5
23.0	23.0				
Robin:					
21.8	23.0	23.3	22.4	23.0	23.0
23.0	22.4	23.9	22.3	22.0	22.6
22.0	22.1	21.1	23.0		
Wren:					
19.8	22.1	21.5	20.9	22.0	21.0
22.3	21.0	20.3	20.9	22.0	20.0
20.8	21.2	21.0			

Obtain the analysis of variance table and test whether or not the mean lengths of the eggs found in the nests of the three species are different. Display the data graphically, and interpret the results.

5.7. Percentage yields from a chemical reaction for changing temperature (factor 1), reaction time (factor 2), and concentration of a certain ingredient (factor 3) are as follows:

Factor 1: x_1	Factor 2: x_2	Factor 3: x_3	Average $\bar{y}_i$ from 5 Experiments
−1	−1	−1	79.7
1	−1	−1	74.3
−1	1	−1	76.7
1	1	−1	70.0
−1	−1	1	84.0
1	−1	1	81.3
−1	1	1	87.3
1	1	1	73.7

Each listed yield is actually the average of five individual independent experiments. The variance of individual measurements can be estimated from the five replications in each cell. It is found that

$$s^2 = \frac{\sum_{i=1}^{8}\sum_{j=1}^{5}(y_{ij}-\bar{y}_i)^2}{8(5-1)} = 40.0$$

a. Estimate the effects of factors 1–3. That is, estimate the coefficients in the regression model

$$y = \beta_0 + \beta_1 x_1 + \beta_2 x_2 + \beta_3 x_3 + \varepsilon$$

Calculate the standard errors of the coefficients and interpret the results. Comment on the nature of the design matrix.

b. Is it possible to learn something about interactions? Consider the interaction effect between factors 1 and 2. Write out the Xmatrix of the regression model $y = \beta_0 + \beta_1 x_1 + \beta_2 x_2 + \beta_3 x_3 + \beta_4 x_1 x_2 + \varepsilon$. Estimate the model and comment on this issue.

5.8. In a study on the effect of coffee consumption on blood pressure, 30 patients are selected at random from among the patients of a medical practice. A questionnaire is administered to each patient to get the following information:

x_1: Average number of cups of coffee consumed/day

x_2: A measure of daily exercise

x_3: Age

x_4: Sex($x_4 = 0$ for males, $x_4 = 1$ for females)

y: Systolic blood pressure during the last visit to the practice

A linear model of the form

$$y = \beta_0 + \beta_1 x_1 + \beta_2 x_2 + \beta_3 x_3 + \beta_4 x_4 + \epsilon, \quad \epsilon \sim N(0, \sigma^2)$$

is considered.

a. Explain carefully the meaning of the parameter β_4.

b. Why is the error term ϵ present in the model?

c. If β_1 is very large, can we conclude from this study that increased coffee consumption **causes** increased blood pressure? Discuss.

d. Another model

$$y = \beta_0 + \beta_1 x_1 + \beta_2 x_2 + \beta_3 x_3 + \beta_4 x_4 + \beta_5 x_1 x_4 + \epsilon$$

is fit to the data. Explain the meaning of the hypothesis $\beta_5 = 0$.

5.9. a. Suppose that y is the average price (in thousands of dollars) of a typical three-bedroom home in a large Canadian city. Fourteen consecutive observations $y_1, y_2, \ldots, y_{14}$ are taken at consecutive 6-month intervals over 7 years. At the beginning of the eighth interval the government implemented steps to slow down the rate of the price increases. A possible model for these data specifies that prices increase linearly with time until the time point 8, at which time the rate of increase (slope) changes. Such a model consists of a pair of straight lines intersecting at time point 8. Formulate this as a linear model $\mathbf{y} = X\boldsymbol{\beta} + \boldsymbol{\epsilon}$. Explicitly define the parameters you are using and write out the X matrix.

b. An alternative model specifies that the government's actions had no effect on prices, and that prices continued to increase at the same constant rate. Set up the appropriate linear model $\mathbf{y} = X\boldsymbol{\beta} + \boldsymbol{\epsilon}$.

c. Suppose you found that the residual sum of squares in (a) is 6.12 and that in (b) is 37.25. What evidence does this provide with regard to whether model (a) is better than model (b)?

5.10. Accident rate data $y_1, \ldots, y_{12}$ were collected over 12 consecutive years $t = 1, 2, \ldots, 12$. At the end of the sixth year, a change in safety regulations occurred. For each of the following situations, set up a linear model of the form $\mathbf{y} = X\boldsymbol{\beta} + \boldsymbol{\epsilon}$. Define X and $\boldsymbol{\beta}$ appropriately.

a. The accident rate y is a linear function of t with the new safety regulations having no effect.

b. The accident rate y is a quadratic function of t with the new regulations having no effect.

c. The accident rate y is a linear function of t. The slope for $t \geq 7$ is the same as for $t < 7$. However, there is a discrete jump in the function at $t = 7$.

d. The accident rate y is a linear function of t. After $t = 7$, the slope changes, with the two lines intersecting at $t = 7$.

5.11. A consumer group conducted an experiment to compare the effectiveness of three commercially available weight-reducing diets, A, B, and C. The group wanted to answer the following questions.

i. Are the three diets achieving similar weight reductions?

ii. Does the weight loss depend on the initial weight, and if so, is this effect the same for the three diets?

Thirty volunteers were randomly assigned to the three diets (10 to each diet). Their weights (in pounds) were recorded at the beginning and after 1 month on the respective diets. The resulting data are given in the file **weightloss** and are shown here:

Diet A		Diet B		Diet C	
X Weight Before	*Y* Weight Loss	*X* Weight Before	*Y* Weight Loss	*X* Weight Before	*Y* Weight Loss
227	14	255	19	206	7
286	16	193	8	222	9
180	−2	186	4	168	2
176	8	145	15	132	0
204	15	219	16	173	−3
155	5	273	19	210	8
303	17	289	25	269	10
146	7	168	6	275	51
215	15	194	12	241	8
187	6	248	21	219	5

a. What regression models would you use to investigate the previous questions? Write out the models in matrix form.

b. In nontechnical language, briefly describe the relationship between diets and weight loss that the models selected in (a) describe.

c. If you wanted to use these results as a basis for recommending one of these diets to other potential dieters, what assumptions would you have to make? What reservations might you have about these assumptions?

d. Use the models to analyze the data. What are your conclusions?

5.12. An exploratory study on the influence of formaldehyde concentration (x_1), catalyst ratio (x_2), curing temperature (x_3), and curing time (x_4) on the wrinkle resistance of cotton cellulose (y) was carried out. Small values of y indicate low wrinkle resistance. The method of data collection consisted of taking 30 samples of cotton cellulose from the last 60 production days, measuring the wrinkle resistance of the sampled items, and looking up the corresponding values of x_1, x_2, x_3, and x_4 for the sample from production records. The data are given in the file **cellulose**.

a. By means of pairwise scatter plots of y versus each explanatory variable, make an assessment of the model to be fitted. Specify your model.

b. Fit the model

$$y = \beta_0 + \beta_1 x_1 + \beta_2 x_2 + \beta_3 x_3 + \beta_4 x_4 + \epsilon$$

Assess the adequacy of this model. Check for any unusual points.
Assess the following hypotheses:

(i) $\beta_1 = 0$;

(ii) $\beta_3 = \beta_4 = 0$;

(iii) $\beta_1 = \beta_2 = \beta_3 = \beta_4 = 0$

c. The investigator believes that there could be interactions between catalyst ratio and curing time and between catalyst ratio and curing temperature. Write down an appropriate model and assess if these hypotheses are supported by the data.

d. Give a model that you believe describes the data set. What manufacturing strategy would you pursue if

i. low wrinkle resistance is preferred?

ii. high wrinkle resistance is preferred?

e. The data obtained are observational (because the investigator took what the process provided in terms of values for x_1, x_2, x_3, and x_4). What restriction does this place on the conclusions that we draw from the model? Can you suggest a way that could get us around this restriction?

5.13. A team of anthropologists and nutrition experts investigated the influence of protein content in the diet on the relationship between age and height of children in an underdeveloped country. Data on height (cm) and age for children fed on a protein-rich diet and for children on a protein-poor diet were obtained. The data are given in the file **diet**.

Protein-rich diet

Age	(x)	0.2	0.5	0.8	1.0	1.0	1.4	1.8
Height	(y)	54	54.3	63	66	69	73	82
Age	(x)	2.0	2.0	2.5	2.5	3.0	2.7	
Height	(y)	83	80.3	91	93.2	94	94	

Protein-poor diet

Age	(x)	0.4	0.7	1.0	1.0	1.5	2.0	2.0
Height	(y)	52	55	61	63.5	66	68.5	67.9
Age	(x)	2.4	2.8	3.0	1.3	1.8	0.2	3.0
Height	(y)	72	76	74	65	69	51	77

a. Plot height versus age on a single graph using different symbols for the two diets.

b. Carry out a test of significance to determine if the linear relationship between height and age is the same for both diets.

5.14. a. Assume that the columns of the matrix X in the model $\mathbf{y} = X\boldsymbol{\beta} + \boldsymbol{\epsilon}$, $\boldsymbol{\epsilon} \sim N(0, \sigma^2 I)$ are orthogonal. Show that $\hat{\beta}_i$ and $\hat{\beta}_j$ are independent.

b. Suppose an extra term is added to the model,

$$\mathbf{y} = X\boldsymbol{\beta} + \mathbf{z}\gamma + \boldsymbol{\epsilon}$$

Assume that $\mathbf{z}$ is orthogonal to the columns of X. Show that the estimate of $\boldsymbol{\beta}$ in the expanded model is the same as the estimate in the original model and prove that it has the same distribution. (Note that X need not have orthogonal columns for this result to hold.)

c. Consider the linear model $\mathbf{y} = X\boldsymbol{\beta} + \boldsymbol{\epsilon}$ in which the first column of X is a column of 1's and the remaining columns are centered about their means. Show that $\hat{\beta}_0 = \bar{y}$. Furthermore, show that if the errors are independent $N(0, \sigma^2)$, the estimator $\hat{\beta}_0$ is distributed independently of $(\hat{\beta}_1, \ldots, \hat{\beta}_p)$.

5.15. In a study to examine the fuel efficiency of an automobile as a function of its engine characteristics, 20 automobiles are considered. The following information is collected for each car: y = fuel efficiency, x_1 = weight (in 1,000 lbs), and x_2 = engine type (A or B). Assume that the first 10 cars have type A engine, and the rest have type B engine. Set up a linear model, carefully defining the X matrix and the $\boldsymbol{\beta}$ vector for each of the following situations:

a. It is believed that expected fuel efficiency depends on the weight of the car and types of the engine.

b. It is suspected that the effect of weight on expected fuel efficiency depends on the different types of engines.

5.16. In an experiment to study the effect of temperature (x) on the yield of a chemical reaction (y), 30 experimental runs were conducted. The level of temperature was carefully controlled at each of five levels, coded as $x = -2, -1, 0, 1, 2$. Two catalysts were used. For each catalyst three runs were taken at each level of temperature, and the yield was measured. The model

$$y = \beta_0 + \beta_1 x + \beta_2 x^2 + \beta_3 z + \epsilon,$$
$$\epsilon \sim N(0, \sigma^2)$$

was considered, where $z = 0$ for catalyst 1 and $z = 1$ for catalyst 2.

a. Carefully interpret the parameter β_3 in this model.

b. The model was fit to the data and the output is summarized below. The residual sum of squares is 25.05, and

Parameter	Estimate	Standard Error
β_0	29.83	0.33
β_1	0.95	0.13
β_2	0.41	0.11
β_3	−0.32	0.36

Is there any evidence of a difference in the two catalysts? Find a 95% confidence interval for β_2.

c. We also know that

$$(X'X)^{-1} = \begin{bmatrix} 0.114 & 0 & -0.023 & -0.067 \\ 0 & 0.017 & 0 & 0 \\ -0.023 & 0 & 0.012 & 0 \\ -0.067 & 0 & 0 & 0.133 \end{bmatrix}$$

i. Explain why $\hat{\beta}_1$ and $\hat{\beta}_3$ are independent random variables.

ii. Find a 95% confidence interval for the expected yield when the standard temperature ($x = 0$) and catalyst 2 are used.

iii. Find a 95% prediction interval for the yield of a new experiment run under standard temperature ($x = 0$) and with catalyst 2.

d. It was thought that the effects of temperature differ for the two catalysts. Accordingly, the model

$$y = \beta_0 + \beta_1 x + \beta_2 x^2 + \beta_3 z + \beta_4 zx + \beta_5 zx^2 + \epsilon$$

was fit to the data, leading to a residual sum of squares of 19.70. Is there any evidence that the effects of temperature differ for the two catalysts?

5.17. Consider the data in Table 2.2. In Eq. (2.37) of Section 2.8 we considered the linear regression of test scores on poverty.

a. Investigate whether the model can be improved by including the square of poverty as an additional regressor variable.

b. Check whether it is necessary to introduce an indicator for students in a college community such as Iowa City.

6 Model Checking

6.1 INTRODUCTION

In previous chapters, we considered the linear regression model and discussed in detail the assumptions that complete its specification. Chapters 2 and 4 described the inference within a specified model, assuming that all assumptions are met. We also illustrated how such models can be used for prediction as well as other purposes.

Computer packages (such as S-Plus, SAS, and Minitab) can fit any model one supplies, but usually they do so without questioning whether or not the model is adequate. The fact that estimation today is so easy has led to an abundance of models being fit to data sets. However, many of these models will be inappropriate, and model checking becomes very important. One needs to make sure that the adopted models are adequate and satisfy all model assumptions.

As shown in Chapter 1, the essence of model building can be represented by the diagram in Figure 6.1. Initially, a model is specified from available data and/or theory. In some circumstances, good theory and prior studies are available. In other cases, no such information is available, and the model needs to be specified from the data at hand; well-chosen plots of the information can greatly help with the model specification. The specified model is then estimated and submitted to extensive diagnostic checking. If the model is found adequate, it can be used for the purposes for which it was designed. However, if the model is found inadequate, the specification must be modified and the model-building cycle continued until a satisfactory structure is obtained.

A number of assumptions are made when specifying a model, and these assumptions need to be checked. A fitted model can be inadequate for several reasons:

i. The functional form of the model may not be adequate. The model may be missing needed variables and nonlinear components, such as squares of covariates and interactions among covariates.

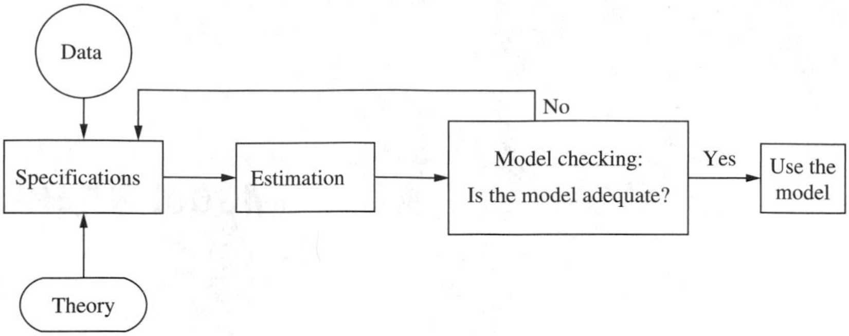

FIGURE 6.1 Model Building Framework

ii. The error specification may be incorrect. In particular, the error variances $V(\epsilon_i)$ may not be constant, the errors may not be normally distributed, and the errors may not be independent.

iii. Unusual observations in the data may have an undue influence on the model fit. There may be outlying data points that have a major impact on the estimates and the conclusions.

In Section 6.2, we examine the residuals as a whole. Residual plots can tell us whether the functional form is misspecified and whether the usual assumptions on the error terms are violated. Residual plots can also draw attention to particular cases that seem "odd" when judged in comparison to the rest of the data; in Section 6.3, we study specific case diagnostics that assess the influence of individual cases on the model results. In Section 6.4, we discuss goodness-of-fit tests that tell us whether our model provides an adequate representation of the functional relationship between the response and the explanatory variables.

6.2 RESIDUAL ANALYSIS

6.2.1 RESIDUALS AND RESIDUAL PLOTS

The residual vector is given by $\boldsymbol{e} = \boldsymbol{y} - \hat{\boldsymbol{\mu}}$, the difference between the observed response and the fitted value; the ith component $e_i = y_i - \hat{\mu}_i$ corresponds to the ith case in the data set. The residual estimates the random component ϵ in the model. Misspecification and departures from the underlying assumptions in the model are reflected in the pattern of the residuals. Hence, a thorough residual analysis and a graphical display of the residuals provide an effective method of discovering model inadequacies.

The discussion in Chapter 4 has shown that the vector of residuals $\boldsymbol{e}$ is orthogonal to the vector space $L(X) = L(\mathbf{1}, \boldsymbol{x}_1, \boldsymbol{x}_2, \ldots, \boldsymbol{x}_p)$ that is spanned by the regressor vectors. We assume that the unit vector, and hence an intercept term, is included in the model. The residuals are orthogonal to the regressor vectors

$\mathbf{1}, \boldsymbol{x}_1, \boldsymbol{x}_2, \ldots, \boldsymbol{x}_p$ and also orthogonal to $\hat{\boldsymbol{\mu}} = X\hat{\boldsymbol{\beta}}$ because the vector of fitted values is in the space $L(X)$. Orthogonality implies that

$$\sum_{i=1}^{n} e_i = \sum_{i=1}^{n} e_i x_{i1} = \cdots = \sum_{i=1}^{n} e_i x_{ip} = \sum_{i=1}^{n} e_i \hat{\mu}_i = 0$$

These residual properties are consequences of the adopted least squares fitting procedure. They hold, irrespective of whether or not the model is adequate.

What happens to the residuals if some model assumptions are violated? For example, what happens to the expected value of the residuals? The expected value of the residual vector can be written as

$$E(\boldsymbol{e}) = E(\boldsymbol{y} - \hat{\boldsymbol{\mu}}) = E[(I - H)\boldsymbol{y}] = (I - H)E(\boldsymbol{y}) \tag{6.1}$$

The residuals have mean vector zero if the "true" expected value $E(\boldsymbol{y})$ is a vector in the space $L(X) = L(\mathbf{1}, \boldsymbol{x}_1, \boldsymbol{x}_2, \ldots, \boldsymbol{x}_p)$. Any element in the linear vector space can be written as $E(\boldsymbol{y}) = X\boldsymbol{\alpha}$, for some $\boldsymbol{\alpha}$, and hence $(I - H)X\boldsymbol{\alpha} = (I - X(X'X)^{-1}X')X\boldsymbol{\alpha} = \mathbf{0}$. For a correctly specified functional relationship, the residuals will have expectation zero. On the other hand, for a misspecified model where $E(\boldsymbol{y})$ is not in the space $L(X)$, $E(\boldsymbol{e}) \neq \mathbf{0}$ (see the appendix, No. 1, for details). Assume, for example, that the true model is given by $E(\boldsymbol{y}) = X\boldsymbol{\beta} + \boldsymbol{u}\gamma$, where $\boldsymbol{u}$ is an additional regressor vector that is not part of $L(X)$, and γ a parameter. Fitting the model without $\boldsymbol{u}$ leads to the residuals $\boldsymbol{e} = (I - H)\boldsymbol{y}$ and

$$E(\boldsymbol{e}) = (I - H)E(\boldsymbol{y}) = (I - H)(X\boldsymbol{\beta} + \boldsymbol{u}\gamma) = \gamma(I - H)\boldsymbol{u} \neq \mathbf{0} \tag{6.2}$$

The vector $(I - H)\boldsymbol{u}$ is not zero because $\boldsymbol{u}$ is not in $L(X)$. Equation (6.2) shows that the residuals obtained from the model without $\boldsymbol{u}$ are related to $\boldsymbol{u}$. Hence, a graph of the residuals obtained from the model without $\boldsymbol{u}$ against the variable $\boldsymbol{u}$ reveals a pattern. Graphs of the residuals against variables that are not part of the specified model but are thought to have an effect on the response should be considered as diagnostic tools. Patterns in such plots indicate that needed terms have been omitted. Also, one is always concerned about whether a linear specification in the regressor variables is sufficient or whether quadratic terms are needed. Since quadratic terms are not part of $L(X)$, scatter plots of the residuals from the linear model against the explanatory variables already part of the model should also be considered. Nonlinear patterns would arise in these graphs if a model that is linear in the covariates is not adequate.

Under the model assumptions, the vector of residuals $\boldsymbol{e} = (I - H)\boldsymbol{y}$ and the vector of fitted values $\hat{\boldsymbol{\mu}}$ are uncorrelated. This was shown in Eq. (4.19). This states that the fitted values should not carry any information on the residuals. Hence, for a correctly specified model, a graph of the residuals against the fitted values should show no patterns. On the other hand, violations of the model assumptions introduce correlations among $\boldsymbol{e}$ and $\hat{\boldsymbol{\mu}}$, and a graph of the residuals against fitted values would reflect an association. Hence, a graph of the residuals against the fitted values should be part of the standard diagnostic-checking repertoire.

If all standard assumptions are met, the residuals $\boldsymbol{e} = \boldsymbol{y} - \hat{\boldsymbol{\mu}}$ have mean zero and covariance matrix $\sigma^2(I - H)$, where $H = X(X'X)^{-1}X'$ is the usual "hat" matrix. The residuals follow a multivariate normal distribution if the errors are normal. This result implies that

$$V(e_i) = \sigma^2(1 - h_{ii}) \quad \text{and} \quad \text{Cov}(e_i, e_j) = -\sigma^2 h_{ij} \quad \text{for} \quad i \neq j \tag{6.3}$$

where h_{ij} is the (i, j)th element of H. Since h_{ii}'s are not necessarily the same for all i, the variances of the residuals, $V(e_i)$, are not identical, although $V(\epsilon_i) = \sigma^2$ is constant. Similarly, although $\text{Cov}(\epsilon_i, \epsilon_j) = 0$, $\text{Cov}(e_i, e_j) = -\sigma^2 h_{ij}$ will not equal zero because h_{ij} is not necessarily zero. The nonconstancy of the variance of the residuals and the slight dependence among the residuals are consequences of the model fitting.

Residuals are often standardized so that they have mean zero and variance one. Since residuals are estimates of the errors, and since the unobserved errors have variance σ^2, one computes

$$e_i^s = \frac{e_i}{s} \tag{6.4}$$

where $s^2 = \boldsymbol{e}'\boldsymbol{e}/(n - p - 1)$ is the usual estimate of σ^2. The residuals in Eq. (6.4) are called the **standardized residuals**. Their variance is approximately one, but only approximately because their definition does not take account of the correct variance of the residuals given in Eq. (6.3). The **studentized residuals** are given by

$$d_i = e_i / s\sqrt{1 - h_{ii}} \tag{6.5}$$

This standardization uses the correct variance of the residual, $V(e_i) = \sigma^2(1 - h_{ii})$. The exact distribution of the studentized residuals (as well as that of the standardized residuals) is complicated because s and e_i are not statistically independent. However, with normal errors, the distribution of the studentized residuals is approximately normal with mean zero and variance one. A histogram or a dot plot of the studentized residuals helps us assess whether one or more of the residuals are unusually large. A studentized residual d_i larger than 2 or 3 in absolute value would make us question whether the model is adequate for that case i. It may be that the response y_i for this particular case is an "outlier" due to a poor design or a recording mistake. However, it could also be that our model is missing an important component.

The properties of the residuals that we have discussed here help us devise useful residual plots for model checking. For adequate models, we do not expect any systematic patterns in the residuals. Hence, plots of the residuals against (i) the case order if cases have been ordered by time, (ii) the explanatory variables in the model, (iii) other variables not in the model but considered important, and (iv) the fitted values should all show random scatter. Since the mean of the residuals is zero ($\bar{e} = 0$, as long as a constant is included in the model), the residuals should lie within a horizontal band around zero and should not exhibit any interpretable patterns. Usually, one works with standardized or studentized residuals, and then the horizontal bands at ± 2 have special meaning. Approximately 95% of these residuals should be within ± 2, and almost all of them should be within ± 3. Patterns in the residuals indicate that the functional form of the model is not correct, that important variables have been omitted, and that perhaps the error variance is not constant.

Figure 6.2 shows various residual plots for illustration. If a model is adequate, then a plot of the residuals e_i against i should exhibit the random pattern depicted in Figure 6.2(a).

A similar plot [Figure 6.2(b)] should result if the residuals e_i are plotted against the fitted values, against any of the p explanatory variables, or against any other regressor variable that is not in the model. Figures 6.2(c)–6.2(g) show departures from random scatter. The plot in Figure 6.2(c) indicates that the addition of a linear term in another regressor variable can improve the model. Figure 6.2(d) shows that the model lacks a quadratic term of the included regressor x; the inclusion of such a term will improve the model. Figures 6.2(e) and 6.2(f) show that the variance of the residuals increases with the regressor variable x and the level. Later, we discuss how to modify the model in such situations. If the pattern in the residual plot is similar to that in Figure 6.2(f), a transformation of the response y is necessary. This will be discussed in Section 6.5.

6.2.2 ADDED VARIABLE PLOTS

An added variable plot is useful when deciding whether a new regressor variable (that currently is not in the model) should be included. It turns out to be a more powerful graph than the plot of the residuals against the new regressor variable.

Let $\boldsymbol{e}$ be the residual vector from a fit of $\boldsymbol{y}$ on X. Let $\boldsymbol{u}$ be the vector of observations on a new regressor variable that is not part of X, and suppose that $\boldsymbol{e}_u$ is the residual vector of the regression of $\boldsymbol{u}$ on X. Then a plot of $\boldsymbol{e}$ against $\boldsymbol{e}_u$ is called an **added variable plot**. If this plot indicates random scatter without an apparent relationship, then there is no need to include $\boldsymbol{u}$ in the model. Systematic patterns in the plot indicate that the variable $\boldsymbol{u}$ should be included.

How can one justify this plot? Assume that $\boldsymbol{u}$ is part of the model $\boldsymbol{y} = X\boldsymbol{\beta} + \boldsymbol{u}\gamma + \boldsymbol{\epsilon}$; γ is the regression parameter that corresponds to $\boldsymbol{u}$. A regression of $\boldsymbol{y}$ on X alone (without the regressor $\boldsymbol{u}$) leads to the residuals $\boldsymbol{e} = (I - H)\boldsymbol{y}$, where $H = X(X'X)^{-1}X'$ is the usual hat matrix. The residuals of the regression of $\boldsymbol{u}$ on X are given by $\boldsymbol{e}_u = (I - H)\boldsymbol{u}$. Note that the mean of either residual sequence is zero. Hence, the slope in the regression of $\boldsymbol{e}$ on $\boldsymbol{e}_u$ is given by

$$\tilde{\gamma} = \boldsymbol{e}_u'\boldsymbol{e}/\boldsymbol{e}_u'\boldsymbol{e}_u = \frac{\boldsymbol{u}'(I - H)\boldsymbol{y}}{\boldsymbol{u}'(I - H)\boldsymbol{u}} \tag{6.6}$$

Now consider the complete model $\boldsymbol{y} = X\boldsymbol{\beta} + \boldsymbol{u}\gamma + \boldsymbol{\epsilon}$, and the regression of $\boldsymbol{y}$ on both X and $\boldsymbol{u}$. The least squares estimate of the parameter vector $(\boldsymbol{\beta}, \gamma)$ is

FIGURE 6.2
Residual Plots

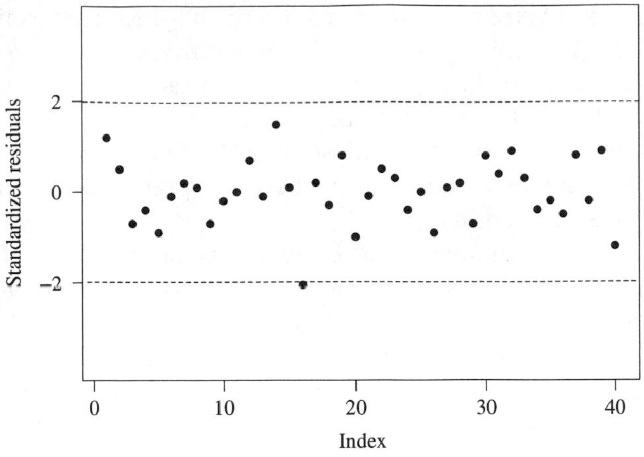

(a)

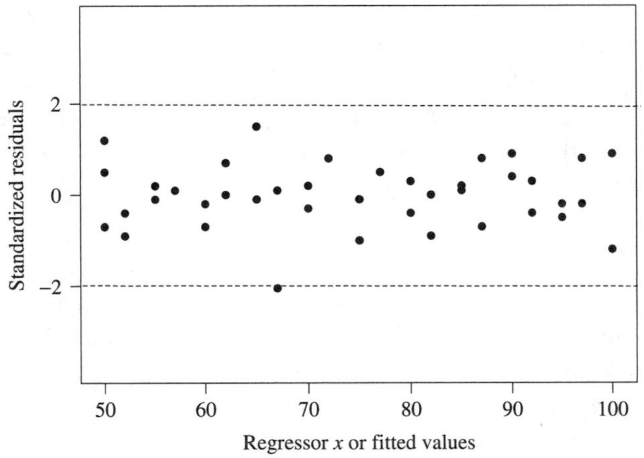

(b)

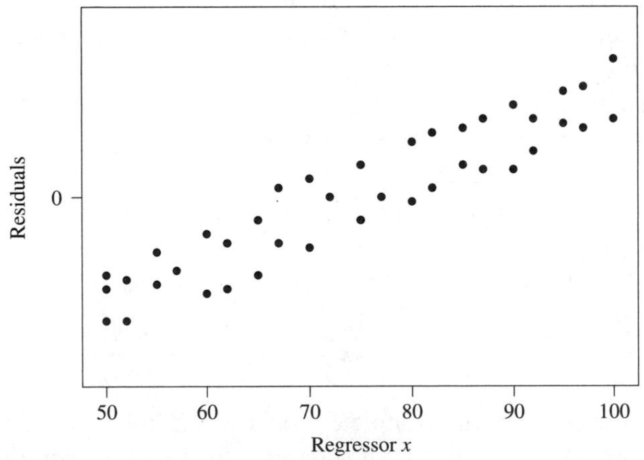

(c)

FIGURE 6.2
(Continued)

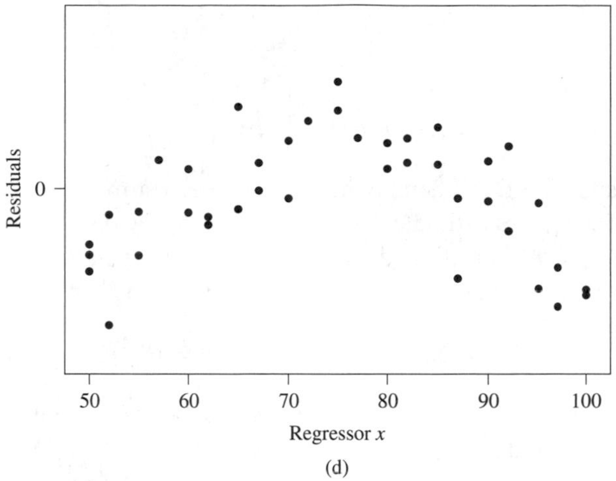

(d)

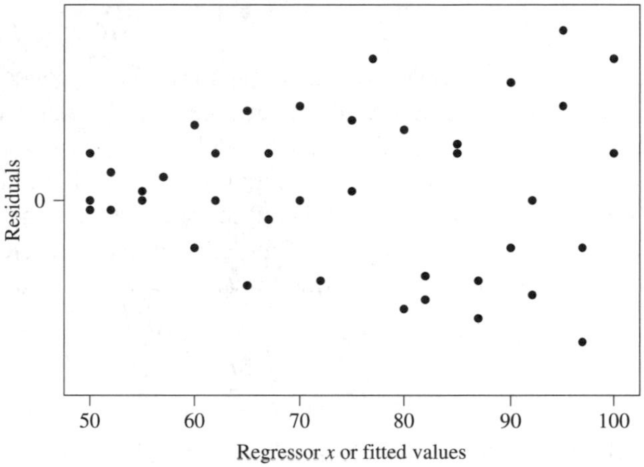

(e)

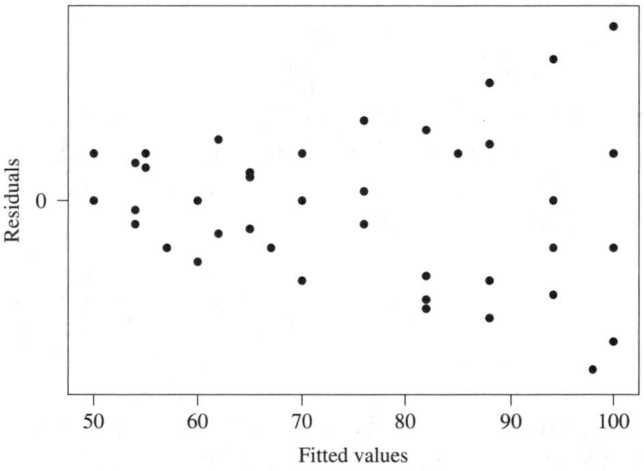

(f)

given by

$$\begin{bmatrix} \hat{\boldsymbol{\beta}}_* \\ \hat{\gamma} \end{bmatrix} = \begin{bmatrix} X'X & X'\boldsymbol{u} \\ \boldsymbol{u}'X & \boldsymbol{u}'\boldsymbol{u} \end{bmatrix}^{-1} \begin{bmatrix} X'\boldsymbol{y} \\ \boldsymbol{u}'\boldsymbol{y} \end{bmatrix} = \begin{bmatrix} A_{11} & A_{12} \\ A_{21} & A_{22} \end{bmatrix}^{-1} \begin{bmatrix} X'\boldsymbol{y} \\ \boldsymbol{u}'\boldsymbol{y} \end{bmatrix} \tag{6.7}$$

We have used $\hat{\boldsymbol{\beta}}_*$ to denote the least squares estimate of $\boldsymbol{\beta}$ to distinguish it from the least squares estimate $\hat{\boldsymbol{\beta}}$ in the regression of $\boldsymbol{y}$ on X alone. Using the result on the inverse of a partitioned matrix given in the appendix, we find that the estimate of γ (the coefficient that corresponds to $\boldsymbol{u}$) is

$$\hat{\gamma} = -B_{22}^{-1}B_{21}X'\boldsymbol{y} + B_{22}^{-1}\boldsymbol{u}'\boldsymbol{y} = B_{22}^{-1}[-B_{21}X'\boldsymbol{y} + \boldsymbol{u}'\boldsymbol{y}] \tag{6.8}$$

where $B_{22} = \boldsymbol{u}'\boldsymbol{u} - \boldsymbol{u}'X(X'X)^{-1}X'\boldsymbol{u} = \boldsymbol{u}'(I-H)\boldsymbol{u}$ and $B_{21} = \boldsymbol{u}'X(X'X)^{-1}$. Hence, we find that

$$\hat{\gamma} = \frac{1}{\boldsymbol{u}'(I-H)\boldsymbol{u}}[-\boldsymbol{u}'H\boldsymbol{y} + \boldsymbol{u}'\boldsymbol{y}] = \frac{\boldsymbol{u}'(I-H)\boldsymbol{y}}{\boldsymbol{u}'(I-H)\boldsymbol{u}} = \tilde{\gamma} \tag{6.9}$$

the slope in the scatter plot of e against e_u in Eq. (6.6). Hence, a linear relationship and a nonzero slope $\tilde{\gamma}$ in the added variable plot of e against e_u indicates that the variable $\boldsymbol{u}$ should be included in the model.

One can also show that the strength of the association in the added variable plot gives evidence on whether the effect of the added variable $\boldsymbol{u}$ is statistically significant. Consider the square of the correlation coefficient between e and e_u as a measure of the strength of the linear association in the added variable plot. That is,

$$\left(r_{e,e_u}\right)^2 = \frac{[\boldsymbol{u}'(I-H)\boldsymbol{y}]^2}{[\boldsymbol{y}'(I-H)\boldsymbol{y}][\boldsymbol{u}'(I-H)\boldsymbol{u}]} \tag{6.10}$$

Consider the square of the t ratio (or the partial F statistic) of the estimate $\hat{\gamma}$ in the regression model on both X and $\boldsymbol{u}$,

$$(t_{\hat{\gamma}})^2 = \frac{(\hat{\gamma})^2}{s^2 B_{22}^{-1}} = \frac{[\boldsymbol{u}'(I-H)\boldsymbol{y}]^2}{s^2[\boldsymbol{u}'(I-H)\boldsymbol{u}]} \tag{6.11}$$

where $s^2 = \boldsymbol{y}'(I-H_*)\boldsymbol{y}/(n-p-2)$ is the estimate of σ^2 in the model $\boldsymbol{y} = X\boldsymbol{\beta} + \boldsymbol{u}\gamma + \epsilon$ and H_* is the hat matrix calculated with the matrix $[X, \boldsymbol{u}]$. Here, we have used the result in Eq. (6.9) and the expression for B_{22}. With this, we can write Eq. (6.10) as

$$\begin{aligned} \left(r_{e,e_u}\right)^2 &= (t_{\hat{\gamma}})^2 \frac{s^2}{\boldsymbol{y}'(I-H)\boldsymbol{y}} = (t_{\hat{\gamma}})^2 \frac{\boldsymbol{y}'(I-H_*)\boldsymbol{y}}{\boldsymbol{y}'(I-H)\boldsymbol{y}} \Big/ (n-p-2) \\ &\approx (t_{\hat{\gamma}})^2/(n-p-2) \end{aligned} \tag{6.12}$$

The stronger the association is in the added variable plot, the stronger the evidence that the variable $\boldsymbol{u}$ should be included in the model.

6.2.3 CHECKING THE NORMALITY ASSUMPTION

We look at the residuals $e_1, e_2, \ldots, e_n$ to check the assumption that the unobserved errors (the ϵ_i's) in the model are normally distributed. Even better, we look at the studentized residuals $d_1, \ldots, d_n$ because they standardize the residuals for the differing variances. The simplest approach to check for normality is to plot a histogram and check whether it resembles that of a $N(0, 1)$ distribution. Most of the residuals (approximately 95%) should fall between -2 and $+2$, and the histogram should be bell shaped and symmetric about 0.

Another, preferable, approach is to prepare a **normal probability plot**. The basic idea of this plot is simple. First, we order the studentized residuals $d_1, \ldots, d_n$ from the smallest to the largest. Let the result be denoted by $d_{(1)}, \ldots, d_{(n)}$. Note that $d_{(1)}$ will be negative and $d_{(n)}$ positive. Let Z be a $N(0, 1)$ random variable with cumulative distribution function $\Phi(z) = P(Z \leq z)$. If the residuals are from a standard normal distribution, then the expected value of the ith smallest residual $d_{(i)}$ should be the normal percentile (quantile) of order $p_i = (i - 1/2)/n$. That is,

$$E\left(d_{(i)}\right) \simeq \Phi^{-1}(p_i)$$

Hence, a plot of the residuals $d_{(i)}$ versus the implied normal percentiles (normal scores) $\Phi^{-1}(p_i)$, for $i = 1, 2, \ldots, n$, should show points that are scattered around a straight line. Deviations from a straight line pattern, which are easy to detect by the eye, indicate a lack of normality.

Figure 6.3 shows probability distributions and corresponding normal probability plots for four data sets. Figure 6.3(a) is the ideal (straight line); it confirms a normal distribution. Figure 6.3(b) gives the probability plot (S shaped) of a light-tailed (uniform) distribution. The plot (inverted S shape) in Figure 6.3(c) is that of a heavy-tailed distribution. The plot in Figure 6.3(d) comes from a skewed distribution. The plots in Figures 6.3(b)–6.3(d) indicate lack of normality.

It should be noted that some computer programs (such as Minitab) reverse the axes and plot the normal percentiles (i.e., the normal scores) against the residuals. This does not matter because one only checks for linearity. We prefer to plot the residuals against the normal scores to be consistent with the other residual plots that graph the residuals against the case index, the fitted values, etc.

If the normal probability plots are prepared by hand, then it is easier to use a special graph paper called the **normal probability paper**. On this paper one plots $d_{(i)}$ against p_i and the scale on the x axis of the paper is drawn to perform the Φ^{-1} transformation. However, since most software packages supply normal probability plots automatically, we will not discuss its construction further.

6.2.4 SERIAL CORRELATION AMONG THE ERRORS

Standard regression inference assumes that the ϵ's, or equivalently the observations y's, are independent. If a regression model is fit to time series data (i.e., data observed sequentially in time such as monthly, quarterly, or yearly data)

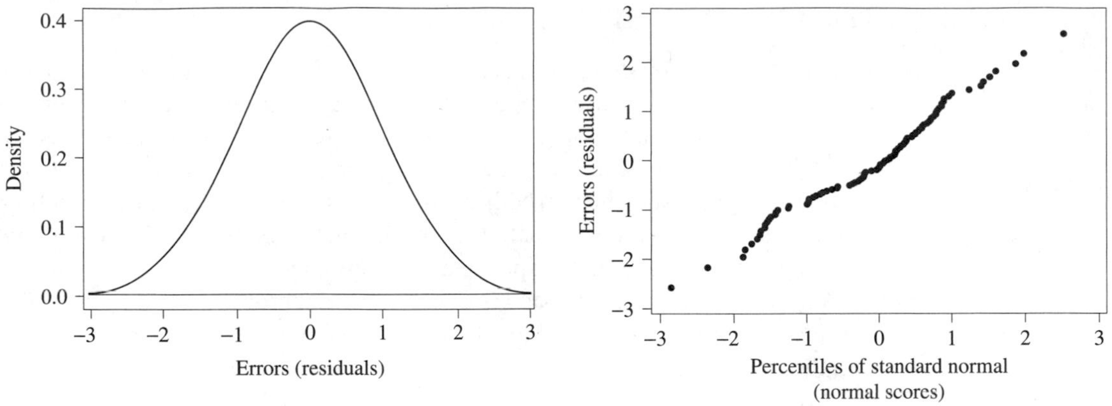

(a) Normal distribution and the ideal normal probability plot

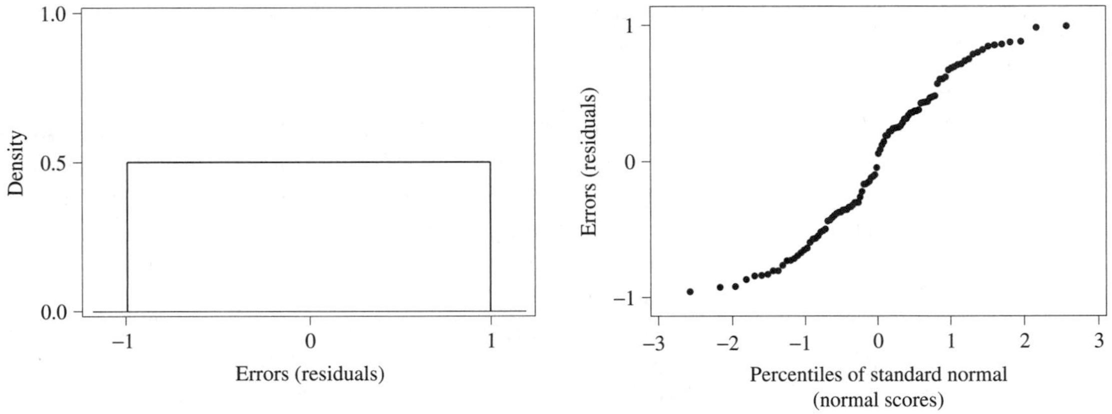

(b) Light-tailed (uniform) distribution

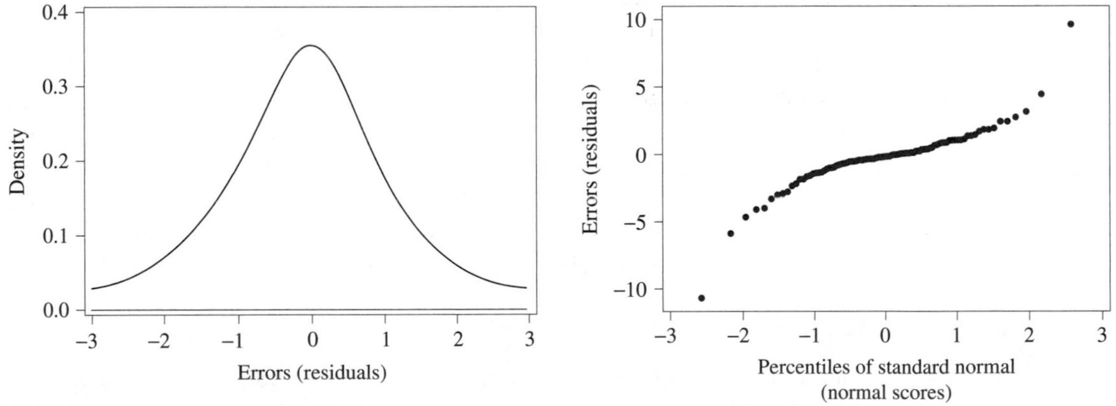

(c) Heavy-tailed distribution

FIGURE 6.3 Normal Probability Plots

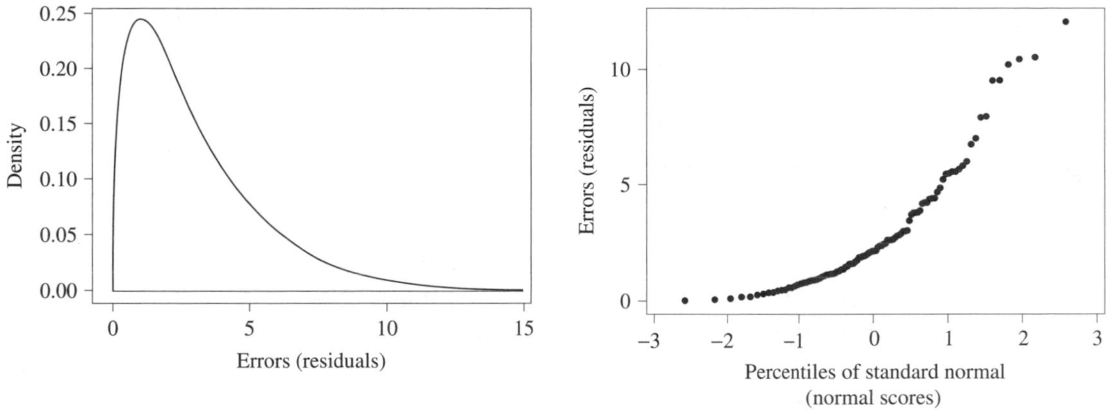

(d) Skewed distribution

FIGURE 6.3 (Continued)

it is likely that the errors are serially correlated. **Serial correlation** or **autocorrelation** indicates that errors at adjacent time periods are correlated. Many business and economic data are positively autocorrelated; that is, a positive error last month implies a similar positive error for this month. A detailed discussion of autocorrelations and models for autocorrelation is given in Chapter 10. There, we also show that standard inference procedures derived under the independence assumption (and that ignore serial correlation) can have a major effect on the standard errors of the regression coefficients and that the associated significance tests may be misleading. Hence, a check for serial correlation among the errors is of special importance if regression models are fit to time series data. Since the case index i reflects time, we replace it by t.

A straightforward approach to check for the serial correlation among the errors of a regression model is to focus on the residuals and calculate the lag k **sample autocorrelation** of the residuals,

$$r_k = \frac{\sum_{t=k+1}^{n} e_t e_{t-k}}{\sum_{t=1}^{n} e_t^2}, \quad k = 1, 2, \ldots \tag{6.13}$$

We always assume that an intercept is in the model, forcing the mean of the residuals to be zero, $\bar{e} = 0$. The autocorrelation in Eq. (6.13) is the sample correlation between e_t and its kth lag, e_{t-k}. Think about e_t as y_t and e_{t-k} as x_t, and write down the usual correlation coefficient between x_t and y_t. Ignoring "end" effects, the correlation coefficient simplifies to the expression in Eq. (6.13). The lag k autocorrelation is always between -1 and $+1$; it measures the association within the same series (residuals) k steps apart. Hence, its name, **auto**correlation. If the errors in the regression model are uncorrelated, then it can be shown that

$$E(r_k) \cong 0 \quad \text{and} \quad V(r_k) \cong \frac{1}{n} \quad \text{for} \quad k > 0$$

In addition, for reasonably large sample size n, the distribution of r_k is approximately normal. A simple check for serial correlation compares r_k with its standard error $\frac{1}{\sqrt{n}}$.

We can graph the autocorrelations r_k as a function of the lag k. We refer to such a graph as the **autocorrelation function** of the residuals. Note that by definition $r_0 = 1$. Two horizontal bands at twice the standard error, $\pm 2/\sqrt{n}$, are usually added to the graph. Sample autocorrelations that are outside the limits are indications of autocorrelation. On the other hand, if all (or most) of the autocorrelations are within these limits, the assumption of independent errors can be adopted. A word of caution must be added here. This approach involves multiple significance tests at the 5% level and not just a single test for which the significance level would be appropriate. Hence, the results should not be taken too literally. Autocorrelations that are barely outside the limits at higher lags and are difficult to explain should not be taken as conclusive evidence of autocorrelation.

The **Durbin–Watson test** statistic examines the lag 1 autocorrelation r_1 in more detail. It is given by

$$D = \frac{\sum_{t=2}^{n}(e_t - e_{t-1})^2}{\sum_{t=1}^{n} e_t^2} = \frac{\sum_{t=2}^{n} e_t^2 + \sum_{t=2}^{n} e_{t-1}^2 - 2\sum_{t=2}^{n} e_t e_{t-1}}{\sum_{t=1}^{n} e_t^2} \approx 2(1 - r_1) \qquad (6.14)$$

In the last step of Eq. (6.14), we have ignored end effects of the sum in the numerator. The distribution of this statistic under independent errors can be derived. It is complicated, and special tables of critical values are needed to carry out the test. For independent errors, the Durbin–Watson test statistic is approximately 2; for correlated errors the test statistic is either smaller or larger than 2. The special tables of critical values tell us how far from 2 the statistic must be before one can conclude that independence is violated. Note that the Durbin–Watson statistic examines just the lag 1 autocorrelation. Autocorrelations at lags higher than 1 are ignored. Hence, one should always supplement this test with a graph of the sample autocorrelation function of the residuals and also examine autocorrelations at higher lags.

Another simple graphical procedure is to plot e_t versus e_{t-k} and assess visually whether there is any apparent relationship between these variables. If the plot indicates a random scatter, then there is no correlation among errors k steps apart. On the other hand, if the plot shows an association, then there is evidence of serial correlation. Such scatter plots for lag $k = 1$ are shown in Figure 6.4. No serial correlation at lag 1 is indicated in Figure 6.4(a), whereas Figures 6.4(b) and 6.4(c) indicate positive and negative lag 1 serial correlations, respectively.

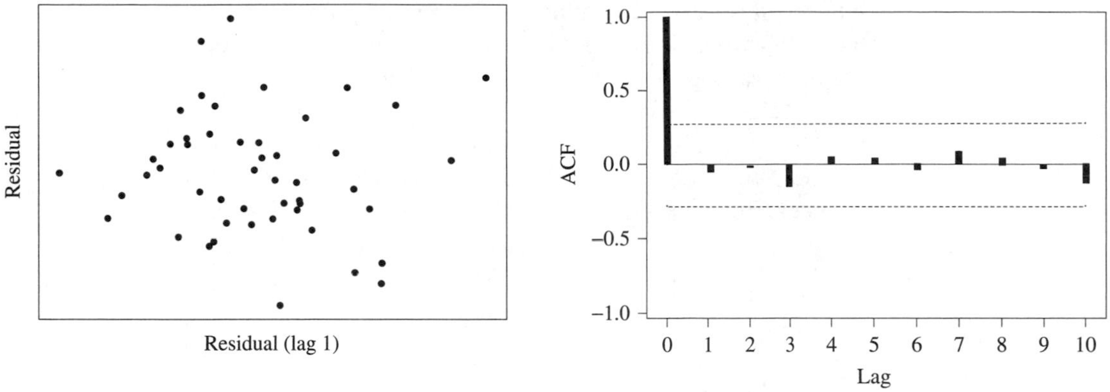

(a) No Autocorrelation

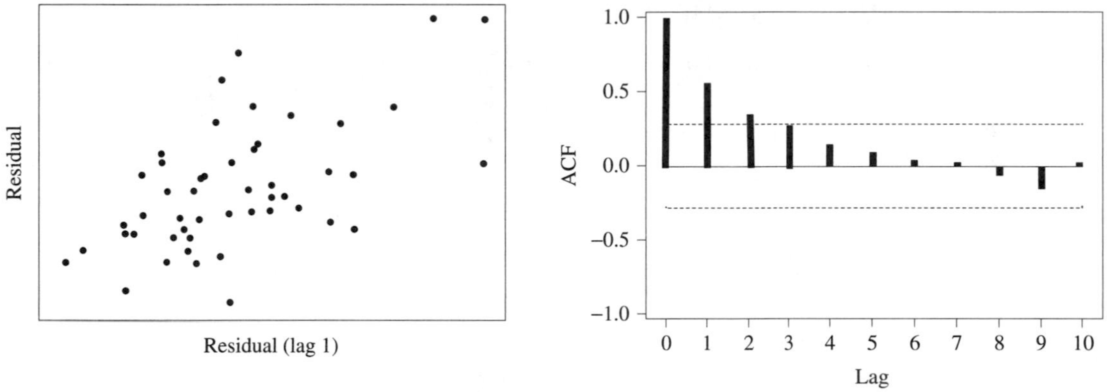

(b) Positive Autocorrelation

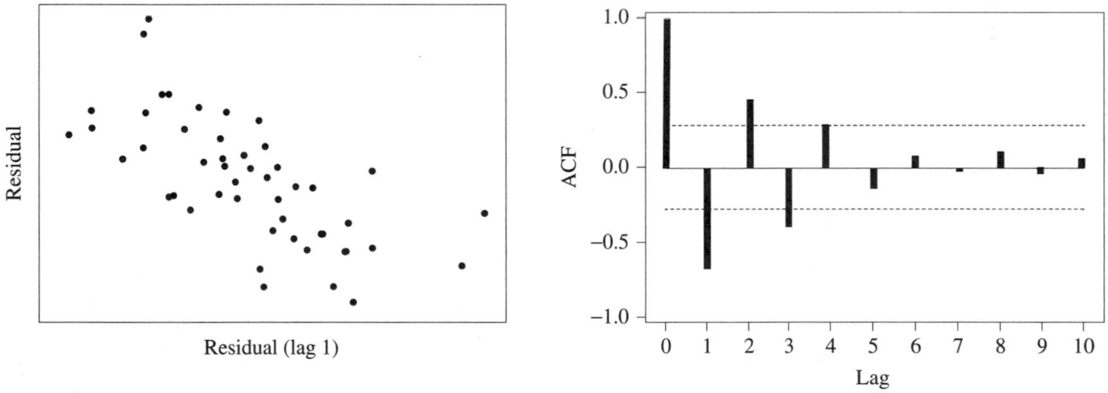

(c) Negative Autocorrelation

FIGURE 6.4 Plots of e_t against e_{t-1} for Three Generated Series and the Associated Autocorrelation Functions

We have not shown scatter plots of e_t versus e_{t-k} for $k > 1$. The sample autocorrelation functions in the right panels of Figures 6.4(b) and 6.4(c) indicate serial correlations at higher lags ($k > 1$).

The calculations of the Durbin–Watson statistic and the autocorrelation function of the residuals are called for if the ordering of the cases is meaningful—that is, if the case index i stands for time or run order. These statistics are not appropriate in the cross-sectional situation because the arrangement of the cases is arbitrary. For example, in a regression that involves data from the 50 U.S. states, it is arbitrary whether the states are ordered alphabetically, by the length of their names, or by the number of vowels. The Durbin–Watson test and the autocorrelation function of the residuals have no place in such analysis. For example, these statistics are not meaningful in the context of the urea formaldehyde foam insulation (UFFI) data.

6.2.5 EXAMPLE: RESIDUAL PLOTS FOR THE UFFI DATA

Residual plots for this data set are given in Figure 6.5. The plots do not show any systematic patterns in the residuals. The normal probability plot indicates an approximate straight line pattern. Thus, it seems that all model assumptions are satisfied.

6.3 THE EFFECT OF INDIVIDUAL CASES

So far, we have considered methods that assess the global adequacy of our fitted model with respect to the form of the model and its error structure. The next step of model criticism examines the question whether all observations arise from the same model. At the same time, we will check whether some of the observations have an unduly large influence on the fit of the model.

6.3.1 OUTLIERS

An outlying case is defined as a particular observation $(y_i, x_{i1}, \ldots, x_{ip})$ that differs from the majority of the cases in the data set. Since several variables are involved in each case, one must distinguish among outliers in the y (response) dimension and outliers in the x (covariate) dimension.

Outliers in the x dimension are cases that have unusual values on one or more of the covariates. Since we do not attempt to model the x variables, it is just their detection that matters. If there is only one dimension, a simple dot diagram or histogram of the values of the single covariate will reveal outliers. If there are two covariates, a scatter plot of one variable against the other can identify unusual cases. Note that a case may not be aberrant when looking at each dimension individually but may be quite unusual if both dimensions are considered simultaneously. Of course, the detection becomes tricky if more than two dimensions are involved.

FIGURE 6.5
Residual Plots for the UFFI Data

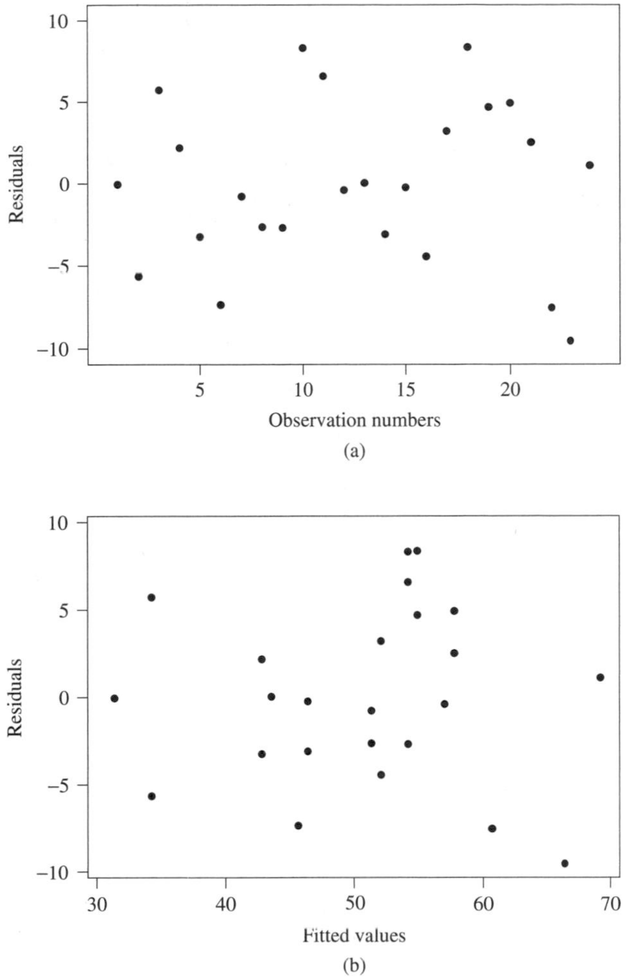

Outliers in the y dimension are linked to the regression model as one tries to explain the response as a function of the covariates. Outliers in the y dimension may be due to many different reasons:

i. The random component in the regression model may be unusually large.

ii. The response y or the covariates x may have been recorded incorrectly. The x variables may be measured correctly, but the y value may be incorrect. Also, the x variables may be recorded incorrectly, in which case the value of y, which would be reasonable for the correct values of x, becomes unreasonable.

iii. It could be that both y and x variables are correct, but that there is another covariate that is missing from the model and that can explain the "strange" observation.

FIGURE 6.5 (Continued)

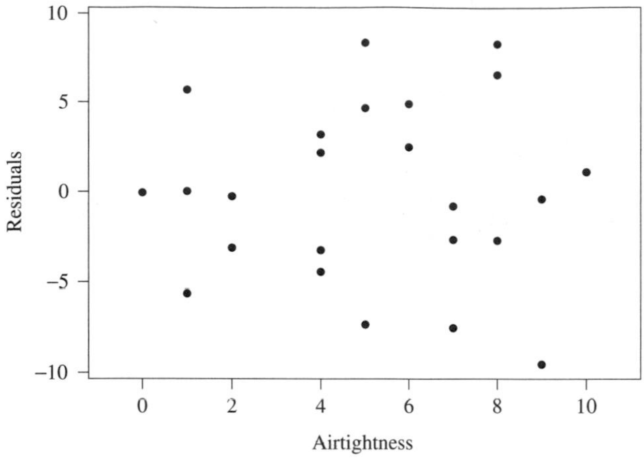

(c)

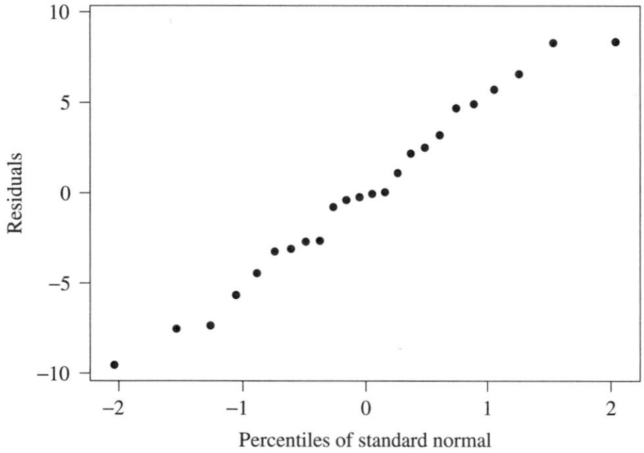

(d)

With the correct tools it is not too difficult to spot outliers. The tools that are discussed in this section will help us do this. However, finding the causes of outliers and deciding what to do with them is a more difficult issue. Careful detective work and consultation with individuals who are involved with the collection of the data are essential. The treatment of outliers that are due to obvious misrecordings is straightforward; we omit such points from the analysis if they cannot be corrected. However, the strategy is more difficult if causes and reasons cannot be found. A conservative approach is to fit the model twice: once with all cases and once with the outlying case removed. If the conclusions are unaffected, then the case in question does not cause difficulties because it does not matter whether the suspect case is kept as part of the data. If the conclusions are changed greatly, any statements originating from the data set must be very tentative. If possible, further data should be collected. Outliers should not automatically be

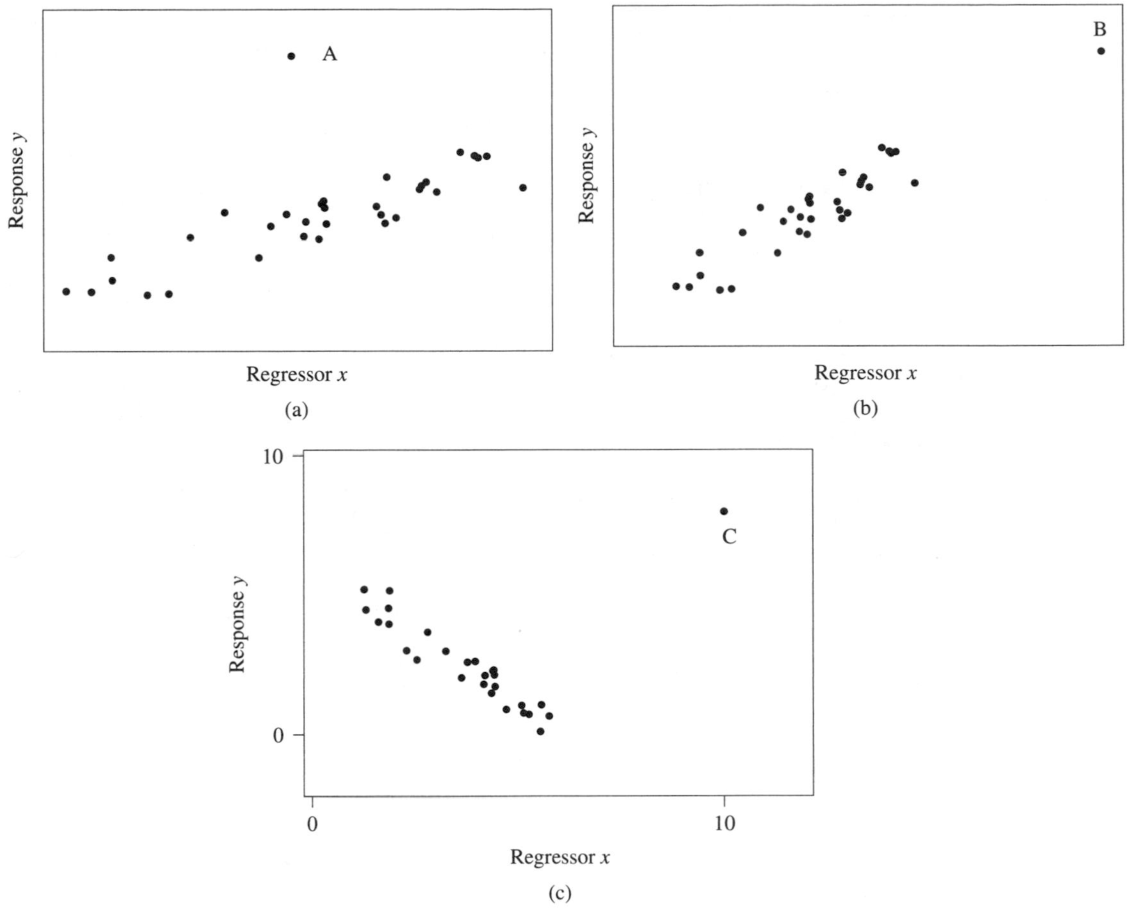

FIGURE 6.6 Scatterplots Showing Three Types of Outliers

thought of as nuisances. Often, unusual outlying cases are very interesting and informative. A study of the causes of unusual values can reveal a great deal about the situation under study.

Let us illustrate the issue of outliers graphically within the context of the simplest situation with one response y and one covariate x. Consider the scatter plots in Figure 6.6. We can spot three unusual observations.

Case A is not unusual with regard to its covariate (x); its x value is in the center of the observed covariate region. It is the response corresponding to this setting of x that is highly unusual. Case A represents an outlier in the y dimension.

Consider the other two cases, B and C. These two cases are unusual with regard to the covariate. Their x values are very different from all others; they are much larger. Cases B and C are outliers with respect to the x dimension. What about the y dimensions for these two cases? Assume that the model—derived from the majority of the cases—is appropriate. Then the response for case B is right on the "model trajectory." Hence, there is no evidence that this point is outlying in the y dimension. Case B is outlying in only the x dimension. Case C is different because its response y is far from the trajectory determined by the majority of the observations. Hence, case C is outlying in the x as well as the y dimension.

Many complicated procedures for detecting outliers have been devised. A simple first step involves the studentized residuals in Eq. (6.5),

$$d_i = e_i / s\sqrt{1 - h_{ii}}$$

which follow an approximate $N(0,1)$ distribution as long as the model assumptions are satisfied. Large values of d_i, for example, $|d_i| > 2.5$, are unexpected and are indicative of aberrant behavior in the response (y) dimension. A histogram or a dot plot of the studentized residuals will show such cases. These cases are also indicated in the normal probability plot as points with unusually low and high values on the y axis that fail to fall within the expected straight line pattern. However, this simple diagnostic tool—graphs of the studentized residuals—is not without problems. As the following discussion shows, a case may be quite peculiar and nevertheless have a small studentized residual.

Case A in Figure 6.6 would certainly be flagged on such a residual graph. Case B, on the other hand, would show a rather unremarkable residual.

6.3.2 LEVERAGE AND INFLUENCE MEASURES

We say that an individual case has a major influence on a given statistical procedure if the conclusions of the analysis are significantly altered when the case is omitted from the analysis. Influence depends on the statistical procedure that one has in mind. The focus could be the fitted line, a particular parameter estimate, or all parameter estimates considered as a group. Consider case C in Figure 6.6. It is an influential case because the fitted line is radically different if this case is omitted from the analysis. However, case B is not influential since the fitted line does not change when it is omitted from the analysis. As another example, consider the graph in Figure 6.7, in which we can see one outlying case with a very different value for the covariate. When fitting a straight line model without the case (Figure 6.7b) the fitted regression line has a negative slope. With the case (Figure 6.7a) the fitted least squares line is "attracted" by the response of the outlying case, the line gets "pulled" up, and the slope becomes positive.

Leverage

Before we can define influence measures, we need to discuss the concept of leverage, the "pull" that individual cases exert. Recall that the vector of fitted values can be written as a linear transformation of the response vector,

$$\hat{\mu} = X(X'X)^{-1}X'\mathbf{y} = H\mathbf{y}$$

with the hat matrix providing the coefficients. The ith fitted value can be written as

$$\hat{\mu}_i = h_{ii}y_i + \sum_{j \neq i} h_{ij}y_j \tag{6.15}$$

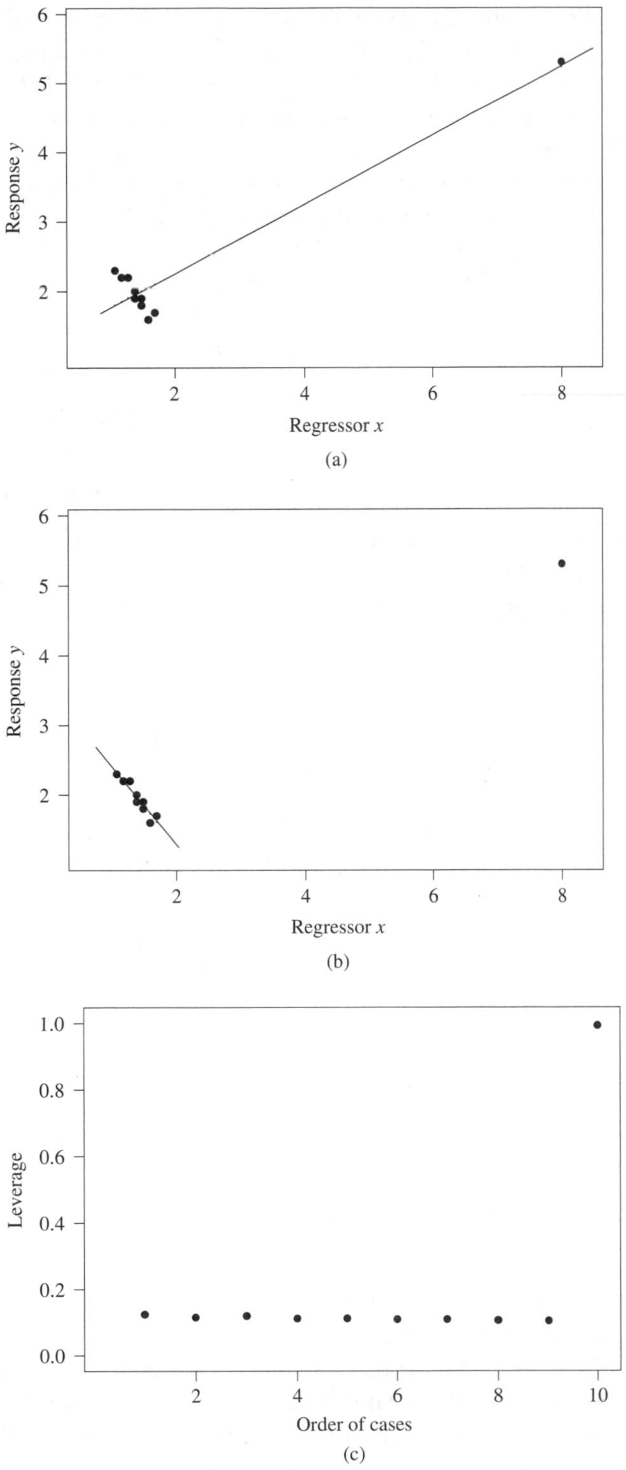

FIGURE 6.7 Plots Illustrating an Influential Case. Data Are in Table 6.1

The weight h_{ii} indicates how heavily y_i contributes to the fitted value $\hat{\mu}_i$. If h_{ii} is large (compared to other h_{ij}'s), then $h_{ii}y_i$ dominates $\hat{\mu}_i$. Recall that the residual e_i has variance $V(e_i) = \sigma^2(1 - h_{ii})$, and hence $h_{ii} \leq 1$. If h_{ii} is close to one, then $V(e_i) \cong 0$ and $\hat{\mu}_i \cong y_i$. This implies that the fitted model will pass very close to the data point $(y_i, x_{i1}, \ldots, x_{ip})$. We say that case i exerts high leverage on the fitted line, in the sense that the fitted value $\hat{\mu}_i$ is attracted to the response of the ith case. We refer to h_{ii}, the ith diagonal element of the hat matrix, as the **leverage** of case i. Large values of h_{ii} indicate large leverage.

Let us establish a few useful properties of leverage:

i. The leverage h_{ii} is a function of the covariates but does not include the response vector $\boldsymbol{y}$.

ii. $\frac{1}{n} \leq h_{ii} \leq 1$. The result $h_{ii} \leq 1$ was shown earlier. The fact that $h_{ii} \geq 1/n$ is shown in Exercise 6.4.

iii. The leverage h_{ii} is small for cases with $x_{i1}, \ldots, x_{ip}$ near the centroid $(\bar{x}_1, \ldots, \bar{x}_p)$ that is determined from all cases. The leverage h_{ii} is large when $(x_{i1}, \ldots, x_{ip})$ is far from the centroid. Take the case of a single explanatory variate, $p = 1$. In this case,

$$h_{ii} = \frac{1}{n} + \frac{(x_i - \bar{x})^2}{\sum_{j=1}^{n}(x_j - \bar{x})^2}$$

The leverage is smallest if $x_i - \bar{x} = 0$, and it is large if x_i is far from $\bar{x}$.

iv.

$$\sum_{i=1}^{n} h_{ii} = \text{tr}[H] = \text{tr}[X(X'X)^{-1}X'] = \text{tr}[X'X(X'X)^{-1}] = \text{tr}[I_{p+1}] = p + 1$$

where tr$[A]$ is the trace of a matrix A.

The average leverage in a model with $(p + 1)$ regression parameters (column dimension of the X matrix) is

$$\bar{h} = \frac{\sum_{i=1}^{n} h_{ii}}{n} = \frac{p+1}{n}$$

A comparison of the leverage of a particular case with its average provides a simple rule for spotting cases with unusual leverage. A case for which the leverage exceeds twice the average, that is $h_{ii} > 2\bar{h} = \frac{2(p+1)}{n}$, is usually considered a high-leverage case. Cases with high leverage need to be identified and examined carefully. High leverage may have many reasons; the case may include misrecorded covariates, or the case may reflect a design point that has been selected very differently from the rest. Low-leverage cases will not influence the fit much because they do not have much pull on the fitted model. High leverage is a prerequisite for making a case a high influence point, but as the following section shows, not every high-leverage case is influential.

TABLE 6.1 LEVERAGE FOR A SMALL DATA SET WITH ONE COVARIATE

x	1.1	1.3	1.2	1.4	1.4	1.5	1.5	1.6	1.7	8.0
y	2.3	2.2	2.2	1.9	2.0	1.9	1.8	1.6	1.7	5.3
Leverage	0.12	0.12	0.12	0.11	0.11	0.11	0.11	0.11	0.10	0.99

Let us consider the $n = 10$ cases in Table 6.1. These data are used in the plots of Figure 6.7. The case with $x = 8$ has the highest leverage,

$$h_{ii} = \frac{1}{10} + \frac{(8 - 2.07)^2}{\sum_{j=1}^{10}(x_j - 2.07)^2} = 0.99$$

The value 8 is far from the average $\bar{x} = 2.07$. The average leverage is $2/10 = 0.20$, and twice the average leverage is 0.40. The leverage of the case with $x = 8$ is 0.99 and is much larger than 0.40; hence, this case is flagged as having large leverage.

Cook's Influence Measure

One way to express influence is to study how the deletion of a case affects the parameter estimates. Suppose we delete the ith case and fit the regression model $\mathbf{y} = X\boldsymbol{\beta} + \boldsymbol{\epsilon}$ to the remaining $n - 1$ cases. Let $\hat{\boldsymbol{\beta}}_{(i)}$ denote the estimate of $\boldsymbol{\beta}$ without the ith case and $\hat{\boldsymbol{\beta}}$ the estimate with all cases. Then the change in $\hat{\boldsymbol{\beta}}$ that is due to deleting the ith case, $\hat{\boldsymbol{\beta}} - \hat{\boldsymbol{\beta}}_{(i)}$, is a good measure of the influence of the ith case on the vector of parameter estimates. We can calculate this change for all $i = 1, 2, \ldots, n$. At first, it appears as if one would need to calculate the least squares estimates for $n + 1$ data sets: the data set with all cases and the n data sets that are missing exactly one of the cases. However, this is not true because one can show (see the appendix) that

$$\hat{\boldsymbol{\beta}}_{(i)} = \hat{\boldsymbol{\beta}} - \left(\frac{e_i}{1 - h_{ii}}\right)(X'X)^{-1}\mathbf{x}_i \tag{6.16}$$

where $\mathbf{x}_i'$ is the ith row of X that corresponds to the deleted case. Hence, the difference in the estimates is given by

$$\hat{\boldsymbol{\beta}} - \hat{\boldsymbol{\beta}}_{(i)} = \frac{e_i}{1 - h_{ii}}(X'X)^{-1}\mathbf{x}_i \tag{6.17}$$

All n differences can be obtained as a by-product of a single regression. All one needs to store is the residuals $\mathbf{e} = \mathbf{y} - X\hat{\boldsymbol{\beta}}$ and the leverages h_{ii}. Note that the expression in Eq. (6.17) represents a vector of $p + 1$ components. Large changes in any component make the ith case influential. It is often useful to condense the vector of changes into a single number. The magnitude of the change in the estimates needs to be assessed by comparing the change to the inherent variability of the estimates. If the inherent variability is large, one should not get overly concerned. We know that an estimate of the covariance matrix of the estimates is given by $V(\hat{\boldsymbol{\beta}}) = s^2(X'X)^{-1}$. Using the inverse of the covariance matrix as weights and standardizing the result by the number of parameters leads to the summary measure

$$D_i = (\hat{\beta}_{(i)} - \hat{\beta})'(X'X)(\hat{\beta}_{(i)} - \hat{\beta})/(p+1)s^2 \tag{6.18}$$

Substituting from Eq. (6.17) into this expression yields

$$D_i = \frac{e_i^2 \mathbf{x}_i'(X'X)^{-1}\mathbf{x}_i}{(1-h_{ii})^2(p+1)s^2} = \frac{h_{ii}d_i^2}{(1-h_{ii})(p+1)} \tag{6.19}$$

where d_i is the studentized residual. The expression in Eq. (6.18) is known as **Cook's D statistic**, and it is standard output of most regression programs.

What does it take to get a sizeable influence measure D_i? One needs both a large leverage h_{ii}, or ratio $h_{ii}/(1-h_{ii})$, and a large studentized residual d_i. Large leverage alone does not do it. If the residual is small (and the response at the high-leverage case is on the overall model trajectory), then the influence will be small also. A large residual alone does not do it either. If the case has no leverage, the influence will be small. Thus, a large leverage and a large studentized residual are needed to make a case influential. Of course, the influence is zero if $\hat{\beta}_{(i)} = \hat{\beta}$.

Cook's D statistic can alternatively be written as

$$D_i = \frac{(X\hat{\beta} - X\hat{\beta}_{(i)})'(X\hat{\beta} - X\hat{\beta}_{(i)})}{(p+1)s^2} = \frac{(\hat{\mu} - \hat{\mu}_{(i)})'(\hat{\mu} - \hat{\mu}_{(i)})}{(p+1)s^2} \tag{6.20}$$

where $\hat{\mu} = X\hat{\beta}$ is the vector of fitted values that is determined from all cases, and $\hat{\mu}_{(i)} = X\hat{\beta}_{(i)}$ is the vector of fitted values with the ith case deleted; the ith element of $\hat{\mu}_{(i)}$ is the "out-of-sample" prediction of y_i.

PRESS Residuals

A related quantity of interest is the **prediction error**

$$e_{(i)} = y_i - \hat{y}_{(i)}, \quad i = 1, 2, \ldots, n \tag{6.21}$$

where $\hat{y}_{(i)} = \mathbf{x}_i'\hat{\beta}_{(i)}$ is the prediction of y_i that is obtained from fitting the model without the ith case. The $e_{(i)}$'s are called the PRESS residuals since the sum of squares of these residuals $\sum_{i=1}^{n} e_{(i)}^2$ is referred to as the prediction error sum of squares (PRESS). Using Eqs. (6.21) and (6.16), we can write

$$\begin{aligned} e_{(i)} &= y_i - \mathbf{x}_i'\hat{\beta}_{(i)} = y_i - \mathbf{x}_i'\left[\hat{\beta} - \frac{e_i(X'X)^{-1}\mathbf{x}_i}{(1-h_{ii})}\right] \\ &= e_i + \frac{e_i h_{ii}}{(1-h_{ii})} = \frac{e_i}{(1-h_{ii})} \end{aligned} \tag{6.22}$$

The computation of the PRESS residuals is easy; only the residuals e_i and the leverages $h_{ii} (i = 1, 2, \ldots, n)$ from the regression on the complete data set are required.

The DEFITS Statistics

Another measure that expresses the effect of deleting the ith case compares the ith fitted value $\hat{\mu}_i$ to the prediction $\hat{y}_{(i)}$ of y_i that is obtained from fitting the model to the data set without the ith case. This can be written as

$$\hat{\mu}_i - \hat{y}_{(i)} = \left(y_i - \hat{y}_{(i)}\right) - (y_i - \hat{\mu}_i) = e_{(i)} - e_i = \frac{e_i}{(1 - h_{ii})} - e_i = \frac{e_i h_{ii}}{(1 - h_{ii})} \qquad (6.23)$$

The change is small if either e_i is small or h_{ii} is near 0. Note that the difference in Eq. (6.23) compares the fitted value $\hat{\mu}_i$ to the prediction $\hat{y}_{(i)}$, whereas the PRESS residual in Eq. (6.21) compares the observation y_i to $\hat{y}_{(i)}$. We define the standardized difference (also referred to as **delta fits**) as

$$\begin{aligned} \text{DEFITS}_i &= \frac{\left(\hat{\mu}_i - \hat{y}_{(i)}\right)}{\left[s^2_{(i)} h_{ii}\right]^{1/2}} = \frac{e_i h_{ii}}{(1 - h_{ii})\left(s^2_{(i)} h_{ii}\right)^{1/2}} \\ &= \left[D_i (p+1) s^2 / s^2_{(i)}\right]^{1/2} \end{aligned} \qquad (6.24)$$

where

$$s^2_{(i)} = \frac{\sum_{j \neq i} \left(y_j - \mathbf{x}'_j \hat{\beta}_{(i)}\right)^2}{(n - p - 2)}$$

is the unbiased estimator of σ^2 without the ith case. We show in the appendix that it can be written as

$$s^2_{(i)} = \frac{\left[(n - p - 1)s^2 - e_i^2/(1 - h_{ii})\right]}{(n - p - 2)} \qquad (6.25)$$

The DEFITS_i in Eq. (6.24) is a slight variant of D_i. The ratio $s^2/s^2_{(i)}$ will usually be small, and DEFITS_i approximates the square root of $(p + 1)D_i$.

Summary

The **leverage** h_{ii} is used to assess whether a case is unusual with regard to its covariates (the x dimension). One compares the leverage to twice the average leverage, $2(p + 1)/n$, and flags values that exceed this bound. High-leverage cases may have a large influence on the model fit. The **influence** of a case is evaluated by calculating how certain aspects of the model fit change when the case is omitted from the analysis. Usually, one looks at changes in individual components of the vector of estimates, at the Cook's distance as an overall measure involving all components of the parameter vector, and at differences between predicted values that are obtained without the use of the case and observed and fitted values. Cook's distance is by far the most popular measure. The question about a bound for Cook's D arises. How large must Cook's D be before one should start to get concerned? The issue of tests of significance is not straightforward. Furthermore, we believe that influence measures are best suited for comparative purposes, assessing whether the influence of one case is much larger than that of others. Nevertheless, cutoffs have been recommended in the literature. Values of D that are larger than 1 are certainly of great concern. Even values larger than 0.50 should be scrutinized.

Plots of the residuals (ordinary, studentized, and PRESS residuals) are quite useful because they draw attention to outliers in the y dimension. The difference between the ordinary and the PRESS residuals is that PRESS residuals omit the ith case when determining the implied (fitted) value. Studentizing the residuals helps because the residuals can be compared with the bounds ± 2. Studentized residuals, leverages, and Cook's distances are standard output of most computer packages.

6.3.3 EXAMPLE: FORESTRY DATA

The information in Table 6.2 (also given in the file **forestry**) represents a subset of a larger data set on the results of a forestry experiment. Several unimportant

TABLE 6.2 FORESTRY DATA

NA	HT	CAL	HTCAL
101.51	36.5	1.10	40.150
79.54	33.0	1.00	33.000
20.62	22.0	0.30	6.600
53.07	26.0	0.50	13.000
43.02	24.0	0.50	12.000
31.88	24.0	0.40	9.600
26.78	21.0	0.40	8.400
29.93	26.0	0.40	10.400
18.90	16.5	0.36	5.940
54.30	34.5	0.89	30.705
51.30	23.0	0.47	10.810
35.50	20.0	0.47	9.400
13.30	17.0	0.37	6.290
25.10	19.0	0.51	9.690
18.70	17.5	0.42	7.350
21.50	16.5	0.33	5.445
14.80	16.5	0.31	5.115
21.20	19.0	0.48	9.120
21.60	16.0	0.39	6.240
20.70	17.0	0.40	6.800
30.90	17.0	0.46	7.820
32.50	19.0	0.44	8.360
11.18	17.0	0.37	6.290
186.00	51.0	0.70	35.700
163.00	41.0	0.73	29.930
130.50	37.0	0.65	24.050
139.00	44.0	0.69	30.360
132.00	43.0	0.66	28.380
171.00	36.0	0.70	25.200
155.00	40.0	0.62	24.800
93.00	41.5	0.59	24.485
161.00	45.5	0.80	36.400
87.00	36.5	0.53	19.345
127.00	41.0	0.64	26.240
140.00	39.0	0.64	24.960

TABLE 6.3 S-PLUS STATEMENTS AND OUTPUT: MODEL (6.26) RESIDUAL ANALYSIS FOR FORESTRY DATA (READ ACROSS)

```
# Reading in the data as a matrix 'forest' from the file 'forestry.dat'
> forest<-matrix(scan('forestry.dat',multi.line = T),byrow = T,ncol = 4,nrow = 35)
> na<-forest[,1]
# Deleting column 1 from the matrix 'forest'
> xv<-forest[,-1]
> forestfit<-lsfit(xv,na)
# Residual Analysis: Leverages, Cook's distancs etc (some part is deleted)
> fordiag<-ls.diag(forestfit)
> fordiag
$hat
[1]   0.4484* 0.3087* 0.1005 0.0782 0.0761 0.0547 0.0470 0.0704 0.1013 0.1578
[11]  0.0610  0.0633  0.0848 0.1016 0.0585 0.1406 0.1748 0.0736 0.0824 0.0666
[21]  0.0714  0.0534  0.0848 0.2843 0.0738 0.0659 0.1032 0.0923 0.0610 0.0881
[31]  0.1236  0.2565  0.1358 0.0829 0.0726

$stud.res
[1] -0.630  -0.581  -1.061  0.141 -0.158 -0.806 -0.369 -1.397  0.192 -2.492*
[11] 0.453   0.382  -0.181  0.172  0.059  0.272 -0.110 -0.069  0.494  0.237
[21] 0.854   0.421  -0.286  0.044  1.544  0.990 -0.444 -0.495  3.915* 1.505
[31]-2.162   0.120  -1.041 -0.206  0.955

$cooks
[1]   0.0823  0.0385  0.0313 0.0004 0.0005 0.0095 0.0017 0.0358 0.0011 0.2491*
[11]  0.0034  0.0025  0.0008 0.0009 0.0001 0.0031 0.0007 0.0001 0.0056 0.0010
[21]  0.0141  0.0026  0.0019 0.0002 0.0455 0.0173 0.0058 0.0064 0.1703* 0.0525
[31]  0.1473  0.0013  0.0424 0.0010 0.0179
```

explanatory variables have been deleted. The objective of the experiment was to build a model to predict total needle area (NA) of a seedling from its caliper (CAL; a measure of trunk size), height (HT), and a derived measure HTCAL (product of HT and CAL) that is related to the volume of the trunk.

The first step is to fit a linear model

$$\mathrm{NA} = \beta_0 + \beta_1\mathrm{CAL} + \beta_2\mathrm{HT} + \beta_3\mathrm{HTCAL} + \epsilon \tag{6.26}$$

The fit produces fitted values, residuals, and various other quantities, such as leverages and Cook's D. Some of these are listed in Table 6.3, where "hat" refers to leverages (h_{ii}), "stud.res" to studentized residuals, and "cooks" to Cook's D statistics. The following observations can be made:

i. The leverages h_{11} and h_{22} are quite large (0.4484 and 0.3087, respectively; entries with an asterisk). Note that the average of the leverages $\bar{h} = (p+1)/n = 4/35 = 0.1143$ and $2\bar{h} = 0.2286$. Thus, h_{11} and h_{22} are larger than $2\bar{h}$. Looking at the raw data, this seems to be due to the large values for CAL. Upon further checking, it was found that these two values had been measured incorrectly. The first two observations were deleted in subsequent runs.

ii. The studentized residuals for cases 10 (−2.492) and 29 (3.915) are unusually large, as are their Cook's distances (entries with an asterisk).

Upon checking, it was found that the CAL measurement on case 10 was incorrect, and hence this case was also deleted. The measurements on case 29 were checked, found to be correct, and hence left unchanged.

The residuals e_i are plotted in Figure 6.8 against the fitted values and the three independent variables. A normal probability plot of the studentized residuals is also shown. Because this data set is cross-sectional and since the ordering of the cases is arbitrary, there is no reason to investigate a potential lack of independence and calculate the autocorrelation function of the residuals.

The following observations can be made:

i. From the plot of the residuals against the fitted values $\hat{\mu}_i$, it appears that the variance is larger for seedlings with larger needle area. This suggests fitting a model in terms of the logarithm of NA. We will explain this in more detail in Section 6.5.

ii. Plots of the residuals against the explanatory variables reveal no obvious patterns.

iii. The normal probability plot of the residuals has a single large value that fails to fall within the straight line pattern; the aberrant point corresponds to case 29.

Based on these diagnostics, we changed the model and considered

$$\ln(\text{NA}) = \alpha_0 + \alpha_1\text{CAL} + \alpha_2\text{HT} + \alpha_3\text{HTCAL} + \epsilon \tag{6.27}$$

Note that the regression coefficients α_0, α_1, α_2, and α_3 in this model are different from the β's in the model (6.26) since ln(NA) is considered in Eq. (6.27). Cases 1, 2, and 10 were deleted from the data set. Fitting the model (6.27) leads to the results in Table 6.4 (on p. 199). Only part of the output is shown here.

We also generated the usual residual plots (see Figure 6.9 on p. 197). The results in Table 6.4 and the plots in Figure 6.9 indicate the following:

i. All leverages are relatively small. Case 29 (case 32 in the earlier run) has the highest leverage of 0.4249. However, no problem with the measurements could be found. The studentized residual corresponding to this observation is small. A further run without this observation produced little change. Also, Cook's influence measure (0.1154) is unremarkable.

ii. The plot of the residuals against the fitted values shows that the log transformation has reduced the dependency of the variance on the level.

iii. Plots of the residuals against the three explanatory variables show no apparent patterns. The model form seems acceptable. The normal probability plot of the studentized residuals reveals no serious departures from normality.

FIGURE 6.8
Residual Plots from Model (6.26): Forestry Data

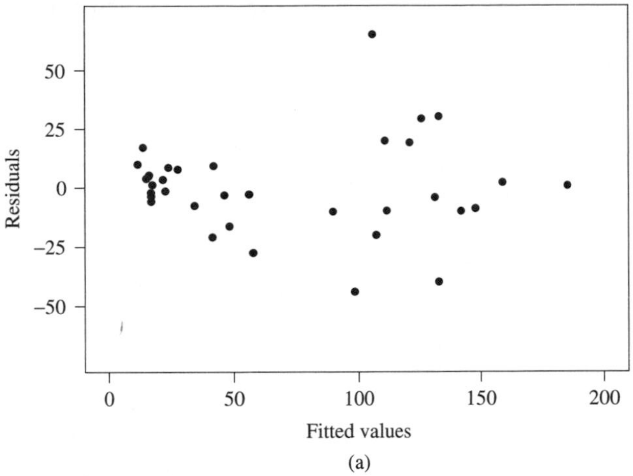

(a)

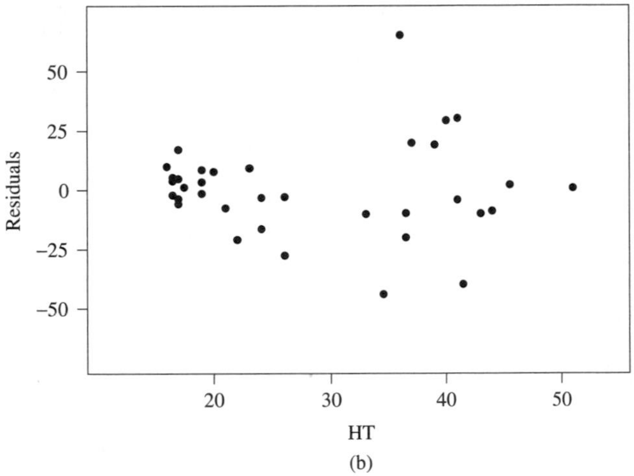

(b)

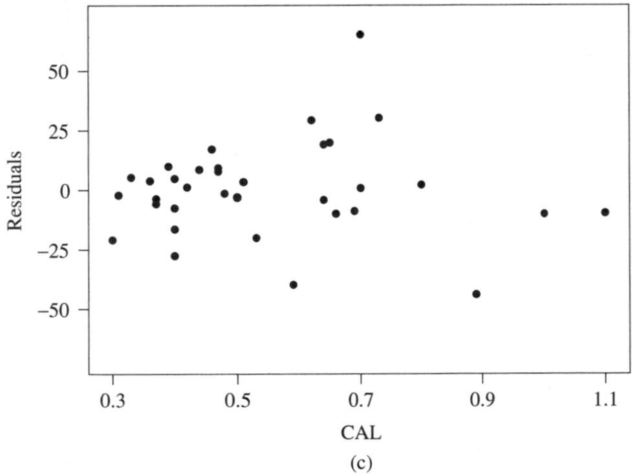

(c)

FIGURE 6.8 (Continued)

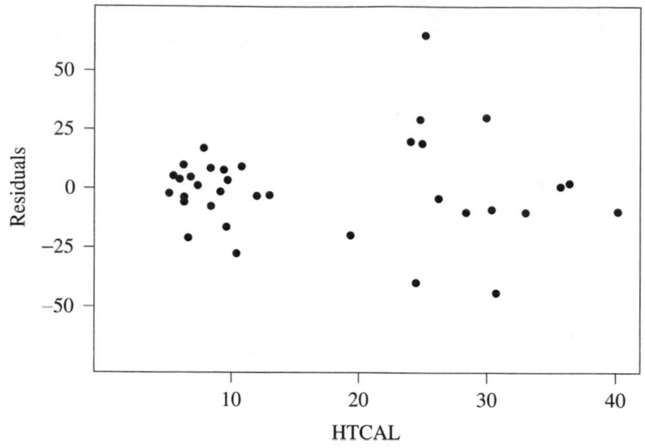

(d)

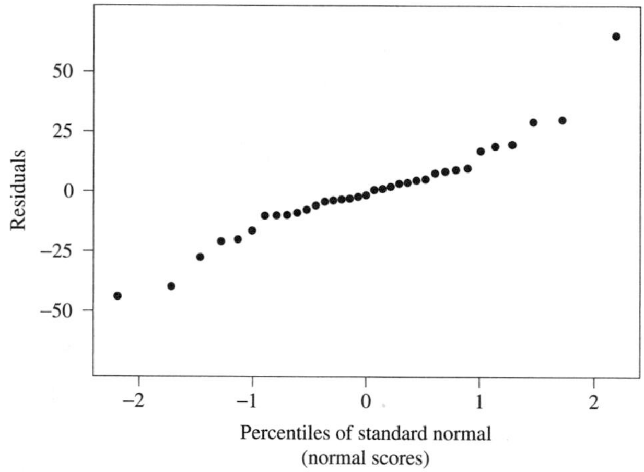

(e)

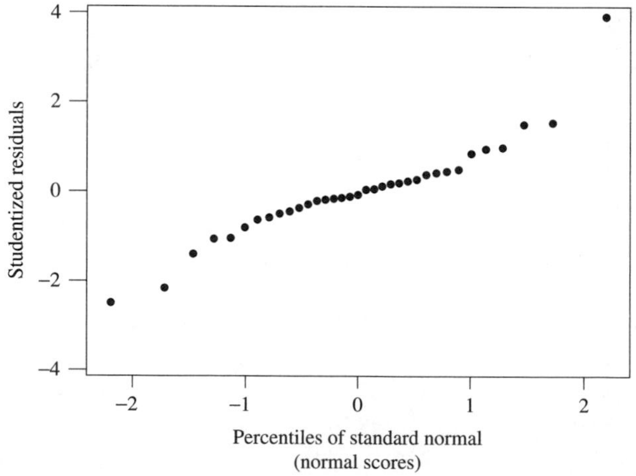

(f)

FIGURE 6.9
Residual Plots from Model (6.27): Adjusted Forestry Data

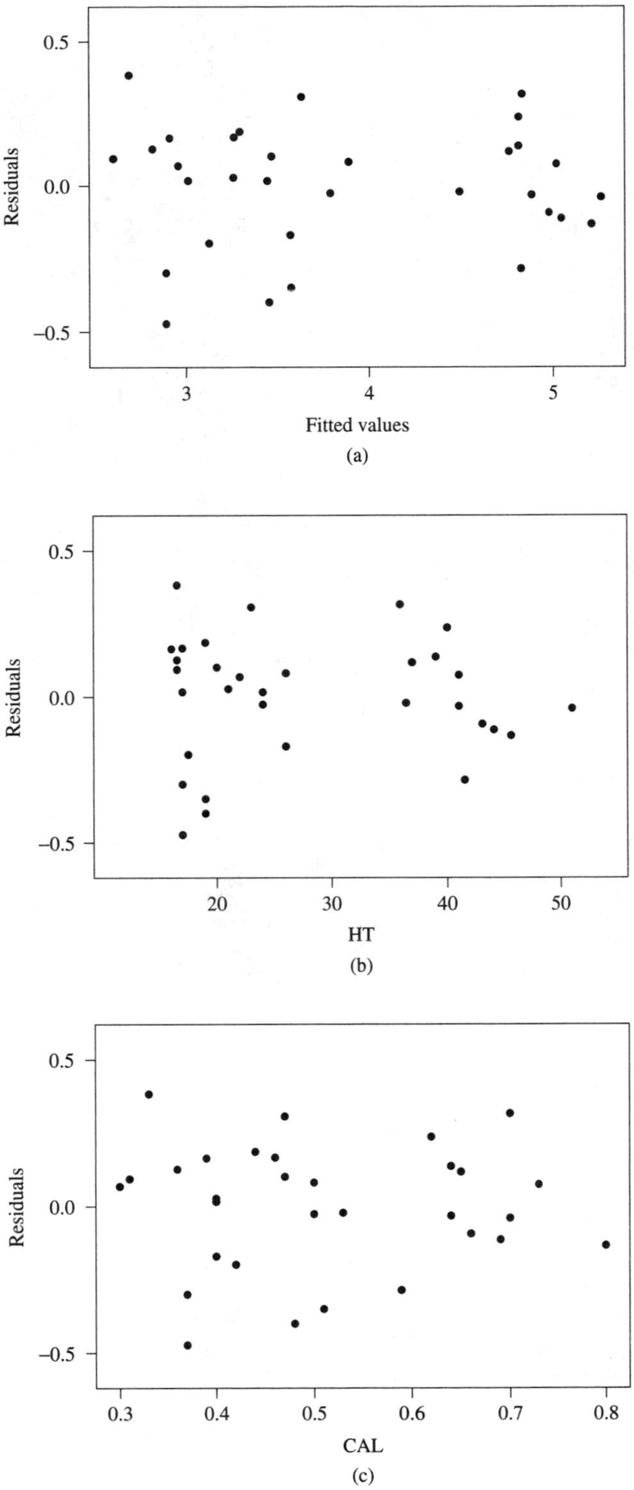

FIGURE 6.9 (Continued)

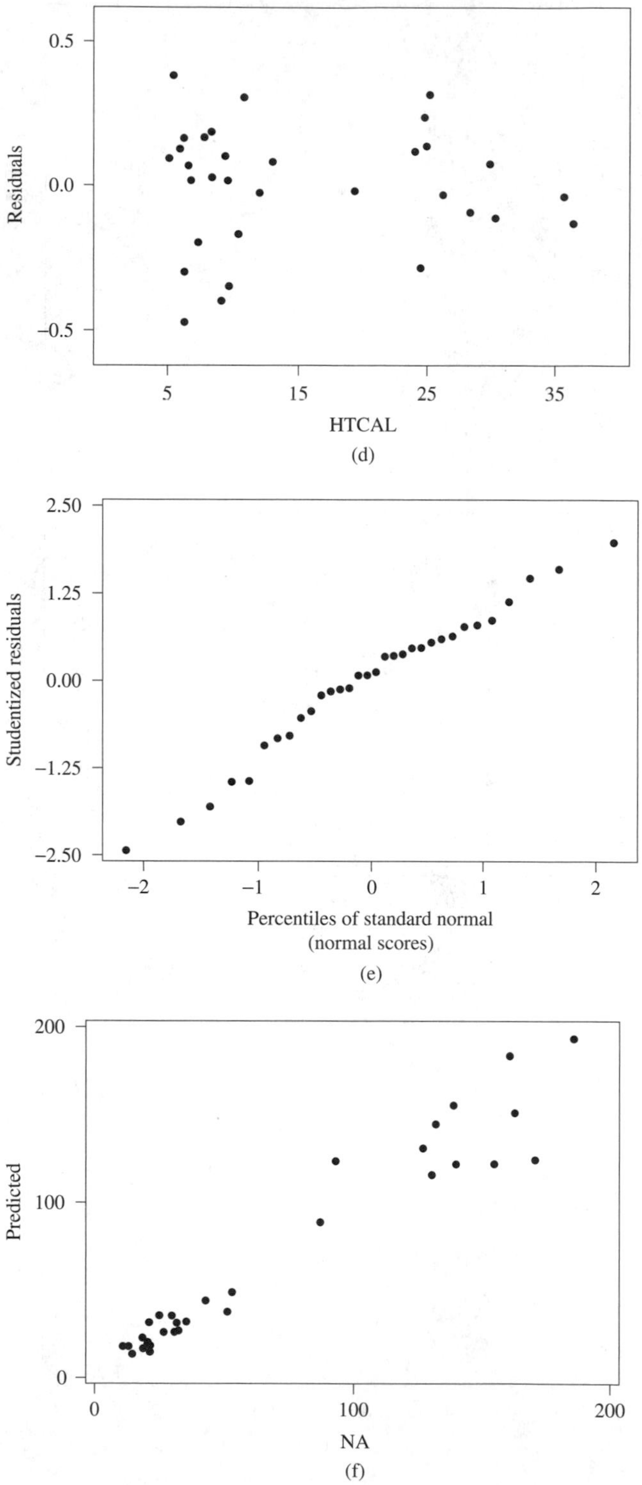

TABLE 6.4 PARTS OF S-PLUS OUTPUT: MODEL (6.27) ADJUSTED FORESTRY DATA

$hat

[1]	0.2258	0.0830	0.0852	0.0881	0.0532	0.1341	0.1033	0.0644	0.0798	0.0859
[11]	0.1655	0.0662	0.1509	0.1960	0.1066	0.0882	0.0703	0.1096	0.0608	0.0859
[21]	0.2958	0.1447	0.0814	0.1090	0.0926	0.1454	0.0923	0.1598	0.4249	0.1922
[31]	0.0839	0.0753								

$stud.res

[1]	0.350	0.383	−0.121	0.079	0.128	−0.823	0.602	1.470	0.474	−1.448
[11]	−1.805	−0.925	1.982	0.471	−2.020	0.778	0.081	0.799	0.869	−2.433
[21]	−0.208	0.360	0.553	−0.533	−0.434	1.598	1.134	−1.437	−0.785	−0.104
[31]	−0.149	0.640								

$cooks

[1]	0.0092	0.0034	0.0004	0.0002	0.0002	0.0265	0.0107	0.0357	0.0050	0.0474
[11]	0.1495	0.0152	0.1580	0.0139	0.1096	0.0148	0.0001	0.0199	0.0123	0.1183
[21]	0.0047	0.0057	0.0069	0.0089	0.0050	0.1028	0.0324	0.0946	0.1154	0.0007
[31]	0.0005	0.0085								

Coefficients:

	Value	Std. Error	t value	Pr(> |t|)
(Intercept)	−0.4465	0.6044	−0.7388	0.4662
ht	0.1051	0.0240	4.3823	0.0001
cal	6.0217	1.3007	4.6296	0.0001
htcal	−0.1082	0.0398	−2.7204	0.0111

Residual standard error: 0.2204 on 28 degrees of freedom
Multiple R-Squared: 0.9471
F-statistic: 167.1 on 3 and 28 degrees of freedom, the p-value is 0

iv. The t ratios in Table 6.4 indicate that on the log scale, all three explanatory variables are important.

v. The plot of $\exp\{\hat{\mu}\}$ against observed y (NA) in the last panel of Figure 6.9 shows that the model fits the data reasonably well and suggests that the model should produce reasonable predictions.

vi. One area for further exploration is suggested by the fact that the cases separate into two groups. In the original numbering, cases 24–35 are made on much larger trees. One should check whether the same model is appropriate for both groups of trees.

6.4 ASSESSING THE ADEQUACY OF THE FUNCTIONAL FORM: TESTING FOR LACK OF FIT

A formal test of model adequacy can be performed if one has repeated observations at some of the constellations of the explanatory variables. For simplicity, we consider the case of a single regressor variable first. Suppose that our data set includes replicated observations at some of the considered x values:

$$\begin{array}{ll} x_1: & y_{11}, y_{12}, \ldots, y_{1n_1} \\ x_2: & y_{21}, y_{22}, \ldots, y_{2n_2} \\ \vdots & \vdots \\ x_k: & y_{k1}, y_{k2}, \ldots, y_{kn_k} \end{array} \tag{6.28}$$

This means that at setting x_i, n_i response observations $y_{i1}, y_{i2}, \ldots, y_{in_i} (i = 1, 2, \ldots, k)$ are available. We assume that these are genuine replications and not just additional measurements on the same experiment.

We are interested in modeling the relationship between y and x. In regression, one typically uses residuals from a parametric model fit to obtain an estimate of σ^2. If the model is incorrect, then this estimate is not appropriate. With repeated observations at some of the covariate constellations, one has the oppportunity to calculate an estimate of σ^2 that does not depend on the assumed model. We can use the resulting information to construct a test of model adequacy.

The data in (6.28) resemble observations from k groups and can be characterized as

$$y_{ij} = \mu_i + \epsilon_{ij}, \quad i = 1, 2, \ldots, k, \ j = 1, 2, \ldots, n_i \tag{6.29}$$

with $E(\epsilon_{ij}) = 0$ and $V(\epsilon_{ij}) = \sigma^2$. This can be written in the linear model form

$$\boldsymbol{y} = X\boldsymbol{\beta} + \boldsymbol{\epsilon}$$

where $\boldsymbol{y}$ is the $n \times 1$ vector of responses ($n = \sum_{i=1}^{k} n_i$), $\boldsymbol{\beta} = (\mu_1, \mu_2, \ldots, \mu_k)'$ is the vector of unknown means, X is the $n \times k$ design matrix with ones and zeros representing the k groups, and $\boldsymbol{\epsilon}$ is the vector of errors satisfying the usual assumptions. This formulation is identical to the one in the k sample problem in Chapter 5 (Section 5.3).

The least squares estimate of the group means μ_i is given by

$$\hat{\boldsymbol{\beta}} = (\hat{\mu}_1, \ldots, \hat{\mu}_k)' = (\bar{y}_1, \bar{y}_2, \ldots, \bar{y}_k)'$$

where $\bar{y}_i = \sum_{j=1}^{n_i} y_{ij}/n_i$, $i = 1, 2, \ldots, k$ are the sample group means. The residual sum of squares is given by

$$S(\hat{\boldsymbol{\beta}}) = \sum_{i=1}^{k} \sum_{j=1}^{n_i} (y_{ij} - \bar{y}_i)^2 \tag{6.30}$$

It has $n - k$ degrees of freedom because we are estimating k parameters. This sum of squares is referred to as the **pure error sum of squares** (PESS). It is called pure error because it does not depend on any functional representation of the μ_i. Note that this is the same as the within sum of squares discussed in Section 5.3.

Now suppose that we entertain the parametric model $\mu_i = \beta_0 + \beta_1 x_i$ and fit the parameters in $y_{ij} = \beta_0 + \beta_1 x_i + \epsilon_{ij}$ by least squares. Minimizing $S(\boldsymbol{\beta}) = \sum_{i=1}^{k} \sum_{j=1}^{n_i} (y_{ij} - \beta_0 - \beta_1 x_i)^2$ with respect to β_0 and β_1, we obtain estimates $\hat{\beta}_0$ and $\hat{\beta}_1$ and can calculate the residual sum of squares $S(\hat{\boldsymbol{\beta}}_A)$. It is larger than $S(\hat{\boldsymbol{\beta}}) = \text{PESS}$ since the minimization is restricted. It involves only two parameters,

β_0 and β_1, compared to the k parameters (means) that led to $S(\hat{\beta})$. $S(\hat{\beta}_A)$ has $(n-2)$ degrees of freedom. The **additional sum of squares** is given by

$$S(\hat{\beta}_A) - S(\hat{\beta}) = \sum_{i=1}^{k}\sum_{j=1}^{n_i}(y_{ij} - \hat{\beta}_0 - \hat{\beta}_1 x_i)^2 - \sum_{i=1}^{k}\sum_{j=1}^{n_i}(y_{ij} - \bar{y}_i)^2 \quad (6.31)$$

It is useful to write $S(\hat{\beta}_A)$ in a slightly different form,

$$\begin{aligned} S(\hat{\beta}_A) &= \sum_{i=1}^{k}\sum_{j=1}^{n_i}(y_{ij} - \bar{y}_i + \bar{y}_i - \hat{\beta}_0 - \hat{\beta}_1 x_i)^2 \\ &= \sum_{i=1}^{k}\sum_{j=1}^{n_i}(y_{ij} - \bar{y}_i)^2 + \sum_{i=1}^{k} n_i(\bar{y}_i - \hat{\beta}_0 - \hat{\beta}_1 x_i)^2 \\ &= S(\hat{\beta}) + \sum_{i=1}^{k} n_i(\bar{y}_i - \hat{\beta}_0 - \hat{\beta}_1 x_i)^2 \end{aligned} \quad (6.32)$$

The sum of the cross-products is zero. The additional sum of squares,

$$S(\hat{\beta}_A) - S(\hat{\beta}) = \sum_{i=1}^{k} n_i(\bar{y}_i - \hat{\beta}_0 - \hat{\beta}_1 x_i)^2 \quad (6.33)$$

involves the squared distances between the group means and the linear fit. The weights n_i correspond to the number of observations at x_i. The additional sum of squares measures the lack of fit of the linear model. Hence, it is referred to as the **lack-of-fit sum of squares** (LFSS). It has $(n-2)-(n-k) = k-2$ degrees of freedom. The full model involves the k means, whereas the restricted model parameterizes the k means with two parameters as $\mu_i = \beta_0 + \beta_1 x_i$. The test of the restriction $\mu_i = \beta_0 + \beta_1 x_i$ is given by the F statistic

$$\begin{aligned} F &= \frac{\text{additional sum of squares}/(k-2)}{\text{residual sum of squares from the unrestricted model}/(n-k)} \\ &= \frac{[S(\hat{\beta}_A) - S(\hat{\beta})]/(k-2)}{S(\hat{\beta})/(n-k)} = \frac{\text{LFSS}/(k-2)}{\text{PESS}/(n-k)} \end{aligned} \quad (6.34)$$

If this statistic is small, then we cannot reject the parametric model. If it is large, we reject the model $\mu_i = \beta_0 + \beta_1 x_i$. The test in Eq. (6.34) assesses the adequacy of the model. Hence, it is also called a goodness-of-fit test. Under the hypothesis that the restriction is true (i.e., the true model is linear) this ratio is distributed as $F(k-2, n-k)$. Large values of F provide evidence against the hypothesis of linearity.

Example: Chemical Yield

An experiment was conducted to study the relationship between the yield from a chemical reaction (y) and the reaction temperature (x). All other factors were held constant. Table 6.5 lists data obtained from 12 runs.

A previous study with temperatures ranging from 60 to 90 °C suggested that the response (yield) is approximately linear in temperature,

$$y = \beta_0 + \beta_1 x + \epsilon \quad (6.35)$$

TABLE 6.5 CHEMICAL YIELD DATA

Reaction Temperature (°C), x	Yield (gs), y	$\bar{y}_i$
60	51	51.0
70	82,78	80.0
80	91,96	93.5
90	98,89,99	95.3
100	82,83	82.5
110	54,52	53.0

TABLE 6.6 ANOVA TABLE FOR CHEMICAL YIELD ($\hat{\mu} = 93.29 - 0.16x$)

Source	df	SS	MS	F
Model	1	69.31	69.31	0.20
Residual	10	3,393.60	339.36	
Lack-of-Fit	4	3,309.93	827.48	59.34
Pure Error	6	83.67	13.95	

The fitted regression line is given by

$$\hat{\mu} = 93.29 - 0.16x$$

and the analysis of variance (ANOVA) table is shown in Table 6.6.

The top part of Table 6.6 (with $F = 0.20$) gives the wrong impression that the variation in y cannot be explained by x. However, this F test assumes that the linear model in Eq. (6.35) is correct. The adequacy of this model can be checked with a lack-of-fit test, as we have replicated observations at some of the temperatures. The replications in the design give us the opportunity to test the model adequacy.

The PESS is given by

$$\begin{aligned} S(\hat{\beta}) = \text{PESS} &= (82-80)^2 + (78-80)^2 + (91-93.5)^2 + (96-93.5)^2 \\ &\quad + (98-95.3)^2 + (89-95.3)^2 + (99-95.3)^2 \\ &\quad + (82-82.5)^2 + (83-82.5)^2 + (54-53)^2 + (52-53)^2 \\ &= 83.67 \end{aligned}$$

and its degrees of freedom are $1+1+2+1+1 = 6$. Alternatively, one can get these degrees of freedom from the number of observations ($n = 12$) minus the number of groups ($k = 6$).

We already estimated the linear model and found that $S(\hat{\beta}_A) = 3{,}393.60$. Hence, the sum of squares due to lack of fit is

$$\text{LFSS} = S(\hat{\beta}_A) - S(\hat{\beta}) = \text{SSE} - \text{PESS} = 3{,}393.60 - 83.67 = 3{,}309.93$$

The degrees of freedom for lack of fit are 4 because there are six separate covariate constellations and two parameters in the linear model. These quantities are shown

FIGURE 6.10
Chemical Yield Data

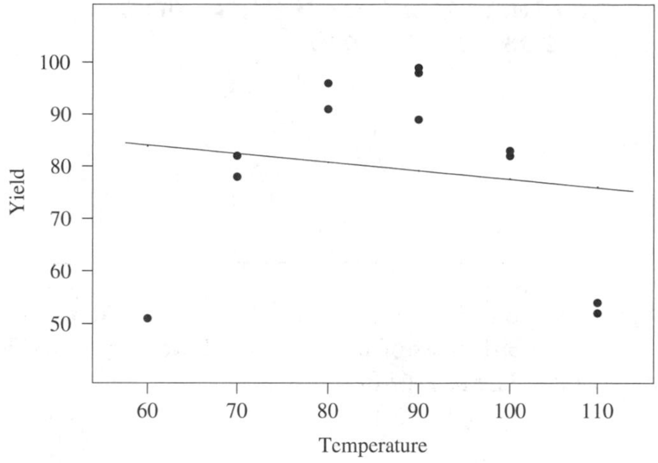

in the bottom half of Table 6.6. The F ratio

$$F = \frac{3{,}309.93/4}{83.67/6} = 59.34$$

is huge, much larger than any reasonable percentile of the $F(4, 6)$ distribution. The probability value

$$P(F(4, 6) \geq 59.34) \leq 0.001$$

is tiny. Hence, we reject the linear model. We have found serious lack-of-fit, which makes the earlier F test, $F = 0.20$, meaningless. One needs to develop a better model before one can test for a relationship with x.

One could have seen this from the graph of y against x. The graph in Figure 6.10 shows convincingly that a model linear in x is not appropriate. The numerical (lack-of-fit test) analysis alone does not show the nature of inadequacy. A scatter plot of the data reveals that the inadequacy comes from the curvlinear nature of the data in the range considered. We are trying to fit a straight line through a set of points that clearly indicate a quadratic pattern. Although it may have been appropriate to fit a linear model (in x) over the temperature range 60–90°C, such a model is certainly not appropriate over the wider range of 60–110°C.

Figure 6.10 leads us to modify the initial linear model in x and consider a quadratic one,

$$y = \beta_0 + \beta_1 x + \beta_2 x^2 + \epsilon \tag{6.36}$$

The fitted model is $\hat{\mu} = -422.98 + 12.17x - 0.07x^2$, and the ANOVA table is given in Table 6.7.

As before, a lack-of-fit test for the model Eq. (6.36) can be performed. The residual sum of squares from the quadratic model is $S(\hat{\beta}_A) = 89.01$; it has $12 - 3 = 9$ degrees of freedom. The pure error sum of squares is still $S(\hat{\beta}) = 83.67$ because we have the same constellations with replications on x and x^2. The sum

TABLE 6.7 ANOVA TABLE FOR CHEMICAL YIELD ($\hat{\mu} = -422.98 + 12.17x - 0.07x^2$)

Source	df	SS	MS	F
Model	2	3,373.90	1,686.95	170.57
Residual	9	89.01	8.89	
Lack-of-Fit	3	5.34	1.78	0.13
Pure Error	6	83.67	13.95	

of squares due to lack of fit is SSLF $= S(\hat{\beta}_A) - S(\hat{\beta}) = 89.01 - 83.67 = 5.34$, with $9 - 6 = 3$ degrees of freedom. The lower part of Table 6.7 shows this decomposition. The lack-of-fit test statistic

$$F = \frac{5.34/3}{83.67/6} = 0.13$$

is very small and not significant. Hence, there is no evidence to contradict the quadratic model.

Over the range 60–90°C, the linear model in x provides a reasonable approximation. However, if that model is used to predict the yield beyond the range from 60 to 90°C, for example at 110°C, the predictions will fall apart. Extrapolations to situations outside the range considered by the experiment can be very dangerous.

6.4.1 LACK-OF-FIT TEST WITH MORE THAN ONE INDEPENDENT VARIABLE

When there are two or more regressor variables, replicate measurements need to be in agreement on all regressors. We require replications at covariate constellations. The data should look like the following:

Constellation at $x_{11}, x_{12}, \ldots, x_{1p}$: n_1 replications $y_{11}, y_{12}, \ldots, y_{1n_1}$

Constellation at $x_{21}, x_{22}, \ldots, x_{2p}$: n_2 replications $y_{21}, y_{22}, \ldots, y_{2n_2}$

$\vdots$ $\vdots$

Constellation at $x_{k1}, x_{k2}, \ldots, x_{kp}$: n_k replications $y_{k1}, y_{k2}, \ldots, y_{kn_k}$

Corresponding to the ith constellation $(x_{i1}, x_{i2}, \ldots, x_{ip})$ we assume that we have n_i observations $y_{i1}, \ldots, y_{in_i}$ available.

The calculation of the pure error sum of squares $S(\hat{\beta}) = S(\hat{\mu}_1, \hat{\mu}_2, \ldots, \hat{\mu}_k)$ is unchanged. The restricted model is now the model with $(p+1) < k$ coefficients, $y_{ij} = \beta_0 + \beta_1 x_{i1} + \cdots + \beta_p x_{ip} + \epsilon_{ij}$. Least squares estimates can be calculated and the residual sum of squares $S(\hat{\beta}_A)$ can be obtained. $S(\hat{\beta}_A)$ has $(n - p - 1)$ degrees of freedom. The lack-of-fit sum of squares, $S(\hat{\beta}_A) - S(\hat{\beta})$, has $(k - p - 1)$ degrees of freedom. The lack-of-fit statistic (also called the goodness-of-fit test),

$$F = \frac{[S(\hat{\beta}_A) - S(\hat{\beta})]/(k - p - 1)}{S(\hat{\beta})/(n - k)} \tag{6.37}$$

follows an $F(k - p - 1, n - k)$ distribution under the hypothesis that the parametric model is appropriate. Large values of F lead us to reject this model.

6.5 VARIANCE-STABILIZING TRANSFORMATIONS

The plot of the residuals (e_i) against the fitted values ($\hat{\mu}_i$) in Figure 6.8 for the forestry data set (Table 6.2) shows a "funnel" shape indicating that the constant variance assumption is violated. In the subsequent analysis of the data, we considered a logarithmic transformation of the response y. Figure 6.9 shows that this transformation is successful in making the variance constant. In situations such as these, we need a certain transformation to **stabilize** the variance. In this section, we illustrate how one can select the correct transformation of the response. We consider the general regression model

$$y_i = f(\mathbf{x}_i; \boldsymbol{\beta}) + \epsilon_i = \eta_i + \epsilon_i \tag{6.38}$$

where $\eta_i = E(y_i) = f(\mathbf{x}_i; \boldsymbol{\beta})$ is the mean of the response. In addition, we assume that the variance of ϵ_i is related to the mean η_i such that

$$V(y_i) = V(\epsilon_i) = [h(\eta_i)]^2\sigma^2 \tag{6.39}$$

where h is a known function.

We like to find a transformation $g(y_i)$ of the response y_i such that the variance of $g(y_i)$ is constant. For this we approximate the function $g(y_i)$ by a first-order Taylor series around η_i:

$$g(y_i) \approx g(\eta_i) + (y_i - \eta_i)g'(\eta_i) \tag{6.40}$$

where $g'(\eta_i)$ is the first derivative of $g(y_i)$ evaluated at η_i. Then the variance of the transformed variable $g(y_i)$ can be approximated as

$$V[g(y_i)] \approx V[g(\eta_i) + (y_i - \eta_i)g'(\eta_i)] = [g'(\eta_i)]^2 V(y_i) = [g'(\eta_i)]^2[h(\eta_i)]^2\sigma^2 \tag{6.41}$$

To stabilize the variance, we need to choose the transformation $g(.)$ such that

$$g'(\eta_i) = 1/h(\eta_i) \tag{6.42}$$

Often, we find that these transformations, in addition to stabilizing the variance, lead to simplifications in the functional form of the regression model.

We now consider two special cases that arise quite often in practical applications.

Example 1 $h(\eta_i) = \eta_i$; the standard deviation of the response is proportional to the mean level. Then the transformation $g(\eta_i)$ has to satisfy $g'(\eta_i) = 1/\eta_i$. This means that $g(\eta_i) = \ln(\eta_i)$, where ln is the natural logarithm. Hence, in cases in which the standard deviation of y is proportional to its mean, one should consider the logarithmic transformation of y and regress $\ln(y_i)$ on the explanatory variables. This is what we did in the forestry example of Section 6.3.

Example 2 $h(\eta_i) = \eta_i^{1/2}$; the variance is proportional to the mean level. In this case, $g'(\eta_i) = \eta_i^{-1/2}$ and $g(\eta_i) = 2\eta_i^{1/2}$. The square root transformation $y^{1/2}$ stabilizes the variance.

6.5.1 BOX–COX TRANSFORMATIONS

We now outline a special class of transformations called the **Box–Cox transformations** or the **power transformations** (Box and Cox, 1964),

$$g(y_i) = \left(y_i^\lambda - 1\right)/\lambda \tag{6.43}$$

If $\lambda = 1$, no transformation is needed and we analyze the original data. If $\lambda = -1$, we analyze the reciprocal $1/y_i$. If $\lambda = 1/2$, we analyze $y_i^{1/2}$ (Example 2). It can be shown that $\text{limit}_{\lambda\to 0}[(y_i^\lambda - 1)/\lambda] = \ln(y_i)$; we analyze $\ln(y_i)$ if $\lambda = 0$ (Example 1). The two examples that we considered previously are special cases of this class.

Box and Cox (1964) show that the maximum likelihood estimate of λ minimizes SSE(λ), where SSE(λ) is the residual sum of squares from fitting the regression model with transformed response

$$y_i^{(\lambda)} = \left(y_i^\lambda - 1\right)/\lambda \bar{y}_g^{\lambda-1}$$

$\bar{y}_g = \left[\prod_{i=1}^n y_i\right]^{1/n}$ is the geometric mean of the y_i's. If $\lambda = 0$, we take

$$y_i^{(\lambda=0)} = \text{limit}_{\lambda\to 0} y_i^{(\lambda)} = \bar{y}_g \ln(y_i)$$

To illustrate the use of Box–Cox transformations, we consider the gas consumption data in Chapter 1 (Table 1.4). This data set is also analyzed in Chapter 4. The dependent varable is y = fuel efficiency (in miles per gallon). Although there are six regressor variables, for simplicity we consider only one of them, x = weight of the automobile. We estimate the regression parameters and obtain the residual sum of squares [SSE(λ)] for several values of λ in the model

$$\frac{y_i^\lambda - 1}{\lambda \bar{y}_g^{\lambda-1}} = \beta_0 + \beta_1 x_i + \epsilon_i \tag{6.44}$$

The results are shown in Table 6.8. The minimum of SSE(λ) is at $\lambda = -0.75$.

Abraham and Ledolter (1983) showed, with a finer grid on λ, that the actual optimum is $\hat{\lambda} = -0.65$. One can also calculate confidence intervals for λ (Box

TABLE 6.8 RESIDUAL SUM OF SQUARES SSE(λ): GAS CONSUMPTION DATA

λ	SSE(λ)
1.00	292.58
0.50	257.74
0.25	245.47
0.00 (ln y_i)	236.34
−0.25	230.23
−0.50	227.07
−0.75	226.07
−1.00	229.67
−1.25	235.63
−1.50	244.98

and Cox, 1964). In fact, the value $\lambda = -1.0$ is in the 95% confidence interval. This value leads to a reciprocal transformation $z = 1/y$ (fuel consumption in gallons per traveled mile or per 100 miles if multiplied by 100) considered in Chapter 4 (see Section 4.3.1).

6.5.2 SEVERAL USEFUL RESULTS FOR LOGARITHMIC TRANSFORMATIONS

Logarithm and Elasticity

Expanding $\ln(x + \Delta x)$ in a Taylor series around x gives for small Δx,

$$\ln(x + \Delta x) \cong \ln(x) + \frac{\Delta x}{x}$$

Elasticity is defined as the percentage change in the response y that results from a percentage change in the input variable x. That is,

$$\text{Elasticity} = \frac{\Delta y/y}{\Delta x/x}$$

For example, an elasticity of +3 implies that a 1% change in the input x results in a 3% change in the response y.

Estimation of Elasticities in Log–Log Regressions: Consider a regression of the logarithm of y on the logarithm of x. That is,

$$E[\ln(y)] = \beta_0 + \beta_1 \ln(x)$$

The slope β_1 in this model represents the elasticity. Consider a change in the input of magnitude Δx. Then the resulting change in y is given by

$$\begin{aligned} E[\ln(y + \Delta y) - \ln(y)] &= \beta_0 + \beta_1 \ln(x + \Delta x) - [\beta_0 + \beta_1 \ln(x)] \\ &= \beta_1[\ln(x + \Delta x) - \ln(x)] \end{aligned}$$

Applying the fact given above leads to

$$E[\ln(y) + \frac{\Delta y}{y} - \ln(y)] \cong \beta_1[\ln(x) + \frac{\Delta x}{x} - \ln(x)]$$

or $\beta_1 \cong \dfrac{E[\Delta y/y]}{\Delta x/x}$. The slope in the log–log regression measures the elasticity.

Logarithm and Proportional Increase in a Time Series

Suppose $y_t\,(t = 1, 2, \ldots,)$ are time series observations that increase proportionally. That is, $y_t = y_{t-1}(1 + r_t)$, where $100r_t$ represents the percentage increase of the series. Then $\ln(y_t) = \ln(y_{t-1}) + \ln(1 + r_t)$. Expanding $\ln(1 + r_t)$ in a Taylor series around 1, we obtain $\ln(1 + r_t) \cong \ln(1) + r_t = r_t$. Thus, $\ln(y_t) \cong \ln(y_{t-1}) + r_t$. Hence, the proportional changes in a time series can be expressed as successive differences of the logarithms of the series. That is, $r_t \cong \ln(y_t) - \ln(y_{t-1})$.

Appendix

1. EFFECTS OF MISSPECIFICATION

Assume that the true linear model is $\boldsymbol{y} = X\boldsymbol{\beta} + U\boldsymbol{\gamma} + \boldsymbol{\epsilon}$, where X and U are $n \times (p+1)$ and $n \times q$ matrices. Suppose that $\hat{\boldsymbol{\beta}} = (X'X)^{-1}X'\boldsymbol{y}$ is the least squares estimate from the incorrect model $\boldsymbol{y} = X\boldsymbol{\beta} + \boldsymbol{\epsilon}$, and $\boldsymbol{e} = (I - H)\boldsymbol{y}$ is the corresponding residual, where $H = I - X(X'X)^{-1}X'$. Then

$$E(\hat{\boldsymbol{\beta}}) = (X'X)^{-1}X'E(\boldsymbol{y}) = (X'X)^{-1}X'(X\boldsymbol{\beta} + U\boldsymbol{\gamma}) = \boldsymbol{\beta} + (X'X)^{-1}X'U\boldsymbol{\gamma}$$

The least squares estimate obtained under the incorrect model is biased, and the bias amounts to $(X'X)^{-1}X'U\boldsymbol{\gamma}$. The bias disappears if $X'U = O$, a matrix of zeros. Furthermore,

$$E(\boldsymbol{e}) = (I - H)E(\boldsymbol{y}) = (I - H)[X\boldsymbol{\beta} + U\boldsymbol{\gamma}] = (I - H)U\boldsymbol{\gamma} \neq \mathbf{0}$$

The first term in the previous equation is zero since $(I - H)X = (I - X(X'X)^{-1}X')X = 0$, an $n \times (p+1)$ matrix of zeros. The second term is a nonzero vector because U is not in $L(X)$. Hence, the expected value of the residual vector from the incorrect model is not $\mathbf{0}$.

2. SOME USEFUL MATRIX RESULTS

Suppose that α is a scalar, $\boldsymbol{u}$ and $\boldsymbol{v}$ are $n \times 1$ vectors, and A is an $n \times n$ invertible matrix.

i. Consider the partitioned matrix $A = \begin{bmatrix} A_{11} & A_{12} \\ A_{21} & A_{22} \end{bmatrix}$, where A_{11} and A_{22} are square matrices whose inverses exist. Define $B_{22} = A_{22} - A_{21}A_{11}^{-1}A_{12}$, $B_{12} = A_{11}^{-1}A_{12}$, and $B_{21} = A_{21}A_{11}^{-1}$. Then the inverse of A is given by

$$A^{-1} = \begin{bmatrix} A_{11}^{-1} + B_{12}B_{22}^{-1}B_{21} & -B_{12}B_{22}^{-1} \\ -B_{22}^{-1}B_{21} & B_{22}^{-1} \end{bmatrix}$$

ii.

$$\left(I - \alpha\boldsymbol{u}\boldsymbol{v}'\right)^{-1} = I + \frac{\alpha\boldsymbol{u}\boldsymbol{v}'}{(1 - \alpha\boldsymbol{v}'\boldsymbol{u})} \tag{6.45}$$

iii.

$$\left(A - \boldsymbol{u}\boldsymbol{v}'\right)^{-1} = A^{-1} + \frac{A^{-1}\boldsymbol{u}\boldsymbol{v}'A^{-1}}{\left(1 - \boldsymbol{v}'A^{-1}\boldsymbol{u}\right)} \tag{6.46}$$

Note that I is the identity matrix of order n, and $(1 - \alpha\boldsymbol{v}'\boldsymbol{u})$ and $(1 - \boldsymbol{v}'A^{-1}\boldsymbol{u})$ are scalars. The results in (i)–(iii) can be shown by confirming that $AA^{-1} = A^{-1}A = I$.

3. A USEFUL RESULT FOR $\hat{\beta}_{(i)}$

Consider the linear model $\boldsymbol{y} = X\boldsymbol{\beta} + \boldsymbol{\epsilon}$ and the parameter estimate $\hat{\boldsymbol{\beta}} = (X'X)^{-1} X'\boldsymbol{y}$ seen in Chapters 4–6. Let $X_{(i)}$ denote the X matrix with the ith row $\boldsymbol{x}_i'$ deleted, and let $\boldsymbol{y}_{(i)}$ and $\boldsymbol{\epsilon}_{(i)}$ denote respectively the vectors $\boldsymbol{y}$ and $\boldsymbol{\epsilon}$ without the ith element. Then we can write without loss of generality

$$X = \begin{pmatrix} X_{(i)} \\ \boldsymbol{x}_i' \end{pmatrix} \quad \text{and} \quad \boldsymbol{y} = \begin{pmatrix} \boldsymbol{y}_{(i)} \\ y_i \end{pmatrix} \tag{6.47}$$

Without the ith observation the model and the estimate of $\boldsymbol{\beta}$ become

$$\boldsymbol{y}_{(i)} = X_{(i)}\boldsymbol{\beta} + \boldsymbol{\epsilon}_{(i)}, \quad \text{and} \quad \hat{\boldsymbol{\beta}}_{(i)} = \left(X_{(i)}'X_{(i)}\right)^{-1} X_{(i)}'\boldsymbol{y}_{(i)} \tag{6.48}$$

From Eq. (6.47) we can write

$$X'X = X_{(i)}'X_{(i)} + \boldsymbol{x}_i\boldsymbol{x}_i' \quad \text{and} \quad X'\boldsymbol{y} = X_{(i)}'\boldsymbol{y}_{(i)} + \boldsymbol{x}_i y_i \tag{6.49}$$

Then we have

$$\begin{aligned}
\hat{\boldsymbol{\beta}}_{(i)} &= (X'X - \boldsymbol{x}_i\boldsymbol{x}_i')^{-1}(X'\boldsymbol{y} - \boldsymbol{x}_i y_i) \\
&= \left[(X'X)^{-1} + \frac{(X'X)^{-1}\boldsymbol{x}_i\boldsymbol{x}_i'(X'X)^{-1}}{1 - \boldsymbol{x}_i'(X'X)^{-1}\boldsymbol{x}_i}\right][X'\boldsymbol{y} - \boldsymbol{x}_i y_i] \\
&= \hat{\boldsymbol{\beta}} - (X'X)^{-1}\boldsymbol{x}_i y_i + \left[\frac{(X'X)^{-1}\boldsymbol{x}_i\boldsymbol{x}_i'\hat{\boldsymbol{\beta}} - (X'X)^{-1}\boldsymbol{x}_i h_{ii} y_i}{1 - h_{ii}}\right] \\
&= \hat{\boldsymbol{\beta}} - \frac{(X'X)^{-1}\boldsymbol{x}_i}{(1 - h_{ii})}[y_i(1 - h_{ii}) - \boldsymbol{x}_i'\hat{\boldsymbol{\beta}} + h_{ii} y_i] \\
&= \hat{\boldsymbol{\beta}} - \frac{(X'X)^{-1}\boldsymbol{x}_i}{(1 - h_{ii})}(y_i - \boldsymbol{x}_i'\hat{\boldsymbol{\beta}}) \\
&= \hat{\boldsymbol{\beta}} - \frac{(X'X)^{-1}\boldsymbol{x}_i e_i}{(1 - h_{ii})}
\end{aligned} \tag{6.50}$$

4. A USEFUL EXPRESSION FOR $s_{(i)}^2$

We can write

$$\sum_{j \neq i} (y_j - \boldsymbol{x}_j'\hat{\boldsymbol{\beta}}_{(i)})^2 = (\boldsymbol{y}_{(i)} - X_{(i)}\hat{\boldsymbol{\beta}}_{(i)})'(\boldsymbol{y}_{(i)} - X_{(i)}\hat{\boldsymbol{\beta}}_{(i)}) \tag{6.51}$$

where $\mathbf{y}_{(i)}$ and $X_{(i)}$ are defined previously. In addition, let $\mathbf{x}_i'$ be the ith row of X. Hence, the right-hand side of Eq. (6.51) is equal to

$$\begin{aligned}
&(\mathbf{y} - X\hat{\beta}_{(i)})'(\mathbf{y} - X\hat{\beta}_{(i)}) - (y_i - \mathbf{x}_i'\hat{\beta}_{(i)})^2 \\
&= [\mathbf{y} - X\hat{\beta} + X(\hat{\beta} - \hat{\beta}_{(i)})]'[\mathbf{y} - X\hat{\beta} + X(\hat{\beta} - \hat{\beta}_{(i)})] - (y_i - \mathbf{x}_i'\hat{\beta}_{(i)})^2 \\
&= [\mathbf{e} + X(\hat{\beta} - \hat{\beta}_{(i)})]'[\mathbf{e} + X(\hat{\beta} - \hat{\beta}_{(i)})] - e_{(i)}^2 \\
&= \mathbf{e}'\mathbf{e} + (\hat{\beta} - \hat{\beta}_{(i)})'X'X(\hat{\beta} - \hat{\beta}_{(i)}) - e_{(i)}^2 \quad \text{(the other terms are zero since } X'\mathbf{e} = \mathbf{0}) \\
&= \mathbf{e}'\mathbf{e} + (p+1)D_i s^2 - e_{(i)}^2 \qquad (D_i \text{ is Cook's distance}) \\
&= \mathbf{e}'\mathbf{e} + \frac{h_{ii}e_i^2}{(1-h_{ii})^2} - \frac{e_i^2}{(1-h_{ii})^2} \qquad \text{(using Eq. 6.19 and Eq. 6.22)} \\
&= \mathbf{e}'\mathbf{e} - \frac{e_i^2}{(1-h_{ii})} = (n-p-1)s^2 - e_i^2/(1-h_{ii})
\end{aligned}$$

Hence,

$$s_{(i)}^2 = \frac{[(n-p-1)s^2 - e_i^2/(1-h_{ii})]}{(n-p-2)}$$

EXERCISES

6.1. A research team studies the influence of body weight (g) and heart weight (mg) on the kidney weight (mg) of rats. Ten rats were selected over a range of body weights, and the following results were recorded. The data are given in the file **kidney**.

Kidney weight (mg) y	Body weight (g) x_1	Heart weight (mg) x_2
810	34	210
480	43	223
680	35	205
920	33	225
650	34	188
650	26	149
650	30	172
620	31	164
740	27	188
600	28	163

a. Consider the following models:

$$y = \beta_0 + \beta_1 x_1 + \epsilon$$
$$y = \beta_0 + \beta_2 x_2 + \epsilon$$
$$y = \beta_0 + \beta_1 x_1 + \beta_2 x_2 + \epsilon$$

Which of these models are appropriate? Discuss.

b. Determine if there are any unusual data points. If there are, does their removal have an effect on the fitting results for the models in (a)? If there are cases with large effects, how would you present the results to the research team?

6.2. The cloud point of a liquid is a measure of the degree of crystallization in a stock that can be measured by the refractive index. It has been suggested that the percentage of I-8 in the base stock is a good predictor of cloud point. Data were collected on stocks with different (and well-known) percentage of I-8. These are given in the file **cloud**.

a. Fit the model $y = \beta_0 + \beta_1 x + \epsilon$ to the data. Test the model for lack of fit.

b. Fit the model $y = \beta_0 + \beta_1 x + \beta_2 x^2 + \epsilon$. Test the model for lack of fit. Comment on the results and compare them to the ones in (a).

6.3. Suppose we have a single explanatory variable x. Fit the linear model

$\boldsymbol{y} = \mathbf{1}\beta_0 + \boldsymbol{x}\beta_1 + \boldsymbol{\epsilon}$, and obtain the vector of residuals $\boldsymbol{e}$. Suppose that the true model is the quadratic model

$$\boldsymbol{y} = \mathbf{1}\beta_0 + \boldsymbol{x}\beta_1 + \boldsymbol{x}_2\beta_2 + \boldsymbol{\epsilon}$$

where $\boldsymbol{x}_2' = (x_1^2, x_2^2, \ldots, x_n^2)$.

Show that $E(\boldsymbol{e}) = \beta_2(I - H)\boldsymbol{x}_2$, where H is the hat matrix from the fitted linear model.

6.4. Consider the linear model $\boldsymbol{y} = X\boldsymbol{\beta} + \boldsymbol{\epsilon}$, with X an $n \times (p+1)$ matrix with rank $(p+1)$, and $\boldsymbol{\epsilon}$ a vector of uncorrelated errors with mean $\mathbf{0}$ and covariance matrix $\sigma^2 I$. Let $\hat{\boldsymbol{\mu}} = X\hat{\boldsymbol{\beta}}$, where $\hat{\boldsymbol{\beta}}$ is the vector of least squares estimates.

a. Find the mean vector and the covariance matrix of $\hat{\boldsymbol{\mu}}$.

b. Show that $\frac{1}{n}\sum_{i=1}^{n} V(\hat{\mu}_i) = \frac{(p+1)}{n}\sigma^2$.

Hint: Find the trace of $V(\hat{\boldsymbol{\mu}})$; use the fact that trace of AB = trace of BA if the products are defined.

c. Let $H = (h_{ij})$ be any $n \times n$ symmetric idempotent matrix: $H' = H$ and $HH = H$. Show that the diagonal elements h_{ii} must lie between zero and one.
Hint: Consider $\boldsymbol{a}_i'H$, where $\boldsymbol{a}_i$ is a $n \times 1$ vector with all components 0 except for the ith element, which is 1.

d. Assume that the linear model includes a constant term. Then the diagonal elements h_{ii} of the hat matrix $H = X(X'X)^{-1}X'$ satisfy $h_{ii} \geq \frac{1}{n}$.

Hint: Parameterize the model by centering the regressor variables $(x_{ij} - \bar{x}_j)$, for $j = 1, 2, \ldots, p$.

e. Consider the linear model $\boldsymbol{y} = X\boldsymbol{\beta} + \boldsymbol{\epsilon}$, where the X matrix has **rank less than** $p+1$. Then $X'X\boldsymbol{\beta} = X'\boldsymbol{y}$ has infinitely many solutions for $\boldsymbol{\beta}$. Suppose that $\hat{\boldsymbol{\beta}}$ and $\tilde{\boldsymbol{\beta}}$ are two solutions and let $\hat{\boldsymbol{\mu}} = X\hat{\boldsymbol{\beta}}$ and $\tilde{\boldsymbol{\mu}} = X\tilde{\boldsymbol{\beta}}$ be the corresponding fitted values. Show that $\hat{\boldsymbol{\mu}} = \tilde{\boldsymbol{\mu}}$. This shows that both solutions of the normal equations will produce the same fitted values and residuals.

6.5. a. Suppose I is the $r \times r$ identity matrix, $\boldsymbol{w}$ and $\boldsymbol{v}$ are $r \times 1$ column vectors, and α is a constant. Show by direct multiplication that

$$(I + \alpha\boldsymbol{v}\boldsymbol{w}')^{-1} = I - \left(\frac{\alpha}{1 + \alpha\boldsymbol{v}'\boldsymbol{w}}\right)\boldsymbol{v}\boldsymbol{w}'$$

b. Use the result in (a) to obtain an expression for $(A + \boldsymbol{w}\boldsymbol{w}')^{-1}$ in terms of A^{-1} and $\boldsymbol{w}$.

c. Suppose we use least squares to fit the model

$$\boldsymbol{y} = X\boldsymbol{\beta} + \boldsymbol{\epsilon}$$

to data from n subjects. Data $(y_{n+1}, x_{n+1,1}, \ldots, x_{n+1,p})$ become available on one more case so that the model becomes

$$\begin{pmatrix} \boldsymbol{y} \\ \cdots\cdot \\ y_{n+1} \end{pmatrix} = \begin{pmatrix} X \\ \cdots\cdot \\ \boldsymbol{w}' \end{pmatrix}\boldsymbol{\beta} + \begin{pmatrix} \boldsymbol{\epsilon} \\ \cdots\cdot \\ \epsilon_{n+1} \end{pmatrix}$$

where $\boldsymbol{w}' = (1, x_{n+1,1}, \ldots, x_{n+1,p})$, or

$$\boldsymbol{y}_1 = X_1\boldsymbol{\beta} + \boldsymbol{\epsilon}_1$$

i. Find an expression for $(X_1'X_1)^{-1}$ in terms of $(X'X)^{-1}$ and $\boldsymbol{w}$.

ii. Find an expression for $\hat{\boldsymbol{\beta}}_1 = (X_1'X_1)^{-1}X_1'\boldsymbol{y}_1$ in terms of $\hat{\boldsymbol{\beta}} = (X'X)^{-1}X'\boldsymbol{y}$.

This provides a simple way of updating the least squares estimate as more data become available. It is used in deletion diagnostics.

6.6. Consider the multiple regression model $\boldsymbol{y} = X\boldsymbol{\beta} + \boldsymbol{\epsilon}$, where X consists of the k columns $\boldsymbol{x}_1, \ldots, \boldsymbol{x}_k$. Prove that $\hat{\beta}_k$ can be obtained by the following three steps:

Step 1. Regress $\boldsymbol{y}$ on $\boldsymbol{x}_1, \ldots, \boldsymbol{x}_{k-1}$ and denote the vector of residuals by $\boldsymbol{r}$.

Step 2. Regress $\boldsymbol{x}_k$ on $\boldsymbol{x}_1, \ldots, \boldsymbol{x}_{k-1}$ and denote the vector of residuals by $\boldsymbol{u}$.

Step 3. Fit the model $\boldsymbol{r} = \beta_k\boldsymbol{u} + \boldsymbol{\epsilon}$. The resulting estimate $\hat{\beta}_k$ is identical to the estimate $\hat{\beta}_k$ in $\boldsymbol{y} = X\boldsymbol{\beta} + \boldsymbol{\epsilon}$.

Hint: Use $\tilde{X}$ to denote the first $k-1$ columns of X. Then $\hat{\boldsymbol{\beta}}$ satisfies $X'X\hat{\boldsymbol{\beta}} = X'\boldsymbol{y}$, where

$$X'X = \begin{pmatrix} \tilde{X}'\tilde{X} & \tilde{X}'\boldsymbol{x}_k \\ \boldsymbol{x}_k'\tilde{X} & \boldsymbol{x}_k'\boldsymbol{x}_k \end{pmatrix} \text{ and } X'\boldsymbol{y} = \begin{pmatrix} \tilde{X}'\boldsymbol{y} \\ \boldsymbol{x}_k'\boldsymbol{y} \end{pmatrix}.$$

6.7. Explain why the following statements are true or false:

a. A residual plot of e_i against $\hat{\mu}_i$ is more informative than that of e_i against y_i.

b. An outlier should always be rejected.

c. Consider the model

$$y_i = \beta x_i + \epsilon_i, \ i = 1, 2, \ldots, 10$$

with $\sum_{i=1}^{10} x_i^2 = 1$. The leverage of the first case is x_1^2.

6.8. In a certain regression model relating y to variables x_1 and x_2, the X matrix is given by

$$X = \begin{bmatrix} 1 & -1 & -1 \\ 1 & -1 & -1 \\ 1 & +1 & -1 \\ 1 & +1 & -1 \\ 1 & -1 & +1 \\ 1 & -1 & +1 \\ 1 & +1 & +1 \\ 1 & +1 & +1 \end{bmatrix}$$

For the model $\mathbf{y} = X\boldsymbol{\beta} + \epsilon$, determine the degrees of freedom for

a. residual sum of squares.

b. regression sum of squares.

c. pure error sum of squares.

d. lack-of-fit sum of squares.

6.9. Discuss the following statements and explain why they are true or false.

a. Increasing the number of predictor variables will never decrease the R^2.

b. Multicollinearity affects the interpretation of the regression coefficients.

c. The variance inflation factor of $\hat{\beta}_j$ depends on the R^2 of the regression of the response variable y on the regressor variable x_j.

d. A high leverage point is always highly influential.

e. Standardized residuals are always smaller than the ordinary residuals.

6.10. Consider the (production) function

$$y = \beta_0 (x_1)^{\beta_1} (x_2)^{\beta_2}$$

You are interested in the coefficients (elasticities) β_1 and β_2. Indicate which of the following statements is *true.*
One can obtain estimates of β_1 and β_2 by

a. Regressing y on x_1 and x_2.

b. Regressing $\log(y)$ on x_1 and x_2.

c. Regressing y on $\log(x_1)$ and $\log(x_2)$.

d. Regressing $\log(y)$ on $\log(x_1)$ and $\log(x_2)$.

e. None of the above.

6.11. Indicate which of the following statements about the Durbin–Watson statistic is *true.*

a. A Durbin–Watson statistic of zero indicates that all regressors are insignificant in predicting the response variable.

b. A Durbin–Watson statistic of 2 indicates the residuals are serially correlated.

c. The Durbin–Watson statistic measures how influential an observation is in determining the regression line.

d. The Durbin–Watson statistic can be used to test the normality of the residuals.

e. A Durbin–Watson statistic cannot be negative.

6.12. Consider the Florida county ($n = 67$) votes for Gore and Buchanan in the 2000 presidential election. A scatter plot of Buchanan votes (y) on Gore votes (x) is given below. A simple linear regression model leads to the following diagnostics. Consider the four points labeled A through D. Assign them the correct case numbers.

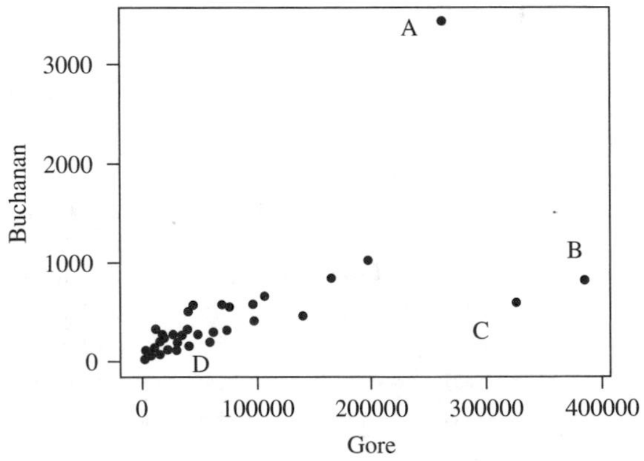

Case	County	GORE	BUCH	Res	StanRes	Leverage	CookDist
1	BAY	18850	248	90.832	0.279	0.017	0.001
2	BRADFORD	3072	65	−26.911	−0.083	0.019	0.000
3	BROWARD			−888.823	−3.317	0.334	2.754
4	DADE			−877.700	−3.058	0.235	1.438
5	PALM BEACH			2215.452	7.333	0.153	4.843
6	PASCO	69550	570	203.139	0.624	0.017	0.003
7	....						

6.13. Consider consecutive weekly stock closings of your choice and use the most recent 52 weeks as your sample.

a. Graph the observations against time.

b. Estimate the coefficients in the linear regression on time, $y_t = \beta_0 + \beta_1 x_t + \varepsilon_t$, where $x_t = t$. Investigate the shortcomings of the fitted model. In particular, assess whether the errors are independent by calculating the autocorrelations of the residuals and the Durbin–Watson test statistic.

c. Regress the observation at time t on the previous observation at time $t - 1$, $y_t = \beta_0 + \beta_1 y_{t-1} + \varepsilon_t$. Repeat the residual diagnostics in (b).

d. Which of the two models is more appropriate? Which model would you use if you wanted to predict the next observation?

6.14. The data are taken from Williams, E. J. *Regression Analysis*. New York: Wiley, 1959. The data are given in the file **tearpaper**.

Five different pressures were applied during the sheet pressing stage in the manufacture of paper. Pressure is thought to affect the tear factor of the paper (which is the percentage of a standard force necessary to tear the paper). Batches of paper were manufactured under different pressures. From these batches, four sheets were selected at random and the tear strength of each sheet was evaluated. Note that pressure is equally spaced on the logarithmic scale.

Specify an appropriate model for this data set. Estimate the model and perform needed diagnostic checks. Use the replications to test for lack of fit.

x = Pressure	y = Tear Factor
35.0	112
35.0	119
35.0	117
35.0	113
49.5	108
49.5	99
49.5	112
49.5	118
70.0	120
70.0	106
70.0	102
70.0	109
99.0	110
99.0	101
99.0	99
99.0	104
140.0	100
140.0	102
140.0	96
140.0	101

6.15. The data are taken from Bennett, G. W. Determination of anaerobic threshold. *Canadian Journal of Statistics,* 16, 307–310, 1988. The data are given in the file **ventilation**.

The data are from an experiment in kinesiology. A subject performed a standard exercise at gradually increasing levels, and oxygen uptake and expired ventilation (which is related to the rate of exchange of gases in the lungs) were measured. The expired ventilation (y) and the oxygen uptake (x) are related nonlinearly.

Graph expired ventilation against oxygen uptake. Repeat the graph for various appropriate transformations, and develop a model that relates the transformed variables. Consider the Box–Cox family of transformations. Estimate the appropriate transformation.

x = Oxygen Uptake	y = Expired Ventilation
574	21.9
592	18.6
664	18.6
667	19.1
718	19.2
770	16.9
927	18.3
947	17.2
1,020	19.0
1,096	19.0
1,277	18.6
1,323	22.8
1,330	24.6
1,599	24.9
1,639	29.2
1,787	32.0
1,790	27.9
1,794	31.0
1,874	30.7
2,049	35.4
2,132	36.1
2,160	39.1
2,292	42.6
2,312	39.9
2,475	46.2
2,489	50.9
2,490	46.5
2,577	46.3
2,766	55.8
2,812	54.5
2,893	63.5
2,957	60.3
3,052	64.8
3,151	69.2
3,161	74.7
3,266	72.9
3,386	80.4
3,452	83.0
3,521	86.0
3,543	88.9
3,676	96.8
3,741	89.1
3,844	100.9
3,878	103.0
4,002	113.4
4,114	111.4
4,152	119.9
4,252	127.2
4,290	126.4
4,331	135.5
4,332	138.9
4,390	143.7
4,393	144.8

6.16. The data are taken from Robertson, J. D., and Armitage, P. Comparison of two hypertensive agents. *Anaesthesia,* 14, 53–64, 1959. The data are given in the file **recovery**.

Hypertensive drugs are used routinely to lower a patient's blood pressure, and such drugs are administered continuously during surgery. Since surgery times vary, the total amount of the drug that is administered varies from case to case. Also, patients react differently to such drugs, and hence blood pressure during surgery varies across patients. The sooner blood pressure rises to normal levels, the better. The recovery time (i.e., the time it takes for a patient's systolic blood pressure to return to normal) is an important variable.

The following table lists, for a sample of 53 patients, the recovery time, the logarithm of the administered dose, and the average systolic blood pressure while the drug is being administered.

Discuss how recovery time is related to the dose and the blood pressure that is achieved during surgery. Fit appropriate regression models, check for model violations, and interpret the results. Explore the usefulness of transformations on the response.

x_1 = Log Dose	x_2 = Blood Pressure	y = Recovery Time
2.26	66	7
1.81	52	10
1.78	72	18
1.54	67	4
2.06	69	10
1.74	71	13
2.56	88	21
2.29	68	12
1.80	59	9
2.32	73	65
2.04	68	20
1.88	58	31
1.18	61	23
2.08	68	22
1.70	69	13
1.74	55	9
1.90	67	50
1.79	67	12
2.11	68	11
1.72	59	8
1.74	68	26
1.60	63	16
2.15	65	23
2.26	72	7
1.65	58	11
1.63	69	8
2.40	70	14
2.70	73	39
1.90	56	28
2.78	83	12
2.27	67	60
1.74	84	10

x_1 = Log Dose	x_2 = Blood Pressure	y = Recovery Time
2.62	68	60
1.80	64	22
1.81	60	21
1.58	62	14
2.41	76	4
1.65	60	27
2.24	60	26
1.70	59	28
2.45	84	15
1.72	66	8
2.37	68	46
2.23	65	24
1.92	69	12
1.99	72	25
1.99	63	45
2.35	56	72
1.80	70	25
2.36	69	28
1.59	60	10
2.10	51	25
1.80	61	44

6.17. The data are taken from Brown, B. M. and Maritz, J. S. Distribution-free methods in regression. *Australian Journal of Statistics,* 24, 318–331, 1982. The data are given in the file **rigidity**.

Measurements on 50 varieties of timber are made on their rigidity, elasticity, and air-dried density. The objective is to predict rigidity as a function of elasticity and air-dried density. Pay careful attention to the case diagnostics.

y = Rigidity	x_1 = Elasticity	x_2 = Density
1,000	99.0	25.3
1,112	173.0	28.2
1,033	188.0	28.6
1,087	133.0	29.1
1,069	146.0	30.7
925	91.0	31.4

y = Rigidity	x_1 = Elasticity	x_2 = Density
1,306	188.0	32.5
1,306	194.0	36.8
1,323	195.0	37.1
1,379	177.0	38.3
1,332	182.0	39.0
1,254	110.0	39.6
1,587	203.0	40.1
1,145	193.0	40.3
1,438	167.0	40.3
1,281	188.0	40.6
1,595	238.0	42.3
1,129	130.0	42.4
1,492	189.0	42.5
1,605	213.0	43.0
1,647	165.0	43.0
1,539	210.0	46.7
1,706	224.0	49.0
1,728	228.0	50.2
1,703	209.0	50.3
1,897	240.0	50.3
1,822	248.0	51.3
2,129	261.0	51.7
2,053	245.0	52.8
1,676	186.0	53.8
1,621	188.0	53.9
1,990	252.0	54.9
1,764	222.0	55.1
1,909	244.0	55.2
2,086	274.0	55.3
1,916	276.0	56.9
1,889	254.0	57.3
1,870	238.0	58.3
2,036	264.0	58.6
2,570	189.0	58.7
1,474	223.0	59.5
2,116	245.0	60.8
2,054	272.0	61.3
1,994	264.0	61.5
1,746	196.0	63.2
2,604	268.0	63.3
1,767	205.0	68.1
2,649	346.0	68.9
2,159	246.0	68.9
2,078	237.5	70.8

6.18. The data are taken from Cook, R. D., and Weisberg, S. *Residuals and Influence in Regression*. London: Chapman & Hall, 1982. The data are given in the file **liver**.

In an experiment to investigate the amount of drug retained in the liver of a rat, 19 rats were weighed and dosed. The dose was approximately 40 mg/1 kg of body weight; it can be expected that the weight of the liver is strongly correlated with body weight. After a fixed length of time the rat was sacrificed, the liver weighed, and the percentage of dose in the liver determined.

a. Check whether liver weight and body weight are strongly correlated.

b. Explain the percentage of dose in the liver as a function of body weight, liver weight, and administered dose. Pay careful attention to model diagnostics.

Body Weight	Liver Weight	Dose	y = Dose in Liver
176	6.5	0.88	0.42
176	9.5	0.88	0.25
190	9.0	1.00	0.56
176	8.9	0.88	0.23
200	7.2	1.00	0.23
167	8.9	0.83	0.32
188	8.0	0.94	0.37
195	10.0	0.98	0.41
176	8.0	0.88	0.33
165	7.9	0.84	0.38
158	6.9	0.80	0.27
148	7.3	0.74	0.36
149	5.2	0.75	0.21
163	8.4	0.81	0.28
170	7.2	0.85	0.34
186	6.8	0.94	0.28
146	7.3	0.73	0.30
181	9.0	0.90	0.37
149	6.4	0.75	0.46

6.19. The data are taken from Snapinn, S. M. and Small, R. D. Tests of significance using regression models for ordered categorical data. *Biometrics,* 42, 583–592, 1966. The data are given in the file **phosphorus**.

Chemical determinations of inorganic phosphorus (x_1) and a component of organic phosphorus (x_2) in the soil are used to explain the plant-available phosphorus (y) of corn grown in the soil. The units are parts per million. Data for 18 soils are shown.

Develop appropriate regression models. Pay careful attention to case diagnostics.

x_1 = Inorganic Phosphorus	x_2 = Organic Phosphorus	y = Plant Phosphorus
0.4	53	64
0.4	23	60
3.1	19	71
0.6	34	61
4.7	24	54
1.7	65	77
9.4	44	81
10.1	31	93
11.6	29	93
12.6	58	51
10.9	37	76
23.1	46	96
23.1	50	77
21.6	44	93
23.1	56	95
1.9	36	54
26.8	58	168
29.9	51	99

6.20. The data are taken from Weiner, B. *Discovering Psychology*. Chicago: Science Research Association, 1977. The table lists the average vocabulary size of children at various ages. The data are given in the file **vocabulary**.

a. Obtain a scatter plot of the information.
b. The objective is to predict vocabulary as a function of age. Check for outliers and pay careful attention to the case diagnostics.

x = Age	y = Vocabulary
1.0	3
1.5	22
2.0	272
2.5	446
3.0	896
3.5	1,222
4.0	1,540
4.5	1,870
5.0	2,072
6.0	2,562

6.21. The data are taken from Jerison, H. J. *Evolution of the Brain and Intelligence*. New York: Academic Press, 1973. The data are given in the file **brainweight**.

Data on body weight and brain weight of animals are given. Construct a regression model that relates the two variables. Transformations may be needed before a regression model can be fit. Identify animals that differ from the common pattern that is established by the majority of the animals.

Species	x = Body Weight	y = Brain Weight
Mountain beaver	1.35	8.1
Cow	465.00	423.0
Grey wolf	36.33	119.5
Goat	27.66	115.0
Guinea pig	1.04	5.5
Diplodocus	11,700.00	50.0
Asian elephant	2,547.00	4,603.0
Donkey	187.10	419.0

Species	x = Body Weight	y = Brain Weight
Horse	521.00	655.0
Potar monkey	10.00	115.0
Cat	3.30	25.6
Giraffe	529.00	680.0
Gorilla	207.00	406.0
Human	62.00	1,320.0
African elephant	6,654.00	5,712.0
Triceratops	9,400.00	70.0
Rhesus monkey	6.80	179.0
Kangaroo	35.00	56.0
Hamster	0.12	1.0
Mouse	0.023	0.4
Rabbit	2.50	12.1
Sheep	55.50	175.0
Jaguar	100.00	157.0
Chimpanzee	52.16	440.0
Brachiosaurus	87,000.00	154.5
Rat	0.28	1.9
Mole	0.122	3.0
Pig	192.00	180.0

6.22. The data are taken from Hill, W. J. and Wiles, R. A. Plant experimentation (PLEX). *Journal of Quality Technology,* 7, 115–122, 1975. The data are given in the file **chemyield**.

The data are from an experiment on a large continuous chemical operation consisting of several separate reaction steps. The variable of interest is percentage yield (y). Factors believed to affect yield are the concentration of reactant A in the solvent S, the ratio of reactant B to reactant A, and the temperature in the reactor. Only coded variables are given because the paper does not reveal the units of measurements. The 15 runs in this experiment were carried out in random order over a period of 10 days, with 8 hours at each operating condition. Note that replicates at some of the operating conditions allow for a test of lack of fit.

Analyze the information, construct an appropriate regression model, and interpret the results. Check the adequacy of the model by constructing residual diagnostics and the appropriate lack of fit test.

Concentration of A	Ratio of B to A	Temperature	% Yield	
−1	−1	−1	75.4	
1	−1	−1	73.9	
−1	1	−1	76.8	
1	1	−1	72.8	
−1	−1	1	75.3	75.3
1	−1	1	71.4	
−1	1	1	76.5	77.2
1	1	1	72.3	
0	0	0	74.4	74.5
−2	0	0	79.0	78.4
2	0	0	69.2	

6.23. The data are taken from Snee, R. D. An alternative approach to fitting models when re-expression of the response is useful. *Journal of Quality Technology,* 18, 211–225, 1986. The data are given in the file **stopping**.

The table shows the stopping distances (feet) for cars traveling at the indicated speeds (miles per hour).

Find a model that explains the stopping distance in terms of the traveled speed. Consider variance-stabilizing transformations and carry out lack-of-fit tests (this is made possible by the replications).

x = Speed	y = Stopping Distance
4	4
5	2
5	8
5	8
5	4

x = Speed	y = Stopping Distance
7	6
7	7
8	9
8	8
8	13
8	11
9	5
9	5
9	13
10	8
10	17
10	14
12	11
12	21
12	19
13	18
13	27
13	15
14	14
14	16
15	16
16	19
16	14
16	34
17	29
17	22
18	47
18	29
18	34
19	30
20	48
21	55
21	39
21	42
22	35
24	56
25	33
25	59
25	48
25	56
26	39
26	41
27	78
27	57
28	64
28	84
29	68
29	54
30	60
30	101
30	67
31	77
35	85
35	107
36	79
39	138
40	110
40	134

6.24. The data are taken from Ryan, T. A., Joiner, B. L., and Ryan, B. F. The *Minitab Student Handbook*. Boston: Duxbury, 1985. The data are given in the file **volumetrees**.

Measurements on the volume (cubic feet), height (feet), and diameter at breast height (inches, measured at 54 inches above the ground) of 31 black cherry trees in the Allegheny National Forest in Pennsylvania are listed.

Develop a model that relates the volume of a tree to its diameter and height. A study of the volume of a (tapered) cylinder will suggest an appropriate model specification. Construct scatter plots of volume against diameter and height. Consider appropriate transformations. Fit appropriate regression models, obtain and interpret the estimates of the coefficients, calculate the ANOVA table, and discuss the adequacy of the model fit. Consider appropriate residual diagnostics for checking model adequacy. Use your model(s) to obtain a 95% confidence interval for the mean volume of a tree with diameter 11 inches and height 70 feet. Discuss whether it is also reasonable to obtain a confidence interval for the mean volume of a tree with diameter 11 inches and height 95 feet.

x_1 = Diameter	x_2 = Height	y = Volume
8.3	70	10.3
8.6	65	10.3
8.8	63	10.2
10.5	72	16.4
10.7	81	18.8
10.8	83	19.7
11.0	66	15.6
11.0	75	18.2
11.1	80	22.6
11.2	75	19.9
11.3	79	24.2
11.4	76	21.0
11.4	76	21.4
11.7	69	21.3
12.0	75	19.1
12.9	74	22.2
12.9	85	33.8
13.3	86	27.4
13.7	71	25.7
13.8	64	24.9
14.0	78	34.5
14.2	80	31.7
14.5	74	36.3
16.0	72	38.3
16.3	77	42.6
17.3	81	55.4
17.5	82	55.7
17.9	80	58.3
18.0	80	51.5
18.0	80	51.0
20.6	87	77.0

6.25. The data are taken from Dempster, A. P. *Elements of Continuous Multivariate Analysis*. Reading, MA: Addison-Wesley, 1969. The data are given in the file **volumeeggs**.

The table shows the volume, the length of the longest diameter, and the largest circular cross section of 12 eggs. Find a model that explains the volume. Think about appropriate transformations.

Longest Diameter	Largest Cross Section	y = Volume
2.151	1.889	5.755
2.086	1.859	5.479
2.099	1.874	5.551
2.138	1.874	5.686
2.195	1.866	5.755
2.125	1.851	5.339
2.170	1.851	5.618
2.164	1.866	5.551
2.201	1.843	5.551
2.151	1.835	5.551
2.157	1.851	5.551
2.112	1.866	5.618

6.26. The data are taken from Joglekar, G., Schuenemeyer, J. H., and LaRiccia, V. Lack-of-fit testing when replicates are not available. *American Statistician,* 43, 135–143, 1989. The data are given in the file **windmill**.

Direct current output from a windmill generator is measured against wind velocity (miles per hour). Data for 25 observations are given.

Model DC output as a function of wind velocity. Construct scatter plots of DC output against wind velocity. Pay particular attention to any possible transformations. Consider transformations such as the reciprocal, the square, the square root, and the logarithmic transformations.

Carefully check the adequacy of your model(s). Use your model to obtain a 90% confidence interval for the mean DC output when the wind velocity is 7 miles per hour.

x = Wind Velocity	y = DC Output
2.45	0.123
2.70	0.500
2.90	0.653
3.05	0.558
3.40	1.057
3.60	1.137
3.95	1.144
4.10	1.194
4.60	1.562
5.00	1.582
5.45	1.501
5.80	1.737
6.00	1.822

x = Wind Velocity	y = DC Output
6.20	1.866
6.35	1.930
7.00	1.800
7.40	2.088
7.85	2.179
8.15	2.166
8.80	2.112
9.10	2.303
9.55	2.294
9.70	2.386
10.00	2.236
10.20	2.310

7 Model Selection

7.1 INTRODUCTION

In many empirical model-building contexts we are faced with **observational data** for which we cannot preselect (or control) the levels of the regressor variables. In such observational studies, a large number of potential explanatory variables are measured along with the response. The goal of the study may be to build a model for prediction or, even simpler, to decide which of the potential explanatory variables influence the response. Often, these models are not motivated by any theory but are completely descriptive.

With observational data the analyst lacks the ability to select the levels of the covariates (the explanatory or regressor variables), and this fact makes model building difficult. The situation is different in experimental studies in which the analyst can control the levels of the covariates. By selecting the levels of the covariates in an "optimal" way, the model building can be simplified considerably. In this chapter, we focus on model building with observational data. The covariates have to be taken as they come, irrespective of whether their underlying design is good or bad.

We wish to develop a systematic procedure for selecting the "best" models. This requires a definition of what we mean by best. The definition needs to incorporate the concepts of **model fit** and **model simplicity** (parsimony). We also need to develop a strategy for finding the best models. Our goal is to find procedures that lead to acceptable models and that even regression novices can follow. Even better, "automatic" procedures may be possible that do not require much user input.

In previous chapters, we developed the inference for given regression models of the form

$$y = X\boldsymbol{\beta} + \boldsymbol{\epsilon}, \quad \text{with} \quad \boldsymbol{\epsilon} \sim N(\mathbf{0}, \sigma^2 I)$$

Given the response $\boldsymbol{y}$ and the set of explanatory variables describing the matrix X, it is fairly straightforward to obtain estimates and to carry out various tests of hypotheses. So where is the source of difficulty with model building?

Several difficulties are listed below:

i. In observational studies, the constellations of the covariates may reflect poor combinations.

ii. The real (observational) world and the idealized (model) world are not the same. There may be no true model after all because all models are at best approximations.

iii. The presence of high-leverage cases, outliers, and influential cases may have an impact on the model selection.

iv. Variables may be given in a certain metric that makes model building very difficult.

v. The purpose of the model may be unclear. A model that is good for prediction may not be best in terms of providing the most accurate historic fit (i.e., the most accurate description of the given data). The fit to historic data can always be improved by adding more regressors to the model. However, each additional regressor variable requires the estimation of a parameter. If the covariate is not needed, then the unnecessary estimation error adds variability to the prediction.

A more formal explanation of this follows. We learned in Chapter 6 that the covariance matrix of the vector of fitted values from a linear regression on p covariates is $V(\hat{\mu}) = H\sigma^2$, where $H = X(X'X)^{-1}X'$. Thus, the average variance of the fitted values is $\sum_{i=1}^{n} V(\hat{\mu}_i)/n = \sigma^2 \text{tr}(H)/n = \sigma^2(p+1)/n$. The prediction of the response y_{new} for a new case with covariates $\boldsymbol{x}_{\text{new}}$ is given by $\hat{y}_{\text{new}} = \boldsymbol{x}'_{\text{new}}\hat{\boldsymbol{\beta}}$. The prediction error

$$y_{\text{new}} - \hat{y}_{\text{new}} = \epsilon_{\text{new}} + (\mu_{\text{new}} - \hat{y}_{\text{new}})$$

combines the variability of the new observation and the error in the fitted value. It has average variance $\sigma^2(1 + (p+1)/n)$. Each unnecessary parameter increases the variance of the prediction error by a factor of $1/n$.

Example: Power Plant Data

Various aspects of model building are illustrated with an example from the book *Applied Statistics* by D. R. Cox and E. J. Snell. Table 7.1 lists the construction costs of 32 light water reactor (LWR) power plants, together with characteristics of the plants and details on their construction. The objective of the modeling is to learn which of the covariates influence capital cost so that we can predict the cost of constucting a new plant with certain specified characteristics. The response y and the 10 regressor variables are as follows:

C Cost in dollars $\times 10^{-6}$, adjusted to 1976 base

D Date construction permit issued

T_1 Time between application for permit and issue of permit

T_2 Time between issue of operating license and construction permit

TABLE 7.1 POWER PLANT DATA [DATA FILE: powerplant]

C	D	T_1	T_2	S	PR	NE	CT	BW	N	PT
460.05	68.58	14	46	687	0	1	0	0	14	0
452.99	67.33	10	73	1,065	0	0	1	0	1	0
443.22	67.33	10	85	1,065	1	0	1	0	1	0
652.32	68.00	11	67	1,065	0	1	1	0	12	0
642.23	68.00	11	78	1,065	1	1	1	0	12	0
345.39	67.92	13	51	514	0	1	1	0	3	0
272.37	68.17	12	50	822	0	0	0	0	5	0
317.21	68.42	14	59	457	0	0	0	0	1	0
457.12	68.42	15	55	822	1	0	0	0	5	0
690.19	68.33	12	71	792	0	1	1	1	2	0
350.63	68.58	12	64	560	0	0	0	0	3	0
402.59	68.75	13	47	790	0	1	0	0	6	0
412.18	68.42	15	62	530	0	0	1	0	2	0
495.58	68.92	17	52	1,050	0	0	0	0	7	0
394.36	68.92	13	65	850	0	0	0	1	16	0
423.32	68.42	11	67	778	0	0	0	0	3	0
712.27	69.50	18	60	845	0	1	0	0	17	0
289.66	68.42	15	76	530	1	0	1	0	2	0
881.24	69.17	15	67	1,090	0	0	0	0	1	0
490.88	68.92	16	59	1,050	1	0	0	0	8	0
567.79	68.75	11	70	913	0	0	1	1	15	0
665.99	70.92	22	57	828	1	1	0	0	20	0
621.45	69.67	16	59	786	0	0	1	0	18	0
608.80	70.08	19	58	821	1	0	0	0	3	0
473.64	70.42	19	44	538	0	0	1	0	19	0
697.14	71.08	20	57	1,130	0	0	1	0	21	0
207.51	67.25	13	63	745	0	0	0	0	8	1
288.48	67.17	9	48	821	0	0	1	0	7	1
284.88	67.83	12	63	886	0	0	0	1	11	1
280.36	67.83	12	71	886	1	0	0	1	11	1
217.38	67.25	13	72	745	1	0	0	0	8	1
270.71	67.83	7	80	886	1	0	0	1	11	1

S Power plant net capacity (MWe)

PR Prior existence of an LWR on same site (= 1)

NE Plant constructed in northeast region of the United States (= 1)

CT Use of cooling tower (= 1)

BW Nuclear steam supply system manufactured by Babcock–Wilcox (= 1)

N Cumulative number of power plants constructed by each architect–engineer

PT Partial turnkey plant (= 1); a special feature in the contract that may affect capital cost

One difficulty here is a relatively small sample size ($n = 32$), coupled with a large number of covariates ($p = 10$). Another difficulty is the fact that the covariates do not vary independently. In an observational study such as this, the analyst has no control over the measurements; they have to be taken as they come.

The difficulty arises from the special nature of the space $L(X)$ that is spanned by the regressor columns $\mathbf{1}, \boldsymbol{x}_1, \ldots, \boldsymbol{x}_p$. When estimating the model $\boldsymbol{y} = X\boldsymbol{\beta} + \boldsymbol{\epsilon}$, we always assume that X is of full column rank $p + 1$ or, in other words, that the columns of X are not linearly dependent. A major problem occurs when these vectors are getting "close" to being linearly dependent. Linear dependence means that we can express a regressor vector, for example, $\boldsymbol{x}_j$, as a linear combination of the other regressor vectors. "Almost" linearly dependent means that although the linear combination does not explain $\boldsymbol{x}_j$ perfectly, only a very small part is left unexplained. In observational studies, such almost linear dependence is often the case. One refers to this phenomenon as **multicollinearity**. We discussed its consequences and detection in Chapter 5, Section 5.4.

The plot of T_1 (time between application for permit and issue of permit) against D (date of construction) is shown in Figure 7.1. It indicates that with each calendar year the approval process is taking longer. The covariates T_1 and D are closely related which implies that a model that regresses cost on D and one that regresses cost on T_1 will fit the data equally well. There is little information in the data to distinguish between these two models.

One needs to check for associations among the explanatory variables. One procedure often suggested is an examination of scatter plots and sample correlations, r_{ij}, among all pairs (see Chapter 5, Section 5.4),

$$r_{ij} = \frac{\sum_{\ell=1}^{n} (x_{i\ell} - \bar{x}_i)(x_{j\ell} - \bar{x}_j)}{\sqrt{\sum_{\ell=1}^{n} (x_{i\ell} - \bar{x}_i)^2 \sum_{\ell=1}^{n} (x_{j\ell} - \bar{x}_j)^2}}$$

where $\bar{x}_i$ denotes the average of the measurements on the variable x_i. A matrix of the correlations provides all pairwise linear associations among the explanatory variables.

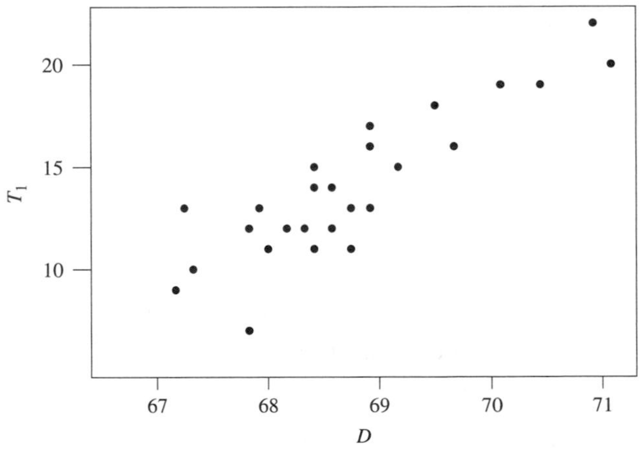

FIGURE 7.1 Plot of T_1 against D: Power Plant Data

TABLE 7.2 CORRELATIONS AMONG REGRESSOR VARIABLES: POWER PLANT DATA

	D	T_1	T_2	S	N
D	1.00	0.86	−0.40	0.02	0.55
T_1		1.00	−0.47	−0.09	0.40
T_2			1.00	0.31	−0.23
S				1.00	0.19
N					1.00

For the power plant example, the matrix of pairwise correlations among the five covariates D, T_1, T_2, S, and N is given in Table 7.2. The correlation between D and T_1 is 0.86; it summarizes the linear association between T_1 and D that we see in Figure 7.1. Note that correlations with and among the other five variables PR, NE, CT, BW, and PT are not shown.

The effect of near collinearity (or multicollinearity) among the explanatory variables is that different methods of analysis (or different analysts) may end up with final models that look very different, but describe the data equally well. If the problem is one of prediction, this difference may be unimportant. However, if the model is to be used to assess which of the explanatory variables are important in their effect on the response, different models can lead to very different conclusions.

Examining pairwise correlations is a step in the right direction. However, such a strategy ignores the joint effect of two or more covariates on another covariate x_j. A better way to check for multicollinearity is to regress x_j on all other covariates and investigate whether the coefficient of multiple determination R^2 from such a regression is large. An R^2 larger than 0.90 is often taken as an indication of multicollinearity; see the discussion of variance inflation factors in Chapter 5, Section 5.4.

A major advantage of designed experiments is that the experimenter can select the values of the covariates $x_1, \ldots, x_p$. Most sound experimental plans make sure that the covariate vectors are far from near collinearity. In fact, many experimental plans select the experiments such that these vectors are orthogonal; see the example in Chapter 5, Section 5.5.

Model Selection Procedures

In this chapter, we consider different strategies for model selection that should be seen in the following context. We are given observations on a response y and q potential explanatory variables, $v_1, \ldots, v_q$. The objective of the strategy is to select a model of the form, $y = \beta_0 + \beta_1 x_1 + \cdots + \beta_p x_p + \epsilon$, where

i. $x_1, \ldots, x_p$ is a subset of the original q regressors, $v_1, \ldots, v_q$;

ii. no important variable is left out of the model; and

iii. no unimportant variable is included in the model.

One straightforward approach is to consider all possible regressions. This is discussed in Section 7.2. Other approaches use certain model selection algorithms that can be operationalized and carried out by computers. Such "automatic" procedures are considered in Section 7.3.

7.2 ALL POSSIBLE REGRESSIONS

Given a set of potential explanatory variables $v_1, \ldots, v_q$, it is certainly feasible to fit all possible regression models to the data. For instance, if there are three regressor variables v_1, v_2, v_3, then there are eight possible models: one model with no variables and just the constant, three models with exactly one variable each; three models with two variables each, and one model with all three variables. Fitting all possible regressions requires us to fit 2^q models if q variables are involved. However, there are very clever and fast algorithms for performing this task even if q is large. The detailed output from such an approach can become very unwieldy, and hence only certain key summary statistics such as the residual sum of squares or simple functions of this quantity such as R^2 are usually shown. In the following section, we discuss these summary statistics and explain how they can be used to narrow the list of models.

7.2.1 R^2, R^2_{adj}, s^2, AND AKAIKE'S INFORMATION CRITERION

Let R^2_p denote the R^2 from a model containing p variables and $(p+1)$ regression coefficients. The intercept is usually part of the model. Recall that

$$R^2_p = 1 - \frac{\text{SSE}_p}{\text{SST}} \tag{7.1}$$

where SSE_p is the residual sum of squares for the p variable model, and $\text{SST} = \sum_{i=1}^{n} (y_i - \bar{y})^2$ is the total sum of squares, which is the same for all models and does not change with p. Everything else equal, one prefers models with larger R^2_p. Note that there will be several models with p variables, and each one will have a different R^2. It makes sense to select the best, or best few, with the largest R^2 from the group of models with p variables. R^2_p increases with p. If you allow more flexibility by adding another variable to the model, then you will automatically decrease (or at least not increase) the error sum of squares, SSE_p. The consequence is that one can get a model with R^2_p close to one by adding increasingly more variables. In the limit, if you have n observations and if the model contains n parameters, $\text{SSE} = 0$ and $R^2 = 1$.

For these reasons, we consider the **adjusted R^2**, which incorporates a penalty for each estimated coefficient,

$$R^2_{\text{adj}.p} = 1 - \frac{\text{SSE}_p/(n-p-1)}{\text{SST}/(n-1)} \tag{7.2}$$

Here the sums of squares are adjusted by their degrees of freedom: SSE_p by $n-p-1$, and SST by $n-1$. Simple algebra shows that

$$R^2_{\text{adj}.p} = 1 - \frac{(n-1)}{(n-p-1)}(1-R_p^2) = 1 - \frac{s_p^2}{\text{SST}/(n-1)} = 1 - \frac{s_p^2}{s_y^2} \quad (7.3)$$

where $s_p^2 = \text{SSE}_p/(n-p-1)$ is the mean square of the residuals, and s_y^2 is the sample variance of the response, without any adjustments by regressor variables. Equation (7.3) shows that $R^2_{\text{adj}.p}$ does not necessarily increase with p any more. If there is no improvement in R^2 by the addition of a variable, the term $(n-1)/(n-p-1)$ actually lowers the adjusted R^2. For this reason, the adjusted R^2 is a better measure for model selection. Models with the largest (or largest few) $R^2_{\text{adj}.p}$ are preferable. Two comments should be made at this point.

i. The adjusted R^2 represents the (proportionate) reduction in the residual variance achieved by the regression model. It compares the variance in the data before any regressors are put in the model (s_y^2) with the variance that remains after the regressors have been incorporated (s_p^2). It examines the reduction in **mean** squares compared to the reduction in **sums** of squares considered by R^2.

ii. An equivalent model selection tool examines the **mean square error** of the residuals s_p^2 and finds models that lead to the smallest (or the smallest few) values.

We previously mentioned that there will be more than one model for each fixed p. Instead of examining all these models, we usually restrict our attention to the best few—for example, the best three or four models—with largest values for R^2 and R^2_{adj} and smallest values for s^2.

Example: Power Plant Data

Preliminary analysis and data plots indicate that it is better to use $y = \ln(C)$ as the response variable. Let us ignore for the moment any issue of outliers. An analysis of the full model with all 10 covariates does not reveal any gross outliers or highly influential points; the only possible problem is case 26 with somewhat elevated Cook's D and studentized residual. Table 7.3 shows the results of fitting all possible regressions; only the best four models of each size (in terms of R^2) are shown. Note that for many sizes p, two or more models give almost the same R^2. For example, in the 4 variable category, the set D, T_2, S, PT ($R^2 = 0.764$) and the set T_1, S, NE, PT ($R^2 = 0.763$) give almost identical R^2. The model with D, S, NE, PT is the leader in the category $p = 4$ ($R^2 = 0.810$, $R^2_{\text{adj}} = 0.781$, $s^2 = 0.0312$). In the 5 variable category, D, S, NE, CT, PT is the best ($R^2 = 0.831$, $R^2_{\text{adj}} = 0.798$, $s^2 = 0.0288$). In the 6 variable group, D, S, NE, CT, N, PT is the best ($R^2 = 0.844$, $R^2_{\text{adj}} = 0.807$, $s^2 = 0.0276$), but the increase in R^2 (or the decrease in s^2) is small. We note that as the number of variables increases from five upwards, the increase in R^2 (decrease in s^2) is very small. Thus, one may wish to settle on a model with five covariates, such as D, S, NE, CT, PT.

TABLE 7.3 POWER PLANT DATA: ALL POSSIBLE REGRESSIONS

p	R^2	R^2_{adj}	C_p	AIC	s^2	Variables
1	0.4545	0.4363	55.91	−78.68	0.0805	PT
1	0.3957	0.3756	64.95	−75.41	0.0891	D
1	0.2064	0.1799	94.09	−66.68	0.1171	T_1
1	0.1963	0.1695	95.64	−66.28	0.1186	S
2	0.6551	0.6314	27.04	−91.36	0.0526	S PT
2	0.5813	0.5525	38.40	−85.15	0.0639	D S
2	0.5656	0.5356	40.82	−83.97	0.0663	D PT
2	0.5529	0.5221	42.77	−83.05	0.0682	N PT
3	0.7584	0.7325	13.16	−100.8	0.0382	D S PT
3	0.7167	0.6863	19.58	−95.64	0.0448	T_1 S PT
3	0.7088	0.6776	20.79	−94.76	0.0460	S N PT
3	0.6989	0.6666	22.32	−93.70	0.0476	S NE PT
4	0.8095	0.7813	7.29	−106.4	0.0312	D S NE PT
4	0.7821	0.7498	11.52	−102.0	0.0357	D S CT PT
4	0.7640	0.7290	14.30	−99.49	0.0387	D T_2 S PT
4	0.7633	0.7282	14.41	−99.40	0.0388	T_1 S NE PT
5	0.8306	0.7980	6.06	−108.1	0.0288	D S NE CT PT
5	0.8216	0.7873	7.44	−106.4	0.0303	D T_2 S NE PT
5	0.8177	0.7826	8.04	−105.8	0.0310	D S NE CT N
5	0.8149	0.7793	8.46	−105.3	0.0315	D S NE N PT
6	0.8441	0.8067	5.97	−108.8	0.0276	D S NE CT N PT
6	0.8376	0.7986	6.98	−107.5	0.0287	D T_2 S NE CT PT
6	0.8367	0.7976	7.11	−107.3	0.0289	D T_2 S PR NE PT
6	0.8335	0.7935	7.61	−106.7	0.0294	D S NE CT BW PT
7	0.8502	0.8065	7.04	−108.0	0.0276	D S NE CT BW N PT
7	0.8497	0.8058	7.12	−107.9	0.0277	D T_2 S NE CT N PT
7	0.8482	0.8039	7.34	−107.6	0.0280	D S PR NE CT N PT
7	0.8472	0.8026	7.50	−107.4	0.0281	D T_2 S PR NE CT PT
8	0.8626	0.8148	7.13	−108.8	0.0264	D T_2 S PR NE CT N PT
8	0.8538	0.8029	8.49	−106.8	0.0281	D S PR NE CT BW N PT
8	0.8525	0.8012	8.68	−106.5	0.0283	D T_2 S NE CT BW N PT
8	0.8506	0.7986	8.98	−106.1	0.0287	D T_1 T_2 S NE CT N PT
9	0.8631	0.8071	9.05	−106.9	0.0275	D T_2 S PR NE CT BW N PT
9	0.8627	0.8065	9.12	−106.8	0.0276	D T_1 T_2 S PR NE CT N PT
9	0.8538	0.7940	10.48	−104.8	0.0294	D T_1 S PR NE CT BW N PT
9	0.8526	0.7923	10.67	−104.6	0.0296	D T_1 T_2 S NE CT BW N PT
10	0.8635	0.7985	11.00	−105.0	0.0287	D T_1 T_2 S PR NE CT BW N PT

Another summary statistic that is often considered for model selection is **Akaike's information criterion (AIC).** It is defined as

$$\text{AIC}_p = n \ln(\text{SSE}_p/n) + 2(p+1) \tag{7.4}$$

Models with smaller values of AIC are preferred. The first term in Eq. (7.4) involves the logarithm of the (biased) maximum likelihood estimate of σ^2, SSE_p/n. We want this component to be small. The second term, $2(p+1)$, represents a

TABLE 7.4 PARTIAL OUTPUT OF MODEL SELECTION: POWER PLANT DATA

Variables	With All Observations	With Case 26 Deleted
$p = 5$: D, S, NE, CT, PT		
R^2	0.8306	0.8505
R^2_{adj}	0.7980	0.8206
s^2	0.0288	0.0251
AIC	−108.1	−105.5
$p = 6$: D, S, NE, CT, N, PT		
R^2	0.8441	0.8654
R^2_{adj}	0.8067	0.8318
s^2	0.0276	0.0235
AIC	−108.8	−110.2
$p = 6$: D, S, NE, CT, PT, PR		
R^2	0.8326	0.8678
R^2_{adj}	0.7924	0.8348
s^2	0.0296	0.0231
AIC	−106.5	−110.8
$p = 6$: D, S, NE, CT, PT, T_2		
R^2	0.8376	0.8540
R^2_{adj}	0.7986	0.8175
s^2	0.0288	0.0255
AIC	−101.7	−103.7

penalty function that increases with the number of estimated parameters. For the $p = 5$ variable model, D, S, NE, CT PT, the Akaike information criterion is AIC $= -108.1$ ($s^2 = 0.0288$), whereas the $p = 6$ variable model, D, S, NE, CT, N, PT, has AIC $= -108.8$ ($s^2 = 0.0276$). These two models are not too different from each other.

Let us further illustrate the use of R^2, R^2_{adj}, and AIC in the context of our example. Since the 26th observation has a fairly large Cook's D (and large studentized residual), we repeat the analysis after deleting the 26th observation. A summary of our results is shown in Table 7.4. We note:

i. The model results for the complete data and the reduced data (without case 26) are quite similar. The performance of each model improves somewhat if the 26th observation is deleted.

ii. The model with D, S, NE, CT, PT, and N is the best in the 6 variable category when all data are used. Best means largest adjusted R^2, smallest s^2, and smallest AIC. Since these measures are related through monotone functions, a model that is best on one is also best on the others. However, the model with D, S, NE, CT, PT, and PR is best in this category ($R^2 = 0.868$, $R^2_{\text{adj}} = 0.835$, $s^2 = 0.0231$) if the 26th observation is deleted.

iii. The model with D, S, NE, CT, PT, and T_2 seems to be a good competitor as well.

The results in Table 7.4 suggest that the gain (in terms of R^2_{adj} or s^2) is very small for models with $p = 6$ over the model with $p = 5$ (D, S, NE, CT, PT). Thus, we select, at least tentatively, this five-variable model. Residual plots for this model do not indicate any systematic patterns. As indicated previously, Cook's D is somewhat large for case 26, which corresponds to the power plant with the largest capacity (S).

7.2.2 C_p STATISTIC

Mallows' C_p statistic (see Mallows, 1973) is another useful summary statistic that helps us choose among candidate models. The largest model that we can fit has q regressors and $(q + 1)$ parameters. Let us denote the mean square error from this model by s^2. We assume that the largest model gives an adequate description, and hence $E(s^2) = \sigma^2$.

Consider a candidate model with p regressors $(p \le q)$ and $(p + 1)$ parameters written as $\mathbf{y} = X_1\boldsymbol{\beta}_1 + \boldsymbol{\epsilon}$ where X_1 contains $\mathbf{1}$ (the column of ones) and the p regressor vectors. If this smaller model is already adequate, then

$$\frac{\text{SSE}_p}{\sigma^2} \sim \chi^2_{n-p-1}$$

Hence,

$$E(\text{SSE}_p) = (n - p - 1)\sigma^2 \quad \text{and} \quad E\left(\frac{\text{SSE}_p}{n - p - 1}\right) = \sigma^2$$

The mean square error from this model, $\text{SSE}_p/(n - p - 1)$, is an unbiased estimator of σ^2 only if the model is adequate. If it is not, the mean square error is inflated. That is, $E(\text{SSE}_p) > (n - p - 1)\sigma^2$ and $E(\text{SSE}_p/n - p - 1) > \sigma^2$.

The C_p statistic for a model with p regressor variables and $(p + 1)$ parameters is defined as

$$C_p = \frac{\text{SSE}_p}{s^2} - [n - 2(p + 1)] \tag{7.5}$$

What happens to this measure if the candidate model is adequate? In this case, SSE_p is an estimate of $(n - p - 1)\sigma^2$, s^2 is also an estimate of σ^2, and C_p is an estimate of $\dfrac{(n - p - 1)\sigma^2}{\sigma^2} - [n - 2(p + 1)] = p + 1$. Hence,

$$E(C_p) \cong p + 1 \tag{7.6}$$

The approximation (instead of strict equality) arises because the expected value of a ratio is not exactly equal to the ratio of the expectations.

On the other hand, for a candidate model that is not yet large enough to be adequate, $E(\text{SSE}_p) > (n - p - 1)\sigma^2$, and the C_p statistic will be larger than $p + 1$. This result suggests the following strategy: Calculate the C_p statistic for each candidate model. This gives us q values for C_1; $q(q - 1)/2$ values for C_2; ...; 1 value for C_q. Within each group of p variables we prefer low values of C_p because this indicates low bias. We graph C_p against $p + 1$, the number of parameters, and add a line through the points (0,0) and $(q+1, q+1)$. Note that for the largest model with q regressors, $C_q = q + 1$. We search for the simplest model (with smallest p) that gives us an acceptable model; that is, we search for a C_p value close to $p+1$ (close to the line). Good candidate models are those with few variables and $C_p \cong p+1$. Once we have found such a model, there is no need to employ a more complicated model that involves more than p variables.

Example: Power Plant Data

The C_p statistics for all possible models in the power plant example [with $y = \ln(C)$ as the response] are shown in Table 7.3 . Only the four models with lowest C_p values are shown in each category. A C_p plot is given in Figure 7.2. The following models seem to be good candidates for a final model, and their C_p values are shown as follows:

$p+1$	C_p	Variables					
6	6.06	D	S	NE	CT	PT	
7	5.97	D	S	NE	CT	PT	N
7	6.98	D	S	NE	CT	PT	T_2

The two six-variable models (with seven parameters) are essentially the same, except that one contains N and the other T_2. Both models contain the regressors D, S, NE, CT, PT of the best five-variable model. The model with $p = 5$ gives a very acceptable C_p statistic; the value 6.06 is very close to $p + 1 = 6$. Smaller models with $p \le 4$ are not acceptable; the best model with four regressors gives $C_4 = 7.29$, which is quite a bit larger than $4 + 1 = 5$. Hence, it appears worthwhile to consider a model with five regressors. However, should we use a model with six regressors? The C_p analysis indicates that this is not necessary. Alternatively, one

FIGURE 7.2
C_p Plot for Power Plant Data

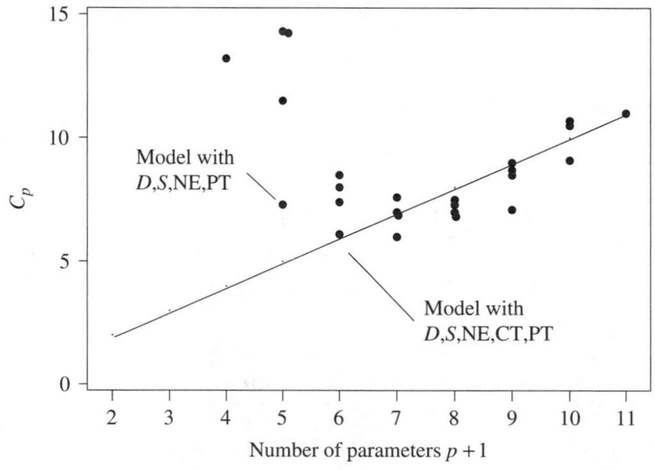

TABLE 7.5 SUMMARY OF FINAL MODEL WITH AND WITHOUT CASES 19 AND 26

	All Data	26 out	19 out	19, 26 out
s^2	0.029	0.025	0.022	0.019
Intercept	−5.40	−8.74	−5.17	−7.91
D	0.156	0.203	0.154	0.192
S	0.00087	0.00098	0.00075	0.00085
NE	0.197	0.174	0.223	0.202
CT	0.115	0.164	0.146	0.184
PT	−0.348	−0.297	−0.314	−0.276

TABLE 7.6 C_p RESULTS OF FOUR MODELS WITH AND WITHOUT CASES 19 AND 26

Variables in						$p+1$	All Data	26 out	19 out	19, 26 out
D	S	NE	CT	PT		6	6.06[a]	8.64	3.60[a]	4.03[a]
D	S	NE	CT	PT	N	7	5.97[a]	7.88	5.41	5.55
D	S	NE	CT	PT	T_2	7	6.98	10.00	4.71[a]	5.31[a]
D	S	NE	CT	PR	N	7	7.75	7.44[a]	>11.00	5.77

[a] best of its size.

can check whether the coefficients for the added regressor N (or regressor T_2) are significant. In this case, we find that they are not.

Here, everything has gone well. Too well you might say. All procedures have led to essentially the same five-variable model. As a final check, we examine a complete analysis of the five-variable model. The results are summarized in Table 7.5.

Cases 26 and 19 have elevated studentized residuals and Cook's D. Plant 26 has an unusually high cost (y) and also the largest capacity (S); plant 19 has the largest cost. Case 26 also has the largest leverage, 0.425. To examine the effects of deleting one or both of these cases, we refit the model three more times. The results are shown in Table 7.5.

No unduly large residuals or high leverage points can be found after omitting cases 19 and 26. Repeating the regressions for the four previous models, on each of the four data sets, leads to the C_p values shown in Table 7.6.

Note that if we decide to delete observation 26 alone, we might derive a different final model with PR and N included but PT omitted. In fitting this model, we find that there is strong evidence that all of the β's differ from 0 except for PR (probability value = 0.09).

7.2.3 PRESS STATISTIC

In chapter 6, we defined PRESS residuals as

$$e_{(i)} = y_i - \hat{y}_{(i)}, \quad i = 1, 2, \ldots, n \tag{7.7}$$

where $\hat{y}_{(i)} = \boldsymbol{x}_i' \hat{\boldsymbol{\beta}}_{(i)}$ is the prediction of y_i that is calculated with the least squares estimate resulting from the data set that does not include the ith case. The sum of these squared residuals is taken as the PRESS model selection criterion,

$$\text{PRESS}_p = \sum_{i=1}^{n} e_{(i)}^2 = \sum_{i=1}^{n} \left(\frac{e_i}{1 - h_{ii}} \right)^2 \tag{7.8}$$

using the result in Eq. (6.22) of Chapter 6, $e_{(i)} = \dfrac{e_i}{1 - h_{ii}}$. We can express the measure as a function of the ordinary residuals, e_i, and the leverages, h_{ii}, of the original regression. Models with smaller PRESS statistics are preferred.

Example: Power Plant Data Continued

We show the PRESS statistics for four models considered previously:

Variables	PRESS Statistic
All 10 Variables	1.5175
D, S, NE, CT, PT	1.1631
D, S, NE, CT, PT, N	1.2168
D, S, NE, CT, PT, T_2	1.2664

Again, the model with the five variables D, S, NE, CT, PT surfaces as our best choice. Models with more than five variables inflate the prediction errors.

7.3 AUTOMATIC METHODS

Ideally, one would like to devise an automatic model selection procedure that determines the best model from a list of q potential covariates. Three procedures are described in this section.

7.3.1 FORWARD SELECTION

The algorithm starts with the simplest model and adds variables as necessary. The algorithm proceeds as follows:

1. Fit the q models with a single covariate, $y = \beta_0 + \beta_1 v_k + \epsilon$, $k = 1, \ldots, q$. Set $x_1 = v_k$, where v_k is the variable that has the **most significant** regression coefficient. We look at the t ratio for testing $\beta_1 = 0$ or, equivalently, at the Fstatistic. If the most significant regressor is not significant enough (i.e., its probability value is larger than a preset significance level α), the algorithm stops. There is no need to include any of the variables, and the model that includes just the constant is appropriate. If the smallest probability value is smaller than the preset α, we include this variable in the model and go to Step 2.

2. Lock in the covariate you have found in step 1, and repeat the procedure in step 1 with models that include two regressors,

$$y = \beta_0 + \beta_1 x_1 + \beta_2 v_k + \epsilon, \quad k = 1, \ldots, q, \quad v_k \neq x_1$$

Set $x_2 = v_k$, where v_k is the variable that is most significant. We establish significance by looking at the partial t test of $\beta_2 = 0$ (or, equivalently, the partial F test that compares the full model with the restricted one found in step 1). If the probability value associated with the partial test is larger than the preset α, the procedure stops. If it is smaller than α, the variable v_k is added to the model.

3. Continue this algorithm until no remaining v_k generates a probability value that is smaller than the preset significance level α.

With this algorithm, once a variable enters the model it remains in the model. One has to specify the significance level α. Values such as 0.05 and 0.10 are typically used. One refers to it as "alpha to enter."

Example: Power Plant Data

The following output summarizes the results of forward selection on the nuclear power plant data. Here, we use the full data set in Table 7.1 ($n = 32$) and consider $y = \ln(C)$ as the response.

Case 1: Preset Significance Level $\alpha = 0.5$ Note that such $\alpha = 0.5$ is quite large, certainly larger than the commonly used values 0.05 or 0.10. It will make it very easy for variables to enter, and once variables enter, they never leave. The variables enter in the order PT, S, D, NE, CT, N, BW, PR, and T_2. The final model with nine regressors gives the results shown in Table 7.7.

TABLE 7.7 FINAL MODEL: FORWARD SELECTION, $\alpha = 0.5$

Source	df	Sum of Squares	Mean Square	F
Model	9	3.822	0.425	15.42
Error	22	0.606	0.028	

Variable	Estimated Coefficient	Probability Value for Testing $\beta_i = 0$
PT	−0.216	0.0982
S	0.0008	0.0001
D	0.243	0.0002
NE	0.261	0.0028
CT	0.114	0.1097
N	−0.012	0.1240
BW	0.026	0.7834
PR	−0.103	0.2059
T_2	0.0054	0.2335

TABLE 7.8 FINAL MODEL: FORWARD SELECTION, $\alpha = 0.15$

Source	df	Sum of Squares	Mean Square	F
Model	5	3.678	0.736	25.50
Error	26	0.750	0.029	

Variable	Estimated Coefficient	Probability Value for Testing $\beta_i = 0$
PT	−0.347	0.0013
S	0.0008	0.0001
D	0.156	0.0002
NE	0.197	0.0113
CT	0.115	0.0839

Note that many partial t tests in the final model are not significant. This model includes too many covariates; this is a consequence of the very large α.

Case 2: Preset Significance Level $\alpha = 0.15$ Setting $\alpha = 0.15$ leads to a simpler model with fewer covariates. They are included in order PT, S, D, NE, and CT. The final model is shown in Table 7.8.

i. From these two cases, we learn that the preset significance level α matters a great deal. A large preset significance level α leads to large models with many regressor variables. Consequently, one does not want an α that is too large because this would lead to many unnecessary variables. The probability values in the output of Table 7.7 show that several variables could be omitted. On the other hand, α should not be too small because this would make it too difficult for a regressor to enter the model. Values such as $\alpha = 0.10$ and $\alpha = 0.05$ are good compromises.

ii. The estimated coefficients and the probability values for the hypothesis $\beta_i = 0$ in the two models are different. For example, the probability value for PT in case 1 is 0.0982, whereas it is 0.0013 in case 2. This is because these tests are **partial** tests and the two sets of regressor variables in the models are not the same.

iii. In Table 7.7 (preset value = 0.5), the probability value for N at the final stage is 0.12. However, at the sixth iteration when N was considered for entry, the probability value was not 0.12; in fact, it was some value larger than 0.15 but less than 0.5. Thus, it entered the model in case 1. Note that N did not enter the model in case 2 since the probability value at that stage was larger than the preset value 0.15.

7.3.2 BACKWARD ELIMINATION

1. Start by fitting the very largest model

$$y = \beta_0 + \beta_1 v_1 + \cdots + \beta_q v_q + \epsilon$$

Consider dropping the variable v_k, which has the **least significant** regression coefficient. Calculate the partial t tests (or partial F tests), and determine the coefficient with the smallest t ratio and largest probability value. If this probability value is smaller than some preset significance level, then you cannot simplify the model further and you stop the algorithm. If it is larger than the preset α, you omit this regressor from the model and go to step 2.

2. Repeat the procedure in step 1 with the simplified model. That is, fit the model that does not include the dropped v_k,

$$y = \beta_0 + \beta_1 v_1 + \cdots + \beta_{k-1} v_{k-1} + \beta_{k+1} v_{k+1} + \cdots + \beta_q v_q + \epsilon$$

Find the least significant regression coefficient. If the probability value of this coefficient is smaller than a preset α, then you stop because you cannot simplify the model further. If the probability value is larger, you remove this variable from the model and continue until the maximum probability value for any variable left in the model is less than the preset value.

With this method, once a variable is omitted from the model it does not reenter. The preset significance level is called the "alpha to drop."

Example: Power Plant Data Continued

We use the preset significance level $\alpha = 0.10$. Here, the variables leave in the order T_1, BW, PR, T_2, and N, to produce a model that includes PT, S, D, NE, and CT. Using this model, we get the same final model results as in Table 7.8.

In general, the forward selection and backward elimination procedures need not lead to the same model. The outcomes from these procedures depend on the preset significance values. In the power plant example, forward selection with a preset significance level, α to enter $= 0.15$, and the backward elimination with an α to drop of 0.10 (or 0.15) lead to the same set of variables.

7.3.3 STEPWISE REGRESSION

This method combines forward selection and backward elimination.

1. Start as in forward selection using the specified significance level α to enter.
2. At each stage, once a variable has been included in the model, check all other variables in this model for their partial significance. Remove the least significant regressor variable for which the probability value for testing the hypothesis $\beta_k = 0$ is greater than the preset significance level α to drop.
3. Continue until no variables can be added and none removed, according to the specified criteria.

In this procedure, a variable may enter and leave the model several times during the execution of the algorithm. The procedure depends on the two alphas. The most commonly used values for the preset significance levels are between 0.15 and 0.05.

Example: Power Plant Data Continued

We use preset significance levels α to enter $= 0.15$ and α to drop $= 0.15$. The procedure terminates with PT, S, D, NE, and CT. In fact, no variables were removed along the way. The model summary is identical to the one in Table 7.8.

This example shows these procedures at their best. All three algorithms lead to the same conclusion: a model that involves the five explanatory variables PT, S, D, NE, and CT. However, we have previously seen that several other quite reasonable models describe the data equally well but involve other variables.

All "automatic" algorithms should be used with caution. In situations in which there is an appreciable degree of multicollinearity among the explanatory variables, the three methods may lead to quite different final models. In such situations, it is preferable to examine all possible regressions because such an analysis can show that several different models perform quite similarly (in terms of R^2, s^2, C_p).

Most observational studies will have some degree of multicollinearity. Hence, one should be cautious with automatic model selection procedures.

EXERCISES

7.1. In an experiment involving one dependent variable (y) and four explanatory variables x_1, x_2, x_3, and x_4, all possible regressions are fit to a data set consisting of $n = 13$ cases. A constant term is routinely included in all models. The results are summarized as follows:

Regressors in Model	Residual Sum of Squares
None	4,073.6
x_1	1,898.5
x_2	1,359.5
x_3	2,909.1
x_4	1,325.8
x_1, x_2	86.9
x_1, x_3	1,840.6
x_1, x_4	112.1
x_2, x_3	623.1
x_2, x_4	1,303.3
x_3, x_4	263.6
x_1, x_2, x_3	72.2
x_1, x_2, x_4	72.0
x_1, x_3, x_4	76.2
x_2, x_3, x_4	110.7
x_1, x_2, x_3, x_4	71.8

a. What model will result from automatic backward elimination with significance level (α to drop) 0.05?

b. What model will result from automatic forward selection with a significance level (α to enter) 0.1?

c. What model will result from automatic stepwise regression with significance levels α to enter $= \alpha$ to drop $= 0.1$?

d. Compare the value of the C_p statistic for the model you found in (a) with that of the model that includes all four x's.

e. In the one-variable models, x_2 and x_4 seem to be important. However, the model with x_1 and x_2 and the model with x_1 and x_4 are the best in the two-variable group, and not the model with x_2 and x_4. Explain.

f. Consider the regression model with the variables x_1, x_2, x_3, and x_4. Test the hypothesis $\beta_1 = \beta_3 = 0$.

7.2. Consider the data given in the file **hald**. It contains the variables y, x_1, x_2, x_3, x_4.

a. For each of the following criteria, indicate which set of independent variables is best for predicting y.

i. R^2

ii. C_p

b. Using (i) backward elimination, (ii) forward selection, and (iii) stepwise regression, find the best sets of independent variables.

7.3. A company studies its marketing and production processes in order to better predict production overhead costs (y_1), direct production costs (y_2), and marketing costs (y_3). It selects as predictor variables direct labor input (x_1), production quantity (x_2), sales quantity (x_3), and the change in production from the last period (x_4). Data on these variables for the past 15 months are in the file **market**.

a. For each of the three response variables, select the best model(s) for prediction. Do these models change with the selected response variable?

b. Assess which of the input factors are most important for influencing (i) overhead costs and (ii) direct production costs.

7.4. Explain why the following statements are true or false.

a. All criteria for the selection of the best regression equation lead to the same set of regressor variables.

b. Addition of a variable to a regression equation does not decrease R^2.

c. Addition of a variable to a regression model always decreases the residual mean square.

7.5. The data are taken from Woodley, W. L., Biondini, R., and Berkeley, J. Rainfall results 1970–75: Florida Area Cumulus Experiment. *Science,* 195, 735–742; 1977. The data are given in the file **rainseeding**.

These particular data come from an experiment during the summer of 1975 that investigated the usefulness of silver iodide to increase rainfall. Experiments were carried out on 24 days that were judged suitable for seeding. Suitability was judged on the basis of a suitability criterion (SC) that had to be at least 1.5 (with larger values indicating better suitability). On each day, the decision to seed or not to seed was made at random ($A = 1$ if seeding occurred; A $= 0$ if no seeding). The response is the amount of rain (in cubic meters $\times 10^7$) that fell on the target area during a 6-hr period during that day. In addition, the data set includes the following covariates:

- *Time:* Number of days after the first day of the experiment (June 1, 1975)
- *Echo coverage:* The percentage cloud cover in the experimental area, determined from radar measurements
- *Echo motion:* An indicator whether the radar echo was moving (1) or stationary (2)
- *Prewetness:* The total rainfall in the target area 1 hr before seeding (in cubic meters $\times 10^7$)

Investigate appropriate models that relate the amount of rainfall to the explanatory variables. Use model selection procedures

Seeding Action	Time	Suitability Criterion	Echo Coverage	Echo Motion	Pre-wetness	$y =$ Rainfall
0	0	1.75	13.4	2	0.274	12.85
1	1	2.70	37.9	1	1.267	5.52
1	3	4.10	3.9	2	0.198	6.29
0	4	2.35	5.3	1	0.526	6.11
1	6	4.25	7.1	1	0.250	2.45
0	9	1.60	6.9	2	0.018	3.61
0	18	1.30	4.6	1	0.307	0.47
0	25	3.35	4.9	1	0.194	4.56
0	27	2.85	12.1	1	0.751	6.35
1	28	2.20	5.2	1	0.084	5.06
1	29	4.40	4.1	1	0.236	2.76
1	32	3.10	2.8	1	0.214	4.05
0	33	3.95	6.8	1	0.796	5.74
1	35	2.90	3.0	1	0.124	4.84
1	38	2.05	7.0	1	0.144	11.86
0	39	4.00	11.3	1	0.398	4.45
0	53	3.35	4.2	2	0.237	3.66
1	55	3.70	3.3	1	0.960	4.22
0	56	3.80	2.2	1	0.230	1.16
1	59	3.40	6.5	2	0.142	5.45
1	65	3.15	3.1	1	0.073	2.02
0	68	3.15	2.6	1	0.136	0.82
1	82	4.01	8.3	1	0.123	1.09
0	83	4.65	7.4	1	0.168	0.28

(all possible regressions, backward elimination and stepwise regression). Assess the effectiveness of cloud seeding, after having adjusted your analysis for important covariates. Check for unusual cases, and determine the sensitivity of your results to these cases.

7.6. The data are taken from Vandaele, W. Participation in illegitimate activities: Erlich revisited. In: *Deterrence and Incapacitation* (Blumstein, A., Cohen, J., and Nagin, D., Eds.,). Washington, DC: National Academy of Sciences, pp. 270–335, 1978. The data are given in the file **crimerate**.

Data on crime-related statistics for 47 U.S. states in 1960 are given. The data set includes

- Crime rate: Number of offenses known to police per 1,000,000 population
- Age: Age distribution—Number of males aged 14–24 per 1,000 of total state population
- *S*: Binary variable distinguishing southern states (1) from the rest of the states
- Ed: Mean number of years of schooling x 10 of the population, 25 years or older
- PE: Police expenditures—Per capita expenditure on police protection by state and local government in 1960
- PE-1: Police expenditures—Per capita expenditure on police protection by state and local government in 1959
- LF: Labor force participation rate per 1,000 civilian urban males in the age group 14–24
- *M*: The number of males per 1,000 females
- Pop: The state population size in 100,000
- NW: The number of nonwhites per 1,000
- UE1: Unemployment rate of urban males per 1,000 in the age group 14–24
- UE2: Unemployment rate of urban males per 1,000 in the age group 35–39
- Wealth: Median value of transferable goods and assets or family income (units 10 dollars)
- IncIneq: Income inequality—Number of families per 1,000 earning below one-half of the median income

Crime Rate	Age	S	Ed	PE	PE-1	LF	M	Pop	NW	UE1	UE2	Wealth	Inc Ineq
79.1	151	1	91	58	56	510	950	33	301	108	41	394	261
163.5	143	0	113	103	95	583	1,012	13	102	96	36	557	194
57.8	142	1	89	45	44	533	969	18	219	94	33	318	250
196.9	136	0	121	149	141	577	994	157	80	102	39	673	167
123.4	141	0	121	109	101	591	985	18	30	91	20	578	174
68.2	121	0	110	118	115	547	964	25	44	84	29	689	126
96.3	127	1	111	82	79	519	982	4	139	97	38	620	168
155.5	131	1	109	115	109	542	969	50	179	79	35	472	206
85.6	157	1	90	65	62	553	955	39	286	81	28	421	239
70.5	140	0	118	71	68	632	1,029	7	15	100	24	526	174
167.4	124	0	105	121	116	580	966	101	106	77	35	657	170
84.9	134	0	108	75	71	595	972	47	59	83	31	580	172
51.1	128	0	113	67	60	624	972	28	10	77	25	507	206
66.4	135	0	117	62	61	595	986	22	46	77	27	529	190
79.8	152	1	87	57	53	530	986	30	72	92	43	405	264
94.6	142	1	88	81	77	497	956	33	321	116	47	427	247
53.9	143	0	110	66	63	537	977	10	6	114	35	487	166
92.9	135	1	104	123	115	537	978	31	170	89	34	631	165
75.0	130	0	116	128	128	536	934	51	24	78	34	627	135
122.5	125	0	108	113	105	567	985	78	94	130	58	626	166
74.2	126	0	108	74	67	602	984	34	12	102	33	557	195
43.9	157	1	89	47	44	512	962	22	423	97	34	288	276
121.6	132	0	96	87	83	564	953	43	92	83	32	513	227
96.8	131	0	116	78	73	574	1,038	7	36	142	42	540	176
52.3	130	0	116	63	57	641	984	14	26	70	21	486	196
199.3	131	0	121	160	143	631	1,071	3	77	102	41	674	152
34.2	135	0	109	69	71	540	965	6	4	80	22	564	139
121.6	152	0	112	82	76	571	1,018	10	79	103	28	537	215
104.3	119	0	107	166	157	521	938	168	89	92	36	637	154
69.6	166	1	89	58	54	521	973	46	254	72	26	396	237
37.3	140	0	93	55	54	535	1,045	6	20	135	40	453	200
75.4	125	0	109	90	81	586	964	97	82	105	43	617	163
107.2	147	1	104	63	64	560	972	23	95	76	24	462	233
92.3	126	0	118	97	97	542	990	18	21	102	35	589	166
65.3	123	0	102	97	87	526	948	113	76	124	50	572	158
127.2	150	0	100	109	98	531	964	9	24	87	38	559	153
83.1	177	1	87	58	56	638	974	24	349	76	28	382	254
56.6	133	0	104	51	47	599	1,024	7	40	99	27	425	225
82.6	149	1	88	61	54	515	953	36	165	86	35	395	251
115.1	145	1	104	82	74	560	981	96	126	88	31	488	228
88.0	148	0	122	72	66	601	998	9	19	84	20	590	144
54.2	141	0	109	56	54	523	968	4	2	107	37	489	170
82.3	162	1	99	75	70	522	996	40	208	73	27	496	224
103.0	136	0	121	95	96	574	1,012	29	36	111	37	622	162
45.5	139	1	88	46	41	480	968	19	49	135	53	457	249
50.8	126	0	104	106	97	599	989	40	24	78	25	593	171
84.9	130	0	121	90	91	623	1,049	3	22	113	40	588	160

Use model selection procedures (all possible regressions and stepwise regression) to find adequate models that relate the crime rate to the explanatory variables. Check your models (outliers and influential cases, multicollinearity), and interpret the estimated coefficients.

7.7. The data are taken from Brownlee, K. A. *Statistical Theory and Methodology in Science and Engineering,* 2nd ed. London: Wiley, 1965. The data are given in the file **stackloss**.

These data arise in the production of nitric acid in the process of oxidizing ammonia. The response variable, stack loss, is the percentage of the ingoing ammonia that escapes unabsorbed. Key process variables are the airflow, the cooling water inlet temperature (in degrees C), and the acid concentration (in percent).

Construct a regression model that relates the three predictor variables to the response, stack loss. Check the adequacy of the fitted model.

Air Flow	Cooling Temp	Acid Percent	$y =$ Stack Loss
80	27	58.9	4.2
80	27	58.8	3.7
75	25	59.0	3.7
62	24	58.7	2.8
62	22	58.7	1.8
62	23	58.7	1.8
62	24	59.3	1.9
62	24	59.3	2.0
58	23	58.7	1.5
58	18	58.0	1.4
58	18	58.9	1.4
58	17	58.8	1.3
58	18	58.2	1.1
58	19	59.3	1.2
50	18	58.9	0.8
50	18	58.6	0.7
50	19	57.2	0.8
50	19	57.9	0.8
50	20	58.0	0.9
56	20	58.2	1.5
70	20	59.1	1.5

7.8. The data are taken from a study by Gorman, J. W., and Toman, R. J. Selection of variables for fitting equations to data. *Technometrics,* 8, 27–51, 1966. The article also gives a detailed description of the experiment. The data are given in the file **asphalt**.

Data on pavement durability are given here. We list measurements on the change in rut depth (y) of 31 experimental asphalt pavements that were prepared under different conditions, as specified by the levels of the following five design variables: viscosity of the asphalt, percentage of asphalt in the

$y =$ Change in Rut Depth	Viscosity	% Asphalt Surface	% Asphalt Base	% Fines Surface	% Voids Surface	Run Indicator
6.75	2.80	4.68	4.87	8.4	4.916	−1
13.00	1.40	5.19	4.50	6.5	4.563	−1
14.75	1.40	4.82	4.73	7.9	5.321	−1
12.60	3.30	4.85	4.76	8.3	4.865	−1
8.25	1.70	4.86	4.95	8.4	3.776	−1
10.67	2.90	5.16	4.45	7.4	4.397	−1
7.28	3.70	4.82	5.05	6.8	4.867	−1
12.67	1.70	4.86	4.70	8.6	4.828	−1
12.58	0.92	4.76	4.84	6.7	4.865	−1
20.60	0.68	5.16	4.76	7.7	4.034	−1
3.58	6.00	4.57	4.82	7.4	5.450	−1
7.00	4.30	4.61	4.65	6.7	4.853	−1
26.20	0.60	5.07	5.10	7.5	4.257	−1
11.67	1.80	4.66	5.09	8.2	5.144	−1
7.67	6.00	5.42	4.41	5.8	3.718	−1
12.25	4.40	5.01	4.74	7.1	4.715	−1
0.76	88.00	4.97	4.66	6.5	4.625	1
1.35	62.00	5.01	4.72	8.0	4.977	1
1.44	50.00	4.96	4.90	6.8	4.322	1
1.60	58.00	5.20	4.70	8.2	5.087	1
1.10	90.00	4.80	4.60	6.6	5.971	1
0.85	66.00	4.98	4.69	6.4	4.647	1
1.20	140.00	5.35	4.76	7.3	5.115	1
0.56	240.00	5.04	4.80	7.8	5.939	1
0.72	420.00	4.80	4.80	7.4	5.916	1
0.47	500.00	4.83	4.60	6.7	5.471	1
0.33	180.00	4.66	4.72	7.2	4.602	1
0.26	270.00	4.67	4.50	6.3	5.043	1
0.76	170.00	4.72	4.70	6.8	5.075	1
0.80	98.00	5.00	5.07	7.2	4.334	1
2.00	35.00	4.70	4.80	7.7	5.705	1

surface course, percentage of asphalt in the base course, percentage of fines in the surface course, and percentage of voids in the surface course. The last variable is an indicator variable that separates the results of 16 pavements tested in one set of runs from 15 tested in the second run. Note that asphalt viscosity is considerably higher in the second set of runs.

An objective of the experiment was to determine the important factors that affect the change in rut depth. Develop a regression model that explains the change in rut depth as a function of the explanatory variables. Check the adequacy of the model and interpret the model results.

7.9. The data, taken from H. Strasser (Vienna University of Economics and Business Administration, 2003), include information on 109 Austrian school children. Measurements on the following variables were taken: gender (0, male; 1, female), age (in months), IQ, Math1 (assessing mathematics computation), Math2 (assessing mathematics problem solving), Read1 (assessing reading speed), Read2 (assessing reading comprehension). Data for the first five children are given below; the complete data set is given in the file **achievement**.

Gender	Age	IQ	Math1	Math2	Read1	Read2
1	121	99	12	11	27	17
0	124	83	13	4	12	15
1	103	117	5	8	30	26
1	127	83	8	6	30	12
0	115	109	7	4	26	27
.	.	.	.	.	.	

Analyze the information. Explore relationships among the various scores. Discuss whether age, gender, and IQ are important predictors of mathematics and reading abilities.

7.10. The data are taken from Kieschnick, R., and McCullough, B. D. Regression analysis of variates observed on (0,1): Percentages, proportions, and fractions. *Statistical Modelling: An International Journal,* 3, 193–213, 2003. The data are given in the file **election 2000**.

The dependent variable is the fraction of a state's total counted vote that was for President George Bush in the 2000 presidential election. Independent variables include the unemployment rate, the total population, the proportion of males, the proportion of males older than 18 years, the proportion of the population older than 65 years, the proportion of the rural (nonmetro) population, the proportion of the population below the poverty rate, the total number of households, and the proportion of households earning more than \$50,000, \$75,000, or \$100,000.

Develop appropriate regression models (using all possible regressions and stepwise regression) and check their adequacy. Pay particular attention to checking the model assumptions.

States	Total votes	% Votes for Bush	Unemployment Rate	Total Population	% Male	% Male Age >18	% Population >65	% of nonmetro Population	% Population below Poverty Level	Number of Households	% House Income >$50,000	% House Income >$75,000	% House Income >$100,000
Alabama	1,672,551	56.47	4.6	4,447,100	48.27	47.26	13.0	32.68	16.1	1,472,906	14.3	4.3	2.0
Alaska	285,560	58.62	6.6	626,932	51.69	51.83	5.7	58.48	9.4	85,359	36.3	16.8	6.9
Arizona	1,532,016	51.02	3.9	5,130,632	49.92	49.42	13.0	14.03	13.9	1,571,330	11.0	4.2	1.8
Arkansas	921,781	51.31	4.4	2,673,400	48.80	47.94	14.0	53.66	15.8	836,388	10.0	2.9	1.6
California	10,965,856	41.65	4.9	33,871,648	49.82	49.26	10.6	3.31	14.2	9,709,296	23.1	10.0	4.4
Colorado	1,741,368	50.75	2.7	4,301,261	50.35	50.00	9.7	18.83	9.3	1,566,037	15.6	6.0	2.4
Connecticut	1,459,525	38.44	2.3	3,405,565	48.43	47.51	13.8	8.55	7.9	1,024,529	37.2	15.4	6.3
Delaware	327,622	41.90	4.0	783,600	48.56	47.64	13.0	19.99	9.2	236,751	31.3	14.3	5.8
District of Columbia	201,894	8.95	5.8	572,059	47.09	46.27	12.2	0.00	20.2	262,959	21.4	10.8	6.4
Florida	5,963,110	48.85	3.6	15,982,378	48.80	48.05	17.6	7.16	12.5	5,613,696	18.4	7.0	3.5
Georgia	2,596,804	54.67	3.7	8,186,453	49.19	48.45	9.6	30.78	13.0	2,152,584	18.4	6.9	4.0
Hawaii	367,951	37.46	4.3	1,211,537	50.25	49.85	13.3	27.68	10.7	232,147	22.5	10.7	5.8
Idaho	501,621	67.17	4.9	1,293,953	50.12	49.62	11.3	66.59	11.8	274,713	15.6	4.5	2.2
Illinois	4,742,123	42.58	4.4	12,419,293	48.95	48.16	12.1	15.12	10.7	4,563,760	19.1	7.3	3.5
Indiana	2,199,302	56.65	3.2	6,080,485	49.06	48.27	12.4	27.80	9.5	1,829,081	15.0	4.2	1.7
Iowa	1,315,563	48.22	2.6	2,926,324	49.06	48.29	14.9	54.68	9.1	1,201,408	14.3	4.2	1.8
Kansas	1,072,216	58.04	3.7	2,688,418	49.42	48.72	13.3	43.42	9.9	1,052,913	16.3	4.9	2.6
Kentucky	1,544,187	56.50	4.1	4,041,769	48.88	48.05	12.5	51.18	15.8	1,573,260	11.9	3.5	1.6
Louisiana	1,765,656	52.55	5.5	4,468,976	48.40	47.40	11.6	24.59	19.6	1,675,045	13.1	5.5	3.7
Maine	651,817	43.97	3.5	1,274,923	48.67	47.84	14.4	59.66	10.9	536,617	21.2	6.7	3.0
Maryland	2,025,480	40.18	3.9	5,296,486	48.29	47.31	11.3	7.28	8.5	1,768,635	27.6	12.8	5.6
Massachusetts	2,702,984	32.50	2.6	6,349,097	48.19	47.23	13.5	1.51	9.3	2,304,767	35.0	16.5	7.4
Michigan	4,232,711	46.14	3.6	9,938,444	49.03	48.24	12.3	17.80	10.5	3,770,974	20.2	7.3	3.2
Minnesota	2,438,685	45.50	3.3	4,919,479	49.52	48.88	12.1	29.60	7.9	1,876,964	19.3	6.1	2.9
Mississippi	994,926	57.62	5.7	2,844,658	48.29	47.26	12.1	67.94	19.9	973,833	11.7	3.4	1.8

(*Continued*)

States	Total votes	% Votes for Bush	Unemployment Rate	Total Population	% Male	% Male Age >18	% Population >65	% of nonmetro Population	% Population below Poverty Level	Number of Households	% House Income >$50,000	% House Income >$75,000	% House Income >$100,000
Missouri	2,359,892	50.42	3.5	5,595,211	48.61	47.73	13.5	32.18	11.7	2,240,337	15.8	5.5	2.2
Montana	410,997	58.44	4.9	902,195	49.82	49.29	13.4	76.76	14.6	336,099	11.4	3.1	1.4
Nebraska	697,019	62.25	3.0	1,711,263	49.29	48.61	13.6	47.42	9.7	599,558	15.3	5.4	2.2
Nevada	608,970	49.52	4.1	1,998,257	50.96	50.79	11.0	12.54	10.5	563,602	15.3	4.6	2.1
New Hampshire	569,081	48.07	2.8	1,235,786	49.19	48.48	12.0	37.66	6.5	445,399	31.2	11.6	5.5
New Jersey	3,187,226	40.29	3.8	8,414,350	48.53	47.62	13.2	0.00	8.5	2,433,813	30.6	14.1	6.4
New Mexico	598,605	47.85	4.9	1,819,046	49.16	48.45	11.7	43.10	18.4	498,566	11.8	3.6	1.7
New York	6,821,999	35.23	4.6	18,976,457	48.21	47.20	12.9	7.92	14.6	5,862,138	26.6	12.9	7.0
North Carolina	2,911,262	56.03	3.6	8,049,313	48.98	48.27	12.0	32.45	12.3	2,530,739	15.7	5.5	2.5
North Dakota	288,256	60.66	3.0	642,200	49.90	49.44	14.7	55.78	11.9	281,483	9.9	2.7	1.1
Ohio	4,705,457	49.97	4.1	11,353,140	48.56	47.67	13.3	18.84	10.6	4,514,331	17.0	5.3	2.1
Oklahoma	1,234,229	60.31	3.0	3,450,654	49.14	48.40	13.2	39.19	14.7	1,269,978	11.6	3.5	1.6
Oregon	1,533,968	46.52	4.9	3,421,399	49.60	49.03	12.8	29.15	11.6	1,268,514	11.4	3.8	1.6
Pennsylvania	4,913,119	46.43	4.2	12,281,054	48.29	47.34	15.6	15.39	11.0	4,256,258	17.3	5.7	2.3
Rhode Island	409,112	31.91	4.1	1,048,319	48.05	47.03	14.5	8.15	11.9	366,684	22.6	8.4	3.0
South Carolina	1,382,717	56.84	3.9	4,012,012	48.59	47.70	12.1	30.04	14.1	1,552,751	13.4	4.1	1.7
South Dakota	316,269	60.30	2.3	754,844	49.62	49.01	14.3	65.43	13.2	262,967	11.0	2.9	1.2
Tennessee	2,076,181	51.15	3.9	5,689,283	48.69	47.81	12.4	32.39	13.5	2,077,975	14.7	4.7	2.1
Texas	6,407,637	59.30	4.2	20,851,820	49.65	49.03	9.9	15.15	15.4	9,392,827	12.0	4.6	2.0
Utah	770,754	66.83	3.2	2,233,169	50.10	49.47	8.5	23.77	9.4	419,854	14.7	5.1	2.0
Vermont	294,308	40.70	2.9	608,827	49.01	48.27	12.7	67.33	9.4	190,533	19.3	6.4	3.1
Virginia	2,739,447	52.47	2.2	7,078,515	49.06	48.37	11.2	21.90	9.6	1,965,893	22.3	9.4	4.7
Washington	2,487,433	44.58	5.2	5,894,121	49.77	49.24	11.2	16.88	10.6	1,672,825	16.6	5.9	2.3
West Virginia	648,124	51.92	5.5	1,808,344	48.61	47.84	15.3	57.66	17.9	725,040	9.7	2.6	1.1
Wisconsin	2,598,607	47.61	3.5	5,363,675	49.39	48.74	13.1	32.13	8.7	2,228,433	14.7	4.8	2.0
Wyoming	218,351	67.76	3.9	493,782	50.30	49.92	11.7	70.00	11.4	181,138	8.7	2.8	1.0

8 Case Studies in Linear Regression

In this chapter, we apply what we have learned about regression to the analysis of several data sets. We have selected one project on the educational achievement of Iowa students. One project deals with the price of Bordeaux wine, whereas another tries to predict the auction price of livestock. A fourth project addresses the prediction of U.S. presidential elections. The scope of these projects is broad enough so that we can illustrate various aspects of regression modeling, including model selection, model estimation, and diagnostic checking. Although we do not cover all details, we give the reader suggestions of what other analyses one could try. We conclude the chapter by presenting guidelines and examples of reader-driven projects in which students of this text select the projects and collect their own data for analysis.

8.1 EDUCATIONAL ACHIEVEMENT OF IOWA STUDENTS

Data on average test scores for 325 school districts in the State of Iowa are given in the file **iowastudent**. The data set from the 2000–2001 school year includes average scores on the mathematics, reading, and science portions of the Iowa Test of Basic Skills for grades 4, 8, and 11. These tests are administered each year to monitor the progress of Iowa students. The data set also includes information on the size of the school district (number of students in the district), average teacher salary (in $), and average teacher experience (in years).

Here, we address two types of issues:

1. Achievement issues: How are test scores related across fields (math, reading, and science), how are achievements related across grades (4th, 8th, and 11th grade), and how are test scores related to the size of the district?
2. Salary issues: Do teacher salaries depend on the size of the district and teacher experience?

TABLE 8.1 PAIRWISE CORRELATION COEFFICIENTS: IOWA EDUCATIONAL ACHIEVEMENT

Correlations among math, reading, and science scores at various grades

	4th math	8th math	11th math	4th read	8th read	11th read	8th sci
8th math	0.415						
11th math	0.256	0.290					
4th read	**0.663**	0.378	0.232				
8th read	0.369	**0.691**	0.260	0.386			
11th read	0.238	0.340	**0.616**	0.241	0.327		
8th sci	0.188	0.521	0.186	0.240	0.530	0.274	
11th sci	0.199	0.230	0.462	0.219	0.269	**0.646**	0.312

Correlations among discipline averages (averaged over years)

	Math avg	Read avg
Read avg	**0.773**	
Sci avg	0.494	**0.617**

Correlations among grade averages (averaged over disciplines)

	4th avg	8th avg
8th avg	0.385	
11th avg	0.318	0.385

8.1.1 ANALYSIS OF ACHIEVEMENT SCORES

Pairwise correlations among math, reading, and science scores for 4th, 8th, and 11th graders are listed in Table 8.1. Scatter plots are not shown here, but you can check that the relationships in most graphs are well described by linear models. Correlations that exceed 0.60 are set in bold type. Reading and math scores at all three grade levels and grade 11 reading and science scores are correlated most strongly. School districts that score high in reading tend to score high on math also, indicating that "good" school districts tend to be good in both areas.

Correlations of test scores across grades are also shown in Table 8.1. Although it is true that test scores are correlated across grades, the correlation across grades is considerably weaker than the correlation across disciplines.

Are test scores related to the size of the district? In the following analysis, we divide school districts into three size groups: large districts (more than 2,000 students), medium-sized districts (between 1,000 and 2,000 students), and small districts (less than 1,000 students).

We use the one-way classification analysis of variance (ANOVA) model in Section 5.3 with average district test score as response and three parameters to represent the mean scores of the three groups. That is,

$$y_i = \mu_1 x_{i1} + \mu_2 x_{i2} + \mu_3 x_{i3} + \varepsilon_i = \beta_0 + \beta_1 x_{i2} + \beta_2 x_{i3} + \varepsilon_i \qquad (8.1)$$

where x_{i1}, x_{i2}, x_{i3} are indicator variables that are one if an observation comes from a small, medium, or large district; see the discussion in Chapter 5. The model can be parameterized without an intercept and three parameters μ_1, μ_2, and μ_3, which are the means of the three groups, or it can be written as a regression model with an intercept and two parameters that relate the size effects to group 1; that is, $\beta_0 = \mu_1$, $\beta_1 = \mu_2 - \mu_1$, and $\beta_2 = \mu_3 - \mu_1$.

The results are given in Table 8.2. For 4th and 8th grades there is little support for size differences among the test scores. However, there is strong evidence that test scores for 11th graders increase with the size of the district. Large school

TABLE 8.2 ANOVA FOR EDUCATIONAL ACHIEVEMENT SCORES AND SIZE OF THE SCHOOL DISTRICT. MINITAB OUTPUT

ANOVA: 4th grade averages

```
Analysis of Variance for 4th average
Source       DF          SS          MS         F        P
Size          2       151.2        75.6      0.76    0.467
Error       319     31597.9        99.1
Total       321     31749.2
                                           Individual 95% CIs For Mean
                                           Based on Pooled StDev
Level         N        Mean       StDev    -+---------+---------+---------+-----
small       222      71.092      10.124                (----*-----)
medium       67      69.716       9.893    (---------*--------)
large        33      72.121       8.817          (------------*-------------)
                                           -+---------+---------+---------+-----
Pooled StDev =        9.953              67.5      70.0      72.5      75.0
```

ANOVA: 8th grade averages

```
Analysis of Variance for 8th average
Source       DF          SS          MS         F        P
Size          2        53.7        26.8      0.40    0.671
Error       261     17525.2        67.1
Total       263     17578.9
                                           Individual 95% CIs For Mean
                                           Based on Pooled StDev
Level         N        Mean       StDev    ---+---------+---------+---------+---
small       178      73.623       8.417        (-----*-----)
medium       55      73.576       7.329    (----------*----------)
large        31      75.011       8.339        (-------------*--------------)
                                           ---+---------+---------+---------+---
Pooled StDev =        8.194                72.0      74.0      76.0      78.0
```

ANOVA: 11th grade averages

```
Analysis of Variance for 11th average
Source       DF          SS          MS         F        P
Size          2       838.7       419.3      4.28    0.015
Error       250     24489.1        98.0
Total       252     25327.8
```

(Continued)

TABLE 8.2 (Continued)

```
                                                Individual 95% CIs For Mean
                                                Based on Pooled StDev
Level          N         Mean       StDev       ----------+---------+---------+------
small        167       73.608      10.729       (----*----)
medium        54       75.642       8.792          (--------*--------)
large         32       79.000       6.432                   (----------*-----------)
                                                ----------+---------+---------+------
Pooled StDev =          9.897                             75.0      78.0      81.0
```

ANOVA: 11th grade math averages

```
Analysis of Variance for 11th math average
Source        DF           SS          MS        F        P
Size           2        362.5       181.3     1.84    0.160
Error        314      30871.2        98.3
Total        316      31233.8
                                                Individual 95% CIs For Mean
                                                Based on Pooled StDev
Level          N         Mean       StDev       -----+---------+---------+---------+-
small        215       78.316      10.551          (----*-----)
medium        68       78.632       9.431       (---------*--------)
large         34       81.824       5.744                (------------*-------------)
                                                -----+---------+---------+---------+-
Pooled StDev =          9.915                        77.5      80.0      82.5      85.0
```

ANOVA: 11th grade reading averages

```
Analysis of Variance for 11th reading average
Source        DF           SS          MS          F        P
Size           2         1042         521       3.63    0.028
Error        310        44516         144
Total        312        45558
                                                Individual 95% CIs For Mean
                                                Based on Pooled StDev
Level          N         Mean       StDev       --+---------+---------+---------+----
small        212        71.12       13.05        (---*----)
medium        68        73.71       10.13           (--------*-------)
large         33        76.61        7.33                (-----------*-----------)
                                                --+---------+---------+---------+----
Pooled StDev =          11.98                     70.0      73.5      77.0      80.5
```

ANOVA: 11th grade science averages

```
Analysis of Variance for 11th science average
Source        DF           SS          MS       F        P
Size           2          963         482    2.64    0.073
Error        253        46183         183
Total        255        47146
                                                Individual 95% CIs For Mean
                                                Based on Pooled StDev
Level          N         Mean       StDev       --------+---------+---------+--------
small        169        72.85       14.54       (-----*-----)
medium        54        76.30       11.18             (---------*---------)
large         33        77.76       11.25              (------------*------------)
                                                --------+---------+---------+--------
Pooled StDev =          13.51                           73.5      77.0      80.5
```

districts score significantly higher than small and medium-sized districts. The test of H_0: $\mu_1 = \mu_2 = \mu_3$, or $\beta_1 = \beta_2 = 0$, leads to the F statistic $=$ 4.28 with probability value 0.015. Examining 11th grade math, reading, and science scores separately, we find that the size of the district matters most for achievement on the reading portion of the test (F ratio $=$ 3.63, probability value $=$ 0.028). The reason why 11th graders score higher in large districts may have to do with the additional educational opportunities large districts can provide.

Table 8.2 lists the output from the one-way ANOVA command of the Minitab statistical software package. Minitab refers to the square root of the mean square error of the regression model in Eq. (8.1) as the pooled standard deviation. That is,

$$\text{Pooled StDev} = \sqrt{\text{SSE}/\text{df}_{\text{Error}}}$$

where SSE is the error sum of squares, and df_{Error} are its degrees of freedom. It is an estimate of $\sigma = \sqrt{V(\varepsilon_i)}$. The 95% confidence interval for the group mean μ_i in this output is calculated from

$$\bar{y}_i \pm t(0.975; \text{df}_{\text{Error}}) \frac{\text{Pooled StDev}}{\sqrt{n_i}}$$

where $\bar{y}_i$ is the sample mean of the n_i observations in group i. These confidence intervals are shown in Table 8.2 as dashed lines.

8.1.2 ANALYSIS OF TEACHER SALARIES

Let us consider the average teacher salary in the school district as the response variable y. Does size of the school district matter? The box plots in Figure 8.1 show that average salaries increase with size of the district, but that the variability

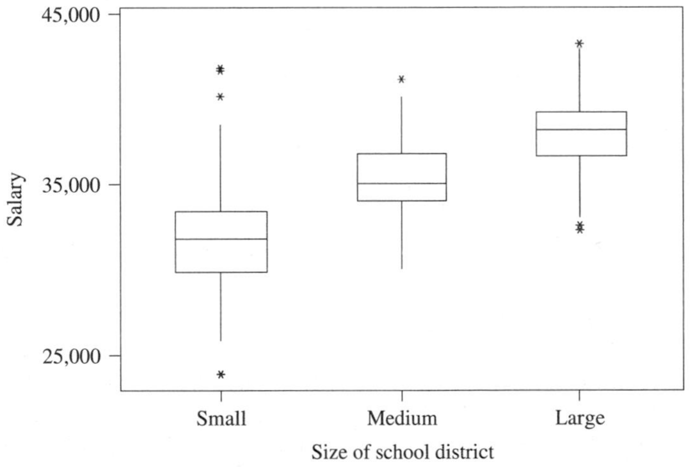

FIGURE 8.1 Box plots of teacher salaries and size of the district. MINITAB uses Q1 − 1.5 (Q3 − Q1) and Q3 + 1.5 (Q3 − Q1) for the endpoints of the lower and upper whiskers; Q1 and Q3 are the first and third quartiles. Observations beyond the endpoints of the whiskers are considered outliers and are denoted by asterisks. A test for the equality of the variances of the three groups was also considered but was found insignificant at the 0.05 significance level

TABLE 8.3 ANOVA FOR AVERAGE TEACHER SALARY AND SIZE OF THE SCHOOL DISTRICT. MINITAB OUTPUT

ANOVA: Salary versus size

```
Analysis of Variance for salary
Source      DF          SS           MS       F       P
Size         2   1.475E+09    737470964  103.30   0.000
Error      322   2.299E+09      7138790
Total      324   3.774E+09
                                        Individual 95% CIs For Mean
                                        Based on Pooled StDev
Level        N        Mean        StDev  ---+---------+---------+---------+---
small      223       31799         2820  (-*-)
medium      68       35326         2249                 (---*--)
large       34       37834         2425                           (---*----)
                                         ---+---------+---------+---------+---
Pooled StDev =        2672              32000     34000     36000     38000
```

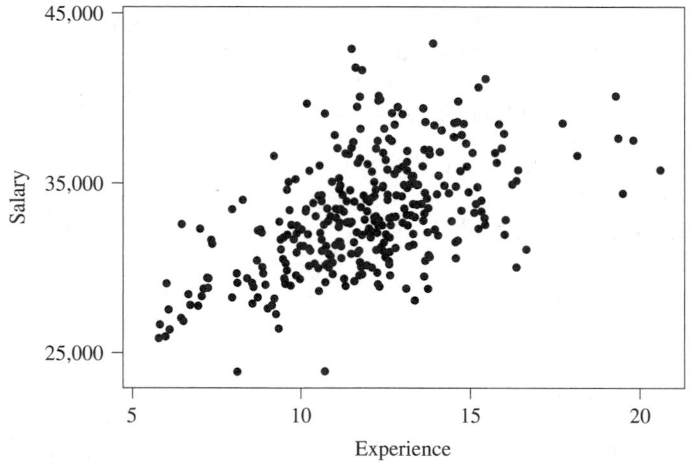

FIGURE 8.2 Scatter Plot of Teacher Salaries against Teacher Experience

(i.e., the width of the boxes) is approximately constant across the three size groups. This is important because the regression (ANOVA) model in Eq. (8.1) with average teacher salary as the response assumes constant error variance.

The results of the one-way ANOVA model in Eq. (8.1) for teacher salaries and the three district size groups are given in Table 8.3. The results show quite convincingly that teacher compensation increases with the size of the district. This effect may have to do with the cost of living. Large school districts are mostly located in urban settings, and it is more expensive to live in urban areas than in small rural towns.

Does experience have an effect on salaries? One would expect that teachers with more experience get paid more, and districts with larger average experience have higher average salaries. The scatter plot in Figure 8.2 confirms

TABLE 8.4 REGRESSION OF AVERAGE TEACHER SALARY ON SIZE OF THE SCHOOL DISTRICT AND TEACHER EXPERIENCE

```
The regression equation is
salary = 24450 + 631 experience + 3053 med size + 5522 large size
```

Predictor	Coef	SE Coef	T	P
Constant	24450.1	581.4	42.06	0.000
Experience	630.98	48.34	13.05	0.000
Medium	3052.8	301.8	10.11	0.000
Large	5522.4	400.1	13.80	0.000

S = 2163   R-Sq = 60.2%   R-Sq(adj) = 59.8%

Analysis of Variance

Source	DF	SS	MS	F	P
Regression	3	2271932689	757310896	161.88	0.000
Residual Error	321	1501699566	4678192		
Total	324	3773632255			

Diagnostics: Cases with large standardized residuals

Case	School District	Exper	Size	Salary	Fit	Residual
79	(Davies County)	16.34	medium	30051	37813	−7762
113	(Sioux Center)	11.82	small	41656	31908	9748
115	(Hudson)	11.63	small	41805	31788	10017
230	(Kingsley-Pierson)	12.31	small	40151	32217	7934
325	(Lineville-Clio)	10.72	small	23912	31214	−7302

this hypothesis. However, we already learned that size of the district matters. Adding the size of the district to the model leads to the regression

$$y_i = \beta_0 + \beta_1 \text{Experience} + \beta_2 x_{i2} + \beta_3 x_{i3} + \varepsilon_i \tag{8.2}$$

The indicators for size were explained previously. The small district group becomes the reference in our comparison. The parameter β_2 measures, for fixed experience, the difference in the average salaries of medium-size and small school districts. The parameter β_3 measures, for fixed experience, the difference in average salaries of large and small school districts. The difference $\beta_3 - \beta_2$ measures, for fixed experience, the difference in average salaries of large and medium-size school districts. The parameter β_1 measures, for given size of the school district, the benefit of an additional year of experience on average pay.

The regression summary in Table 8.4 shows that there is approximately a \$3,000 difference in the average salaries of small and medium-size districts and a \$2,500 difference in the average salaries of medium-size and large districts. Each additional year of experience "costs" the district (and earns the teachers) approximately \$630.

Leverages of six cases are larger than three times the average leverage, 0.037; the largest leverage is 0.047. However, the largest Cook's statistic (0.07569) is rather unremarkable. The histogram of the standardized residuals in Figure 8.3 shows that there are several large residuals. The five cases with standardized

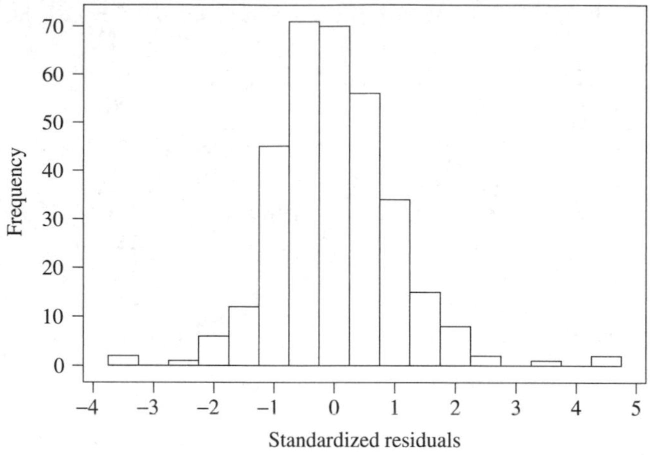

FIGURE 8.3
Histogram of Standardized Residuals from the Regression Model (8.2)

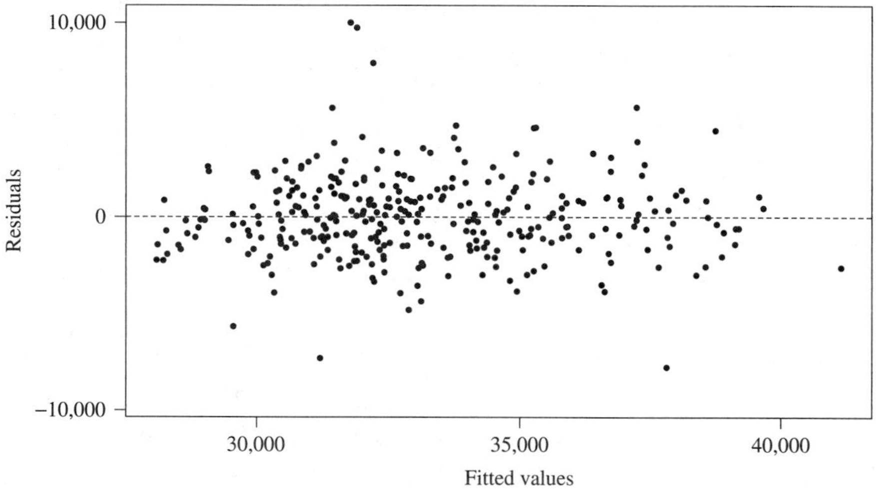

FIGURE 8.4
Scatter Plot of the Residuals against the Fitted Values, Regression Model (8.2)

residuals outside ±3 are listed in Table 8.4. The scatter plot of residuals against fitted values in Figure 8.4 fails to reveal major problems with model (8.2).

8.1.3 CONCLUDING COMMENTS

The average test scores of a school district depend on many factors, such as

- the intellectual ability of the incoming students;
- the support students get from their parents; and
- the instruction that is provided by the school and the teachers.

Of course, all three factors are difficult to measure. One could try to get a measure of student "ability" such as an average intelligence score on the student cohort that enters each district. However, because of the sensitive nature of such data, it may be almost impossible to obtain this information. One could try to assess—through surveys—the role families play in student education. The average number of hours children are tutored or the number of extracurricular activities students are exposed to could be used as proxies. Quality of instruction is very difficult to measure. One could try to obtain proxies such as the average amount of money a school district spends on each child.

Economic factors, such as the average income or the poverty rate of the school district's county, are often used to "explain" educational achievement. Regression results will find strong relationships among test scores and poverty; we already saw this in the example in Section 2.8. However, the mechanisms of the relationship are unclear because wealth (or poverty) of a school district affects all three factors listed previously. Economic conditions will attract smarter students to certain school districts (by families moving into desirable areas), affect the amount of outside help parents provide to their children, and impact the resources the district can provide. A claim that good test scores are due to excellence in instruction may be premature. The ability and composition of the student body and the family support children receive also play a major role.

8.2 PREDICTING THE PRICE OF BORDEAUX WINE

The following data are discussed in Ashenfelter, Ashmore, and Lalonde (1995). For additional discussion, see *Barron's* (December 30, 1996, pp. 17–19) and Chapter 6 of Fair (2002).

Traditionally, the quality of a Bordeaux vintage is first evaluated by experts in March of the following year. These first ratings, however, are rather unreliable because a 4-month old wine is a rather foul mixture of fermenting grape juice and little like the magnificent stuff it can become years later. Wouldn't it be wonderful to be able to rate the quality (and hence predict the price) of the most recent Bordeaux vintage immediately, compared to having to wait several months before a first, and usually inaccurate, assessment of its quality can be made?

Price data are obtained from the London market, a main market for fine wines. Price information from 1990–1991 auctions representing six chateaus (Latour, Lafite, Cheval Blanc, Pichon-Lalande, Cos d'Estournel, and Montrose) are averaged for each vintage, and the average price is expressed as a fraction of the price of the 1961 vintage (which was truly outstanding).

The theory is that the quality of the wine, and hence vintage price, depends on weather variables—such as the average temperature during the growing season (April–September, in degrees centigrade), the amount of rain during the harvest season (August and September, in total millimeters), and the amount of rain in the preceding October–March period—and the age of the vintage. It is thought that conditions for the vintage are best when the growing season is warm, August and September are dry, and the previous winter was wet. Furthermore, because of storage expenses older wines should cost more than younger ones.

Data for the years 1952–1980 are listed in Table 8.5. The 1962 price of 0.331 in column 2, for example, implies that the price of the 1962 vintage amounted to

TABLE 8.5 PRICE AND GROWING CONDITIONS OF BORDEAUX WINE, 1952–1980[a]

Vintage	Price (1961 = 1)	Average Temperature, April–September (°C)	Rainfall, August–September (ml)	Rainfall Previous, October–March (ml)	Age (1983 = 0)
1952	0.368	17.12	160	600	31
1953	0.635	16.73	80	690	30
1954	*	*	*	*	29
1955	0.446	17.15	130	502	28
1956	*	*	*	*	27
1957	0.221	16.13	110	420	26
1958	0.180	16.42	187	582	25
1959	0.658	17.48	187	485	24
1960	0.139	16.42	290	763	23
1961	1.000	17.33	38	830	22
1962	0.331	16.30	52	697	21
1963	0.168	15.72	155	608	20
1964	0.306	17.27	96	402	19
1965	0.106	15.37	267	602	18
1966	0.473	16.53	86	819	17
1967	0.191	16.23	118	714	16
1968	0.105	16.20	292	610	15
1969	0.117	16.55	244	575	14
1970	0.404	16.67	89	622	13
1971	0.272	16.77	112	551	12
1972	0.101	14.98	158	536	11
1973	0.156	17.07	123	376	10
1974	0.111	16.30	184	574	9
1975	0.301	16.95	171	572	8
1976	0.253	17.65	247	418	7
1977	0.107	15.58	87	821	6
1978	0.270	15.82	51	763	5
1979	0.214	16.17	122	717	4
1980	0.136	16.00	74	578	3

[a] The data are stored in the file **wine**.

33.1% of the price of the 1961 vintage. The prices for the 1954 and 1956 vintages are missing. Prices for these two vintages could not be established because the 1954 and 1956 vintages were poor and very little wine was sold.

Scatter diagrams of price against each of the four predictor variables suggest that the relationship between price and these predictor variables is not linear and that a logarithmic transformation of the response may be beneficial (these graphs are not shown here; we encourage you to check this conclusion). Scatter diagrams of the logarithm of price against the four predictor variables are shown in Figures 8.5a–8.5d. The results of fitting the linear model

$$\text{Ln(Price)} = \beta_0 + \beta_1\,\text{Temp} + \beta_2\,\text{Rain} + \beta_3\,\text{PRain} + \beta_4\,\text{Age} + \varepsilon \tag{8.3}$$

FIGURE 8.5
Logarithm of Price against (a) Temperature, (b) Rainfall, (c) Previous Rain, and (d) Age and (e) Plot of Residuals against Fitted Values, Model (8.3)

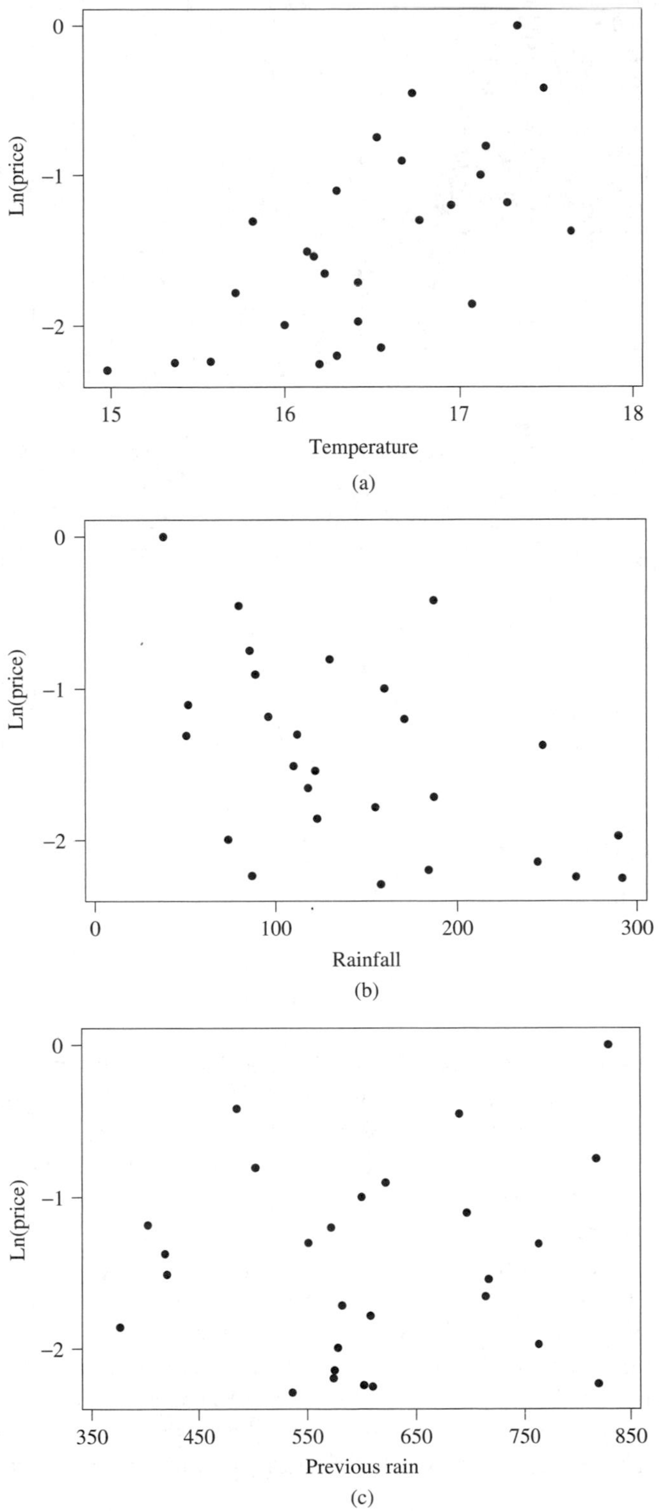

FIGURE 8.5 (Continued)

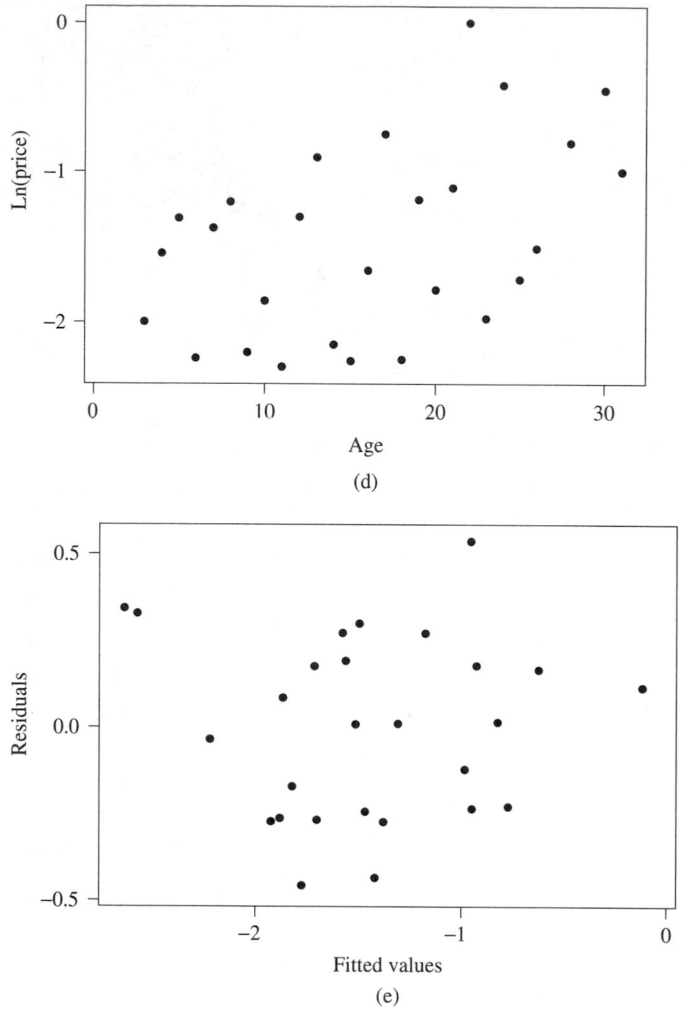

are shown in Table 8.6. The coefficients have the anticipated signs and are statistically significant, indicating that it is not possible to simplify the model. Back-transformation of Eq. (8.3) through exponentiation shows that the regression coefficients are related to percentage changes in price when changing covariates by one unit. The positive coefficient on age (0.0239) implies that the average price increases by $100(e^{0.0239} - 1) = 2.4\%$ annually (assuming, of course, that all other covariates are unchanged). The model in Eq. (8.3) explains about 83% of the variation in the response. The plot of the residuals against the fitted values in Figure 8.5e is rather unremarkable; it shows no gross violations of the fitted model. Also, none of the leverages and none of the Cook's distances are unusually large; the largest residual (case 6 for 1959) barely exceeds 2 standard deviations.

The fitting results and the model diagnostics show that Eq. (8.3) represents a fairly respectable model. The model lends support to the theory that the price of a

TABLE 8.6 MINITAB REGRESSION OUTPUT OF MODEL (8.3)

Regression Analysis: Ln(Price) versus Temp, Rain, PRain, Age

```
The regression equation is
Ln(Price) = - 12.2 + 0.617 Temp - 0.00387 Rain + 0.00117 PRain + 0.0239 Age

Predictor           Coef      SE Coef         T        P
Constant         -12.159        1.686     -7.21    0.000
Temp             0.61699      0.09502      6.49    0.000
Rain          -0.0038659    0.0008062     -4.80    0.000
PRain          0.0011710    0.0004814      2.43    0.024
Age             0.023901     0.007155      3.34    0.003

S = 0.2861          R-Sq = 82.8%      R-Sq(adj) = 79.7%

Analysis of Variance

Source            DF         SS         MS        F          P
Regression         4     8.6795     2.1699    26.51      0.000
Residual Error    22     1.8004     0.0818
Total             26    10.4799

Unusual Observations
Obs             Temp   Ln(Price)         Fit    SE Fit    Residual    St Resid
  6             17.5     -0.4186     -0.9550    0.1124      0.5364       2.04R

R denotes an observation with a large standardized residual
```

Bordeaux wine (i.e., its quality) depends on the growing conditions. Temperature and last year's rainfall are beneficial, whereas excess rain during the growing season is detrimental. The costs of storing the wine are reflected in the positive coefficient of age.

However, the real test is the model performance when using the model to predict the prices for vintages that are not part of the sample data that were used to construct the model. How successful is the regression approach in obtaining out-of-sample predictions? Can we use the model to predict the price of a new vintage?

Ray C. Fair, in his 2002 book *Predicting Presidential Elections and Other Things* (Chapter 6), lists the growing conditions for the years 1987–1991. They are given in Table 8.7. The age variable is a counting variable that is set at 0 in 1983, and hence its values for 1987 and beyond are negative. Note that it does not matter which year is taken as 0 (here, 1983); in a linear model such as Eq. (8.3), all that matters is that the variable changes by the same constant amount.

We use model (8.3) to get out-of-sample price predictions for 1987–1991. Using the new growing conditions, we calculate

$$\begin{aligned}\text{Ln(Price)} &= -12.159 + 0.617\text{Temp} - 0.00387\text{Rain}\\ &\quad + 0.00117\text{PRain} + 0.0239\text{Age}\end{aligned}$$

TABLE 8.7 GROWING CONDITIONS FOR THE PREDICTION SET, 1987–1991[a]

Vintage	Average Temperature	Rainfall	Previous Rainfall	Age (1983 = 0)	Prediction for Ln(Price)	Prediction for Price	Actual Price
1987	16.98	115	452	−4	−1.69301	0.184	0.135
1988	17.10	59	808	−5	−1.00952	0.364	0.271
1989	18.60	82	443	−6	−0.62425	0.536	0.432
1990	18.70	80	468	−7	−0.54945	0.578	0.568
1991	17.70	183	570	−8	−1.46909	0.230	0.142

[a] Predictions and actual prices are shown in columns 7 and 8.

and

$$\text{Price} = \exp(-12.159 + 0.617\text{Temp} - 0.00387\text{Rain} \\ + 0.00117\text{PRain} + 0.0239\text{Age})$$

The predictions of price are given in column 7 of Table 8.7.

Fair also lists price information that he obtained from an East Coast wine distributor. His data imply the following average prices for the 1961 and the 1987–1991 vintages: \$258.33 (1961), \$35.00 (1987), \$70.00 (1988), \$111.67 (1989), \$146.67 (1990), and \$36.67 (1991). From these numbers, he calculates the price of the 1987–1991 vintages relative to 1961 as $35.00/258.33 = 0.135$, $70.00/258.33 = 0.271$, $111.67/258.33 = 0.432$, $146.67/258.33 = 0.568$, and $36.67/258.33 = 0.142$, respectively. These are the entries in the last column of Table 8.7. A comparison of the last two columns in this table addresses the accuracy of the out-of-sample predictions. The model is correct in identifying the worst vintage (1987) and the best vintage (1990). In fact, it ranks the prices on all five vintages correctly. One can calculate percentage absolute errors, such as $100(0.184 - 0.135)/0.135 = 36.3\%$ for 1987. The mean absolute percentage error for the five periods is approximately 32%.

8.3 FACTORS INFLUENCING THE AUCTION PRICE OF IOWA COWS

Table 8.8 lists the results of livestock sales by Wapello Livestock Sales in Wapello, Iowa. This data set is part of a 2001 student project by Jay Heindel at the University of Iowa. It contains the selling price of a cow as well as various characteristics of the animal that is sold. We have also listed an explanation of how these factors can be expected to influence the price. A random sample of 115 sales over a period of 19 weeks (mid-September to the end of January 2000) is analyzed. Explanatory factors include

- Age of the animal: Cows in the mid-range may be more valuable because they have shown the ability to produce calves but are still young enough for subsequent breeding.

TABLE 8.8 AUCTION PRICE (y) AND FACTORS (x) THAT MAY HELP EXPLAIN THE PRICE[a]

Price y ($)	Age (years)	Indicator for Bred	Indicator for Angus	Frame (Large)	Weight (100 lb)	Indicator for Conditioned	Indicator for Registered
1,000	3	1	1	1	10.15	1	0
1,250	3	1	1	1	11.00	1	0
980	5	1	0	0	11.15	1	0
1,015	4	0	1	0	11.00	1	0
995	5	1	0	1	10.00	0	0
825	7	1	0	0	9.80	0	0
850	6	1	0	0	10.25	0	1
1,150	2	1	1	0	10.50	1	1
1,150	2	1	1	1	10.75	1	1
1,200	3	1	1	1	11.75	1	0
1,200	2	1	1	1	11.60	1	0
1,000	6	1	1	0	11.00	1	0
1,000	7	0	1	0	10.00	0	0
1,050	6	1	0	0	10.00	0	1
1,075	4	1	1	1	9.90	0	1
1,165	5	1	1	0	11.35	1	0
780	2	0	1	0	8.85	0	0
800	2	0	1	0	9.50	0	1
1,180	3	0	0	1	11.45	1	1
1,000	4	0	0	1	11.45	0	1
1,200	2	1	0	1	11.50	1	1
1,000	3	1	1	0	10.00	1	0
1,025	3	1	1	0	9.75	1	0
1,175	2	0	1	1	11.35	1	1
800	5	1	0	0	12.00	1	0
915	4	0	1	0	11.85	1	1
1,185	6	1	1	1	12.00	0	0
1,020	5	1	1	0	11.25	0	0
775	7	1	1	0	12.00	1	0
850	7	1	0	1	12.00	1	1
1,200	2	1	1	0	10.00	1	1
1,200	3	1	1	0	10.15	1	1
775	8	1	0	0	12.50	1	0
775	7	1	0	0	11.85	1	0
1,200	3	1	0	1	11.85	0	1
1,135	4	1	0	0	12.00	1	0
1,000	7	1	0	0	11.50	1	0
1,185	3	0	0	1	12.00	0	1
1,155	3	1	0	0	11.85	1	0
1,155	2	0	0	1	12.00	1	1
1,175	2	1	1	1	11.75	1	0
1,200	2	1	0	1	11.50	1	0
1,165	3	1	1	0	11.65	1	0
1,000	6	1	1	0	12.25	1	0
1,200	2	0	1	1	10.25	0	0
1,175	2	1	0	0	10.25	0	1

(*Continued*)

TABLE 8.8 (Continued)

Price y ($)	Age (years)	Indicator for Bred	Indicator for Angus	Frame (Large)	Weight (100 lb)	Indicator for Conditioned	Indicator for Registered
1,000	5	0	0	0	10.00	0	1
1,125	5	0	0	0	10.00	1	0
1,150	4	1	1	1	12.25	1	0
1,125	2	0	0	1	11.25	1	1
875	6	0	0	0	10.00	0	0
1,000	6	1	0	1	11.25	0	0
1,185	3	1	1	0	10.00	0	0
1,185	4	1	1	0	10.00	1	1
1,000	4	1	1	0	11.00	1	0
950	4	1	1	1	10.25	0	0
875	6	1	0	0	11.25	1	0
775	6	0	0	0	10.15	0	0
1,100	2	0	1	0	9.85	0	1
1,125	2	0	1	0	10.35	1	1
1,200	2	0	0	0	10.65	1	1
1,175	4	1	1	1	12.30	1	0
1,000	4	1	0	1	11.75	0	0
1,000	3	1	0	1	11.85	0	0
1,000	6	1	1	0	11.25	0	0
1,000	6	1	0	1	11.25	1	0
985	5	0	1	0	10.50	0	0
1,150	3	1	1	0	11.75	1	1
1,150	4	1	1	0	11.50	1	0
1,075	7	1	0	0	11.45	1	0
1,050	6	0	0	0	11.50	0	0
885	6	0	0	0	10.00	0	0
1,200	4	1	1	1	11.50	1	1
1,150	3	1	1	0	11.50	1	1
1,000	4	1	0	0	10.85	0	0
1,075	2	1	1	0	10.85	0	1
980	4	0	0	0	12.50	1	0
980	3	0	0	0	12.45	1	0
1,000	6	1	1	1	12.00	1	0
1,085	3	1	0	0	11.25	1	0
1,175	5	1	0	0	11.55	1	0
1,150	5	1	0	1	11.15	0	1
1,175	3	1	0	0	11.15	1	0
1,000	2	0	1	0	10.00	1	0
875	7	1	1	1	11.65	1	0
995	5	1	1	1	11.55	1	1
1,000	3	1	1	1	10.15	1	1
1,250	3	1	1	1	11.00	1	1
1,150	2	1	0	1	11.00	1	0
1,150	5	1	0	1	10.85	1	0
1,200	5	1	0	1	10.95	1	1
980	8	1	1	0	12.75	1	0
995	3	0	0	0	10.00	0	0

TABLE 8.8 (Continued)

Price y ($)	Age (years)	Indicator for Bred	Indicator for Angus	Frame (Large)	Weight (100 lb)	Indicator for Conditioned	Indicator for Registered
1,000	3	1	1	1	10.35	1	0
1,165	3	1	0	1	10.35	1	0
1,175	5	0	1	0	11.25	1	1
1,000	6	1	1	1	11.20	1	0
775	5	0	1	1	10.00	0	0
1,000	3	1	1	1	11.65	1	0
1,015	4	1	0	1	11.50	1	0
1,200	4	1	1	1	11.50	1	0
1,175	6	1	1	1	11.75	1	1
1,095	4	0	1	1	11.75	1	0
1,100	3	1	1	0	12.35	1	1
980	3	0	0	0	10.00	0	0
1,110	5	0	1	1	10.35	0	0
1,200	2	1	1	1	11.00	1	0
1,185	4	0	1	1	11.25	1	0
1,185	3	0	1	0	11.00	1	0
1,170	4	0	1	0	11.25	1	0
1,150	3	1	0	0	11.10	1	0
1,000	4	1	1	1	11.00	1	1
975	6	1	0	0	10.00	0	0
1,200	4	0	1	0	11.00	1	0
1,185	3	1	1	0	11.00	1	0

[a] The data are stored in the file **cows**.

- Weight of the animal: Cows are intended for the production of calves, and thcy arc not being sold for meat. Hence, the weight of the cow may not be the most deciding factor because it does not tell about the cow's ability to produce healthy marketable calves.
- Whether the cow has been bred (i.e., currently carrying a calf), indicating that a "free" calf comes with the sale.
- Frame size of the animal: A large size may avoid birthing problems in the future.
- Whether the cow is registered (recorded through a breed organization as a legitimate bloodline of a particular breed).
- Whether the cow is in good condition (i.e., well fed and "filled out").
- Whether an Angus cow is involved: Angus cattle may be more valuable because consumers are willing to pay more for their leaner meat.

Scatter plots of price against the explanatory variables age and weight are shown in Figure 8.6. Box plots of weight against the categorical variables (whether the cow has been bred, is conditioned, its frame size, whether it is an Angus cow,

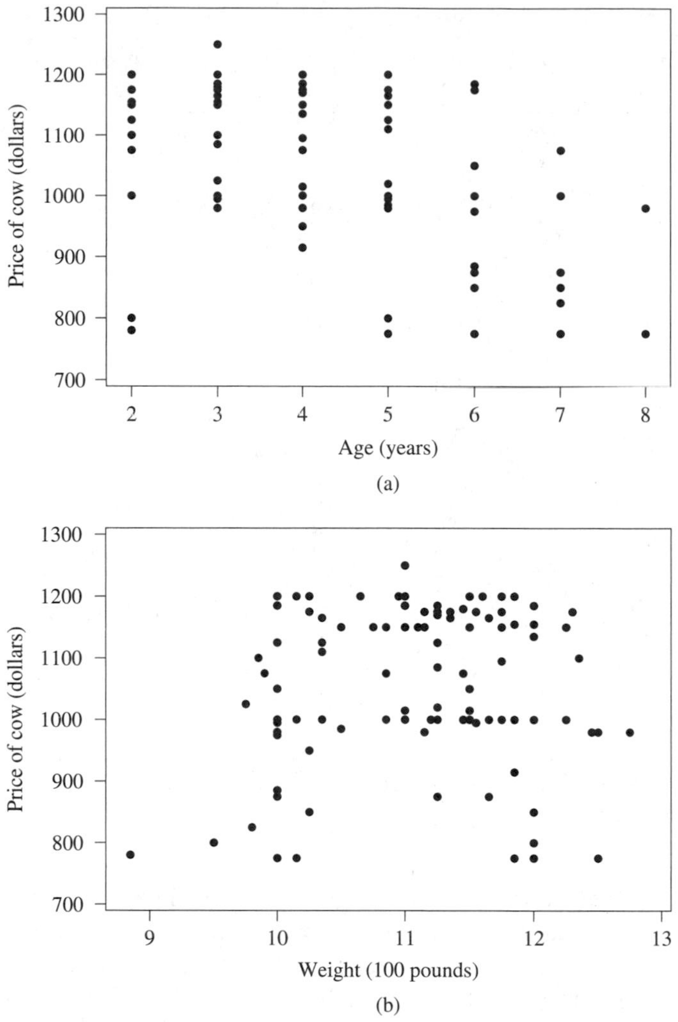

FIGURE 8.6 Scatter Plots of Auction Price against Age and Weight of the Cow

and registered) were also drawn but are not shown here. The graphs in Figure 8.6 suggest that one should allow for quadratic effects of age and weight. It appears that very young and old cows fetch less money than cows in the mid-range; the same applies for weight.

We start by fitting all possible regressions using the seven original explanatory variables plus the two constructed ones (squares of age and weight). The results are shown in Table 8.9. The results indicate that a model with three explanatory variables appears to give an acceptable representation. The Mallows C_p criteria of the two identified models (in boldface type) are quite acceptable; for an acceptable model with three regressor variables we expect a C_p of approximately 4. There appears to be no bias with the three-regressor models. Weight and its square, and the age of the cow (or, alternatively, its square), seem to

TABLE 8.9 BEST SUBSETS REGRESSION OF PRICE ON THE NINE EXPLANATORY VARIABLES

Vars	R-Sq	R-Sq(adj)	C-p	S	Age	Age**2	Weight	Weight**2	Breed	Angus	Frame	Condition	Register	
1	29.7	29.1	34.6	106.40		X								
1	27.7	27.1	38.7	107.89	X									
2	36.2	35.1	23.1	101.80		X	X							
2	35.6	34.4	24.5	102.32		X		X						
3	**46.4**	**44.9**	**4.1**	**93.753**		**X**	**X**	**X**						
3	**46.2**	**44.8**	**4.5**	**93.913**	**X**		**X**	**X**						(age, weight, weight**2)
4	47.5	45.6	3.8	93.220		X	X	X		X				
4	47.3	45.4	4.1	93.337	X		X	X		X				
5	48.4	46.0	3.9	92.823		X	X	X		X			X	
5	48.1	45.7	4.6	93.119	X		X	X		X			X	
6	48.9	46.1	4.8	92.782		X	X	X	X	X			X	
6	48.8	45.9	5.1	92.904		X	X	X		X		X	X	
7	49.2	45.9	6.3	92.973		X	X	X	X	X		X	X	
7	49.0	45.7	6.7	93.142		X	X	X	X	X	X		X	
8	49.3	45.5	8.0	93.289		X	X	X	X	X	X	X	X	
8	49.2	45.4	8.3	93.401	X	X	X	X	X	X		X	X	
9	49.3	45.0	10.0	93.722	X	X	X	X	X	X	X	X	X	

[a] Only the two best models for each group are shown. The placement of the "X" symbol indicates which variables are included.

matter most. Since a linear term of age is easier to interpret than the square of age without the linear component, we consider the model with weight, weight square, and age in more detail. With this model, we explain approximately 46.2% of the variability in price. The standard deviation of the errors is 93.9 (dollars). This needs to be compared with the standard deviation of the selling price by itself, without taking advantage of the cow's characteristics; this turns out to be 126.4 (dollars). Although this is not a striking reduction, the data do not provide us with a better model. The buyers bidding on cows must use other factors that are not recorded.

Detailed estimation results for this model are shown at the top of Table 8.10. Would the additional information that the cow in question is an Angus cow make a difference? We add the indicator for Angus beef and run the extended

TABLE 8.10 ESTIMATION RESULTS FOR TWO REGRESSION MODELS

Regression Analysis: Price versus Age, Weight, Weight2**

The regression equation is
Price = − 6745 − 42.0 Age + 1423 Weight − 63.1 Weight**2

Predictor	Coef	SE Coef	T	P
Constant	−6745	1509	−4.47	0.000
Age	−41.997	5.553	−7.56	0.000
Weight	1422.6	275.8	5.16	0.000
Weight**2	−63.07	12.58	−5.02	0.000

S = 93.91   R-Sq = 46.2%   R-Sq(adj) = 44.8%

Analysis of Variance

Source	DF	SS	MS	F	P
Regression	3	841123	280374	31.79	0.000
Residual Error	111	978981	8820		
Total	114	1820104			

Regression Analysis: Price versus Age, Weight, Weight2, Angus**

The regression equation is
Price = − 6887 − 40.4 Age + 1444 Weight − 64.0 Weight**2 + 27.6 Angus

Predictor	Coef	SE Coef	T	P
Constant	−6887	1503	−4.58	0.000
Age	−40.431	5.612	−7.20	0.000
Weight	1443.9	274.5	5.26	0.000
Weight**2	−64.02	12.51	−5.12	0.000
Angus	27.59	17.91	1.54	0.126

S = 93.34   R-Sq = 47.3%   R-Sq(adj) = 45.4%

Analysis of Variance

Source	DF	SS	MS	F	P
Regression	4	861812	215453	24.73	0.000
Residual Error	110	958292	8712		
Total	114	1820104			

regression model. The results in the bottom of Table 8.10 show that, as expected, the effect of Angus beef is positive, indicating that an Angus cow of equal weight and age would bring \$27.6 more on average. However, the partial t ratio, $27.59/17.91 = 1.54$, is not quite statistically significant. The probability value of a one-sided test of the hypothesis that Angus cattle do bring in more money is $0.126/2 = 0.063$; note that the two-sided probability value in the table needs to be cut in half. It is close to the commonly used 5% significance level. Strictly speaking, with a significance level of 5% one cannot reject the null hypothesis that Angus cattle are not priced higher. However, the p value is close to 0.05, and the results do suggest that Angus cattle may fetch a somewhat higher price.

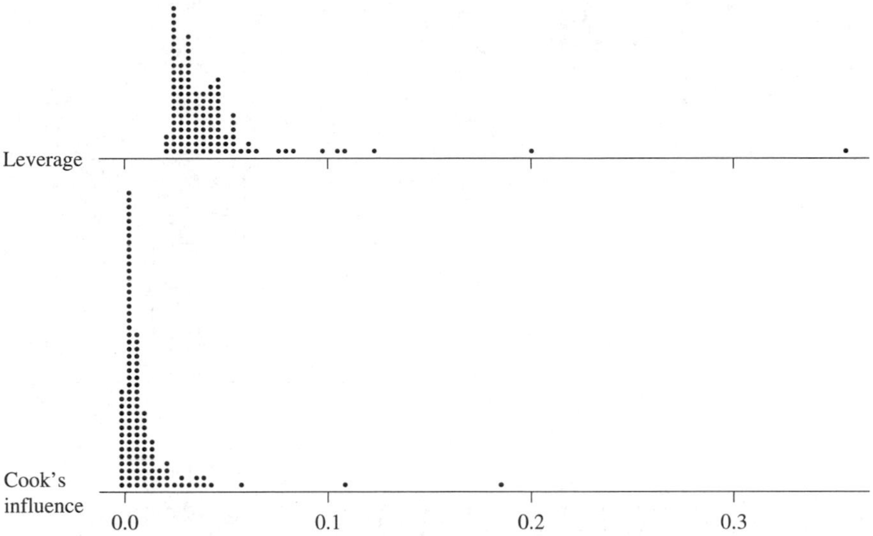

FIGURE 8.7
Dot Plots of Leverages and Cook's Influence Measures from the Model with Age, Weight, Square of Weight, and Indicator for Angus Cattle

What is the optimal weight for a cow? One can take the derivative of the model equation with respect to weight and determine the optimum. The optimum is given by $1,444/[(2)(64)] = 11.28$, and since the second derivative [the coefficient of $(\text{weight})^2$ is -64] is negative, it is indeed a maximum. This means that the optimal weight is 1,128 pounds.

What about the diagnostics of the model that includes age, weight, the square of weight, and the indicator for Angus beef? Do we see serious problems? It should be mentioned that we do not know about the time arrangement of the 115 cases in Table 8.8. All we know is that the data are a sample of 115 auction sales within a 6-month period. Since we cannot assume that the arrangement is sequential in time, the Durbin–Watson statistic is not meaningful in this context. Are there high-leverage cases, and do some cases exert high influence on the regression estimates? Leverages and Cook's influence measures are calculated for the 115 cases and they are graphed in Figure 8.7. The highest leverage is 0.353; it originates from case 17, a cow with the smallest weight (885 pounds). The leverage is quite unusual and certainly much larger than three times the average leverage (the average leverage is $5/115 = 0.0435$). However, this case does not exert much influence on the regression coefficients. The largest Cook's influence measure is 0.184; it comes from case 92, a cow with the largest weight (1,275 pounds). However, it is not unusually large (our usual warning limit is 0.50 or higher) to raise suspicion. None of the cases exert unusual influence on the fitted regression. The histogram of the (standardized) residuals is shown in Figure 8.8. None of the residuals are unusually large, and the shape of the histogram suggests that the normal distribution assumption is adequate. The approximate linear appearance of the normal probability plot (not shown) also confirms that the usual error assumptions are met.

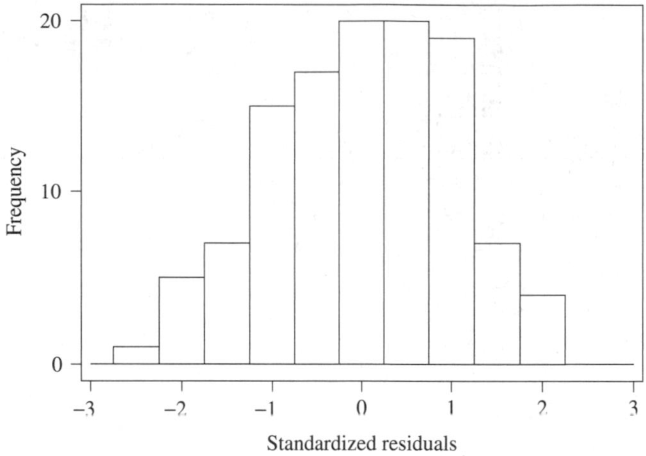

FIGURE 8.8 Histogram of the Standardized Residuals from the Model with Age, Weight, Square of Weight, and Indicator for Angus Cattle

8.4 PREDICTING U.S. PRESIDENTIAL ELECTIONS

8.4.1 A PURELY ECONOMICS-BASED MODEL PROPOSED BY RAY FAIR

Ray Fair, an economics professor at Yale University, uses the state of the economy prior to the election to predict the incumbent vote share in presidential elections. By incumbent vote share we mean the vote share for the candidate of the party occupying the White House during the election. There is a certain attraction to such forecasting models because the explanatory economic variables measure the state of the economy several months prior to the actual election. They are readily available for making real-time predictions of the election outcome.

The data in Table 8.11 are taken from Fair's book, *Predicting Presidential Elections and Other Things* (2002). Chapters 1, 3, and 4 and references in the chapter notes are relevant to our discussion. For detailed discussion and definition of the variables, see Fair's Chapter 3.

- Vote share, the response variable, represents the incumbent share of the two-party vote. By taking the two-party vote share and not the incumbent share of the total vote, one assumes that third-party candidates take about the same amount from each party. This is a reasonable assumption for most elections, except for the 1924 election. There is evidence that La Follette (of the Progressive Party) took more voters from Davis (the Democrat) than from Coolidge (the Republican). It is estimated that 76.5% of the votes for La Follette would have gone to Davis, whereas only 23.5% of the La Follette votes would have gone to Coolidge. This information is incorporated into the listed 1924 incumbent share.
- Growth rate represents the per capita growth rate of real gross domestic product (GDP) in the first three quarters (9 months) of the election year.
- Inflation rate is the average (absolute) inflation rate during the 15 quarters prior to the election (i.e., all the quarters of the administration except the

TABLE 8.11 INCUMBENT VOTE SHARE AND ITS DETERMINANTS: FAIR'S MODEL[a]

Year	Party in Power	Election Outcome	Incumbent Vote Share (%)	Growth Rate (%)	Inflation Rate (%)	Good News Quarters	Duration Value	War	President Running	Party Variable
1916	D	President Wilson beats Hughes	51.7	2.2	4.3	3	0.00	0	1	1
1920	D	Cox loses to Harding	36.1	−11.5	16.5	5	1.00	1	0	1
1924	R	Pres. Coolidge beats Davis and LaFollette	58.2	−3.9	5.2	10	0.00	0	1	0
1928	R	Hoover beats Smith	58.8	4.6	0.2	7	1.00	0	0	0
1932	R	Pres. Hoover loses to Roosevelt	40.8	−14.9	7.1	4	1.25	0	1	0
1936	D	Pres. Roosevelt beats Landon	62.5	11.9	2.4	9	0.00	0	1	1
1940	D	Pres. Roosevelt beats Willkie	55.0	3.7	0.0	8	1.00	0	1	1
1944	D	Pres. Roosevelt beats Dewey	53.8	4.1	5.7	14	1.25	1	1	1
1948	D	Pres. Truman beats Dewey	52.4	1.8	8.7	5	1.50	1	1	1
1952	D	Stevenson loses to Eisenhower	44.6	0.6	2.3	6	1.75	0	0	1
1956	R	Pres. Eisenhower beats Stevenson	57.8	−1.5	1.9	5	0.00	0	1	0
1960	R	Nixon loses to Kennedy	49.9	0.1	1.9	5	1.00	0	0	0
1964	D	Pres. Johnson beats Goldwater	61.3	5.1	1.2	10	0.00	0	1	1
1968	D	Humphrey loses to Nixon	49.6	4.8	3.2	7	1.00	0	0	1
1972	R	Pres. Nixon beats McGovern	61.8	6.3	4.8	4	0.00	0	1	0
1976	R	Ford loses to Carter	48.9	3.7	7.7	4	1.00	0	0	0
1980	D	Pres. Carter loses to Reagan	44.7	−3.8	8.1	5	0.00	0	1	1
1984	R	Pres. Reagan beats Mondale	59.2	5.4	5.4	7	0.00	0	1	0
1988	R	G. Bush beats Dukakis	53.9	2.1	3.3	6	1.00	0	0	0
1992	R	Pres. G. Bush loses to Clinton	46.5	2.3	3.7	1	1.25	0	1	0
1996	D	Pres. Clinton beats Dole	54.7	2.9	2.3	3	0.00	0	1	1

[a] The data are stored in the file **election(Fair)**.

last one). Inflation and deflation (i.e., negative inflation) are treated symmetrically. Deflation is assumed to be just as bad as inflation.

- Good news is the number of quarters out of the 15 quarters prior to the election in which the per capita growth rate exceeds 3.2%.
- Duration takes the value 0 if the incumbent party has been in office for only one consecutive term. It takes the value 1.00 for two consecutive terms, 1.25 for three consecutive terms, 1.50 for four consecutive terms, and 1.75 for five consecutive terms.
- President running variable takes the value 1 if the president is running for reelection; otherwise, the value is 0. Vice presidents who become president during the administration are also given the value 1 if they run for president. The exception is Ford, who is given a 0 because he was not part of the 1972 ticket.
- Party takes the value 1 if the incumbent party is Democratic and 0 if the incumbent party is Republican. It measures the "pure party" effect.
- War: Because of World Wars I and II, the 1920, 1944, and 1948 elections are treated differently. Fair includes a war variable that takes the value 1 for years 1920, 1944, and 1948, and 0 otherwise.

Scatter plots of the incumbent vote share against growth rate, inflation rate, good news variable, and duration are shown in Figures 8.9a–8.9d. The incumbent vote share increases with growth rate and the number of good news quarters, and it decreases with inflation rate and duration. The patterns in these figures are expected because candidates of the incumbent party tend to be elected if the economy is strong. Figure 8.9e shows an interaction diagram of incumbent vote share against incumbent party and whether the incumbent president is running. The graph shows that for Democrats the incumbent effect is weak unless the president is running for reelection.

The summary results from fitting all possible regressions with the listed predictors and the interaction (product) between the party and the president running variables are given in Table 8.12. The model with the five predictors—growth rate, inflation rate, duration, party, and the war variable—leads to an acceptable C_p statistic. The model explains 90.3% of the variation. Although this may seem like a very good fit, the standard deviation of the unexplained component is 2.6%, indicating that the 95% prediction error margins are approximately $\pm 5.2\%$. Also note that we are fitting a regression model with six parameters to just $n = 21$ cases. The detailed fitting results for this model are shown in Table 8.13. Growth rate (with positive coefficient) and inflation rate (with negative coefficient) have the expected signs. The negative effect of duration indicates that incumbency wears out if the incumbent party has occupied the White House for a long time. The negative coefficient for party indicates that Democrats in power tend to do worse. The war variable is significant and positive, indicating that in times of war voters appear to rally around the incumbent party.

FIGURE 8.9
Scatter Plots of Incumbent Vote Share against (a) Economic Growth Rate, (b) Inflation Rate, (c) Good News Variable, and (d) Duration Value and (e) Interaction Diagram—Mean Incumbent Vote Share against Incumbent Party and President Running

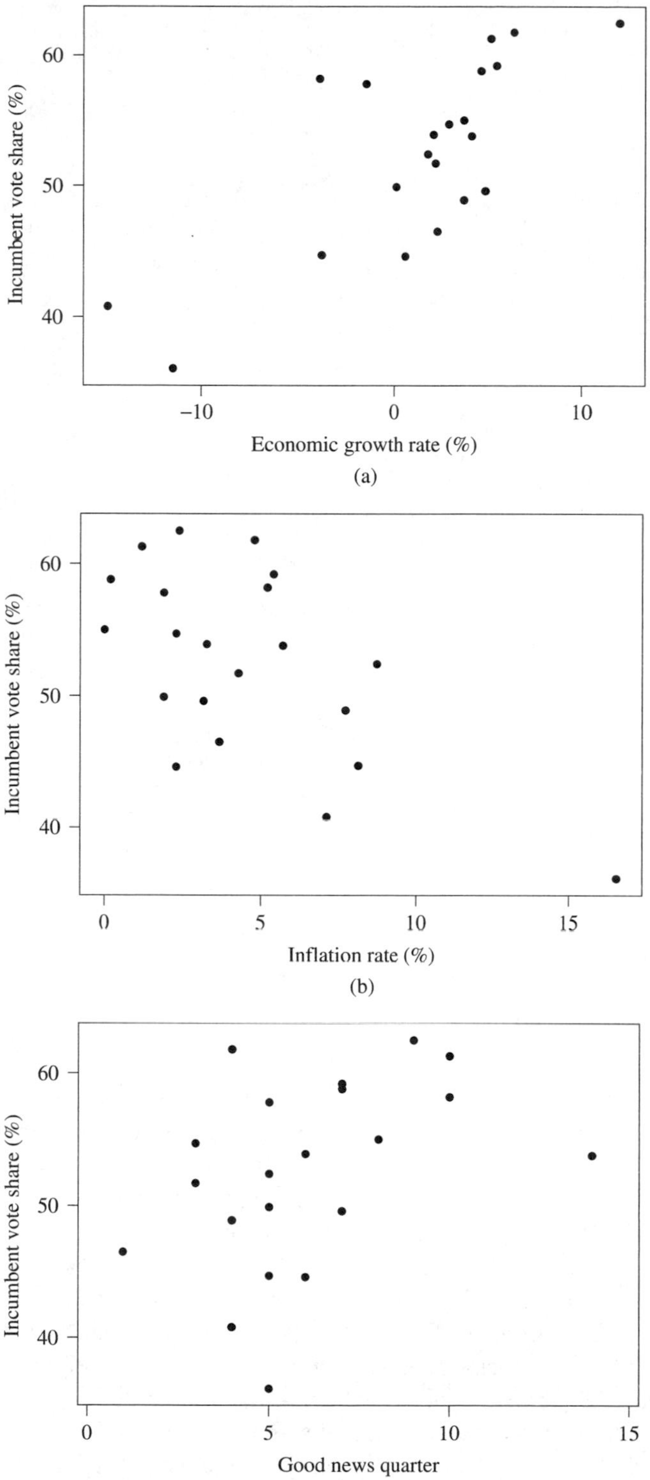

FIGURE 8.9 (Continued)

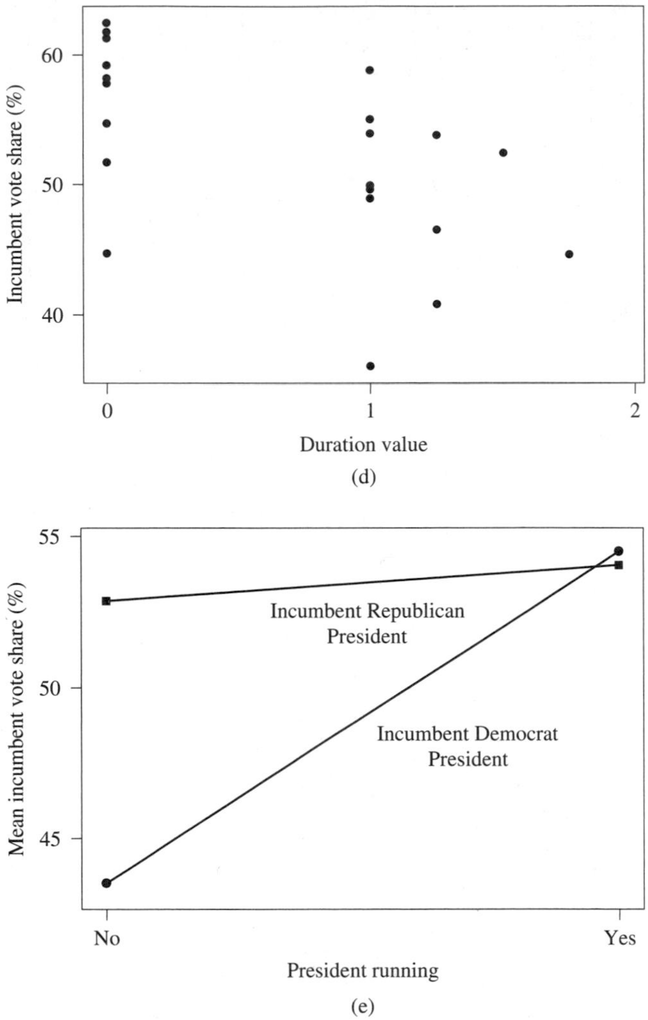

An interaction term for incumbent party and incumbent president running was quite visible in the interaction plot in Figure 8.9e. However, the final model does not include such a term. This is not necessarily a contradiction. The interaction graph gives a marginal view of the data and "collapses" the data over all other variables except the two considered in the graph. It so happens that the five predictors in the model explain this interaction and that conditional on these five variables, the need for an interaction has disappeared.

The Durbin–Watson test is appropriate here because we estimate the regression on time series data. Its value, DW $= 1.82$, is quite close to 2 and does not indicate problems with serial correlation at lag 1. The variance inflation factors are unremarkable and smaller than commonly-used cutoff values; there is not an undue amount of multicollinearity. A scatter plot of the standardized residuals

TABLE 8.12 BEST SUBSETS REGRESSION OF INCUMBENT VOTE SHARE ON ALL LISTED PREDICTOR VARIABLES

Vars	R-Sq	R-Sq(adj)	C-p	S	Growth R	Inflation	Good News	Duration	War	President	Party	Pres*party
1	54.6	52.3	56.9	4.9290	X							
1	38.5	35.3	83.1	5.7375		X						
2	71.0	67.7	32.3	4.0526	X			X				
2	64.7	60.8	42.5	4.4665	X					X		
3	76.5	72.3	25.3	3.7521	X		X	X				
3	76.2	72.0	25.8	3.7777	X			X			X	
4	85.1	81.4	13.3	3.0804	X		X	X			X	
4	81.7	77.1	18.9	3.4161	X	X		X	X			
5	**90.3**	**87.0**	**6.8**	**2.5687**	**X**	**X**		**X**	**X**		**X**	
5	87.5	83.3	11.4	2.9126	X		X	X		X	X	
6	92.3	89.0	5.6	2.3703	X	X	X	X	X		X	
6	90.4	86.3	8.6	2.6403	X	X		X	X	X	X	
7	92.6	88.7	7.0	2.4025	X	X	X	X	X	X	X	
7	92.4	88.2	7.4	2.4457	X	X	X	X	X		X	X
8	92.6	87.7	9.0	2.5001	X	X	X	X	X	X	X	X

against the fitted values is shown in Figure 8.10a; the scatter plot shows no appreciable patterns. The normal probability plot of the standardized residuals is shown in Figure 8.10b; the linear pattern confirms the normality of the residuals. Dot diagrams of leverages and Cook's distances are given in Figure 8.10c; no case exerts an undue influence on the regression results.

Now that we have obtained an acceptable regression model, let us use it to predict the vote share of the incumbent party (represented by candidate Gore) in the 2000 presidential election. In 2000, the growth rate was 2.2%, inflation was at 1.7%, and the growth rates of 7 of the 15 quarters during the 1996–2000 term exceeded 3.2%. The duration variable has value 1 because the Democrats (the incumbent party) have been in power for two consecutive terms. The war indicator is zero and the party indicator is 1 for Democrat. The prediction from the regression equation is

$$\begin{aligned}\text{Incumbent vote share} &= 62.6 + 0.496\,\text{growth rate} - 1.18\,\text{inflation}\\ &\quad - 6.78\,\text{duration value} - 4.64\,\text{party} + 11.0\,\text{war}\end{aligned}$$

TABLE 8.13 ESTIMATION RESULTS FOR THE REGRESSION MODEL

```
The regression equation is
Incumbent Vote Share (%) = 62.6 + 0.496 Growth Rate - 1.18 Inflation
       - 6.78 Duration Value - 4.64 Party + 11.0 War
```

Predictor	Coef	SE Coef	T	P	VIF
Constant	62.636	1.728	36.25	0.000	
Growth R	0.4963	0.1308	3.80	0.002	1.8
Inflatio	−1.1788	0.2673	−4.41	0.001	3.0
Duration	−6.780	1.089	−6.22	0.000	1.4
Party	−4.638	1.272	−3.65	0.002	1.3
War	10.979	2.672	4.11	0.001	2.8

S = 2.569   R-Sq = 90.3%   R-Sq(adj) = 87.0%

Analysis of Variance

Source	DF	SS	MS	F	P
Regression	5	918.73	183.75	27.85	0.000
Residual Error	15	98.97	6.60		
Total	20	1017.71			

Durbin-Watson statistic = 1.82

or

$$\text{Incumb vote share} = 62.6 + 0.496(2.2) - 1.18(1.7) - 6.78(1) - 4.64(1) + 11.0(0)$$
$$= 50.306$$

The actual vote share for Gore in the election was 50.3%. The prediction is right on target. However, note that quite a bit of "luck" was involved. The standard error of the prediction error, calculated from Eq. (4.29) in Section 4.3, amounts to 2.79, and a 95% prediction interval for the incumbent vote share extends from 44.3 to 56.3. Here, we have used the 97.5 percentile of the t distribution with 15 degrees of freedom. A Gore result of, for example, 46 or 54% would not have been out of the ordinary either. Of course, proponents of prediction models tend to "tout their horns" if they get so close to the true value, and they credit this to the quality of their model. On the other hand, if they are not very close to the actual value, they point to the large prediction intervals to excuse their miss.

A Comment about the War Variable

Fair includes a war variable that takes the value 1 for years 1920, 1944, and 1948, and 0 otherwise. Fair argues that inflation and "good news" are irrelevant in these elections. He does not use the actual values of inflation and the good news variable for these three elections but instead sets them equal to zero. He claims that this implies that voters do not take into account past inflation and good news when deciding to vote during the three war-dominated periods.

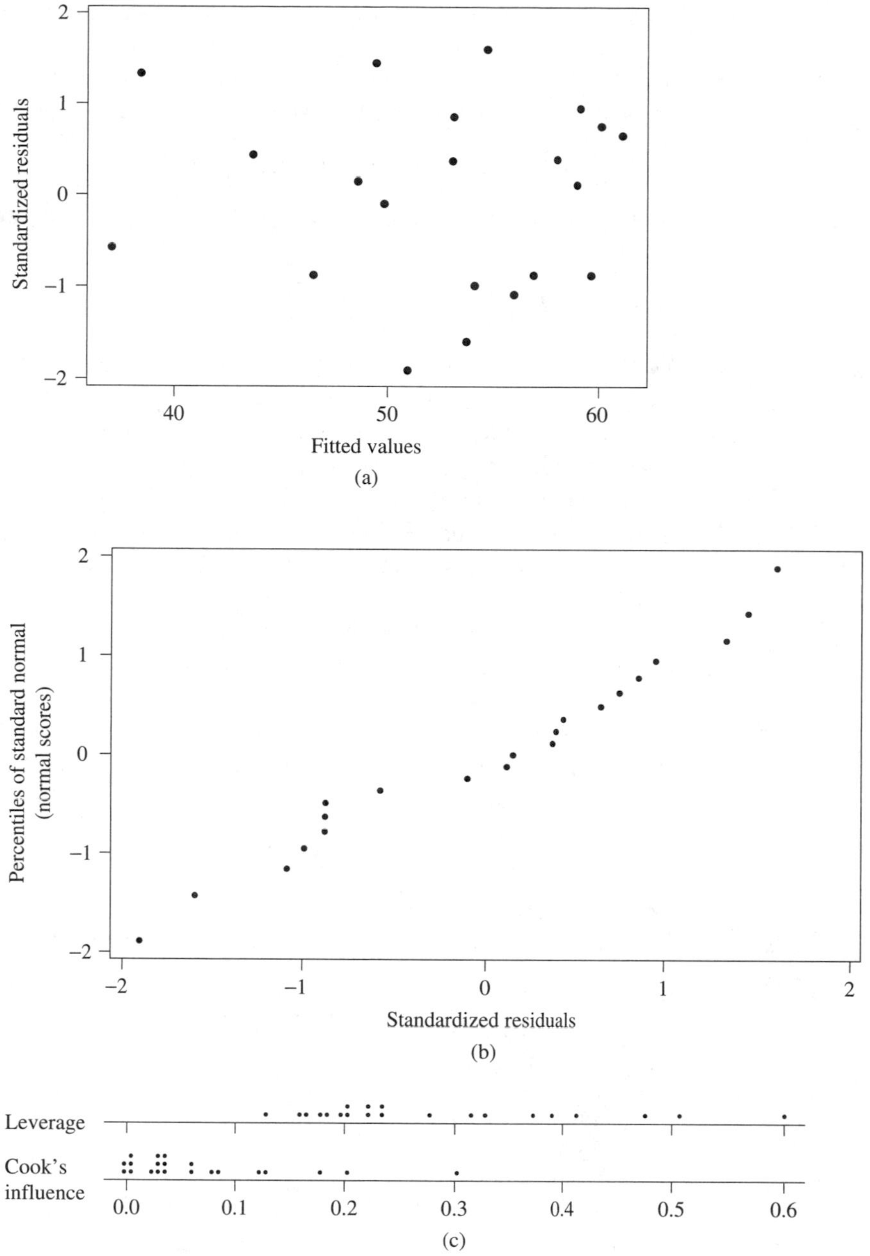

FIGURE 8.10 (a) Scatter plot of standardized residuals against fitted values, (b) normal probability plot of standardized residuals, and (c) dot plots of leverages and Cook's influence measures. Regarding Fig. 8.10(b). Note that the normal probability plot created by Minitab has switched the axes. Minitab plots the percentiles of the standard normal distribution (the normal scores) against the residuals, compared to the graphs in Chapter 6, which plot the residuals against the normal scores

However, this argument is not entirely correct. The decision to plug in zeros for inflation and good news is arbitrary. A better approach uses the original values of the regressor variables but allows for three separate indicator variables—one for each war year (1920, 1944, and 1948). This specification implies that the war years need separate adjustments, but it does not tie the adjustments to specific variables. We have estimated the regression model with the four explanatory variables in

Table 8.13 (growth rate, inflation, duration value, and party) and three separate war indicators. The results show that the war effects for 1920, 1944 and 1948 are about the same. We compare the fit of the full model with the three war indicators to the fit of the restricted model that uses the same indicator for all three war years. The F statistic for testing this restriction is $F = \dfrac{(98.97 - 84.91)/2}{84.91/(21 - 8)} = 1.08$. Its probability value $P(F \geq 1.08) = 0.362$ indicates that the three war years can be treated the same.

8.4.2 PREDICTION MODELS PROPOSED BY POLITICAL SCIENTISTS

Political scientists refer to Fair's model as an "economy incumbency model" because it uses economic variables to predict the incumbent vote share. They criticize his model for failing to incorporate measures of public opinion that are available prior to the election. There is a large literature in political science on the prediction of presidential elections. The book by Campbell and Garand (2000) gives a good summary of several competing models. This book also lists the data for all elections from 1948 through 1996 that were used in the estimation of these models as well as the model predictions for 2000. Data for five popular models (the models by Campbell, Abramowitz, Lewis-Beck and Tien, Holbrook, and Lockerbie) are included in Exercise 8.1. Here, we focus on the model by Michael S. Lewis-Beck and Charles Tien. They use data on the economy and survey responses to several questions on the Gallup poll of the last July prior to the election. Their data are listed in Table 8.14. Note that their model does not include the inflation rate, a significant factor in Fair's model.

- *Incumbent vote* is the percentage of the two-party vote received by the candidate of the president's party.

TABLE 8.14 INCUMBENT VOTE SHARE AND ITS DETERMINANTS[a]

Year	Incumbent Vote	July Popularity	Peace and Prosperity	Future Problems	Leading Indicators	GNP Change	Second Term
1948	52.37	39	*	46.67	0.00	2.42	0
1952	44.60	32	82.41	*	3.88	0.07	0
1956	57.76	69	122.84	56.90	0.00	0.26	1
1960	49.91	49	101.80	49.30	−3.08	1.42	0
1964	61.34	74	140.37	60.32	3.96	3.11	1
1968	49.60	40	85.94	46.55	0.00	2.88	0
1972	61.79	56	106.68	57.35	5.06	4.18	1
1976	48.95	45	80.40	34.85	6.07	2.33	0
1980	44.70	21	113.43	47.37	−5.67	−1.38	1
1984	59.17	52	104.51	51.32	0.00	3.95	1
1988	53.90	51	106.12	53.52	3.23	1.91	0
1992	46.55	32	94.81	45.95	2.48	1.46	0
1996	54.66	57	109.30	52.56	1.69	1.85	1

[a] The data are stored in the file **election(Beck&Tien)**.

- *July popularity* refers to the presidential popularity as measured by the Gallup poll in July before the election.
- *Peace and prosperity* is an index created by adding the percentage of two-party respondents who favored the incumbent party on keeping the United States out of war and on keeping the country prosperous (Gallup question).
- *Future problems* is the percentage of two-party respondents who favored the incumbent party on handling the country's most important problems (Gallup question).
- *Leading indicators* is the percentage change in the government's index of leading indicators during the first two quarters of the election year. It is set at zero if the change in one direction was not sustained for at least 3 months.
- *GNP change* is the nonannualized percentage change in GNP (constant dollars) from the fourth quarter of the year before the election to the second quarter of the election year.
- *Second term* is an indicator variable for a party's second consecutive term in the White House. It is coded 1 if the party is heading into its second term and 0 otherwise.

Scatter plots of incumbent vote against July popularity, response on questions regarding peace and prosperity, response on questions regarding future problems, leading indicators, and GNP change are shown in Figure 8.11. They show some relationships, although the associations are relatively weak.

The summary of all possible regressions is shown in Table 8.15. The model with July popularity, GNP change, and an indicator for a second term seems to give an acceptable fit with little bias; Mallow's C_p statistic (value of 4.9) is close to what one would expect for a good model (it should be approximately 4 for a model with three regressors). Detailed fitting results for this model are shown in Table 8.16. Since the four parameters in this model are estimated on only 13 cases, any conclusion from the estimated model must be made with great caution. The signs of the regression coefficients make sense; one can expect that popularity in July and an increase in GNP help increase the vote share. Running for a second term also increases the incumbent vote share. The Durbin–Watson statistic is unremarkable, indicating that there is no problem with serial correlation. The scatter plot of the standardized residuals in Figure 8.12, the normal probability plot of the standardized residuals (not shown), and the leverages and influence measures (not shown) do not reveal any serious problems with this model. The highest leverage originates from the 1980 election, with a leverage of 0.80 (which is approximately 2.5 times higher than the average leverage of $4/13 = 0.31$). However, Cook's influence measure shows that the influence of this case on the parameter estimates is not out of line. The standard deviation of the residuals amounts to approximately 1.7 percentage points, resulting in an approximate 95% prediction interval with half width $\pm 3.4\%$.

FIGURE 8.11
Scatter Plots of Incumbent Vote against (a) July Popularity, (b) Peace and Prosperity Response, (c) Future Problems Response, (d) Index of Leading Indicators, and (e) GNP Change

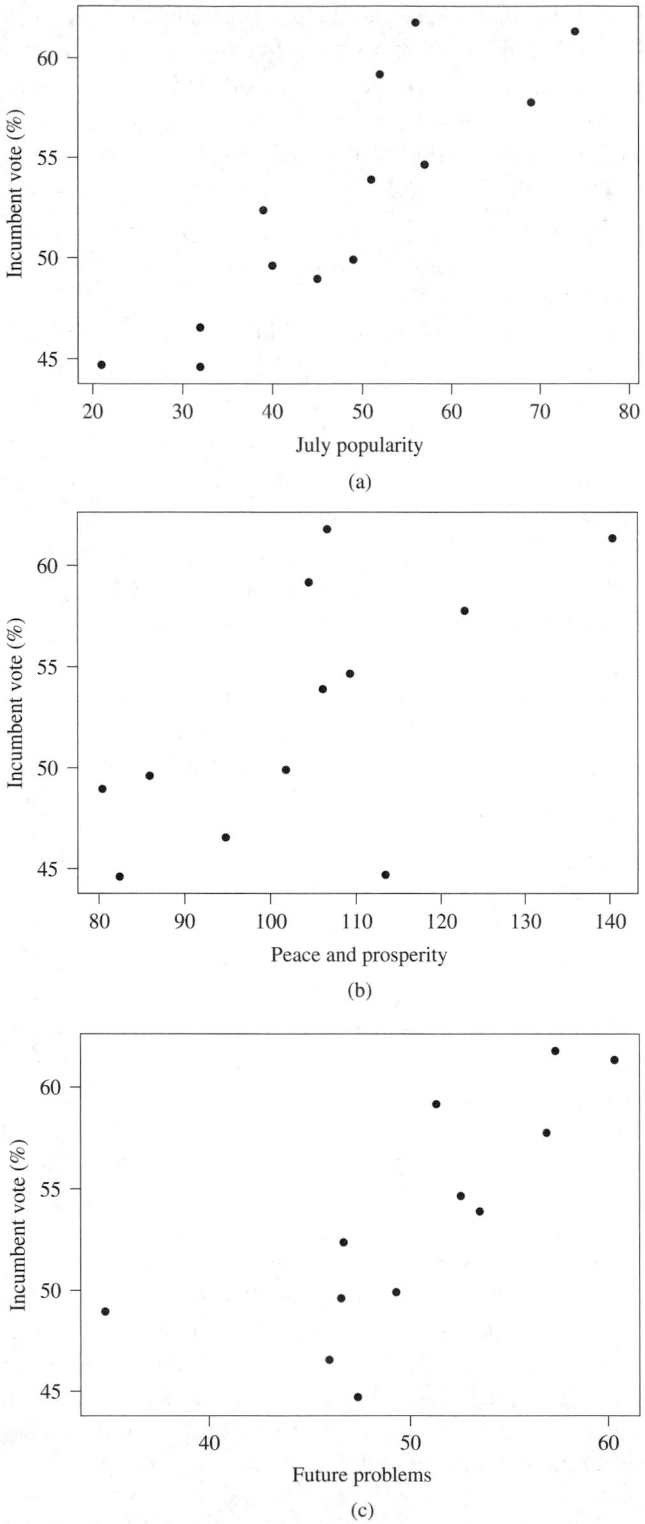

FIGURE 8.11
(Continued)

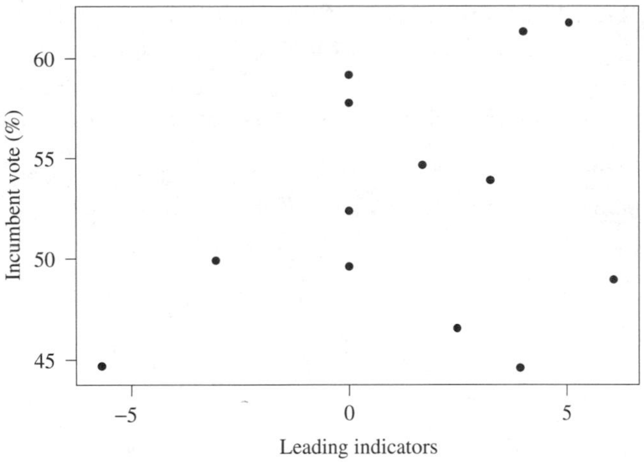

(d)

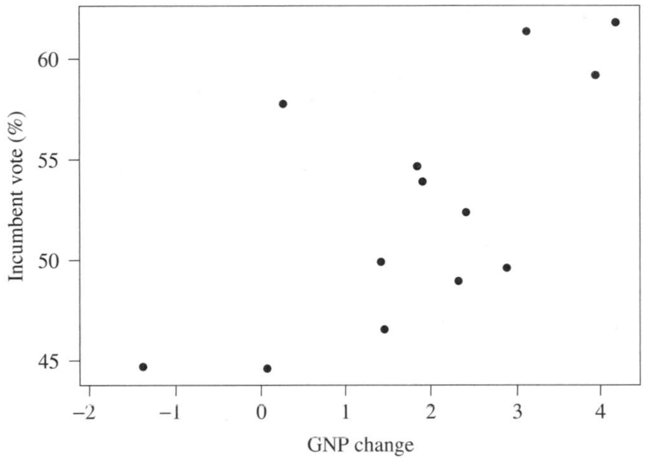

(e)

How does this model perform in predicting Gore's 2000 vote share? July's popularity for the incumbent party's candidate was 59, the GNP change from the fourth quarter of the year before the election to the second quarter of the election year was 2.52, and the indicator for a straight second term is 0. The prediction is given by

$$\text{Incumbent vote} = 38.3 + 0.202\,\text{July popularity} + 1.58\,\text{GNP change} + 4.07\,\text{second term}$$

or

$$\text{Incumbent vote} = 38.3 + 0.202(59) + 1.58(2.52) + 4.07(0) = 54.2$$

with 95% prediction intervals extending from 49.8 to 58.6. Here, we have used the standard deviation of the prediction error in Eq. (4.29) and the 97.5 percentile of

TABLE 8.15 BEST SUBSETS REGRESSION OF INCUMBENT VOTE SHARE ON ALL LISTED PREDICTOR VARIABLES

Best Subsets Regression: Incumbent versus July Popular, Peace and Pr, . . .

Response is Incumbent Vote

11 cases used 2 cases contain missing values.

Vars	R-Sq	R-Sq(adj)	C-p	S	July Pop	Peace an	Future P	Leading	GNP Chan	Second T
1	74.5	71.6	29.4	3.1800	X					
1	58.3	53.7	52.5	4.0647			X			
2	84.3	80.4	17.4	2.6457		X			X	
2	83.5	79.4	18.5	2.7109			X		X	
3	**94.5**	**92.1**	**4.9**	**1.6756**	**X**				**X**	**X**
3	91.8	88.2	8.8	2.0475	X		X		X	
4	96.7	94.6	3.6	1.3904	X		X		X	X
4	95.0	91.7	6.1	1.7248	X	X			X	X
5	97.0	93.9	5.3	1.4721	X	X	X		X	X
5	96.9	93.9	5.4	1.4779	X		X	X	X	X
6	97.2	93.0	7.0	1.5803	X	X	X	X	X	X

the t distribution with 9 degrees of freedom. The prediction intervals are somewhat narrower than the ones obtained by Fair (which extended from 44.3 to 56.3). However, the point prediction misses the true value (50.3) by a larger amount than the point prediction in Fair's model.

8.4.3 CONCLUDING COMMENTS

How successful are these models in predicting the incumbent vote share in presidential elections? We notice that the standard deviation of the forecast errors amounts to approximately 2 to 3 percentage points. This implies that a 52 or 53% (point) prediction is not yet sufficient to justify concluding victory for the incumbent party.

The models are based on very few years of data, and they provide only a very rough glimpse of what can be expected at the next election. Although these models provide a useful yardstick of what can be expected, one should not bet one's fortune on their predictions.

TABLE 8.16 ESTIMATION RESULTS FOR THE MODEL WITH JULY POPULARITY, GNP CHANGE, AND AN INDICATOR FOR A SECOND TERM[a]

```
The regression equation is
Incumbent vote = 38.3 + 0.202 July Popularity + 1.58 GNP
        Change + 4.07 Second Term

Predictor              Coef    SE Coef         T        P    VIF
Constant             38.296      1.700     22.53    0.000
July Popularity     0.20156    0.04363      4.62    0.001    1.7
GNP Change           1.5821     0.3667      4.31    0.002    1.3
Second Term           4.066      1.108      3.67    0.005    1.3

S = 1.717          R-Sq = 93.8%       R-Sq(adj) = 91.8%

Analysis of Variance

Source               DF        SS         MS        F       P
Regression            3    402.49     134.16    45.52   0.000
Residual Error        9     26.53       2.95
Total                12    429.02

Durbin-Watson statistic = 2.04
```

[a] Observe that the summary statistics [s, R-Sq, R-Sq(adj)] are not identical to the summary statistics of the best subset regressions in Table 8.15. This has to do with the 2 years (cases) in which some of the regressor variables had missing values (peace and prosperity in 1948 and future problems in 1952). The best subset regression omits the entire case even if only one regressor variable has a missing value for that case, and hence the analysis in Table 8.15 is based on only 11(= 13 − 2) cases. The model in Table 8.16, on the other hand, uses all 13 cases because its regressor variables do not include the variables peace/prosperity and future problems, which had missing values on 2 years.

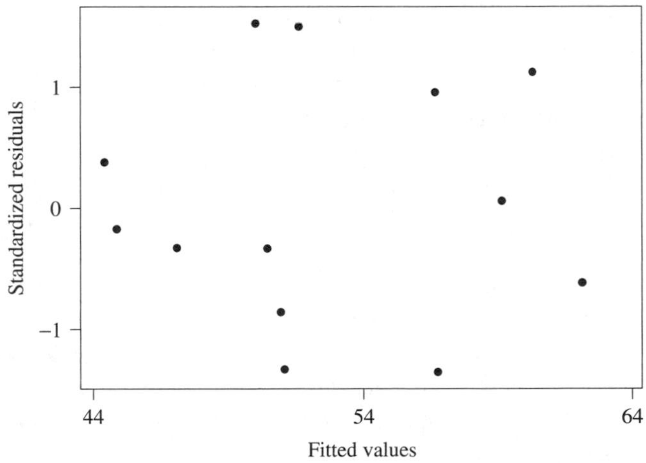

FIGURE 8.12 Scatter Plot of Standardized Residuals against Fitted Values

8.5 STUDENT PROJECTS LEADING TO ADDITIONAL CASE STUDIES

Courses on applied statistics such as regression should incorporate projects in which the student or the reader of this book selects the problems to study, gathers the data, analyzes the information using suitable computer software, and communicates the results in a report. Most textbooks on applied statistics, such as this book, include data. However, the included data sets are typically small to moderately sized, and the data are usually given to the reader as "numbers waiting to be analyzed." You are asked to analyze the data, run specific regression models, and find the best (appropriate) regression models that relate a specified response to certain specified explanatory variables. Such exercises are important because they teach the mechanical aspects of data analysis. However, they are incomplete in that they do not expose you to problem formulation and the difficulties of acquiring relevant data. The message communicated to you is that statistics starts after the data have been collected.

Each course should also contain projects that are relatively unstructured by the instructor. It is your own questions and data that provide the project structure. For such projects, you must generate an interesting problem, figure out what data to gather and where to get them, analyze the information, and put the analysis into words by writing a report on your findings. The data analysis is just one step in this process. The steps of formulating the problem, deciding what data to collect, managing the data acquisition process, and checking its integrity are often the more instructive and challenging parts of the project. This activity teaches you that statistics is more than just analyzing numbers. It engages you in research as you search for solutions to relevant and interesting questions. Active engagement is an important aspect of learning. Much is learned by being involved in writing survey questions and dealing first-hand with such issues as random and nonrandom sampling, nonresponse, and poorly designed questions. Also, when performing your own experiments you think about how to set up the experiment and you struggle with such issues as randomization, blocking, replication, and how your plans are impacted by practical issues. You learn to appreciate the difficulty of obtaining relevant data.

Several acronyms have been used to describe the various steps of the problem-solving cycle. For example, the five-step cycle "DMAIC" consists of defining the problem (D), measure (M), analyze (A), improve (I), and control (C). Deming talks about the Deming–Shewhart wheel PDSA, where P stands for plan, D for do, S for study, and A for act.

It is also very worthwhile to carry out such projects in groups, and we recommend that you form study groups. Our experience shows that group projects work well with committed students who are eager to learn. In such situations, the sharing of information and expertise is very beneficial to learning. Of course, group projects are likely to fail if motivation and group spirit are lacking.

The project output should be a report. The report should start with a short, concise executive summary that describes the problem and the main findings of your study. The write-up should discuss the motivation behind the project, describe the data and the way they were obtained, and discuss the statistical analysis. Furthermore, the report should discuss the appropriateness of the analysis and should reflect on any possible shortcomings. The findings must be interpreted, and the conclusions and implications of the study should be spelled out clearly. Relevant statistical tables and computer output should be put in an appendix. A listing of the raw data (only a subset of the data for large data sets) and a summary of the data definitions and the data sources should be included.

Writing a good research paper is a challenging task. The topic may be interesting and the statistical analysis may be competent, but the project may still fall short due to poor writing. The writing must be well organized and grammatically correct, and it must keep the reader's interest. Fortunately, help on writing is available online because many universities have developed excellent online writing resources. Here are a few particularly good links:

- Purdue University Online Writing Lab: http://owl.english.purdue.edu
- University of Florida—The Reading and Writing Center: http://web.cwoc.ufl.edu/owl
- University of Wisconsin Writing Center: http://www.wisc.edu/writing
- University of Victoria (Canada) Writer's Guide: http://web.uvic.ca/wguide
- Guide to Writing and Grammar: http://webster.commnet.edu/grammar/index.htm

These resources discuss the general structure of research papers and they contain many useful suggestions. They discuss how to cite the work of others and how to avoid plagiarism. They give strategies for clarifying logic and avoiding "deadly" sins, and they help with grammar and style.

An oral presentation, with a subsequent discussion that involves all students in the class, is also very useful. It helps you practice your oral presentation skills, and it teaches you how to respond to questions and criticism. Furthermore, a session for oral presentations exposes you to a wide variety of other topics that have been considered by the various groups.

Your instructor should give you examples of previous projects—good projects as well as bad ones—and should provide suggestions for appropriate topics. Many of the data sets in this book in fact originated from such projects. Successful past projects in our courses have examined the following issues (see Ledolter, 1995):

- "Success" at the university: What are the relationships between college GPA and high school GPA, number of hours studied, etc.? Does gender play a role? Does it help (hurt) your GPA if one lives in a sorority/fraternity? What about the effect of drinking or smoking?

In projects such as these, you have to construct questionnaires and survey your fellow students. You need to think about sampling issues and how to draw representative samples.

- Relationships between preferences (such as listening or buying preferences) and demographic characteristics (such as gender and occupation).
- Sports statistics (football, basketball, and baseball): Find the key variables that explain individual or team performance. Projects that attempt to explain the salaries of baseball players are always fun. Compare the salaries of pitchers, infielders, etc.
- Relationships between CEO compensation and performance; relationship between professor salaries and performance: Much of the needed information can be obtained from Web sites. Salaries at most public universities are readily available. Who wouldn't be interested in finding out the salary of one's professor and learning whether he or she "deserves" it?
- Sales forecasts; marketing applications; applications of statistics to finance and portfolio selection; tracking the performance of investment strategies; predicting economic indicators.
- Effects of legislation on society: For example, investigate the impact of changes in the maximum speed limit on the number and severity of traffic accidents, or study the impact of the motorcycle helmet law on motor cycle accidents that resulted in severe head injuries.
- Experiments with paper helicopters, catapults, rubber balls, and sticky pads. For example, vary certain design characteristics on a simple paper helicopter and study how the settings affect the flying time of the helicopter. The paper by Hunter (1977) is a good reference if you want to conduct such experiments.

Data for projects may be obtained from Internet sources, company data, surveys, statistical reference books, or experiments that you carry out yourself. The *Statistical Data Abstract of the United States* (U.S. Bureau of the Census, U.S. Government Printing Office, Washington, DC, 1879–) is an annual compendium of summary statistics on the political, social, industrial, and economic life of the United States. Four major monthly governmental periodicals provide the majority of the current statistics available on the economy and its operation. The *Survey of Current Business* contains approximately 2500 statistical series on income, expenditures, production, and prices of commodities. Historical figures for the statistical data published in the *Survey of Current Business* are available in a supplement titled *Business Statistics,* published in odd-numbered years. A second source is the *Federal Reserve Bulletin,* which publishes data with emphasis on financial statistics. The third major governmental publication is the *Monthly Labor Review,* which publishes data on work and labor conditions, wage rates, consumer price indices, and the like. The fourth is the *Business Conditions Digest,* which contains several hundred economic time series in a form convenient for forecasters and business analysts.

EXERCISES

8.1. Incumbent Vote Share in Presidential Elections

The model by Campbell includes two predictors: the percentage of support for the inparty candidate in the preference poll conducted by Gallup in early September of the election year and the second-quarter (of the election year) rate of growth in GDP.

Abramowitz models the incumbent vote share as a function of the president's approval rating in the Gallup poll of early July and the annual growth rate of real GDP during the first two quarters of the election year (which are released in August of the election year).

The model developed by Holbrook uses the following predictors: the president's approval rating, a measure of retrospective personal finances indicating whether one is better or worse off financially now than a year ago, and an indicator measuring tenure in office.

Lockerbie uses two measures of change in disposable income, a measure of anticipated financial well-being a year from now and the number of years the incumbent party has controlled the White House.

The data sets and a description of the variables are given below. The data sets are stored in the files **campbell**, **abramowitz**, **holbrook**, and **lockerbie**.

a. Use the data to estimate the various regression models. Assess the significance of the estimated regression coefficients. Investigate whether all listed regressor variables are needed. Simplify the regression models if possible.

b. Check the models for violations of the regression assumptions. Identify high-leverage values and influential observations. Also, check for autocorrelation among the residuals.

c. Discuss how these models can be used for prediction. Discuss how these models can be used for scenario forecasts. That is, develop scenarios for the predictor variables under which the incumbent party candidate has a good chance of holding on to the White House. Your analysis should also incorporate the uncertainty.

d. Discuss in more detail the difference between "fitting models" and "forecasting." Onc could drop a particular year (case) from the regression, estimate the model without the data from that year, and predict the incumbent share for the year that one has omitted. Why would this be different (better or worse) than using all observations and looking at the residuals? Discuss.

e. Summarize the performance of these models, and compare them to Fair's model discussed in Section 8.4.1. Are these worthwhile models?

Data Used by Campbell

Year	Incumbent Party Vote	September Trial Heat	GDP Growth Rate
1948	52.32	45.61	0.91
1952	44.59	42.11	0.27
1956	57.75	55.91	0.64
1960	49.92	50.54	−0.26
1964	61.34	69.15	0.81
1968	49.60	41.89	1.63
1972	61.79	62.89	1.73
1976	48.95	40.00	1.17
1980	44.70	48.72	−2.43
1984	59.17	60.22	1.79
1988	53.90	54.44	0.79
1992	46.55	41.94	0.35
1996	54.74	60.67	1.04

Incumbent vote: Percentage of the two-party vote received by the candidate of the president's party.

September trial heat: Two-party percentage of support for the in-party candidate in the preference poll conducted by Gallup in early September of the election year.

GDP growth rate: Second quarter rate of growth (nonannualized) in the GDP.

Data Used by Abramowitz

Year	Incumbent Vote	Term	Presidential Popularity	GDP Growth Rate
1948	52.3	1	39	1.8
1952	44.6	1	32	−0.2
1956	57.8	0	69	0.5
1960	49.9	1	49	0.4
1964	61.3	0	74	1.1
1968	49.6	1	40	1.3
1972	61.8	0	56	2.2
1976	48.9	1	45	1.1
1980	44.7	0	21	−2.3
1984	59.2	0	52	1.8
1988	53.9	1	51	0.8
1992	46.6	1	31	0.4
1996	54.6	0	56	1.6

Incumbent vote: Percentage of the two-party vote received by the candidate of the president's party.

Term: Binary variable coded 1 if the president's party has held the White House for 8 years or longer and 0 otherwise.

Presidential popularity: President's approval rating in the Gallup poll in early June.

GDP growth rate: Annual growth rate of the real GDP during the first two quarters of the election year.

Data Used by Holbrook

Year	Incumbent Vote	Presidential Popularity	Personal Finances	Tenure in Office
1948	52.4	37.5	116	1
1952	44.6	30.0	94	1
1956	57.8	69.0	110	0
1960	49.9	62.6	108	1
1964	61.3	74.3	120	0
1968	49.6	42.2	114	1
1972	61.8	59.0	129	0
1976	48.9	46.6	103	1
1980	44.7	36.8	79	0
1984	59.2	53.8	121	0
1988	53.9	49.4	111	1
1992	46.5	39.0	97	1
1996	54.6	54.0	114	0

Incumbent vote: Percentage of the two-party vote received by the candidate of the president's party.

Presidential popularity: Average percentage of the public (responding to Gallup polls in the second quarter) who said that they approved of the way the president is handling his job.

Personal finances: Measure of retrospective personal finances based on responses to the Survey of Consumers' question, "Would you say that you (and your family living here) are better off or worse off financially than you were a year ago?"

Tenure in office: Binary variable coded 1 for presidential candidates whose party has held the White House for two or more terms and 0 for candidates whose party has held the White House for one term.

Data Used by Lockerbie

Year	Incumbent Vote	Disposable Income 1	Disposable Income 2	Next Year Better	Tenure
1956	57.80	1.60	3.70	36.00	4
1960	49.90	1.65	2.26	35.00	8
1964	61.30	4.74	1.78	37.00	4
1968	49.60	2.91	3.31	33.00	8
1972	61.79	1.61	2.39	38.00	4
1976	48.90	2.03	0.68	32.00	8
1980	44.70	−1.26	1.59	26.00	4
1984	59.20	2.67	1.67	37.33	4

Year	Incumbent Vote	Disposable Income 1	Disposable Income 2	Next Year Better	Tenure
1988	53.90	1.73	−0.05	36.33	8
1992	46.50	2.10	−1.03	36.33	12
1996	54.74	2.07	2.33	35.67	4

Incumbent vote: Percentage of the two-party vote received by the candidate of the president's party.

Disposable Income 1: Change in per capita real disposable income from the second quarter of the year prior to the election to the second quarter of the election year.

Disposable Income 2: Change in per capita real disposable income from 2 years prior to the election to the year immediately prior to the election.

Next year better: A component of the Index of Consumer Sentiment. This variable is based on the following question: "Now looking ahead—Do you think that a year from now you (and your family living here) will be better off financially, or worse off, or just about the same as now?"

Tenure: Number of years a party has controlled the White House.

8.2. Height and Weight of Boys and of their Mothers and Fathers

This data set is taken from a larger study examining the associations between childhood treatment with methylphenidate (MPH) and adult height and weight [Kramer, J. R., Loney, J., Ponto, L. B., Roberts, M. A., and Grossman, S. Predictors of adult height and weight in boys treated with methylphenidate for childhood behavior problems. *Journal of the American Academy of Child and Adolescent Psychiatry,* 39, 517–524, 2000]. The 93 boys in this study were 6 to 13 years of age, had behavior problems, were referred to a child psychiatry outpatient clinic, treated clinically with MPH for an average of 36 months, and reevaluated between ages 15 and 19. The information for the first five boys and the data description are listed here; the complete data set is contained in the file **height&weight**.

Part 1 In part 1 of this exercise, we ask you to examine the relationship between the height (and weight) of the boys and their age. Pediatricians have access to elaborate nonlinear growth curves, that they obtain from measurements on a very large number of children. With a small data set such as this, it may not be possible to fit elaborate models. However, it may be feasible to obtain useful linear approximations.

a. Consider the measurements taken at referral. Model the relationship between the height and the child's age at referral. Model the relationship between the weight and the child's age. Investigate whether or not the weight at birth has some additional explanatory power.

b. Consider the measurements taken at the follow-up visit. Model the relationship between the height and the child's age at the follow-up visit. Model the relationship between the weight and the child's age.

c. Combine the data set, and use all measurements on weight and height. Ignore the fact that the two measurements on weight (and height) are taken on the same child. Model the relationship between the height and the child's age. Model the relationship between the weight and the child's age.

d. Discuss your findings.

Part 2 The data set also lists the height and weight for mothers and fathers.

a. Investigate, for mothers and fathers separately, relationships among their weight and height.

b. Investigate whether there are relationships between mother's and father's heights and mother's and father's weights. Can you claim that "thin" tends to be attracted by "thin"?

Col 1	Col 2	Col 3	Col 4	Col 5	Col 6	Col 7	Col 8	Col 9	Col 10	Col 11
112	56.5	75.0	199	72	165	9.38	64	135	75	230
89	50.0	64.0	184	62	101	8.41	68	127	70	181
100	53.0	85.0	162	63	92	7.40	68	250	69	190
133	56.8	91.8	206	67	147	7.00	65	138	70	181
111	52.5	67.3	187	69	139	8.88	66	179	72	210

Col 1: Age at referral (months)

Col 2: Height at referral (in.)

Col 3: Weight at referral (lb)

Col 4: Age at follow-up (months)

Col 5: Height at follow-up (in.)

Col 6: Weight at follow-up (lb)

Col 7: Birth weight (lb)

Col 8: Height of mother (in.)

Col 9: Weight of mother (lb)

Col 10: Height of father (in.)

Col 11: Weight of father (lb)

8.3. Modeling Softdrink Sales:

The data in file **softdrink** represent weekly sales and prices of two competing products, brand P and brand C. The sales are recorded in ounces, whereas prices are given in dollars per ounce. Logarithms of the sales of 12-packs for both brands (lnSalesP12 and lnSalesC12) are given, as well as the logarithms of prices for 6-, 12-, and 24-packs (lnPriceP6, lnPriceP12, and lnPriceP24 for brand P and lnPriceC6, lnPriceC12, and lnPriceC24 for brand C).

a. Analysis for brand P
Price effects on sales are measured by regressing the logarithm of sales on the logarithm of prices. Consider the regression model

$$\textbf{M1:}\quad \text{lnSalesP12}_t = \beta_0 + \beta_1 \ln \text{PriceP6}_t + \beta_2 \ln \text{PriceP12}_t + \beta_3 \ln \text{PriceP24}_t + \varepsilon_t$$

The slope coefficients in the log/log model represent price elasticities (see the discussion in Section 6.5). A 1% change in the product's own price translates into a percentage change in sales of magnitude β_2. We expect this elasticity to be negative because sales decrease when prices are increased. On the other hand, the price elasticities for 6- and 24 packs, β_1 and β_3, should be positive. Because of product substitution within the same brand family, we expect increased sales of 12-packs when prices of the other pack sizes are increased.

Estimate the model M1 using regression software of your choice. Interpret the results. Check whether the elasticities have the expected signs. Discuss whether sales of 12-packs are more responsive to price changes in 24-packs than to price changes in 6-packs.

Check the model assumptions, in particular investigate the autocorrelations of the residuals. You will find that there is some autocorrelation in the residuals, and you will revisit this data set in Exercise 10.8. Look for residual outliers, leverage, and Cook's distance (you will find several large ones).

b. Analysis for brand C
Apply the analysis in (a) to the sales of brand C and consider the model

$$\textbf{M2:}\quad \text{lnSalesC12}_t = \alpha_0 + \alpha_1 \ln \text{PriceC6}_t + \alpha_2 \ln \text{PriceC12}_t + \alpha_3 \ln \text{PriceC24}_t + \varepsilon_t$$

Estimate the model. Interpret the estimates. Check the model. Discuss whether or not the results for 12-packs of brand C in model M2 are similar to the results for 12-packs of brand P in model M1.

c. Analysis incorporating prices of both brands

Consider models that do not just include the prices of the considered brand but also the prices of the competing product. Estimate the model

$$\textbf{M3:}\ \ln \text{SalesP12}_t = \beta_0 + \beta_1 \ln \text{PriceP6}_t + \beta_2 \ln \text{PriceP12}_t + \beta_3 \ln \text{PriceP24}_t + \alpha_1 \ln \text{PriceC6}_t + \alpha_2 \ln \text{PriceC12}_t + \alpha_3 \ln \text{PriceC24}_t + \varepsilon_t$$

Interpret the coefficients in this log/log model. Discuss whether the results are much of an improvement over the simpler models M1 and M2. You will find that the price elasticities of the competing brand are essentially zero and not overly significant. This means that sales of brand P are mostly driven by the prices of brand P and not by the prices of brand C.

Repeat the analysis by regressing the log sales of 12-packs of brand C on all six prices.

d. Additional models
Consider the ratio SalesP12/SalesC12 as the response. This is equivalent to analyzing $S/(1 - S)$, where $S =$ SalesP12/[SalesP12 + SalesC12] is the market share of brand P. Estimate the model

$$\textbf{M4:}\ \ln(\text{SalesP12}_t/\text{SalesC12}_t) = \beta_0 + \beta_1 \ln \text{PriceP6}_t + \beta_2 \ln \text{PriceP12}_t + \beta_3 \ln \text{PriceP24}_t + \alpha_1 \ln \text{PriceC6}_t + \alpha_2 \ln \text{PriceC12}_t + \alpha_3 \ln \text{PriceC24}_t + \varepsilon_t$$

Confirm that the elasticities have the expected signs. Confirm that the response decreases with increasing 12-pack price of brand P and decreasing 12-pack price of brand C. The signs of the two price coefficients are different, but their magnitude is approximately the same. Confirm that the same can be said for the price coefficients of the other pack sizes, except that now the signs are reversed. These regression results lead us to a model that contains logarithms of price ratios (i.e., differences of the log prices) as explanatory variables. Consider the model

$$\textbf{M5:}\ \ln(\text{SalesP12}_t/\text{SalesC12}_t) = \beta_0 + \beta_1 \ln \text{PriceRatio6}_t + \beta_2 \ln \text{PriceRatio12}_t + \beta_3 \ln \text{PriceRatio24}_t + \varepsilon_t$$

Interpret the model and compare it to model M4. You will find that the R^2 is not much worse, but the model may be easier to interpret.

8.4. Complete a project as outlined in Section 8.5.

9 Nonlinear Regression Models

9.1 INTRODUCTION

So far, we have assumed that the regression function $\mu = \beta_0 + \beta_1 x_1 + \cdots + \beta_p x_p$ in the model $y = \mu + \varepsilon$ is linear in the parameters. Linearity of the regression function in the parameters $\boldsymbol{\beta}$ is the key here, because this assumption allows us to develop a closed-form solution for the least squares estimator. Linearity of the regression function in the regressor variables is not a relevant issue. The regressor variables $x_1, x_2, \ldots, x_p$ can be any known nonlinear function of the original regressors, such as squares, powers, exponentials, and logarithms of the x's, and the regression function can still be linear in the parameters (see Chapter 1, Section 1.3).

Models that are nonlinear in the parameters can sometimes be transformed into linear representations. For example, the nonlinear model $y = \mu\varepsilon = \alpha x_1^{\beta} x_2^{\gamma} \varepsilon$ can be made linear by applying the logarithm to both sides of the equation, leading to the linear model in the transformed variables, $\ln(y) = \ln(\alpha) + \beta \ln(x_1) + \gamma \ln(x_2) + \ln(\varepsilon)$. A linear regression of $\ln(y)$ on $\ln(x_1)$ and $\ln(x_2)$ gives us estimates of $\alpha^* = \ln(\alpha)$, β, and γ; the estimate of α can be obtained from $\exp(\alpha^*)$. Note that the errors in the original model are multiplicative. The logarithmic transformation makes the errors additive; constant variance and a normal distribution must be assumed for the transformed errors $\ln(\varepsilon)$. The role of transformations in regression was discussed in Section 6.5.

In certain applications the regression models are **intrinsically nonlinear** in the parameters, and these are the models that are discussed in this chapter. Intrinsically nonlinear means that the model cannot be transformed into a linear model. Many nonlinear models arise as solutions of differential equations or, more generally, systems of several differential equations. The differential equations are usually based on scientific theory from fields such as physics, chemistry, or the biological sciences. Biologists are interested in the growth of organisms and they are trying to understand and model the underlying mechanism. Chemists are interested in understanding chemical reactions that involve the interaction of several factors over time. Agricultural and environmental scientists are interested in the growth of crops because they need to understand how tall things grow, how fast they grow, and how growth is affected by various environmental conditions and treatments. Pediatricians are interested in knowing about infant growth, and oncologists study tumor growth. The growth models that arise as solutions to theory-based differential equations are important for several reasons. The coefficients in such models mean something to the specialist, and the coefficients can be parameterized and made dependent on changing conditions and treatments.

Let us illustrate the connection between differential equations and nonlinear models in the context of a simple example. Consider a situation in which the growth rate of a certain organism is proportional to the product of present size and future growth potential. Specifically, let μ_t be size at time t, α be some limiting growth value, and $(\alpha - \mu_t)/\alpha$ the proportional amount of growth at time t yet to be realized. The growth rate is such that

$$\frac{\partial \mu_t}{\partial t} = \gamma \mu_t \frac{\alpha - \mu_t}{\alpha} \tag{9.1}$$

where $\gamma > 0$ is a proportionality coefficient. The solution of this differential equation is given by

$$\mu_t = \frac{\alpha}{1 + \beta \exp(-\gamma t)} \tag{9.2}$$

If you are rusty solving differential equations, you can always check the solution "backwards" and show that the derivative of Eq. (9.2) coincides with the expression in Eq. (9.1). The parameter β in Eq. (9.2) is determined by the initial conditions. Note that the starting value for μ at $t = 0$ is $\alpha/(1 + \beta)$; the limiting value, as t approaches ∞, is α; hence, for growth models, $\beta > 0$. The model in Eq. (9.2) is called the **logistic** or **autocatalytic** growth function.

9.2 OVERVIEW OF USEFUL DETERMINISTIC MODELS, WITH EMPHASIS ON NONLINEAR GROWTH CURVE MODELS

The simplest growth model is the **linear trend model,**

$$\mu_t = \alpha + \gamma t \tag{9.3}$$

The **exponential trend model** is given by

$$\mu_t = \beta \exp(\gamma t) \tag{9.4}$$

The starting value for μ at $t = 0$ is β. The parameter γ represents the growth rate. The level μ_t increases exponentially for positive values of γ. For negative values of γ the model implies exponential decay toward zero. The model is nonlinear in the parameters but can be transformed into a linear model by taking the logarithm on both sides of Eq. (9.4). That is,

$$\ln(\mu_t) = \ln(\beta) + \gamma t$$

Linear and exponential trends imply unbounded growth. However, in most circumstances one expects that growth will not continue beyond a certain level.

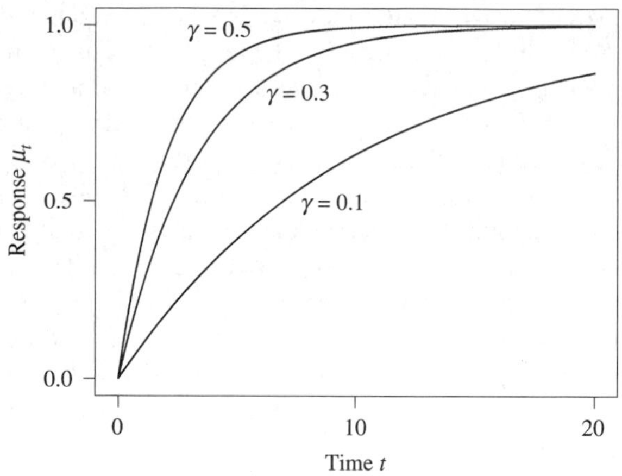

FIGURE 9.1 Modified Exponential Model with $\alpha = \beta = 1$ and Various Values for γ

Trend curves with upper levels, also called **saturation levels**, must be considered. The **"modified" exponential trend model** (also called the **monomolecular** model)

$$\mu_t = \alpha - \beta \exp(-\gamma t) \quad \alpha > 0, \quad 0 < \beta \leq \alpha, \quad \gamma > 0 \tag{9.5}$$

achieves such a saturation. The starting value for μ at $t = 0$ is $\alpha - \beta$; the limiting value, as t approaches ∞, is α. Once these values are set, the parameter $\gamma > 0$ determines the speed with which the function approaches the limiting value. Figure 9.1 shows realizations of this model with $\alpha = \beta = 1$ and several values for γ (0.1, 0.3, 0.5). The increase to the limiting level slows down exponentially.

In certain applications, the growth is described by an early rapid growth, followed by a more mature period of slower growth and concluding with growth bounded by some limit that is ultimately attainable. Such "S-shaped" or "sigmoidal" growth curves can be modeled with the **logistic trend model** in Eq. (9.2),

$$\mu_t = \frac{\alpha}{1 + \beta \exp(-\gamma t)} \quad \alpha > 0, \quad \beta > 0, \quad \gamma > 0 \tag{9.6}$$

The starting value for μ at $t = 0$ is $\alpha/(1 + \beta)$; the limiting value, as t approaches ∞, is α. The parameters α and β determine the initial and the limiting values. Once these values are set, the parameter γ determines the shape of the sigmoidal function. Figure 9.2 shows several realizations of this model for $\alpha = 1$, $\beta = 100$, and several values of γ (0.4, 0.5, 0.6). The logistic growth model implies a characteristic S-shaped response. Such a response function cannot be achieved with the modified exponential model in Eq. (9.5).

Another function with an S-shaped appearance is the **Gompertz growth model**,

$$\mu_t = \alpha \exp[-\beta \exp(-\gamma t)] \quad \alpha > 0, \quad \beta > 0, \quad \gamma > 0 \tag{9.7}$$

The limiting value, as t approaches ∞, is α. The starting value for μ at $t = 0$ is $\alpha \exp(-\beta)$, and with the restrictions on the parameters $0 < \alpha \exp(-\beta) < \alpha$. The parameter $\gamma > 0$ models the shape of the function. Figure 9.3 shows realizations

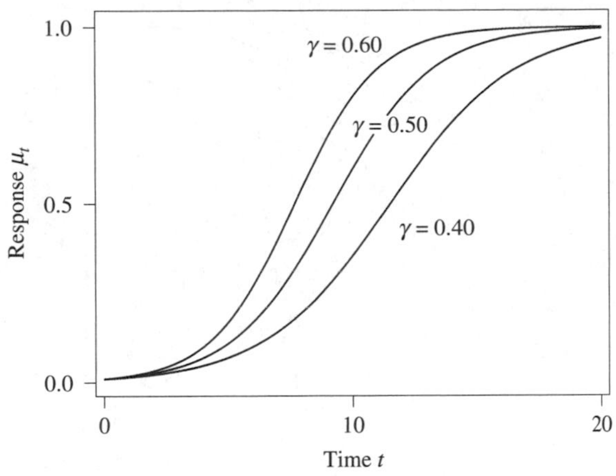

FIGURE 9.2 Logistic Growth Model with $\alpha = 1$, $\beta = 100$, and Various Values of γ

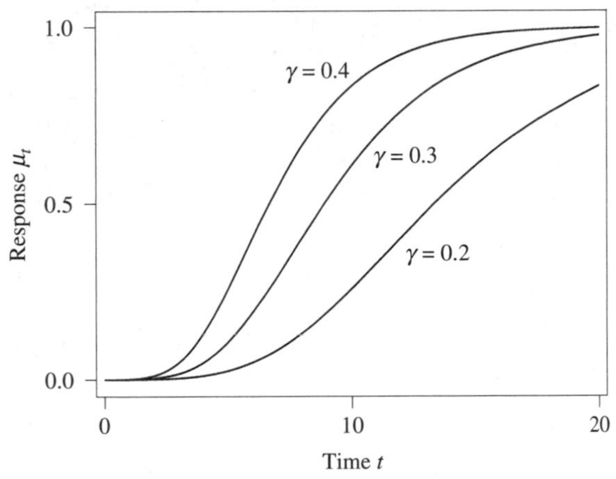

FIGURE 9.3 Gompertz Growth Model with $\alpha = 1$, $\beta = 10$, and Various Values for γ

of this model with $\alpha = 1$, $\beta = 10$, and several values for γ (0.2, 0.3, 0.4). This model is also capable of representing an S-shaped response.

The **Weibull growth model,**

$$\mu_t = \alpha - (\alpha - \beta)\exp[-(\gamma t)^{\delta}] \quad \alpha > 0, \quad \beta < \alpha, \quad \gamma > 0, \quad \delta > 0 \tag{9.8}$$

is another popular model. The starting value for μ at $t = 0$ is β and the limiting value, as t approaches ∞, is α. Several typical realizations from this model are shown in Figure 9.4. For $\delta = 1$, the Weibull model simplifies to the modified exponential model in Eq. (9.5).

The Richards family (Richards, 1959),

$$\mu_t = \alpha[1 + (\delta - 1)\exp(-\gamma(t - \xi))]^{1/(1-\delta)} \tag{9.9}$$

provides another flexible class of models with four parameters. Special cases of this model are the logistic model if $\delta = 2$, the Gompertz model as δ approaches one, and the "modified" exponential (monomolecular) model as $\delta = 0$.

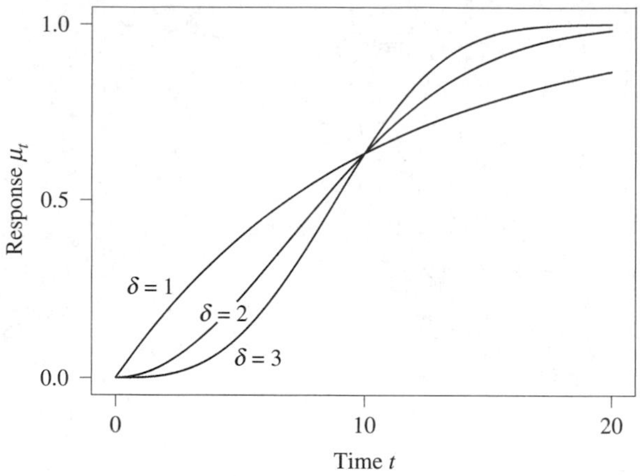

FIGURE 9.4 Weibull Growth Model with $\alpha = 1$, $\beta = 0$, $\gamma = 0.1$, and Various Values for δ

The Morgan–Mercer–Flodin family (Morgan *et al.*, 1975),

$$\mu_t = \alpha - \frac{\alpha - \beta}{1 + (\gamma t)^\delta} \tag{9.10}$$

represents another, and somewhat different, family of growth models. The parameters β and α respectively represent the starting ($t = 0$) and limiting (t approaches ∞) values of the response. The parameter δ determines the shape of the sigmoid function, and $\gamma > 0$ is a scale parameter.

Until now, we have focused on growth models where μ_t increases with time. We conclude this section by discussing an interesting nonlinear model that does not have this increasing property. It arises as the solution of a system of two differential equations and is given by

$$\begin{aligned} \mu_t &= \frac{\gamma_1}{\gamma_1 - \gamma_2}[\exp(-\gamma_2 t) - \exp(-\gamma_1 t)] \quad && \gamma_1 \neq \gamma_2 \\ &= \gamma_1 t \exp(-\gamma_1 t) && \gamma_1 = \gamma_2 \end{aligned} \tag{9.11}$$

This model arises in chemical kinetics where a reactant A decomposes to form the desired product B, which in turn decomposes into an undesired by-product C. Under the assumption of first-order kinetics, the process can be modeled with the following two differential equations:

$$\frac{\partial \mu^*}{\partial t} = -\gamma_1 \mu^* \quad \text{and} \quad \frac{\partial \mu}{\partial t} = \gamma_1 \mu^* - \gamma_2 \mu \tag{9.12}$$

Here, $\gamma_1 > 0$ and $\gamma_2 > 0$ are the reaction rate constants, and μ^* and μ are the concentrations of products A and B. The function in Eq. (9.11) is the solution of the system of the two differential equations in Eq. (9.12). Figure 9.5 shows special cases of the function (9.11).

Note that in all these models the regressor variable is time, and the x axis in all figures represents "time." However, in applications of these models the regressor can also be a variable other than time. For example, in marketing applications the regressor variable can be the amount spent on advertising. The sales effect

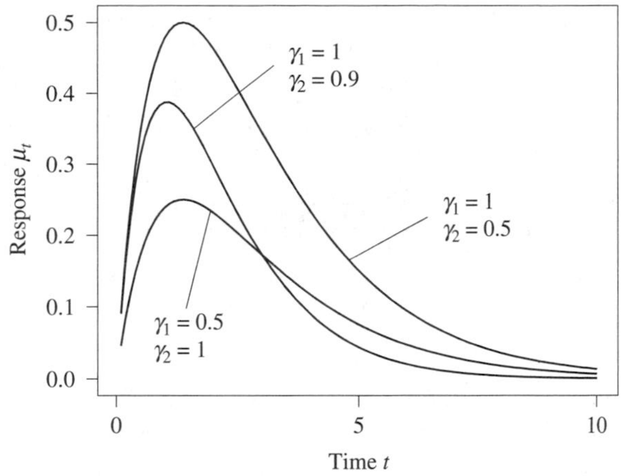

FIGURE 9.5
Model in Eq. (9.11)

of advertising may be characterized by a rapid increase with a moderate level of advertising, a more mature (and approximately linear) effect to advertising in the midrange, and ultimately diminishing and no additional returns when advertising is at very high levels. In chemical or engineering applications, the regressors may be concentrations of certain input variables or process factors such as temperature or pressure.

9.3 NONLINEAR REGRESSION MODELS

The nonlinear functions in Section 9.2 model the "signal" in the observation y. However, no measurement will precisely follow one of these, or any other, simple functional forms. As in the linear regression situation, we need to add a probabilistic error component. In the following, we assume that the errors are additive, independent, and normal with mean zero and constant variance σ^2. Of course, we need to check whether these assumptions are satisfied. In particular, we need to confirm that the errors are indeed uncorrelated if observations are recorded sequentially in time, and that no serial correlation is present. Also, we must check that the errors are normally distributed, and that the variance of the errors is constant and does not depend on the level of the series. The usual regression diagnostics in Chapter 6 apply also to the nonlinear situation.

The nonlinear regression model is given by

$$y_i = \mu_i + \varepsilon_i = \mu(\boldsymbol{x}_i, \boldsymbol{\beta}) + \varepsilon_i \tag{9.13}$$

where $\boldsymbol{x}_i = (x_{i1}, \ldots, x_{im})'$ is the vector of m covariates for the ith case, $\boldsymbol{\beta}$ is a vector of p parameters, and $\mu(\boldsymbol{x}_i, \boldsymbol{\beta})$ is the nonlinear model component. The errors ε_i are independent normal random variables with mean zero and variance σ^2.

The log-likelihood for this model can be written down readily and is given by

$$\ln L(\boldsymbol{\beta}, \sigma^2 \,|\, y_1, y_2, \ldots, y_n) = c - n \ln(\sigma) - \frac{1}{2\sigma^2} \sum_{i=1}^{n} [y_i - \mu(\boldsymbol{x}_i, \boldsymbol{\beta})]^2 \tag{9.14}$$

where $c = -(n/2)\ln(2\pi)$ is a constant that does not depend on the parameters. The maximum likelihood estimator of $\boldsymbol{\beta}$ minimizes the sum of squares

$$S(\boldsymbol{\beta}) = \sum_{i=1}^{n} [y_i - \mu(\boldsymbol{x}_i, \boldsymbol{\beta})]^2 \tag{9.15}$$

and is identical to the least squares estimator of $\boldsymbol{\beta}$. Let us denote it by $\hat{\boldsymbol{\beta}}$. The maximum likelihood estimator of the variance σ^2 is given by

$$\hat{\sigma}^2 = \frac{S(\hat{\boldsymbol{\beta}})}{n} = \frac{\sum_{i=1}^{n} [y_i - \mu(\boldsymbol{x}_i, \hat{\boldsymbol{\beta}})]^2}{n} \tag{9.16}$$

As in the linear model, one adjusts the denominator by the number of estimated parameters and calculates the unbiased least squares estimator of σ^2 from

$$s^2 = \frac{S(\hat{\boldsymbol{\beta}})}{n-p} = \frac{\sum_{i=1}^{n} [y_i - \mu(\boldsymbol{x}_i, \hat{\boldsymbol{\beta}})]^2}{n-p} \tag{9.17}$$

where p is the number of parameters.

So far, this discussion has been identical to our treatment of the linear regression model. The difference here is that the sum of squares $S(\boldsymbol{\beta}) = \sum_{i=1}^{n} [y_i - \mu(\boldsymbol{x}_i, \boldsymbol{\beta})]^2$ is no longer a quadratic function of the parameters, and that it is not possible anymore to write down a closed form expression for the estimator $\hat{\boldsymbol{\beta}}$. **Iterative** estimation schemes must be used to find the value that minimizes the sum of squares.

Our goal in this chapter is to give an introduction, but not a comprehensive treatment, of nonlinear regression modeling. For a comprehensive discussion, we refer the interested reader to the books by Gallant (1987), Bates and Watts (1988), Seber and Wild (1989), and Huet *et al.* (1996). These books go into great detail on how to maximize the likelihood function (or, equivalently, minimize the sum of squares) if the function in Eq. (9.14) is not quadratic in the parameters. In the next section, we give the basic ideas and we discuss the Newton–Raphson and Gauss–Newton algorithms for finding the values that minimize such functions.

9.4 INFERENCE IN THE NONLINEAR REGRESSION MODEL

9.4.1 THE NEWTON–RAPHSON METHOD OF DETERMINING THE MINIMUM OF A FUNCTION

Consider finding the minimum of a function $f(\boldsymbol{\beta})$ with respect to the p-dimensional vector $\boldsymbol{\beta}$. The negative log-likelihood in Eq. (9.14) or the sum of squares in Eq. (9.15), $S(\boldsymbol{\beta}) = \sum_{i=1}^{n} [y_i - \mu(\boldsymbol{x}_i, \boldsymbol{\beta})]^2$, are two such functions of interest. Expand the function $f(\boldsymbol{\beta})$ in a (second-order) Taylor series around its optimum $\boldsymbol{\beta}^*$,

$$f(\boldsymbol{\beta}) \cong f(\boldsymbol{\beta}^*) + \boldsymbol{g}(\boldsymbol{\beta}^*)'(\boldsymbol{\beta} - \boldsymbol{\beta}^*) + \frac{1}{2}(\boldsymbol{\beta} - \boldsymbol{\beta}^*)'G(\boldsymbol{\beta}^*)(\boldsymbol{\beta} - \boldsymbol{\beta}^*) \tag{9.18}$$

where $\mathbf{g}(\boldsymbol{\beta})$ is the column vector of first derivatives of $f(\boldsymbol{\beta})$ with respect to $\boldsymbol{\beta}$. It is the vector with elements $\frac{\partial f(\boldsymbol{\beta})}{\partial \beta_1}, \frac{\partial f(\boldsymbol{\beta})}{\partial \beta_2}, \ldots, \frac{\partial f(\boldsymbol{\beta})}{\partial \beta_p}$. The vector $\mathbf{g}(\boldsymbol{\beta})'$ is the transpose of $\mathbf{g}(\boldsymbol{\beta})$, and the notation $\mathbf{g}(\boldsymbol{\beta}^*)$ expresses the fact that the vector of derivatives is evaluated at $\boldsymbol{\beta} = \boldsymbol{\beta}^*$. $G(\boldsymbol{\beta})$ in Eq. (9.18) is the $p \times p$ matrix of second derivatives of $f(\boldsymbol{\beta})$ with element $\frac{\partial^2 f(\boldsymbol{\beta})}{\partial \beta_i \partial \beta_j}$ in row i and column j. The notation $G(\boldsymbol{\beta}^*)$ indicates that the derivatives are evaluated at $\boldsymbol{\beta} = \boldsymbol{\beta}^*$. The matrix of second derivatives is known as the **Hessian matrix**.

Differentiating the previous equation with respect to the elements in $\boldsymbol{\beta}$ yields

$$\mathbf{g}(\boldsymbol{\beta}) \cong \mathbf{g}(\boldsymbol{\beta}^*) + G(\boldsymbol{\beta}^*)(\boldsymbol{\beta} - \boldsymbol{\beta}^*) = G(\boldsymbol{\beta}^*)(\boldsymbol{\beta} - \boldsymbol{\beta}^*) \tag{9.19}$$

as the vector of first derivatives $\mathbf{g}(\boldsymbol{\beta}^*) = \mathbf{0}$ at the optimum $\boldsymbol{\beta}^*$. Solving Eq. (9.19) for $\boldsymbol{\beta}^*$ leads to

$$\boldsymbol{\beta}^* \cong \boldsymbol{\beta} - [G(\boldsymbol{\beta}^*)]^{-1}\mathbf{g}(\boldsymbol{\beta}) \tag{9.20}$$

For quadratic functions $f(\boldsymbol{\beta})$, $\boldsymbol{\beta}$ can be set equal to any initial value and the optimum $\boldsymbol{\beta}^*$ on the left-hand side of the previous equation is given exactly by the right-hand side. For example, take the scalar case with $p = 1$, and $f(\beta) = a + b\beta + c\beta^2$ and $c > 0$. Then the first and second derivatives are given by $g(\beta) = b + 2c\beta$ and $G(\beta) = 2c$, a positive constant. Note that the β on the right-hand side of the equation cancels and

$$\beta - [G(\beta^*)]^{-1}g(\beta) = \beta - \frac{1}{2c}(b + 2c\beta) = -\frac{b}{2c}$$

which is the value that provides the minimum of $f(\beta) = a + b\beta + c\beta^2$.

For more general functions $f(\boldsymbol{\beta})$, the cancellation of $\boldsymbol{\beta}$ on the right-hand side of the equation no longer happens. However, the equation suggests an iterative procedure in which we revise the current value $\tilde{\boldsymbol{\beta}}$ to obtain a revised value $\boldsymbol{\beta}^*$ according to

$$\boldsymbol{\beta}^* = \tilde{\boldsymbol{\beta}} - [G(\boldsymbol{\beta}^*)]^{-1}\mathbf{g}(\tilde{\boldsymbol{\beta}}) \tag{9.21}$$

The revised value $\boldsymbol{\beta}^*$ can be expected to be closer to the optimum than the current value $\tilde{\boldsymbol{\beta}}$. However, the iterative procedure raises a difficulty because the Hessian matrix is evaluated at the true optimum $\boldsymbol{\beta}^*$, which of course is unknown at the outset. The solution adopted by Newton–Raphson evaluates the Hessian at the current value $\tilde{\boldsymbol{\beta}}$, on the grounds that this will provide a good approximation if the current value is reasonably close to the optimum. The **Newton–Raphson procedure** revises the values according to

$$\boldsymbol{\beta}^* = \tilde{\boldsymbol{\beta}} - [G(\tilde{\boldsymbol{\beta}})]^{-1}\mathbf{g}(\tilde{\boldsymbol{\beta}}) \tag{9.22}$$

The procedure progresses toward a minimum only if the Hessian matrix is positive definite. For concave functions the Hessian can be shown to be positive definite. The procedure breaks down if the matrix G is not invertible, and the iterations may go in the wrong direction (toward a maximum) if the matrix G

is negative definite. However, even with a positive definite Hessian matrix the method may overshoot the minimum and actually increase the objective function. Various safeguards (adding components to make the Hessian positive definite and varying the step size of successive changes) are incorporated in commonly used numerical optimization procedures.

9.4.2 APPLICATION TO NONLINEAR REGRESSION: NEWTON–RAPHSON AND GAUSS–NEWTON METHODS

The Newton–Raphson method can be applied to the minimization of the negative log-likelihood function, $-\ln L(\boldsymbol{\beta})$ in Eq. (9.14). First and second derivatives of $-\ln L(\boldsymbol{\beta})$ need to be calculated and evaluated at the current value $\tilde{\boldsymbol{\beta}}$. Sometimes the matrix of negative second derivatives is replaced by its expectation,

$$E\left[-\frac{\partial^2 \ln L(\boldsymbol{\beta})}{\partial \boldsymbol{\beta} \partial \boldsymbol{\beta}'}\right] = I(\boldsymbol{\beta}) \tag{9.23}$$

$I(\boldsymbol{\beta})$ is called the information matrix. The resulting recursion,

$$\boldsymbol{\beta}^* = \tilde{\boldsymbol{\beta}} + [I(\tilde{\boldsymbol{\beta}})]^{-1} \frac{\partial \ln L(\tilde{\boldsymbol{\beta}})}{\partial \boldsymbol{\beta}} \tag{9.24}$$

is called the **method of scoring**. The scoring calculations are often simpler, and since the information matrix is always positive definite, the problems with nonnegative definite Hessian matrices are avoided.

The least squares (or maximum likelihood with normal errors) estimation of parameters in nonlinear regression models involves the minimization of a sum of squares, $f(\boldsymbol{\beta}) = S(\boldsymbol{\beta}) = \sum_{i=1}^{n} [\varepsilon_i(\boldsymbol{\beta})]^2$, where the errors $\varepsilon_i(\boldsymbol{\beta}) = y_i - \mu_i(\boldsymbol{x}_i, \boldsymbol{\beta})$ involve the parameters $\boldsymbol{\beta}$ in a nonlinear fashion. The vector of first derivatives and the Hessian matrix can be written as

$$\boldsymbol{g}(\boldsymbol{\beta}) = \frac{\partial S(\boldsymbol{\beta})}{\partial \boldsymbol{\beta}} = 2 \sum \left(\frac{\partial \varepsilon_i(\boldsymbol{\beta})}{\partial \boldsymbol{\beta}}\right) \varepsilon_i(\boldsymbol{\beta}) \tag{9.25}$$

$$G(\boldsymbol{\beta}) = \frac{\partial^2 S(\boldsymbol{\beta})}{\partial \boldsymbol{\beta} \partial \boldsymbol{\beta}'} = 2 \sum \left\{ \left(\frac{\partial \varepsilon_i(\boldsymbol{\beta})}{\partial \boldsymbol{\beta}}\right) \left(\frac{\partial \varepsilon_i(\boldsymbol{\beta})}{\partial \boldsymbol{\beta}}\right)' + \frac{\partial^2 \varepsilon_i(\boldsymbol{\beta})}{\partial \boldsymbol{\beta} \partial \boldsymbol{\beta}'} \varepsilon_i(\boldsymbol{\beta}) \right\} \tag{9.26}$$

The vector of derivatives of $\varepsilon_i(\boldsymbol{\beta})$ with respect to the parameters in $\boldsymbol{\beta}$, $\dfrac{\partial \varepsilon_i(\boldsymbol{\beta})}{\partial \boldsymbol{\beta}}$, consists of the elements $\dfrac{\partial \varepsilon_i(\boldsymbol{\beta})}{\partial \beta_1}, \dfrac{\partial \varepsilon_i(\boldsymbol{\beta})}{\partial \beta_2}, \ldots, \dfrac{\partial \varepsilon_i(\boldsymbol{\beta})}{\partial \beta_p}$. The last term in Eq. (9.26), $\dfrac{\partial^2 \varepsilon_i(\boldsymbol{\beta})}{\partial \boldsymbol{\beta} \partial \boldsymbol{\beta}'}$, is the $p \times p$ matrix of second derivatives of $\varepsilon_i(\boldsymbol{\beta})$, with element $\dfrac{\partial^2 \varepsilon_i(\boldsymbol{\beta})}{\partial \beta_j \partial \beta_{j^*}}$ in row j and column j^*. All sums in Eqs. (9.25) and (9.26) go from $i = 1$ through n; we have omitted the limits of the summation in order to simplify the notation.

The Newton–Raphson method revises the estimates according to

$$\boldsymbol{\beta}^* = \tilde{\boldsymbol{\beta}} - \left[\sum \left\{ \left(\frac{\partial \varepsilon_i(\boldsymbol{\beta})}{\partial \boldsymbol{\beta}}\right) \left(\frac{\partial \varepsilon_i(\boldsymbol{\beta})}{\partial \boldsymbol{\beta}}\right)' + \frac{\partial^2 \varepsilon_i(\boldsymbol{\beta})}{\partial \boldsymbol{\beta} \partial \boldsymbol{\beta}'} \varepsilon_i(\boldsymbol{\beta}) \right\}\right]^{-1} \sum \left(\frac{\partial \varepsilon_i(\boldsymbol{\beta})}{\partial \boldsymbol{\beta}}\right) \varepsilon_i(\tilde{\boldsymbol{\beta}}) \tag{9.27}$$

where all first and second derivatives on the right-hand side are evaluated at $\tilde{\boldsymbol{\beta}}$.

Ignoring the second derivatives in the Hessian matrix (one can expect that second derivatives will be small in comparison to first derivatives), the recursions simplify to

$$\boldsymbol{\beta}^* = \tilde{\boldsymbol{\beta}} - \left[\sum\left\{\left(\frac{\partial\varepsilon_i(\boldsymbol{\beta})}{\partial\boldsymbol{\beta}}\right)\left(\frac{\partial\varepsilon_i(\boldsymbol{\beta})}{\partial\boldsymbol{\beta}}\right)'\right\}\right]^{-1}\sum\left(\frac{\partial\varepsilon_i(\boldsymbol{\beta})}{\partial\boldsymbol{\beta}}\right)\varepsilon_i(\tilde{\boldsymbol{\beta}}) \qquad (9.28)$$

This is known as the **Gauss–Newton method**. The method is equivalent to the first-order Taylor series expansion of the nonlinear regression model at the current value $\tilde{\boldsymbol{\beta}}$,

$$\varepsilon_i(\boldsymbol{\beta}) = y_i - \mu(\mathbf{x}_i;\boldsymbol{\beta}) \cong \varepsilon_i(\tilde{\boldsymbol{\beta}}) + \left(\frac{\partial\varepsilon_i(\boldsymbol{\beta})}{\partial\boldsymbol{\beta}}\right)'(\boldsymbol{\beta} - \tilde{\boldsymbol{\beta}})$$

and updating the estimates in the linearized version of the nonlinear model

$$\varepsilon_i(\tilde{\boldsymbol{\beta}}) \cong -\left(\frac{\partial\varepsilon_i(\boldsymbol{\beta})}{\partial\boldsymbol{\beta}}\right)'(\boldsymbol{\beta} - \tilde{\boldsymbol{\beta}}) + \varepsilon_i \qquad (9.29)$$

The vector of first derivatives on the right-hand side of Eq. (9.29), $\dfrac{\partial\varepsilon_i(\boldsymbol{\beta})}{\partial\boldsymbol{\beta}}$, is evaluated at the current estimate $\tilde{\boldsymbol{\beta}}$ and is known; it takes the place of the regressor vector in a standard linear regression model. The least squares estimate of $(\boldsymbol{\beta} - \tilde{\boldsymbol{\beta}})$ can be written down, and by rearranging terms one obtains the updated estimates

$$\boldsymbol{\beta}^* = \tilde{\boldsymbol{\beta}} - \left[\sum\left(\frac{\partial\varepsilon_i(\boldsymbol{\beta})}{\partial\boldsymbol{\beta}}\right)\left(\frac{\partial\varepsilon_i(\boldsymbol{\beta})}{\partial\boldsymbol{\beta}}\right)'\right]^{-1}\sum\left(\frac{\partial\varepsilon_i(\boldsymbol{\beta})}{\partial\boldsymbol{\beta}}\right)\varepsilon_i(\tilde{\boldsymbol{\beta}})$$

These are exactly the recursions given in Eq. (9.28).

The estimation methods—the Newton–Raphson iterations in Eq. (9.27) and the Gauss–Newton recursions in Eq. (9.28)—differ only slightly with respect to the Hessian matrix that is used in the iterations. In almost all situations, they will converge to the same estimates. Let us denote this estimate by $\hat{\boldsymbol{\beta}}$.

9.4.3 STANDARD ERRORS OF THE MAXIMUM LIKELIHOOD ESTIMATES

In the linear regression situation, we were able to derive the exact sampling distribution of the least squares estimates in finite samples. This is no longer possible in the nonlinear situation. However, the inference procedures can be justified from large sample results.

General maximum likelihood theory establishes the asymptotic (when the sample size n is large) normality of the estimates $\hat{\boldsymbol{\beta}}$. It also establishes that the asymptotic covariance matrix of the maximum likelihood estimates is given by the inverse of the information matrix in Eq. (9.23), $I(\boldsymbol{\beta}) = E\left[-\dfrac{\partial^2 \ln L(\boldsymbol{\beta})}{\partial\boldsymbol{\beta}\partial\boldsymbol{\beta}'}\right]$. Evaluating this matrix at the final maximum likelihood estimates gives us an estimate of the asymptotic covariance matrix. That is,

$$V(\hat{\boldsymbol{\beta}}) \cong [I(\hat{\boldsymbol{\beta}})]^{-1} \qquad (9.30)$$

The information matrix can be estimated by the Hessian matrix

$$-\frac{\partial^2 \ln L(\boldsymbol{\beta})}{\partial\boldsymbol{\beta}\partial\boldsymbol{\beta}'} = \frac{1}{2\sigma^2}\frac{\partial^2 S(\boldsymbol{\beta})}{\partial\boldsymbol{\beta}\partial\boldsymbol{\beta}'} = \frac{1}{\sigma^2}\sum\left\{\left(\frac{\partial\varepsilon_i(\boldsymbol{\beta})}{\partial\boldsymbol{\beta}}\right)\left(\frac{\partial\varepsilon_i(\boldsymbol{\beta})}{\partial\boldsymbol{\beta}}\right)' + \frac{\partial^2\varepsilon_i(\boldsymbol{\beta})}{\partial\boldsymbol{\beta}\partial\boldsymbol{\beta}'}\varepsilon_i(\boldsymbol{\beta})\right\}$$

using the result in Eq. (9.26). Hence, the asymptotic covariance matrix of $\hat{\boldsymbol{\beta}}$ is estimated by

$$\begin{aligned} V(\hat{\boldsymbol{\beta}}) &\cong \sigma^2\left[\sum\left\{\left(\frac{\partial\varepsilon_i(\boldsymbol{\beta})}{\partial\boldsymbol{\beta}}\right)\left(\frac{\partial\varepsilon_i(\boldsymbol{\beta})}{\partial\boldsymbol{\beta}}\right)' + \frac{\partial^2\varepsilon_i(\boldsymbol{\beta})}{\partial\boldsymbol{\beta}\partial\boldsymbol{\beta}'}\varepsilon_i(\boldsymbol{\beta})\right\}\right]^{-1} \\ &\cong \sigma^2\left[\sum\left(\frac{\partial\varepsilon_i(\boldsymbol{\beta})}{\partial\boldsymbol{\beta}}\right)\left(\frac{\partial\varepsilon_i(\boldsymbol{\beta})}{\partial\boldsymbol{\beta}}\right)'\right]^{-1} \end{aligned} \tag{9.31}$$

after ignoring the matrix of second derivatives. All derivatives on the right-hand side of Eq. (9.31) are evaluated at the maximum likelihood estimates.

An alternative way of justifying this covariance result is to expand the nonlinear regression function one additional time at the final estimate [see Eq. (9.29) and using $\tilde{\boldsymbol{\beta}} = \hat{\boldsymbol{\beta}}$],

$$\varepsilon_i(\hat{\boldsymbol{\beta}}) \cong -\left(\frac{\partial\varepsilon_i(\boldsymbol{\beta})}{\partial\boldsymbol{\beta}}\right)'(\boldsymbol{\beta} - \hat{\boldsymbol{\beta}}) + \varepsilon_i$$

and to calculate the covariance matrix of the estimates in the last linearization step. Linear regression theory tells us that the covariance matrix of the estimates is

$$V(\hat{\boldsymbol{\beta}}) \cong \sigma^2(X'X)^{-1} = \sigma^2\left[\sum\left(\frac{\partial\varepsilon_i(\boldsymbol{\beta})}{\partial\boldsymbol{\beta}}\right)\left(\frac{\partial\varepsilon_i(\boldsymbol{\beta})}{\partial\boldsymbol{\beta}}\right)'\right]^{-1}$$

which is exactly the result in Eq. (9.31).

The unknown parameter σ^2 in the previous equation is replaced by its least squares estimate in Eq. (9.17). This results in an estimate of the covariance matrix $V(\hat{\boldsymbol{\beta}})$ that can be calculated. The square roots of the diagonal elements in this matrix provide estimates of the standard errors s.e.$(\hat{\beta}_1), \ldots,$ s.e.$(\hat{\beta}_p)$. Off-diagonal elements provide estimates of the covariances among the estimates, which in turn can be used to calculate estimates of the correlation between the parameter estimates.

Large sample theory allows us to proceed with the inference just like we did in the linear regression model. That is, the standard errors can be used to obtain approximate 95% confidence intervals for β_j,

$$\hat{\beta}_j \pm t(1 - \alpha/2; n - p)\text{s.e.}(\hat{\beta}_j)$$

and t ratios $\hat{\beta}_j/\text{s.e.}(\hat{\beta}_j)$ can be used to test H_0: $\beta_j = 0$ against H_1: $\beta_j \neq 0$. The significance is assessed from the percentiles of the t distribution. The additional sum of squares F tests for testing several restrictions among the parameters are still valid under large sample theory.

9.4.4 IMPLEMENTATION ISSUES

Computer software is widely available to find the estimates in nonlinear regression problems. Statistical software packages, such as SPSS, SAS in its routine PROC NLIN, and S-Plus, include reliable estimation routines. For the user, little more than specifying the model and supplying initial starting values is required. Usually, these procedures will work fine.

However, problems with the convergence of these routines may occur. Often, problems are due to poorly specified models and badly chosen starting values. Specifying models that leave certain parameters unidentified (or "almost" unidentified) is a sure roadmap to disaster. The concept of unidentified (or almost unidentified) parameters is akin to multicollinearity (or near multicollinearity). In the linear regression model, one would not think about putting the same regressor into the model twice, because this would leave the individual parameters unidentified; whereas the sum of the two coefficients is identified and estimable, the individual coefficients are not. We learned how to identify such situations by looking at variance inflation factors, and we avoided such situations by simplifying the model specification. The same problem arises in nonlinear models, and in such situations the iterative optimization procedures will experience convergence difficulties. The structure of many nonlinear models is such that a certain degree of multicollinearity is present automatically. Often, the challenge with nonlinear models is to find specifications that are the least subject to the multicollinearity problem.

Nonlinear optimization procedures work best if the magnitudes of the parameters are approximately the same. They often fail if the parameters are very different in size; for example, one parameter is in the range from 0 and 1, whereas the other is in the thousands. One should attempt to reparameterize the model so that the parameters are of approximately equal magnitude.

The iterative estimation procedures will work well as long as we fit appropriately specified models and the functions (sum of squares functions or negative log-likelihoods) are reasonably "well behaved." By that, we mean that the functions are concave and have no local minima that can "trap" iterative algorithms. If the dimension of the parameter vector p is small (one or two), one can always evaluate the function over a grid and look at the whole picture. Of course, this is not feasible if the dimension of β is large. In such cases, one can experiment with different starting values. The model needs to be reconsidered if different starting values do not result in the same final estimate.

9.5 EXAMPLES

9.5.1 EXAMPLE 1: LOSS IN CHLORINE CONCENTRATION

The investigation involves a product that at the time of manufacture must have a fraction 0.50 of available chlorine. The fraction of available chlorine in the product decreases with time. The decrease under controlled conditions is well understood.

TABLE 9.1 DATA ON AVAILABLE CHLORINE AND AGE OF THE PRODUCT [DATA FILE: chlorine]

x = Age: Length of Time Since Produced	y = Available Chlorine (%)
8	0.49, 0.49
10	0.48, 0.47, 0.48, 0.47
12	0.46, 0.46, 0.45, 0.43
14	0.45, 0.43, 0.43
16	0.44, 0.43, 0.43
18	0.46, 0.45
20	0.42, 0.42, 0.43
22	0.41, 0.41, 0.40
24	0.42, 0.40, 0.40
26	0.41, 0.40, 0.41
28	0.41, 0.40
30	0.40, 0.40, 0.38
32	0.41, 0.40
34	0.40
36	0.41, 0.38
38	0.40, 0.40
40	0.39
42	0.39

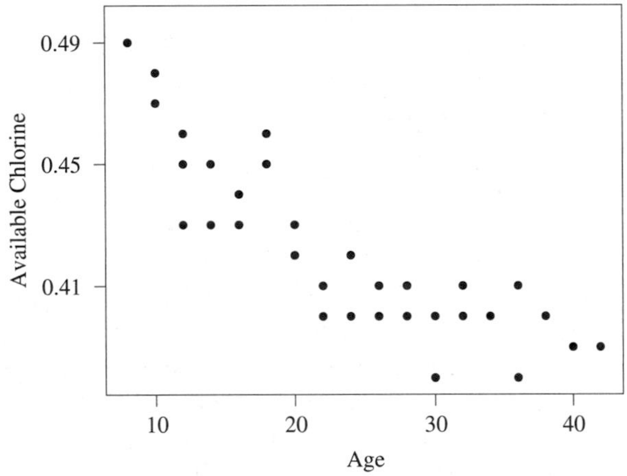

FIGURE 9.6 Scatter Plot of Available Chlorine against Age of the Product

In the 8 weeks before the product reaches the consumer, the fraction of available chlorine declines to approximately 0.49. However, what happens after the 8 weeks is difficult to predict because the handling and storing procedures in warehouses are not always the same.

The data in Table 9.1 are taken from Smith and Dubey (1964). Chlorine concentrations were determined from 44 samples of various ages.

A scatter plot of available chlorine against age is shown in Figure 9.6. Available chlorine decays exponentially from the level 0.49 at age $x = 8$.

TABLE 9.2 SPSS OUTPUT FROM THE NONLINEAR ESTIMATION OF MODEL (9.32): CHLORINE LOSS

Source	DF	Sum of Squares	Mean Square
Regression	2	7.98200	3.99100
Residual	42	5.001680E-03	1.190876E-04
Uncorrected Total	44	7.98700	
(Corrected Total)	43	.03950	

R squared = 1 − Residual SS / Corrected SS = .87338

Parameter	Estimate	Asymptotic Std. Error	Asymptotic 95 % Confidence Interval Lower	Upper
α	.390140032	.005044840	.379959134	.400320931
β	.101632757	.013360412	.074670354	.128595160

Asymptotic Correlation Matrix of the Parameter Estimates

	α	β
α	1.0000	.8879
β	.8879	1.0000

We consider a nonlinear model of the form

$$y_i = \alpha + (0.49 - \alpha)\exp[-\beta(x_i - 8)] + \varepsilon_i \tag{9.32}$$

The model implies a mean chlorine concentration at $x = 8$ of 0.49 and an exponential decay thereafter. The exponential decay flattens out at level α. The model is similar to the "modified" exponential model in Eq. (9.5), except that here we model a decrease (and not an increase) and know the initial response at $x = 8$.

Judging from the graph in Figure 9.6, it is a reasonable assumption that the chlorine concentration flattens out at approximately 0.35; hence, a good starting value for α is given by 0.35. A starting value for β can be obtained as follows: At $x = 38$, $0.40 \approx \alpha + (0.49 - \alpha)\exp(-30\beta)$. Substituting the starting value $\alpha = 0.35$ leads to the equation, $\exp(-30\beta) = 0.05/0.14$ and $\beta = 0.034$.

These values are used as starting values for the nonlinear estimation. The output from the nonlinear regression routine in SPSS is shown in Table 9.2. The estimates are $\hat{\alpha} = 0.39$ and $\hat{\beta} = 0.10$. An estimate of the unexplained variability σ is given by

$$s = \sqrt{\frac{0.00500168}{44 - 2}} = \sqrt{0.000119} = 0.011$$

This amounts to approximately one percentage point. A graph of the fitted values, together with the observations, is shown in Figure 9.7. The figure shows that the model provides a fairly good fit. The coefficient of determination, $R^2 = 1 - \text{SSE}/\text{SST} = 1 - 0.00500168/0.03950 = 0.8734$, indicates that approximately

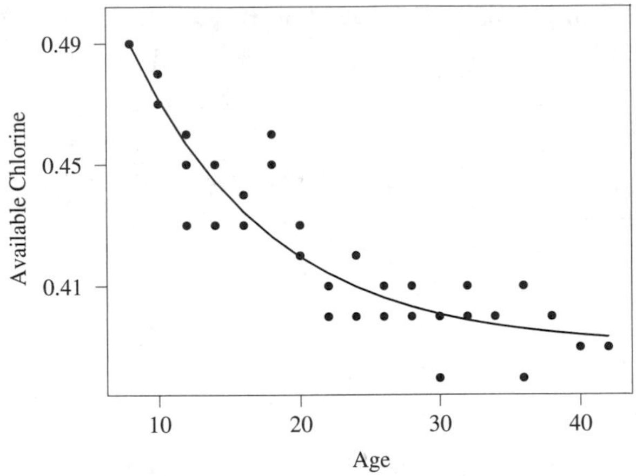

FIGURE 9.7 Observations and Fitted Regression Line: Chlorine Loss

87% of the variation around the mean is explained by the nonlinear regression model.

Because of numerous replications at the values on the covariate, one can also calculate a pure error sum of squares; see Section 6.4. The pure error sum squares from the replications is given by 0.002367 and has 26 degrees of freedom. You can check this by calculating and adding the sums of squares and the degrees of freedom over all groups with replications. Hence, the F test for lack of fit is given by

$$F = \frac{(0.00500168 - 0.002367)/(42 - 26)}{0.002367/26} = 1.81$$

This value needs to be compared to the 95th percentile of the $F(16, 26)$ distribution, which is 2.05. Hence, there is not enough evidence to question the adequacy of the model in Eq. (9.32). Alternatively, one could have calculated the probability value $P(F \geq 1.81) = 1 - 0.9134 = 0.0866$. Since this is larger than the common significance level 0.05, we find no reason to reject the model.

The uncertainty in the estimates is reflected in the confidence intervals. The computer output in Table 9.2 refers to asymptotic standard errors, asymptotic confidence intervals, and asymptotic correlation matrices because it uses an estimate of the asymptotic variance in Eq. (9.31) for these calculations. The 95% confidence interval for β extends from 0.075 to 0.129. We notice a fair amount of correlation (0.89) among the estimates of α and β.

9.5.2 EXAMPLE 2: SHOOT LENGTH OF PLANTS EXPOSED TO GAMMA IRRADIATION

Data on the shoot length of plants emerging from seeds exposed to gamma irradiation are given in Table 9.3. The predictor variable is the dose of gamma irradiation (in kiloRoentgen). The response is the average shoot length of plants (averages of five plants, in centimeters) of a chickpea cultivar emerging from seeds exposed to this dose. The data are taken from Singh *et al.* (1992).

TABLE 9.3 DATA ON PLANT LENGTH AND GAMMA IRRADIATION [DATA FILE: shootlength]

x = Gamma Irradiation	y = Shoot Length
0	8.85
10	9.40
20	9.18
30	8.70
40	7.53
50	6.43
60	5.85
70	4.73
80	3.98
90	3.50
100	3.10
110	2.80

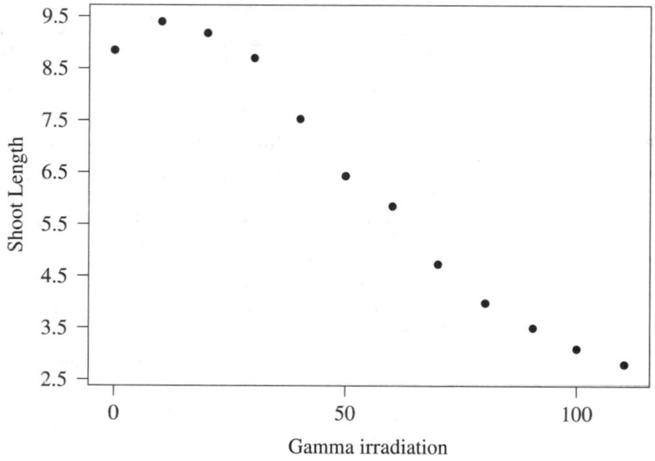

FIGURE 9.8 Scatter Plot of Shoot Length against Gamma Irradiation

A scatter plot of average shoot length against gamma irradiation in Figure 9.8 shows a nonlinear relationship. Singh *et al.* (1992) consider the nonlinear model of the form

$$y = \beta_1 + \frac{\beta_2}{1 + \exp[-\beta_3(x - \beta_4)]} + \varepsilon \tag{9.33}$$

The model is similar to the logistic model in Eq. (9.6), except that it is parametrized differently and that a "background" component β_1 has been included.

Fitting this model with the nonlinear regression routine in SPSS leads to the results in Table 9.4. The table also shows the fitted values and residuals.

The estimate of $V(\varepsilon) = \sigma^2$ is given by $s^2 = 0.63504/8 = 0.0794$. The standard deviation $s = 0.28$ is rather small when compared with the average response

TABLE 9.4 OUTPUT OF THE SPSS NONLINEAR REGRESSION PROGRAM

Source	DF	Sum of Squares	Mean Square
Regression	4	526.19146	131.54786
Residual	8	.63504	.07938
Uncorrected Total	12	526.82650	
(Corrected Total)	11	69.87629	

R squared = 1 − Residual SS / Corrected SS = .99091

Parameter	Estimate	Asymptotic Std. Error	Asymptotic 95 % Confidence Interval Lower	Upper
β_1	2.712748521	.313315157	1.990242473	3.435254569
β_2	−.063851581	.010626341	−.088355967	−.039347194
β_3	6.797488567	.526103337	5.584292096	8.010685038
β_4	56.093286476	2.410491142	50.534683935	61.651889017

Asymptotic Correlation Matrix of the Parameter Estimates

	β_1	β_2	β_3	β_4
β_1	1.0000	−.7800	−.8710	−.4613
β_2	−.7800	1.0000	.8991	.0411
β_3	−.8710	.8991	1.0000	.0514
β_4	−.4613	.0411	.0514	1.0000

Observations, Fitted Values, and Residuals

x = Dose	y = Length	Fitted	Residual
0.00	8.85	9.33	−0.48
10.00	9.40	9.17	0.23
20.00	9.18	8.89	0.29
30.00	8.70	8.43	0.27
40.00	7.53	7.72	−0.19
50.00	6.43	6.76	−0.33
60.00	5.85	5.69	0.16
70.00	4.73	4.69	0.04
80.00	3.98	3.93	0.05
90.00	3.50	3.41	0.09
100.00	3.10	3.10	0.00
110.00	2.80	2.92	−0.12

of 6.2. The nonlinear model in Eq. (9.33) implies monotone decreasing lengths for increasing irradiation; the model is unable to capture the observed "peak" around $x = 10$. One should obtain more data at small gamma irradiation levels to check whether some minor amount of irradiation is actually "helpful" in terms of increasing length, or whether the observation at $x = 10$ is unduly affected by noise. Furthermore, the correlations among the estimates of β_1, β_2, and β_3 make the interpretation of individual coefficients difficult. Although the model fits quite well, one should not put too much emphasis on the interpretation of the individual coefficients.

EXERCISES

9.1. Consider the data on the age (in years) and the leaf area (in square meters) of 12 palm trees. [Source: Rasch, D. The robustness against parameter variation of exact locally optimum designs in nonlinear regression—A case study. *Computational Statistics and Data Analysis,* 20, 441–453, 1995. Also: Sedlacek, G. Zur Quantifizierung und Analyse der Nichtlinearitaet von Regressionsmodellen. *Austrian Journal of Statistics,* 27, 171–190, 1998. The data are given in the file **palmtrees**.

Plot area against age. Consider the Gompertz model in Eq. (9.7), $y = \mu + \varepsilon = \alpha \exp[-\beta \exp(-\gamma \text{Age})] + \varepsilon$, with parameters $\alpha > 0, \beta > 0, \gamma > 0$. Discuss suitable starting values for the parameters. Determine the estimates and assess the adequacy of the model.

Age (years)	Area (m^2)
1	2.02
2	3.62
3	5.71
4	7.13
5	8.33
6	8.29
7	9.81
8	11.30
9	12.18
10	12.67
11	10.62
12	12.01

9.2. Data on the utilization of nitrate in bush beans as a function of light intensity were obtained by J. R. Elliott and D. R. Peirson of Wilfrid Laurier University (Canada). Portions of leaves from three 16-day-old plants were subjected to eight levels of light intensity (in microeinsteins per square meter per second) and the nitrate utilization (in nanomoles per gram per hour) was measured. The data are given in the file **nitrate**.

Nitrate utilization should be zero at zero light intensity and should approach an asymptote as the light intensity increases. The data listed below are analyzed in Bates and Watts (1988); the models that we ask you to consider are taken from their analysis. Note that the experiment was carried out on two different days.

Plot nitrate utilization versus light intensity. Use different symbols for day 1 and day 2 observations.

Bates and Watts (1988) consider several nonlinear models for $y = \mu + \varepsilon$. Fit these models to the data, add the fitted regression line to your scatter plot, and discuss their adequacy.

$$\mu = f(x; \beta_1, \beta_2) = \frac{\beta_1 x}{\beta_2 + x}$$

(Michaelis–Menton model)

$$\mu = f(x; \beta_1, \beta_2) = \beta_1[1 - \exp(-\beta_2 x)]$$

(exponential rise model)

$$\mu = f(x; \beta_1, \beta_2, \beta_3) = \frac{\beta_1 x}{\beta_2 + x + \beta_3 x^2}$$

(quadratic Michaelis–Menton model)

$$\mu = f(x; \beta_1, \beta_2, \beta_3) = \beta_1[\exp(-\beta_3 x) - \exp(-\beta_2 x)]$$

(modified exponential rise model)

Bacon and Watts (1988) also investigate whether the parameters in the Michaelis–Menton models change with the day. They consider an indicator variable z that is 1 if day 2 observations are involved and 0 otherwise, and they study the models

$$\mu = f(x, z; \beta_1, \beta_2, \alpha_1, \alpha_2) = \frac{(\beta_1 + \alpha_1 z)x}{(\beta_2 + \alpha_2 z) + x}$$

$$\mu = f(x, z; \beta_1, \beta_2, \beta_3, \alpha_1, \alpha_2, \alpha_3) = \frac{(\beta_1 + \alpha_1 z)x}{(\beta_2 + \alpha_2 z) + x + (\beta_3 + \alpha_3 z)x^2}$$

Investigate whether these modifications are needed, or whether all or some parameters can be taken constant across both days.

Light Intensity (μE/m^2 sec)	Nitrate Utilization, Day 1 (nmol/g hr)	Nitrate Utilization, Day 2 (nmol/g hr)
2.2	256	549
	685	1,550
	1,537	1,882
5.5	2,148	1,888
	2,583	3,372
	3,376	2,362
9.6	3,634	4,561
	4,960	4,939
	3,814	4,356
17.5	6,986	7,548
	6,903	7,471
	7,636	7,642
27.0	9,884	9,684
	11,597	8,988
	10,221	8,385
46.0	17,319	13,505
	16,539	15,324
	15,047	15,430
94.0	19,250	17,842
	20,282	18,185
	18,357	17,331
170.0	19,638	18,202
	19,043	18,315
	17,475	15,605

9.3. Consider the following models. Discuss how these models can be transformed so that they can be estimated through (linear) least squares.

$$y = \beta_0 (x_1)^{\beta_1} (x_2)^{\beta_2} \varepsilon$$

$$y = \frac{1}{\beta_0 + \beta_1 x + \varepsilon}$$

$$y = \frac{1}{1 + [\exp(\beta_0 + \beta_1 x)]\varepsilon}$$

9.4. Search the engineering/chemistry/biology literature for interesting applications of nonlinear regression. Replicate the estimation if the data are given. Provide a brief report on your findings.

10 Regression Models for Time Series Situations

The standard regression assumptions specify that the errors $\varepsilon_t (t = 1, 2, \ldots, n)$ in the regression model

$$y_t = \beta_0 + \beta_1 x_{t1} + \beta_2 x_{t2} + \cdots + \beta_p x_{tp} + \varepsilon_t$$

are independent, or at least uncorrelated. Independence may be an unreasonable assumption if we estimate the regression model on time series data. In this chapter, we replace the standard case index "i" by "t" in order to emphasize the fact that we deal with time series and not cross-sectional data. That is, we assume that $(y_t, x_{t1}, x_{t2}, \ldots, x_{tp})$ represent the measurements on the response y and the p explanatory variables $x_1, x_2, \ldots, x_p$ at time t—usually months, quarters, or years. For example, (y_t, x_{t1}, x_{t2}), for $t = 1, 2, \ldots, n$, may represent the sales, the price, and the amount spent on advertisement in month t.

10.1 A BRIEF INTRODUCTION TO TIME SERIES MODELS

In this chapter, we assume that observations become available at equally spaced time periods, such as months, quarters, or years. The situation becomes more complicated if time periods are unequally spaced, and we do not address this issue here. In this section, we introduce several simple models that are very useful for characterizing the correlations among time series observations. Correlations among observations k periods apart are referred to as **autocorrelations** or **serial correlations** (see Section 6.2).

10.1.1 FIRST-ORDER AUTOREGRESSIVE MODEL

In our previous discussion on generalized least squares (Section 4.6), we mentioned a situation in which errors were serially (auto) correlated. We parameterized the covariance matrix of the vector of errors $\varepsilon = (\varepsilon_1, \ldots, \varepsilon_t, \ldots, \varepsilon_n)'$ as

$$V(\varepsilon) = \sigma^2 V = \sigma^2 \begin{bmatrix} 1 & \phi & \phi^2 & \cdot & \cdot & \phi^{n-1} \\ \phi & 1 & \phi & \phi^2 & \cdot & \phi^{n-2} \\ \phi^2 & \phi & 1 & \cdot & \cdot & \cdot \\ \cdot & \cdot & \cdot & \cdot & \cdot & \cdot \\ \cdot & \cdot & \cdot & \cdot & 1 & \phi \\ \phi^{n-1} & \phi^{n-2} & \cdot & \cdot & \phi & 1 \end{bmatrix} \tag{10.1}$$

and referred to the associated model as a **first-order autoregressive** representation. Note that the diagonal elements of the matrix V are all one; hence, the errors have equal variance, $V(\varepsilon_t) = \sigma^2$. The matrix V in Eq. (10.1) is in fact a correlation matrix.

All correlations among observations one step apart are the same. The lag 1 autocorrelations are

$$\text{Corr}(\varepsilon_1, \varepsilon_2) = \text{Corr}(\varepsilon_2, \varepsilon_3) = \cdots \text{Corr}(\varepsilon_{t-1}, \varepsilon_t) = \cdots \text{Corr}(\varepsilon_{n-1}, \varepsilon_n) = \phi$$

Correlations must be between -1 and $+1$. Hence, we must restrict the autoregressive parameter to $|\phi| < 1$. Note that we also exclude values of ϕ at the boundary. The case $\phi = 1$ will be covered in the next model when we discuss the random walk.

The matrix V in Eq. (10.1) also implies that all correlations among observations two steps apart (i.e., the lag 2 autocorrelations) are the same and equal to the square of the lag one autocorrelation. That is,

$$\text{Corr}(\varepsilon_1, \varepsilon_3) = \text{Corr}(\varepsilon_2, \varepsilon_4) = \cdots \text{Corr}(\varepsilon_{t-2}, \varepsilon_t) = \cdots \text{Corr}(\varepsilon_{n-2}, \varepsilon_n) = \phi^2$$

The autocorrelations of observations k steps apart are

$$\text{Corr}(\varepsilon_1, \varepsilon_{k+1}) = \text{Corr}(\varepsilon_2, \varepsilon_{k+2}) = \cdots \text{Corr}(\varepsilon_{t-k}, \varepsilon_t) = \cdots \text{Corr}(\varepsilon_{n-k}, \varepsilon_n) = \phi^k$$

We notice two things about these autocorrelations: (i) They depend only on the time lag between the observations, and (ii) they decrease geometrically (or exponentially) with the time lag. We can drop the time index (because the autocorrelations depend only on the time lag) and write the autocorrelations as

$$\rho_1 = \text{Corr}(\varepsilon_{t-1}, \varepsilon_t) = \phi, \quad \rho_2 = \text{Corr}(\varepsilon_{t-2}, \varepsilon_t) = \phi^2, \ldots, \rho_k = \text{Corr}(\varepsilon_{t-k}, \varepsilon_t) = \phi^k \tag{10.2}$$

The autocorrelations, viewed as a function of the lag k, describe the **autocorrelation function**. Note that $\rho_0 = 1$, and $\rho_k = \rho_{-k}$. Hence, the autocorrelation function needs to be shown only for nonnegative k's. We have just seen that the autocorrelation function of errors that follow a first-order autoregressive model exhibits an exponential decay; the farther apart the observations, the weaker the autocorrelation. If the parameter ϕ is large and close to one in absolute value, then the decay is slow; even errors far apart are still correlated. With economic data we usually expect positive values for ϕ; if errors (e.g., in sales) are unusually high in a particular month, then we expect the errors in adjacent months to be large also.

Why do we call this an autoregressive model? This will become clear now as we specify the model that implies this particular autocorrelation structure. Consider the model

$$\varepsilon_t = \phi\varepsilon_{t-1} + a_t \tag{10.3}$$

which "regresses" the correlated error at time t, ε_t, on the previous error ε_{t-1}. The a_t's are the usual "errors" in the regression model, with the typical properties; they have mean zero and are uncorrelated. The autocorrelations among the a_t's are zero at all lags: $\text{Corr}(a_{t-k}, a_t) = 0$ for all $k \neq 0$. Their variance is denoted by σ_a^2. The literature refers to such a sequence of uncorrelated random variables $\{a_t\}$ as a **white noise sequence**. Sometimes, the a_t's are also called random "shocks." Figure 10.1 shows the plot of observations from the model (10.3) with $\phi = 0.5$.

The correlated error at time t, ε_t, can equivalently be written as a linear function of current and previous white noise errors. By repeated substitution in Eq. (10.3), we obtain

$$\begin{aligned}\varepsilon_t &= \phi\varepsilon_{t-1} + a_t = a_t + \phi(\phi\varepsilon_{t-2} + a_{t-1}) = a_t + \phi a_{t-1} + \phi^2\varepsilon_{t-2} \\ &= a_t + \phi a_{t-1} + \phi^2(\phi\varepsilon_{t-3} + a_{t-2}) = \cdots \\ &= a_t + \phi a_{t-1} + \phi^2 a_{t-2} + \phi^3 a_{t-3} + \cdots \end{aligned} \tag{10.4}$$

This expansion is possible because $|\phi| < 1$, which implies that errors in the distant past have little weight. The mean $E(\varepsilon_t) = 0$ because the weights in the expansion (10.4) converge to zero.

The autocorrelations that are implied by this autoregressive model are exactly the ones given by the matrix V in Eq. (10.1). You can see this by multiplying the model in Eq. (10.3) with ε_{t-k} (for $k > 0$),

$$\varepsilon_t\varepsilon_{t-k} = \phi\varepsilon_{t-1}\varepsilon_{t-k} + a_t\varepsilon_{t-k}$$

FIGURE 10.1
Simulations from Three Error Models

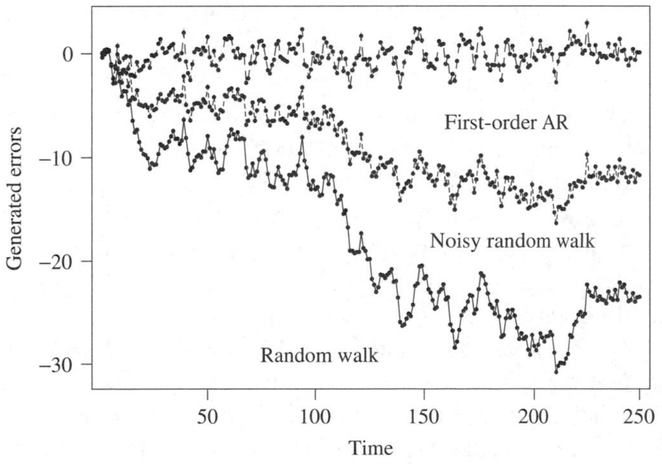

and taking the expectation on both sides of this equation,

$$\text{Cov}(\varepsilon_t, \varepsilon_{t-k}) = \phi\text{Cov}(\varepsilon_{t-1}, \varepsilon_{t-k}) + E(a_t\varepsilon_{t-k}) \tag{10.5}$$

Here we have used the fact that $\text{Cov}(\varepsilon_t, \varepsilon_{t-k}) = E(\varepsilon_t\varepsilon_{t-k})$ since the means $E(\varepsilon_t) = E(\varepsilon_{t-k}) = 0$. The last term in Eq. (10.5) is zero for $k > 0$. This is because ε_{t-k} depends on past errors $a_{t-k}, a_{t-k-1}, a_{t-k-2}, \ldots$, and these components are independent of the current error a_t. Hence,

$$\text{Cov}(\varepsilon_t, \varepsilon_{t-k}) = \phi\text{Cov}(\varepsilon_{t-1}, \varepsilon_{t-k})$$

Dividing both sides of this equation by the time-invariant variance $V(\varepsilon_t)$ leads to

$$\rho_k = \phi\rho_{k-1} \quad \text{for all } k > 0 \tag{10.6}$$

This implies that

$$\rho_k = \phi\rho_{k-1} = \phi\phi\rho_{k-2} = \cdots = \phi^k\rho_0 = \phi^k, \quad k = 1, 2, \ldots \tag{10.7}$$

which are exactly the autocorrelations given by the matrix V in Eq. (10.1). The variance of ε_t can be obtained from the representation in Eq. (10.4):

$$V(\varepsilon_t) = \sigma_a^2[1 + \phi^2 + \phi^4 + \phi^6 + \cdots] = \sigma_a^2/(1 - \phi^2) \tag{10.8}$$

10.1.2 RANDOM WALK MODEL

Consider the first-order autoregressive model in Eq. (10.3), but let the parameter $\phi = 1$. Then

$$\varepsilon_t = \varepsilon_{t-1} + a_t = a_t + a_{t-1} + a_{t-2} + a_{t-3} + \cdots \tag{10.9}$$

We call this the **random walk model**. It expresses the error at time t as a cumulative sum of all random shocks up to time t. We call the random walk model an integrated model because it integrates (sums) the current and previous white noise errors.

The first-order autoregressive model with $|\phi| < 1$ has a fixed level, and in the model without a constant the level is zero. Realizations from the model (10.3) scatter around the fixed level and sample paths do not leave this level for long periods. This "stationary" behavior is different from that of the random walk because the random walk in Eq. (10.9) does not have a fixed level. We call such a model "nonstationary." For nonstationary models, it usually takes a very long time until realizations return to their starting level. Stock prices are good examples of random walks. Looking at charts of daily stock prices, one notices long excursions from any starting point. Also note that the path of a nonstationary time series sequence usually appears quite smooth. Figure 10.1 shows a realization from a random walk.

Differences of a random walk, $w_t = \varepsilon_t - \varepsilon_{t-1} = a_t$, are uncorrelated. Although the ε_t's are nonstationary, wandering "smoothly" with long excursions from any starting point, their first differences w_t are stationary and well behaved. For a random walk, the successive differences are in fact uncorrelated.

10.1.3 SECOND-ORDER AUTOREGRESSIVE MODEL

The first-order autoregressive model is a very simple model for characterizing serial correlation in a time series. However, the model imposes a fairly rigid structure on the autocorrelations. In some circumstances, it may be better to work with a more flexible model that allows more "freedom" for the autocorrelations. The second-order autoregressive model is such an extension. It "regresses" the error at time t on the errors at time $t-1$ and $t-2$:

$$\varepsilon_t = \phi_1 \varepsilon_{t-1} + \phi_2 \varepsilon_{t-2} + a_t \tag{10.10}$$

As in the first-order autoregressive model, it is possible—through repeated substitution—to express ε_t as a linear combination of the current and previous white noise errors a_t's. The coefficients in this expansion depend on the two parameters, ϕ_1 and ϕ_2.

One can derive the autocorrelations that this model implies. The first lag autocorrelation is given by (see Abraham and Ledolter, 1983)

$$\rho_1 = \frac{\phi_1}{1-\phi_2}$$

and the remaining autocorrelations can be obtained from the difference equation,

$$\rho_k = \phi_1 \rho_{k-1} + \phi_2 \rho_{k-2} \quad \text{for } k > 1$$

With ρ_1, one calculates $\rho_2 = \phi_1 \rho_1 + \phi_2$, then $\rho_3 = \phi_1 \rho_2 + \phi_2 \rho_1$, and so on. It can be shown [see standard textbooks on time series analysis, such as Abraham and Ledolter (1983) and Box et al. (1994)] that the autocorrelations of a second-order autoregressive model are combinations of two exponential decays or damped sine waves.

As in the first-order autoregressive model, one needs to put restrictions on the autoregressive parameters; otherwise, the realizations from this model would wander without a fixed level. The parameters must satisfy the constraints

$$\phi_2 + \phi_1 < 1, \quad \phi_2 - \phi_1 < 1, \quad -1 < \phi_2 < 1$$

These are known as the stationarity conditions.

10.1.4 NOISY RANDOM WALK MODEL

The following model provides another useful representation for nonstationary error sequences. The model adds an uncorrelated error (noise) a_t to a fraction of the random walk that we considered previously. It is given by

$$\varepsilon_t = (1-\theta)[a_{t-1} + a_{t-2} + a_{t-3} + \cdots] + a_t = (1-\theta)\sum_{j=1}^{\infty} a_{t-j} + a_t \tag{10.11}$$

The parameter θ that controls the fraction is restricted within -1 and $+1$.

There are several ways of writing this model, and the equivalent representations give useful insights into the nature of the model. Simple manipulation shows that

$$\varepsilon_t = \varepsilon_{t-1} + a_t - \theta a_{t-1} \tag{10.12}$$

Write down ε_t and ε_{t-1} in Eq. (10.11), and convince yourself that their difference equals $a_t - \theta a_{t-1}$. The representation in Eq. (10.12) shows that the noisy random walk expands on the random walk model by adding to the equation a fraction of the previous white noise error $(-\theta a_{t-1})$. The model is capable of characterizing nonstationary error sequences, and it simplifies to the random walk if the parameter $\theta = 0$. Figure 10.1 gives a realization from the model (10.12) with $\theta = 0.5$.

Another representation of the noisy random walk expresses the error at time t as an infinite regression on previous errors,

$$\varepsilon_t = (1-\theta)[\varepsilon_{t-1} + \theta\varepsilon_{t-2} + \theta^2\varepsilon_{t-3} + \cdots] + a_t \tag{10.13}$$

Note that the regression coefficients are all functions of the single parameter θ, and that they decrease geometrically to zero (because the parameter θ is between -1 and $+1$). You can derive the representation in Eq. (10.13) by substituting into the model (10.11) $a_{t-1} = \varepsilon_{t-1} - (1-\theta)[a_{t-2} + a_{t-3} + a_{t-4} + \cdots]$, $a_{t-2} = \varepsilon_{t-2} - (1-\theta)[a_{t-3} + a_{t-4} + a_{t-5} + \cdots]$, and so on.

Successive differences of a noisy random walk, $w_t = \varepsilon_t - \varepsilon_{t-1} = a_t - \theta a_{t-1}$, are linear combinations (or "moving averages") of the present and the previous white noise errors. Since the ε_t's are the result of "integrating" (summing) the differences, the literature refers to the model in Eq. (10.12) as the integrated moving average model of order 1. It is abbreviated as IMA(1,1) because it involves one difference and one moving average term.

Autoregressive models and the integrated models involving differences can be extended. Box and Jenkins discuss **autoregressive integrated moving average** (ARIMA) models that contain autoregressive components (such as the ones we considered), differences (such as the one in the random walk and the noisy random walk), and moving average components (such as the one in the noisy random walk). Time series books will tell you more about these models and their extensions to capture seasonal variation. For our introductory treatment of regression models with time series errors, the models in Eqs. (10.3), (10.9), (10.10), and (10.12) are general enough to illustrate the basic idea.

The first-order integrated moving average (or noisy random walk) model provides a very good representation of a wide range of economic time series. If a simple specific model is assumed on a priori grounds, the first-order integrated moving average model is usually a serious candidate for many economic time series. We typically expect it to provide a better representation than the first-order autoregressive model.

10.1.5 SUMMARY OF TIME SERIES MODELS

In this chapter, we expand the standard regression models by adding autocorrelated errors. Our analysis focuses on the following three models: the first-order autoregressive model, the random walk model, and the noisy random walk model. The autoregressive model describes stationary errors, especially if its parameter ϕ is away from the boundary 1. The other two models describe nonstationary behavior. In order to show the differences among these three models, we have simulated from each of the three models sequences of length 250. Figure 10.1 shows the results for $\phi = 0.5$ and $\theta = 0.5$. The same sequence of standard normal random variables a_t's is used for all three series.

10.1.6 REMARK

In regression modeling, the primary objective is to make inferences about the regression component $E(\mathbf{y}) = \boldsymbol{\mu} = X\boldsymbol{\beta}$. So why is it not sufficient to just adopt a more general covariance matrix $V(\mathbf{y}) = \sigma^2 V$—perhaps reasonably parameterized—and use generalized least squares of Section 4.6 to obtain the estimate of $\boldsymbol{\beta}$? Why is it necessary to study the models that give rise to the various correlation structures?

There are several advantages to an approach that models the autocorrelation in the errors, and these advantages will become clear in the following sections. A model-based approach allows us to (i) get estimates of both the regression parameters and the parameters that govern the autocorrelation in the errors and (ii) derive the optimal forecasts for the response variable.

10.2 THE EFFECTS OF IGNORING THE AUTOCORRELATION IN THE ERRORS

In this section, we illustrate the effects of autocorrelation on standard least squares estimation. In Section 10.5, we discuss the effect of autocorrelation on the forecasts. The consequences of ignoring the autocorrelation in the errors can be very serious, and the assumption of independence among errors should not be taken lightly.

10.2.1 INEFFICIENCY OF LEAST SQUARES ESTIMATION

We assume the multiple linear regression model

$$\mathbf{y} = X\boldsymbol{\beta} + \boldsymbol{\varepsilon} \tag{10.14}$$

with errors that have mean vector zero and covariance matrix $V(\boldsymbol{\varepsilon}) = \sigma^2 V$. The standard least squares estimator

$$\hat{\boldsymbol{\beta}} = (X'X)^{-1}X'\mathbf{y} \tag{10.15}$$

is still unbiased, but now with covariance matrix

$$V(\hat{\boldsymbol{\beta}}) = \sigma^2 (X'X)^{-1}X'VX(X'X)^{-1} \tag{10.16}$$

Our discussion of generalized least squares and the Gauss–Markov result in Section 4.2 shows that the generalized least squares (GLS) estimator

$$\hat{\boldsymbol{\beta}}^{\text{GLS}} = (X'V^{-1}X)^{-1}X'V^{-1}\mathbf{y} \tag{10.17}$$

has the smallest variance among all linear unbiased estimators. This shows that the least squares estimator in Eq. (10.15) is inefficient because its covariance matrix exceeds $V(\hat{\beta}^{GLS}) = \sigma^2(X'V^{-1}X)^{-1}$ by a positive semidefinite matrix.

A Special Case

Let us study the loss in efficiency of the least squares estimator when the errors follow a first-order autoregressive model. The inverse of the correlation matrix V in Eq. (10.1) can be written as $V^{-1} = (1-\phi^2)^{-1}L'L$, where L is the $n \times n$ matrix

$$L = \begin{bmatrix} \sqrt{1-\phi^2} & 0 & 0 & 0 & . & 0 \\ -\phi & 1 & 0 & 0 & . & 0 \\ 0 & -\phi & 1 & 0 & . & 0 \\ . & . & . & . & . & . \\ 0 & 0 & 0 & -\phi & 1 & 0 \\ 0 & 0 & 0 & 0 & -\phi & 1 \end{bmatrix} \tag{10.18}$$

This result can be checked by writing out the elements of the $n \times n$ matrix $(1-\phi^2)^{-1}LV$ and premultiplying the resulting matrix with L'. This leads to $(1-\phi^2)^{-1}L'LV = I$ and proves that $(1-\phi^2)^{-1}L'L$ is the inverse of V.

Consider the simple linear regression without an intercept, $y_t = \beta x_t + \varepsilon_t$. The least squares estimator has variance

$$V(\hat{\beta}) = \sigma^2(X'X)^{-1}X'VX(X'X)^{-1} = \sigma^2 \frac{\sum_{t=1}^{n} x_t^2 + 2\sum_{j=1}^{n-1}\sum_{t=1}^{n-j}\phi^j x_t x_{t+j}}{\left[\sum_{t=1}^{n} x_t^2\right]^2} \tag{10.19}$$

On the other hand, the generalized least squares estimator has variance

$$V(\hat{\beta}^{GLS}) = \sigma^2(X'V^{-1}X)^{-1} = \sigma^2 \frac{1-\phi^2}{\sum_{t=1}^{n} x_t^2 + \phi^2\sum_{t=2}^{n-1} x_t^2 - 2\phi\sum_{t=1}^{n-1} x_t x_{t+1}} \tag{10.20}$$

The loss in efficiency depends on the regressor variable. Consider the special case in which the regressor variable is time; that is, $x_t = t$, for $t = 1, 2, \ldots, n$. The relative efficiencies, $V(\hat{\beta}^{GLS})/V(\hat{\beta})$, for various values of ϕ and sample sizes $n = 50$ and 100, are listed in Table 10.1. The loss in efficiency of the standard least squares estimator is relatively small if the autoregressive parameter ϕ is not too close to the nonstationarity boundary, 1; for ϕ smaller than 0.8, the relative

TABLE 10.1 RELATIVE EFFICIENCIES, $V(\hat{\beta}^{GLS})/V(\hat{\beta})$, FOR VARIOUS VALUES OF ϕ AND SAMPLE SIZES $n = 50$ AND 100

ϕ	0.10	0.20	0.30	0.40	0.50	0.60	0.70	0.80	0.90	0.95	0.99
$n = 50$	0.999	0.998	0.994	0.989	0.982	0.970	0.951	0.915	0.833	0.719	0.331
$n = 100$	0.9997	0.999	0.997	0.994	0.990	0.984	0.973	0.953	0.899	0.819	0.497

efficiency of the least squares estimator is larger than 90%. Only if ϕ is close to the nonstationarity boundary does the loss in efficiency become substantial; the least squares estimator should not be used in such a situation.

10.2.2 SPURIOUS REGRESSION RESULTS WHEN NONSTATIONARY ERRORS ARE INVOLVED

The calculations in Table 10.1 show that the loss in efficiency for the standard least squares estimate when errors follow a stationary model (such as the first-order autoregression discussed previously) is usually relatively small. Problems with nonstationary errors are more worrisome.

In many observational studies with time series data, both the response and the regressor variables are nonstationary, and it is in this situation that the standard regression analysis is misleading. Box and Newbold (1971) and Granger and Newbold (1974) examine the situation in which the response and the explanatory variable follow independent random walks; in this case, both series drift without fixed level (as the graph of a random walk in Figure 10.1 shows), but there is no relationship among the series. One would expect that the standard t ratio, $\hat{\beta}_1/\text{s.e.}(\hat{\beta}_1)$, in the simple linear regression assuming independent errors rejects the null hypothesis $\beta_1 = 0$ in at most 5% of the cases (if the test is conducted at the 5% significance level). This, however, is far from true. Simulations (with sample size $n = 50$) show that the null hypothesis of no relationship between the series is rejected wrongly in approximately 75% of all cases. This is because the standard error, $\text{s.e.}(\hat{\beta}_1) = s/\sqrt{\sum_{t=1}^{n}(x_t - \bar{x})^2}$, grossly underestimates the true variability of the estimate. Consequently, the t ratio $\hat{\beta}_1/\text{s.e.}(\hat{\beta}_1)$ is too large, and the null hypothesis is rejected far too often. The standard analysis errs on the side of finding "spurious" regression relationships; one is led incorrectly to conclude that there is a relationship when none is present.

Equivalently, one could talk about the effect of autocorrelation on the F ratio in the analysis of variance table because in the simple linear regression model the F ratio is the square of the t ratio. The simulations with regressions of two independent random walks show that also the F ratio is way too large. The relationship between the F ratio and the coefficient of determination R^2, $F = (n-2)R^2/(1-R^2)$, implies that also the coefficient of determination from such regressions is too high; the simulations by Granger and Newbold (1974) with independent random walks of lengths $n = 50$ show that the average R^2 is approximately 0.25, with 5% of the simulations leading to an R^2 that exceeds 0.7.

A theoretical explanation for the inflated R^2 when regressing (or correlating) two autocorrelated but independent time series can be given. The variance of the correlation coefficient r_{xy} between two independent but autocorrelated series x and y with autocorrelation functions $\rho_x(k)$ and $\rho_y(k)$ is given by

$$V(r_{xy}) \approx n^{-1} \sum_{j=-\infty}^{\infty} \rho_x(j)\rho_y(j) \tag{10.21}$$

see Box *et al.* (1994). For two first-order autoregressive processes with the same autoregressive parameter and $\rho_x(j)=\rho_y(j)=\phi^{|j|}$, this variance simplifies to

$$V(r_{xy})\approx n^{-1}\frac{1+\phi^2}{1-\phi^2} \tag{10.22}$$

If the autoregressive parameter is large, indicating that the series approach nonstationarity, the ratio $(1+\phi^2)/(1-\phi^2)$ can become quite large. Hence, it is not uncommon to obtain an unusually large sample correlation coefficient, or an unusually large R^2, when dealing with two autocorrelated but independent series. In the simple linear regression of y_t on x_t, the coefficient of determination is the square of the correlation coefficient, $R^2=r_{xy}^2$. Hence, for two uncorrelated first-order autoregressive series x and y,

$$E(R^2)=E\left(r_{xy}^2\right)=V(r_{xy})\approx n^{-1}\frac{1+\phi^2}{1-\phi^2}$$

It is straightforward to replicate and confirm these findings through additional simulations. In Exercise 10.6, we ask you to generate independent random walks of various lengths, run the standard regression analysis, and confirm that one rejects the null hypothesis of no association way too often and that the coefficient of determination R^2 can be quite large.

How can one guard against being tricked into accepting spurious regressions? The answer is simple. One needs to be on the alert and watch for possible autocorrelations in the residuals. If autocorrelations are present, they must be modeled and the standard regression model must be enlarged to take account of the autocorrelation. A model that combines the regression with time series models for the errors must be considered.

A standard correction for serial correlation in the regression errors is to adopt a first-order autoregressive model for the error components in the regression and use the Cochran–Orcutt approach to correct for first-order autoregressive errors. This approach is discussed in the following section. Simulations by Newbold and Davies (1978) show that although this approach is better than taking no correction at all, it is still subject to finding spurious associations if the series in the regression are independent but generated from first-order integrated moving average models. A preferable strategy is to allow for the possibility of alternative time series models for the regression errors and use the data to distinguish among them.

10.3 THE ESTIMATION OF COMBINED REGRESSION TIME SERIES MODELS

10.3.1 A REGRESSION MODEL WITH FIRST-ORDER AUTOREGRESSIVE ERRORS

Consider the regression model with an intercept and p explanatory regressor variables, and assume that the errors follow a first-order autoregressive model. That is,

$$\begin{aligned} y_t &= \mathbf{x}_t'\boldsymbol{\beta}+\varepsilon_t=\beta_0+\beta_1x_{t1}+\cdots+\beta_px_{tp}+\varepsilon_t \\ \varepsilon_t &= \phi\varepsilon_{t-1}+a_t \end{aligned} \tag{10.23}$$

We assume that the random components a_t's $(t = 1, 2, \ldots, n)$ are uncorrelated with mean zero and variance σ_a^2.

Writing the equation for y_{t-1} and subtracting ϕy_{t-1} from y_t results in

$$\begin{aligned} y_t - \phi y_{t-1} &= \mathbf{x}_t'\boldsymbol{\beta} - \phi \mathbf{x}_{t-1}'\boldsymbol{\beta} + a_t \\ &= (1-\phi)\beta_0 + \beta_1(x_{t1} - \phi x_{t-1,1}) + \cdots + \beta_p(x_{tp} - \phi x_{t-1,p}) + a_t \end{aligned} \tag{10.24}$$

The regression of the transformed response $y_t - \phi y_{t-1}$ on the p transformed regressor variables $(x_{t1} - \phi x_{t-1,1}), \ldots, (x_{tp} - \phi x_{t-1,p})$ satisfies the standard regression assumptions. The regression, using the $n - 1$ cases $(t = 2, 3, \ldots, n)$ of the transformed variables, results in the best linear unbiased estimator of $\boldsymbol{\beta}$.

The problem with this approach is that the parameter ϕ is unknown and the transformed model in Eq. (10.24) is no longer linear in the parameters. It involves products of ϕ and the regression parameters $\beta_0, \beta_1, \ldots, \beta_p$.

We need an estimate of ϕ. One approach proceeds as follows. We start with the standard regression in Eq. (10.23) but ignore the autocorrelations and assume independent errors. The estimates of the regression parameters are inefficient because they ignore the autocorrelation (see Section 10.2). Nevertheless, the estimated lag 1 autocorrelation r_1 of the residuals provides an initial estimate of ϕ, and this value can be used to transform the response and the explanatory variables. The regression of the transformed response on the transformed regressors (which is carried out next) results in revised estimates of the regression coefficients $\boldsymbol{\beta}$. The new estimates are more efficient because they incorporate the serial correlation in the errors. With the better estimates we can calculate a better estimate of the errors, $y_t - \mathbf{x}_t'\hat{\boldsymbol{\beta}} = y_t - (\hat{\beta}_0 + \hat{\beta}_1 x_{t1} + \cdots + \hat{\beta}_p x_{tp})$, and from the lag 1 autocorrelation of the new errors we can get an improved estimate of ϕ. The iteration continues; we calculate new transformed variables, get new estimates of the regression parameters, get new residuals and another estimate of ϕ, and so on. One or two iterations are usually sufficient. Standard errors of the regression estimates can be obtained from the last regression of model (10.24).

This is a good and often used approach. It is known as the **Cochran–Orcutt approach** after the researchers who developed it. However, their approach works only for autoregressive errors and not for errors that follow a model with moving average components, such as the nonstationary noisy random walk discussed in Section 10.1.4. In this case, one cannot use an approach that transforms the model equation with correlated errors into one with uncorrelated errors.

The maximum likelihood approach, which is discussed next, works more generally. We now assume that the errors a_t are independent and normal, with mean zero and variance σ_a^2.

Let us first revisit the regression model with first-order autoregressive errors in Eq. (10.24), which we write in transformed form as

$$\begin{aligned} y_t &= \phi y_{t-1} + (1-\phi)\beta_0 + \beta_1(x_{t1} - \phi x_{t-1,1}) + \cdots + \beta_p(x_{tp} - \phi x_{t-1,p}) + a_t \\ &= \mu(\mathbf{x}_t, \boldsymbol{\xi}) + a_t \end{aligned} \tag{10.25}$$

where $\boldsymbol{\xi} = (\phi, \boldsymbol{\beta})'$ is the vector of parameters and $\boldsymbol{x}_t = (y_{t-1}, x_{t1}, \ldots, x_{tp})'$ is the vector of the regressors on the right-hand side of Eq. (10.25). The time index t runs from $t = 2, 3, \ldots, n$. We treat y_1 in the equation with $t = 2$ as an observed value.

From the joint probability density

$$p(a_2, a_3, \ldots, a_n) = \left(\frac{1}{\sqrt{2\pi}\sigma_a}\right)^{(n-1)} \exp\left\{-\frac{1}{2\sigma_a^2}\sum_{t=2}^{n} a_t^2\right\}$$

we obtain the joint probability density of $y_2, y_3, \ldots, y_n$:

$$p(y_2, y_3, \ldots, y_n \mid \boldsymbol{\beta}, \phi, \sigma_a)$$
$$= \left(\frac{1}{\sqrt{2\pi}\sigma_a}\right)^{(n-1)} \exp\left\{-\frac{1}{2\sigma_a^2}\sum_{t=2}^{n}\{y_t - [\phi y_{t-1} + (1-\phi)\beta_0 + \beta_1(x_{t1} - \phi x_{t-1,1}) \right.$$
$$\left. + \cdots + \beta_p(x_{tp} - \phi x_{t-1,p})]\}^2\right\}$$

Treating the joint probability density as a function of the parameters leads to the likelihood and the log-likelihood function

$$l(\boldsymbol{\beta}, \phi, \sigma_a \mid y_2, y_3, \ldots, y_n) = c - (n-1)\ln\sigma_a - \frac{1}{2\sigma_a^2}S(\boldsymbol{\beta}, \phi) \qquad (10.26)$$

where $c = -((n-1)/2)\ln(2\pi)$ is a constant and

$$S(\boldsymbol{\beta}, \phi) = \sum_{t=2}^{n}\{y_t - [\phi y_{t-1} + (1-\phi)\beta_0 + \beta_1(x_{t1} - \phi x_{t-1,1})$$
$$+ \cdots + \beta_p(x_{tp} - \phi x_{t-1,p})]\}^2$$

is the sum of squares of errors. Estimates of $\boldsymbol{\beta}$ and ϕ are obtained by minimizing this sum of squares. Since this sum of squares is conditional on the given starting value y_1, one refers to this sum of squares as the **conditional sum of squares** and to the estimators that minimize this sum of squares as the **conditional least squares** (or conditional maximum likelihood) estimators. The model in Eq. (10.25) is no longer linear in the parameters because it involves products of the parameters $\boldsymbol{\beta}$ and ϕ. Hence, it is not possible to give an explicit expression of the least squares estimates. The estimates must be computed iteratively. Nonlinear least squares procedures (in particular, the Newton–Raphson and Gauss–Newton procedures of Section 9.4) are used for their calculation. The procedures are iterative. They linearize the model, starting from initial estimates and refining the linear expansion as better estimates become available.

From the final nonlinear least squares estimates $\hat{\boldsymbol{\beta}}$ and $\hat{\phi}$, one can obtain the maximum likelihood estimate of σ_a^2:

$$\hat{\sigma}_a^2 = \frac{1}{n-1}S(\hat{\boldsymbol{\beta}}, \hat{\phi})$$

It is common to "adjust" the number of observations in the denominator, $n-1$, by the number of estimated parameters, $p + 1 + 1 = p + 2$, and work with the unbiased estimator of σ_a^2,

$$s_a^2 = \frac{1}{n-p-3} S(\hat{\beta}, \hat{\phi}) \tag{10.27}$$

For statistical inference (such as confidence intervals and tests of hypotheses) one needs standard errors of the estimates. The usual approach is to expand the nonlinear model in Eq. (10.25) in a Taylor series expansion around the final parameter estimates $\hat{\boldsymbol{\xi}}$. This leads to

$$y_t \approx \mu(\boldsymbol{x}_t, \hat{\boldsymbol{\xi}}) + (\boldsymbol{\xi} - \hat{\boldsymbol{\xi}})' \left. \frac{\partial \mu(\boldsymbol{x}_t, \boldsymbol{\xi})}{\partial \boldsymbol{\xi}} \right|_{\boldsymbol{\xi}=\hat{\boldsymbol{\xi}}} + a_t \tag{10.28}$$

In our model, the parameter vector $\boldsymbol{\xi}$ consists of the regression parameters $\boldsymbol{\beta}$ and the autoregressive parameter ϕ. The elements of the column vectors of derivatives of $\mu(\boldsymbol{x}_t, \boldsymbol{\xi})$ in Eq. (10.25) with respect to $\beta_0, \beta_1, \ldots, \beta_p$ and ϕ are given by

$$(1-\phi), (x_{t1} - \phi x_{t-1,1}), \ldots, (x_{tp} - \phi x_{t-1,p}),$$
$$(y_{t-1} - \beta_0 - \beta_1 x_{t-1,1} - \cdots - \beta_p x_{t-1,p})$$

and they are evaluated at the least squares estimates $\hat{\beta}_0, \hat{\beta}_1, \ldots, \hat{\beta}_p, \hat{\phi}$. The vectors of first derivatives are collected in the matrix X, and the covariance matrix of the (iterated) least squares estimates is calculated from

$$V(\hat{\boldsymbol{\xi}}) = \sigma_a^2 (X'X)^{-1} \tag{10.29}$$

An estimate is obtained by replacing σ_a^2 by s_a^2 in Eq. (10.27).

10.3.2 REGRESSION MODEL WITH NOISY RANDOM WALK ERRORS

Let us use this approach to estimate the parameters in the regression model with errors that follow a noisy random walk model,

$$y_t = \boldsymbol{x}_t'\boldsymbol{\beta} + \varepsilon_t = \beta_0 + \beta_1 x_{t1} + \cdots + \beta_p x_{tp} + \varepsilon_t$$
$$\varepsilon_t = \varepsilon_{t-1} + a_t - \theta a_{t-1}$$

or

$$\Delta y_t = \beta_1 \Delta x_{t1} + \cdots + \beta_p \Delta x_{tp} + a_t - \theta a_{t-1} \tag{10.30}$$

where Δ is the differencing operator such that $\Delta y_t = y_t - y_{t-1}$. Note that with differences the intercept β_0 no longer appears in the model.

Assuming that a_1 in

$$a_t = \Delta y_t - \beta_1 \Delta x_{t1} - \cdots - \beta_p \Delta x_{tp} + \theta a_{t-1}, \quad t = 2, 3, \ldots, n$$

is a fixed (nonrandom) constant, we can write the log likelihood function

$$l(\boldsymbol{\beta}, \theta, \sigma_a \mid \Delta y_2, \Delta y_3, \ldots, \Delta y_n) = c - (n-1) \ln \sigma_a - \frac{1}{2\sigma_a^2} S(\boldsymbol{\beta}, \theta) \tag{10.31}$$

where the sum of squares

$$S(\boldsymbol{\beta}, \theta) = \sum_{t=2}^{n} [a_t(\boldsymbol{\beta}, \theta)]^2$$

The errors $a_t(\boldsymbol{\beta}, \theta)$ in the sum of squares are functions of the parameters $\boldsymbol{\beta}$ and θ. They can be calculated recursively from

$$\begin{aligned} a_2 &= \Delta y_2 - \beta_1 \Delta x_{21} - \cdots - \beta_p \Delta x_{2p} + \theta a_1 \\ a_3 &= \Delta y_3 - \beta_1 \Delta x_{31} - \cdots - \beta_p \Delta x_{3p} + \theta a_2 \\ &\ldots. \\ a_n &= \Delta y_n - \beta_1 \Delta x_{n1} - \cdots - \beta_p \Delta x_{np} + \theta a_{n-1} \end{aligned} \tag{10.32}$$

All that is needed is a starting value a_1. One common approach is to start with $a_1 = 0$, a reasonable starting value because the unconditional mean $E(a_1) = 0$. This is known as the conditional least squares (or conditional maximum likelihood) approach. Another approach starts from the mean $E(a_1 \mid \Delta y_2, \ldots, \Delta y_n)$, which can be calculated from the data. The time series literature refers to this second approach as the unconditional least squares approach.

Iterative approaches need to be employed to minimize the sum of squares. With given starting values for the parameters $\boldsymbol{\beta}$ and θ (and starting value a_1), it is straightforward to compute recursively all subsequent errors a_2, a_3, and a_n. From these, we can calculate the resulting sum of squares, $S(\boldsymbol{\beta}, \theta) = \sum_{t=2}^{n} [a_t(\boldsymbol{\beta}, \theta)]^2$. The nonlinear least squares procedures from Chapter 9 can be applied to obtain the estimates and their approximate standard errors.

A Note on Software The combined regression time series models require nonlinear least squares procedures for their estimation. Several computer packages are available for the estimation, including SCA (distributed by Scientific Computing Associates, http://www.scausa.com) and Eviews (distributed by Quantitative Micro Software, http://www.eviews.com).

10.4 FORECASTING WITH COMBINED REGRESSION TIME SERIES MODELS

In this section, we show how to use the regression time series models to predict future values of the response. Our assumption is that we have available data on the response up to and including time period n. The objective is to predict the response at time period $n + r$; that is, we are interested in an r step-ahead prediction of y_{n+r}. We use the notation $y_n(r)$ to denote the r step-ahead forecast; the subscript n stands for the forecast origin; and the number in parentheses, r, stands for the forecast horizon.

Initially (in the following two subsections), we assume that the explanatory variables are under the control of the investigator, and that future values of the explanatory variables are known. This is a reasonable assumption in many applications. For example, when predicting future sales from a model that includes one's own price and advertising expenditures, one knows their values in future periods. However, this would not be a good assumption for variables such as the competitor price or the economic climate. These variables are not under the control of the model user, and they would have to be predicted.

It is tempting to ignore the autocorrelation in the model

$$y_t = \mathbf{x}_t'\boldsymbol{\beta} + \varepsilon_t = \beta_0 + \beta_1 x_{t1} + \cdots + \beta_p x_{tp} + \varepsilon_t$$

and predict the value of the future response y_{n+r} as

$$y_n(r) = \mathbf{x}_{n+r}'\boldsymbol{\beta} = \beta_0 + \beta_1 x_{n+r,1} + \cdots + \beta_p x_{n+r,p}$$

However, as the subsequent discussion shows, this is incorrect.

10.4.1 FORECASTS FROM THE REGRESSION MODEL WITH FIRST-ORDER AUTOREGRESSIVE ERRORS

In order to simplify the notation, we consider the case of a single explanatory variable. The model is given as

$$y_t = \beta_0 + \beta_1 x_t + \varepsilon_t \quad \text{and} \quad \varepsilon_t = \phi\varepsilon_{t-1} + a_t \tag{10.33}$$

Suppose that we have data available for time periods $t = 1, 2, \ldots, n$. The observation at the next time period, $n + 1$, can be written as

$$y_{n+1} = \phi y_n + (1 - \phi)\beta_0 + \beta_1(x_{n+1} - \phi x_n) + a_{n+1}$$

The value of the explanatory variable at time $n + 1$, x_{n+1}, is assumed known. However, the random error a_{n+1} is unknown. Since random errors are assumed independent, a_{n+1} is not predictable from the previous errors $a_n, a_{n-1}, \ldots, a_1$, and $E(a_{n+1} \mid a_n, a_{n-1}, \ldots, a_1) = E(a_{n+1}) = 0$. Hence, the one-step-ahead prediction is

$$y_n(1) = \phi y_n + (1 - \phi)\beta_0 + \beta_1(x_{n+1} - \phi x_n) \tag{10.34}$$

and the one-step-ahead prediction error is $y_{n+1} - y_n(1) = a_{n+1}$, with variance σ_a^2. A 95% prediction interval for the future observation y_{n+1} is given by

$$y_n(1) \pm (1.96)\sigma_a \quad \text{or} \quad \phi y_n + (1 - \phi)\beta_0 + \beta_1(x_{n+1} - \phi x_n) \pm (1.96)\sigma_a$$

The observation two-steps-ahead is

$$\begin{aligned} y_{n+2} &= \phi y_{n+1} + (1 - \phi)\beta_0 + \beta_1(x_{n+2} - \phi x_{n+1}) + a_{n+2} \\ &= \phi[\phi y_n + (1 - \phi)\beta_0 + \beta_1(x_{n+1} - \phi x_n)] + (1 - \phi)\beta_0 \\ &\quad + \beta_1(x_{n+2} - \phi x_{n+1}) + a_{n+2} + \phi a_{n+1} \end{aligned}$$

Future random shocks have mean zero. The two-step-ahead forecast is

$$\begin{aligned} y_n(2) &= \phi[\phi y_n + (1 - \phi)\beta_0 + \beta_1(x_{n+1} - \phi x_n)] + (1 - \phi)\beta_0 + \beta_1(x_{n+2} - \phi x_{n+1}) \\ &= \phi y_n(1) + (1 - \phi)\beta_0 + \beta_1(x_{n+2} - \phi x_{n+1}) \end{aligned} \tag{10.35}$$

The two-step-ahead forecast error is $y_{n+2} - y_n(2) = a_{n+2} + \phi a_{n+1}$, with variance $(1 + \phi^2)\sigma_a^2$.

The r-step-ahead forecast of y_{n+r} can be calculated from the difference equation

$$y_n(r) = \phi y_n(r-1) + (1-\phi)\beta_0 + \beta_1(x_{n+r} - \phi x_{n+r-1}) \tag{10.36}$$

The r-step-ahead forecast error $y_{n+r} - y_n(r) = a_{n+r} + \phi a_{n+r-1} + \cdots + \phi^{r-1} a_{n+1}$ has variance $(1 + \phi^2 + \cdots + \phi^{2(r-1)})\sigma_a^2 = \dfrac{1-\phi^{2r}}{1-\phi^2}\sigma_a^2$. A 95% prediction interval is given by

$$y_n(r) \pm 1.96\sigma_a[(1-\phi^{2r})/(1-\phi^2)]^{1/2} \tag{10.37}$$

If the parameters are estimated from past data, we must replace the parameters β_0, β_1, and ϕ with their estimates. In the prediction interval, we replace σ_a^2 by s_a^2 in Eq. (10.27). However, note that the resulting interval is "optimistic" and too narrow because the substitution approach does not incorporate the uncertainty from the parameter estimation. One can also incorporate this uncertainty, but the analysis becomes tedious.

10.4.2 FORECASTS FROM THE REGRESSION MODEL WITH ERRORS FOLLOWING A NOISY RANDOM WALK

In this case, the observation at time $n+1$ can be written as

$$y_{n+1} = y_n + \beta_1(x_{n+1} - x_n) + a_{n+1} - \theta a_n \tag{10.38}$$

where a_{n+1} is a future error, and its mean $E(a_{n+1} \mid a_n, a_{n-1}, \ldots, a_1) = E(a_{n+1}) = 0$. However, a_n is an error that has been realized already. Assuming that β_1 and θ are known, the forecast is

$$y_n(1) = y_n + \beta_1(x_{n+1} - x_n) - \theta a_n \tag{10.39}$$

The term a_n is the last component in the recursions of Eq. (10.32), which for given values of the parameters is easy to calculate. Note that the model in Eq. (10.38) involves only a single regressor and all terms in Eq. (10.32) involving the regressors $x_2, \ldots, x_p$ can be ignored.

Observations two or more steps beyond the forecast origin ($r \geq 2$),

$$y_{n+r} = y_{n+r-1} + \beta_1(x_{n+r} - x_{n+r-1}) + a_{n+r} - \theta a_{n+r-1}$$

involve only future errors, and expectations of future errors are zero. Hence, the forecasts are given by

$$y_n(r) = y_n(r-1) + \beta_1(x_{n+r} - x_{n+r-1}) \quad \text{for } r \geq 2 \tag{10.40}$$

The r-step-ahead forecast error is $y_{n+r} - y_n(r) = a_{n+r} + (1-\theta)[a_{n+r-1} + \cdots + a_{n+1}]$, with variance $[1 + (r-1)(1-\theta)^2]\sigma_a^2$.

It is instructive to write the one-step-ahead forecast of Eq. (10.39) in a slightly different form. In Section 10.1, we showed that the noisy random walk can be expressed as

$$\varepsilon_t = (1-\theta)[\varepsilon_{t-1} + \theta\varepsilon_{t-2} + \theta^2\varepsilon_{t-3} + \cdots] + a_t$$

This implies that the model $y_{n+1} = \beta_0 + \beta_1 x_{n+1} + \varepsilon_{n+1}$ can be written as

$$y_{n+1} - (1-\theta)[y_n + \theta y_{n-1} + \theta^2 y_{n-2} + \cdots]$$
$$= \beta_1\{x_{n+1} - (1-\theta)[x_n + \theta x_{n-1} + \theta^2 x_{n-2} + \cdots]\} + a_{n+1}$$

The one-step-ahead forecast is

$$\begin{aligned} y_n(1) &= (1-\theta)[y_n + \theta y_{n-1} + \theta^2 y_{n-2} + \cdots] \\ &\quad + \beta_1\{x_{n+1} - (1-\theta)[x_n + \theta x_{n-1} + \theta^2 x_{n-2} + \cdots]\} \\ &= \text{EWMA}(y_n, y_{n-1}, \ldots) + \beta_1[x_{n+1} - \text{EWMA}(x_n, x_{n-1}, \ldots)] \end{aligned} \quad (10.41)$$

where EWMA $(y_n, y_{n-1}, \ldots) = (1-\theta)[y_n + \theta y_{n-1} + \theta^2 y_{n-2} + \cdots]$ and EWMA $(x_n, x_{n-1}, \ldots) = (1-\theta)[x_n + \theta x_{n-1} + \theta^2 x_{n-2} + \cdots]$ are **exponentially weighted averages**. The coefficients in the EWMA decay exponentially to zero, and they define weighted averages as the sum of the weights $(1-\theta)[1 + \theta + \theta^2 + \cdots] = 1$. The forecast in Eq. (10.41) adjusts the exponentially weighted average of past responses by the difference between the new level of the regressor variable and its exponentially weighted average. It is not just the new level, x_{n+1}, that goes into the forecast, but its difference to the historic (exponentially weighted) average.

Parameter estimates are used in place of the unknown values β_1 and θ, and the estimate s_a^2 replaces the error variance σ_a^2.

10.4.3 FORECASTS WHEN THE EXPLANATORY VARIABLE MUST BE FORECAST

What happens if the future value of the explanatory variable is not given but must be forecast? Future economic climate, for example, will not be known to the modeler.

Let us assume that we have available forecasts $x_n(r), r = 1, 2, \ldots$. Let us suppose that the forecast error $x_{n+r} - x_n(r)$ is independent of the future random errors $a_{n+1}, \ldots, a_{n+r}$ in the model, and that it has mean zero (i.e., the forecast is unbiased) and variance $\sigma_x^2(r)$. The variance expresses the accuracy of the forecast of the explanatory variable; usually, it is not negligible because the prediction of the regressor variable is often difficult.

Consider the regression with first-order autoregressive errors, and assume that the parameters ϕ, β_0, and β_1 are known. Replacing the unknown value x_{n+1} with its forecast $x_n(1)$ leads to the forecast

$$y_n(1) = \phi y_n + (1-\phi)\beta_0 + \beta_1[x_n(1) - \phi x_n]$$

and the one-step-ahead forecast error

$$y_{n+1} - y_n(1) = a_{n+1} + \beta_1[x_{n+1} - x_n(1)]$$

with variance $\sigma_a^2 + \beta_1^2\sigma_x^2(1)$. The uncertainty in predicting the future value of the regressor variable inflates the variance, and it may be a large part of the uncertainty of the forecast.

Forecasts more than one step ahead are given by

$$y_n(r) = \phi y_n(r-1) + (1-\phi)\beta_0 + \beta_1[x_n(r) - \phi x_n(r-1)]$$

Simple substitution leads to the resulting r-step-ahead forecast error

$$y_{n+r} - y_n(r) = a_{n+r} + \phi a_{n+r-1} + \cdots + \phi^{r-1} a_{n+1} + \beta_1[x_{n+r} - x_n(r)]$$

Its variance is given by $(1+\phi^2+\cdots+\phi^{2(r-1)})\sigma_a^2 + \beta_1^2\sigma_x^2(r) = \dfrac{1-\phi^{2r}}{1-\phi^2}\sigma_a^2 + \beta_1^2\sigma_x^2(r)$.

10.5 MODEL-BUILDING STRATEGY AND EXAMPLE

Our strategy to guard against problems from autocorrelations in the errors is to combine regression models with simple time series models for the errors. Of course, in the beginning one does not know which error model to consider and one needs to adopt an iterative strategy.

We illustrate the modeling strategy in the following example. We analyze the sales and advertising expenditures for a dietary weight control product over 36 consecutive months (Table 10.2). The data are taken from Blattberg and Jeuland (1981). The scatter plot of sales on advertising in Figure 10.2 shows a linear association between the response (Sales) and the explanatory variable (Advertising), suggesting the regression model

$$y_t = \beta_0 + \beta_1 x_t + \varepsilon_t \tag{10.42}$$

TABLE 10.2 SALES AND ADVERTISING EXPENSES FOR 36 CONSECUTIVE MONTHS[a]

Month	Sales	Advert	Month	Sales	Advert
1	12.0	15	19	30.5	33
2	20.5	16	20	28.0	62
3	21.0	18	21	26.0	22
4	15.5	27	22	21.5	12
5	15.3	21	23	19.7	24
6	23.5	49	24	19.0	3
7	24.5	21	25	16.0	5
8	21.3	22	26	20.7	14
9	23.5	28	27	26.5	36
10	28.0	36	28	30.6	40
11	24.0	40	29	32.3	49
12	15.5	3	30	29.5	7
13	17.3	21	31	28.3	52
14	25.3	29	32	31.3	65
15	25.0	62	33	32.2	17
16	36.5	65	34	26.4	5
17	36.5	46	35	23.4	17
18	29.6	44	36	16.4	1

[a] The data are given in the file **salesadvert**.

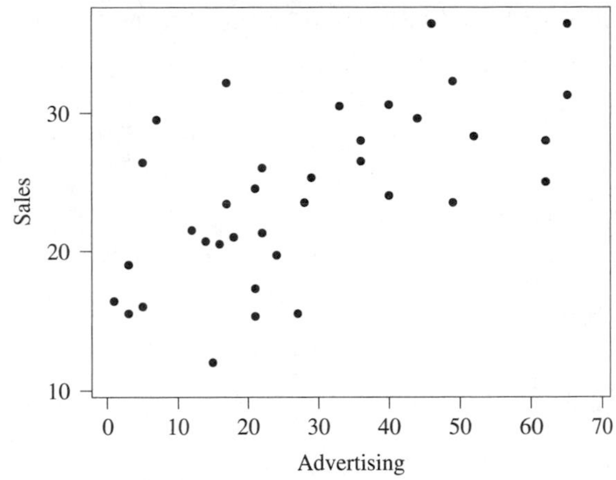

FIGURE 10.2 Scatter Plot of Sales against Advertising

TABLE 10.3 MINITAB REGRESSION OUTPUT FOR THE MODEL $y_t = \beta_0 + \beta_1 x_t + \varepsilon_t$

The regression equation is
Sales = 18.3 + 0.208 Advert

Predictor	Coef	SE Coef	T	P
Constant	18.323	1.489	12.31	0.000
Advert	0.20786	0.04378	4.75	0.000

S = 4.863   R-Sq = 39.9%   R-Sq(adj) = 38.1%

Analysis of Variance

Source	DF	SS	MS	F	P
Regression	1	533.20	533.20	22.54	0.000
Residual Error	34	804.15	23.65		
Total	35	1337.35			

Durbin-Watson statistic = 1.25

The regression output in Table 10.3 shows that approximately 40% of the variation is explained by the regression model ($R^2 = 0.399$). Since time series observations are involved, one needs to check whether the residuals are autocorrelated. The Durbin–Watson test statistic (DW = 1.25) indicates a problem with autocorrelations; it is much smaller than the desired value of 2. The first five autocorrelations of the residuals are shown in Table 10.4. The approximate standard error of a lag k autocorrelation is given by $(n)^{-1/2}$, and with $n = 36$ the standard error is 0.17. The estimated lag 1 autocorrelation is approximately two times its standard error, indicating that there may be serial correlation among the errors. The autocorrelations decay with the lag, suggesting a first-order autoregressive model as an appropriate representation. The series are short, and although 36 observations are sufficient to detect gross violations of the independence assumption, they are usually not sufficient to "fine-tune" the error model. Luckily, the particular choice of the error model usually matters little as long as the model allows for autocorrelation.

TABLE 10.4 FIRST FIVE LAG AUTOCORRELATIONS OF THE RESIDUALS FROM THE MODEL $y_t = \beta_0 + \beta_1 x_t + \varepsilon_t$

```
ACF of residuals
        -1.0 -0.8 -0.6 -0.4 -0.2  0.0  0.2  0.4  0.6  0.8  1.0
          +----+----+----+----+----+----+----+----+----+----+
1 0.317                             XXXXXXXXX
2 0.162                             XXXXX
3 0.196                             XXXXXX
4 0.096                             XXX
5 0.086                             XXX
```

TABLE 10.5 ESTIMATION RESULTS FOR THE THREE REGRESSION MODELS:

Model 1: $y_t = \beta_0 + \beta_1 x_t + \varepsilon_t$, with $\varepsilon_t = \phi\varepsilon_{t-1} + a_t$
Model 2: $y_t = \phi y_{t-1} + \beta_0 + \beta_1 x_t + \beta_2 x_{t-1} + a_t$
Model 3: $y_t = \phi y_{t-1} + \beta_0 + \beta_1 x_t + a_t$

	Model 1 (Eq. 10.43)	Model 2 (Eq. 10.44)	Model 3 (Eq. 10.45)
Estimates[a]			
Constant	22.08 (1.85)	9.08 (2.38)	7.45 (2.47)
ϕ	0.567 (0.128)	0.364 (0.115)	0.528 (0.102)
β_1	0.095 (0.037)	0.130 (0.031)	0.146 (0.033)
β_2		0.097 (0.039)	
R^2	0.588	0.728	0.672
Residuals			
Durbin–Watson	1.70	1.62	2.03
Autocorrelations			
r_1	0.11	0.16	−0.05
r_2	−0.09	−0.15	−0.14
r_3	−0.03	0.14	0.06
r_4	−0.14	0.03	−0.09
r_5	−0.07	−0.11	−0.02

[a] Standard errors in parentheses.

Combining the regression model with the autoregressive errors leads to the new model

$$\begin{aligned} y_t &= \beta_0 + \beta_1 x_t + \varepsilon_t \quad \text{with} \quad \varepsilon_t = \phi\varepsilon_{t-1} + a_t, \quad \text{or} \\ y_t &= \phi y_{t-1} + \beta_0(1-\phi) + \beta_1(x_t - \phi x_{t-1}) + a_t \end{aligned} \tag{10.43}$$

We use the SCA (Scientific Computing Associates) software to estimate the model. The estimation results and the first five autocorrelations of the residuals (the estimates of a_t) are shown under model 1 in Table 10.5. The results indicate that the first-order autoregressive component has been successful in removing the autocorrelation among the errors; all residual autocorrelations are now within one standard error, and no obvious patterns (like the one we saw in Table 10.4) are visible. The R^2 has increased from 0.40 to 0.59. Also note that the estimates of β_0 and β_1 have changed.

The model

$$y_t = \phi y_{t-1} + \beta_0 + \beta_1 x_t + \beta_2 x_{t-1} + a_t \tag{10.44}$$

is slightly more general than the model in Eq. (10.43) because it relaxes the constraint that $\beta_2 = -\beta_1\phi$. This model is also easier to estimate because it can be estimated by standard linear least squares software. This is model 2 in Table 10.5. The estimate $\hat{\beta}_2 = 0.097$ is positive and not negative as implied by the results of model (10.43). This indicates that the constraint imposed by model 1 may not be appropriate; also, the increase in R^2 from 0.59 in model 1 to 0.73 in model 2 is considerable. Again, the estimates of β_0 and β_1 are different from those of models (10.42) and (10.43).

Another alternative is to consider

$$y_t = \phi y_{t-1} + \beta_0 + \beta_1 x_t + a_t \tag{10.45}$$

a model that includes a lag on the response but not a lag on the explanatory variable. The results are given under model 3 in Table 10.5. The reduction in R^2 (from 0.73 in model 2 to 0.67 in model 3) is not very large. Nevertheless, there is evidence that the lag on advertising may be needed because the estimate $\hat{\beta}_2 = 0.097$ in model 2, with standard error s.e.$(\hat{\beta}_2) = 0.039$, is significant.

This analysis shows that the inclusion of the lagged response y_{t-1} in the regression equation goes a long way toward correcting the autocorrelation problem. The question whether one should use a regression model with lagged response and lagged explanatory variables, or a combined regression time series model that restricts the coefficients, is an issue that can be settled from the data. In this example, the regression model (10.44),

$$y_t = \phi y_{t-1} + \beta_0 + \beta_1 x_t + \beta_2 x_{t-1} + a_t$$

provides a good description of the data. The advantage of this model is that it can be estimated by standard linear least squares. Forecasts from this model can be obtained as in Section 10.4. The r-step-ahead forecast of y_{n+r} can be calculated from the difference equation

$$y_n(r) = \phi y_n(r-1) + \beta_0 + \beta_1 x_{n+r} + \beta_2 x_{n+r-1}$$

10.6 COINTEGRATION AND REGRESSION WITH TIME SERIES DATA: AN EXAMPLE

When modeling economic time series, it is often the case that the response y_t and the regressor x_t are nonstationary with changing levels, but that their differences $y_t - y_{t-1}$ and $x_t - x_{t-1}$ are stationary. One calls such nonstationary series (first-order) **integrated** because each series can be thought of as a sum of stationary differences. Random walks, for example, are first-order integrated series.

If the relationship between two nonstationary (integrated) series y_t and x_t is such that the residuals ε_t in the regression

$$y_t = \beta_0 + \beta_1 x_t + \varepsilon_t \tag{10.46}$$

are stationary, then one refers to the two nonstationary series y_t and x_t as **cointegrated**. The errors in the regression (10.46) need not be independent but can follow any stationary time series model, such as the first-order autoregressive model in Eq. (10.3). Stationarity of the errors ε_t is important because it implies that the unexplained component in the regression relationship linking the two nonstationary series is predictable with a time-invariant level. Although the two series drift without fixed levels, the error in the regression relationship is stationary.

Assume that the integrated series y_t and x_t in Eq. (10.46) are independent. In Section 10.2, we discussed the dangers of regressing (correlating) two independent integrated random walk series. Because of the independence between y_t and x_t, the errors in the regression in Eq. (10.46) are also integrated and nonstationary. This is because $\beta_1 x_t$ does not enter the model and $\varepsilon_t = y_t - \beta_0$ is a nonstationary series. This nonstationarity in the errors causes the problems of spurious regression.

For nonstationary but cointegrated series y_t and x_t, on the other hand, the errors ε_t in Eq. (10.46) are stationary. Although there are still problems with the efficiency of least squares estimates if the autocorrelation in the errors is ignored, the problems are not nearly as serious as in the case of nonstationary errors. Also, one can always adjust for the autocorrelation by incorporating time series structure into the errors of the regression.

Many economic laws arise from cointegration relationships. Although the series are nonstationary individually, the errors in the regression are in fact stationary. We illustrate this fact with the U.S. hog series studied by Quenouille (1968). The data set consists of annual data on the price of hogs, the number of hogs, the price of corn, the supply of corn, and the farm wage rate and covers the years from 1867 through 1948. Quenouille logarithmically transforms each variate and then linearly codes the logs, so as to produce numbers of comparable magnitude in the different series. Box and Tiao (1977), in their analysis of this data set, suggest that the price of hogs and the wage rate be shifted backward in time by 1 year. The data set in Table 10.6 lists the transformed and time-shifted observations.

Time sequence plots of the five series are shown in Figure 10.3. The graphs show that individually all five series are integrated. The series are nonstationary without fixed levels, and their autocorrelations (not shown) decay slowly.

Here, we model the price of the hogs as a function of the remaining four variables. We expect the price of hogs to increase with decreasing supply of hogs, increasing price of corn, and increasing farm wage rate. Despite the nonstationary nature of the individual series, we expect that the relationship among the five series is stable, and that the errors in the regression are stationary. In other words, we expect that the relationship is cointegrated.

The results of the regression

$$y_t = \beta_0 + \beta_1 x_{t1} + \beta_2 x_{t2} + \beta_3 x_{t3} + \beta_4 x_{t4} + \varepsilon_t \qquad (10.47)$$

TABLE 10.6 ANNUAL NUMBER OF HOGS, PRICE OF HOGS, PRICE OF CORN, SUPPLY OF CORN, AND FARM WAGE RATE FROM 1867 THROUGH 1947[a]

Year	Hog Price	Hog Supply	Corn Price	Corn Supply	Farm Wage Rate
1867	509	538	944	900	719
1868	663	522	841	964	716
1869	751	513	911	893	724
1870	739	529	768	1,051	732
1871	598	565	718	1,057	740
1872	556	594	634	1,107	748
1873	594	600	735	1,003	756
1874	667	584	858	1,025	748
1875	776	554	673	1,161	740
1876	754	553	609	1,170	732
1877	689	595	604	1,181	744
1878	498	637	457	1,194	756
1879	643	641	612	1,244	778
1880	681	647	642	1,232	799
1881	728	634	849	1,095	799
1882	829	629	733	1,244	799
1883	751	638	672	1,218	799
1884	704	662	594	1,289	801
1885	633	675	559	1,313	803
1886	663	658	604	1,251	806
1887	709	629	678	1,205	806
1888	763	625	571	1,352	806
1889	681	648	490	1,361	810
1890	627	682	747	1,218	813
1891	667	676	651	1,368	810
1892	804	655	645	1,278	806
1893	782	640	609	1,279	771
1894	707	668	705	1,208	771
1895	653	678	453	1,404	780
1896	639	692	382	1,427	789
1897	672	710	466	1,359	799
1898	669	727	506	1,371	820
1899	729	712	525	1,423	834
1900	784	708	595	1,425	848
1901	842	705	829	1,234	863
1902	886	680	654	1,443	884
1903	784	682	673	1,401	906
1904	770	713	691	1,429	928
1905	783	726	660	1,470	949
1906	877	729	643	1,482	960
1907	777	752	754	1,417	971
1908	810	766	813	1,409	982
1909	957	720	790	1,417	987
1910	970	682	712	1,455	991
1911	903	743	831	1,394	1,004
1912	995	743	742	1,469	1,013
1913	1,022	730	847	1,357	1,004

(Continued)

TABLE 10.6 (Continued)

Year	Hog Price	Hog Supply	Corn Price	Corn Supply	Farm Wage Rate
1914	998	723	850	1,402	1,013
1915	928	753	830	1,452	1,053
1916	1,073	782	1,056	1,385	1,149
1917	1,294	760	1,163	1,464	1,248
1918	1,346	799	1,182	1,388	1,316
1919	1,301	808	1,180	1,428	1,384
1920	1,134	779	805	1,487	1,190
1921	1,024	770	714	1,467	1,179
1922	1,090	777	865	1,432	1,228
1923	1,013	841	911	1,459	1,238
1924	1,119	823	1,027	1,347	1,246
1925	1,195	746	846	1,447	1,253
1926	1,235	717	869	1,406	1,253
1927	1,120	744	928	1,418	1,253
1928	1,112	791	924	1,426	1,255
1929	1,129	771	903	1,401	1,223
1930	1,055	746	777	1,318	1,114
1931	787	739	507	1,411	982
1932	624	773	500	1,467	929
1933	612	793	716	1,380	978
1934	800	768	911	1,161	1,013
1935	1,104	592	816	1,362	1,045
1936	1,075	633	1,019	1,178	1,100
1937	1,052	634	714	1,422	1,097
1938	1,048	649	687	1,406	1,090
1939	891	699	754	1,412	1,100
1940	921	786	791	1,390	1,188
1941	1,193	735	876	1,424	1,303
1942	1,352	782	962	1,487	1,422
1943	1,243	869	1,050	1,472	1,498
1944	1,314	923	1,037	1,490	1,544
1945	1,380	774	1,104	1,458	1,582
1946	1,556	787	1,193	1,507	1,607
1947	1,632	754	1,334	1,372	1,629

[a] The data are given in the file **hogs**.

assuming independent errors $\varepsilon_t = a_t$, are shown in Table 10.7. The time sequence plot of the residuals in Figure 10.4 and the autocorrelations of the residuals in Table 10.7 show that the residuals are stationary, and that the relationship in Eq. (10.47) is indeed cointegrated. There is evidence of autocorrelation in the residuals. We notice a somewhat large positive autocorrelation at lag 1 ($r_1 = 0.27$, with approximate standard error $1/\sqrt{81} = 0.11$), implying a Durbin–Watson statistic (DW = 1.36) smaller than the desired value of 2.

Here, we model the autocorrelation with a first-order moving average process and consider

$$\begin{aligned} y_t &= \beta_0 + \beta_1 x_{t1} + \beta_2 x_{t2} + \beta_3 x_{t3} + \beta_4 x_{t4} + \varepsilon_t \\ \varepsilon_t &= a_t - \theta a_{t-1} \end{aligned} \tag{10.48}$$

FIGURE 10.3
Time Sequence Plots of the Annual (a) Price of Hogs, (b) Number of Hogs, (c) Price of Corn, (d) Supply of Corn, and (e) Farm Wage Rate from 1867 through 1947

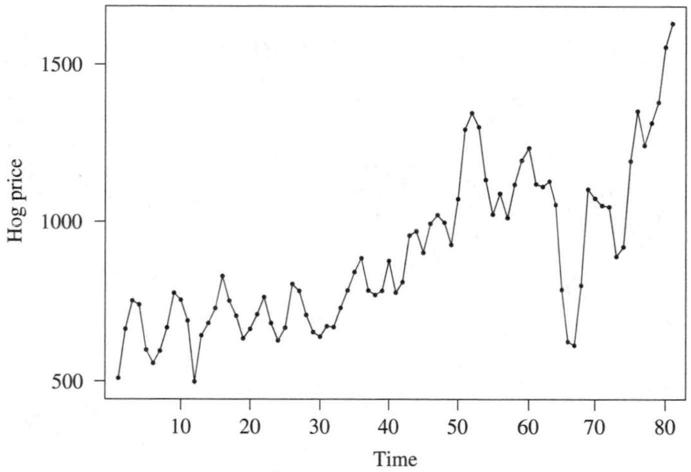

(a)

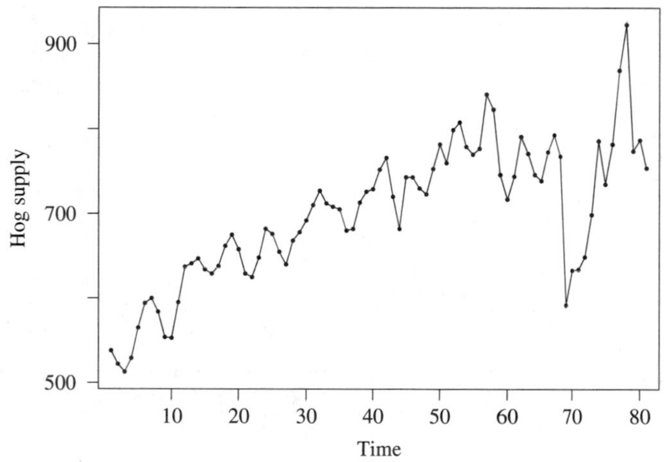

(b)

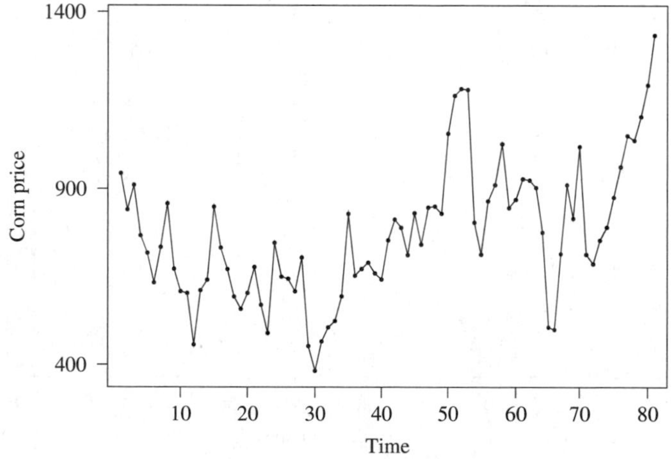

(c)

FIGURE 10.3 (Continued)

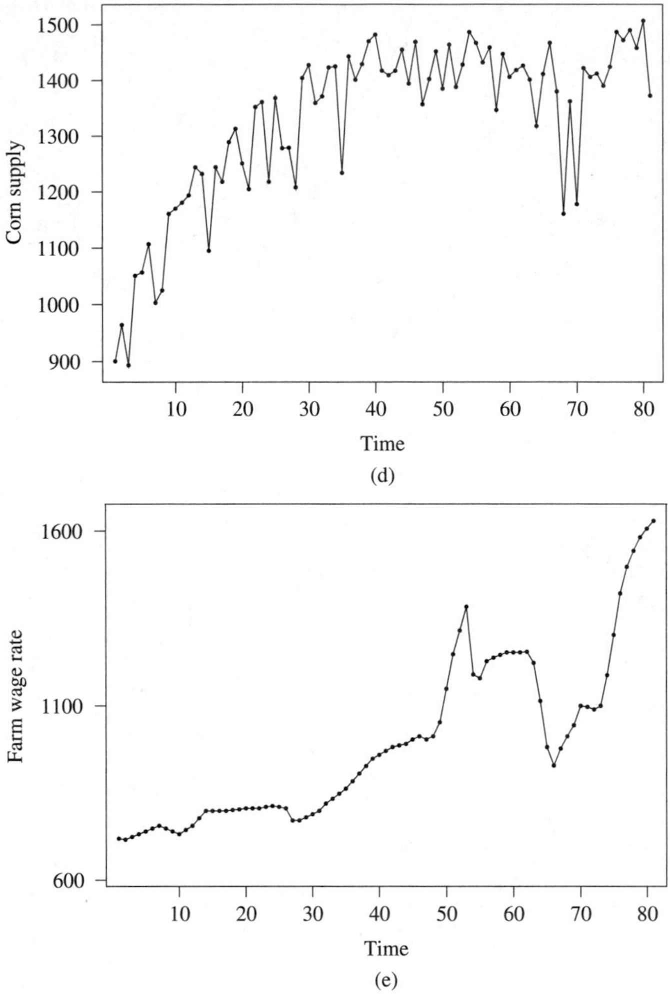

FIGURE 10.4 Time Sequence Plot of the Residuals from the Regression Model in Eq. (10.47)

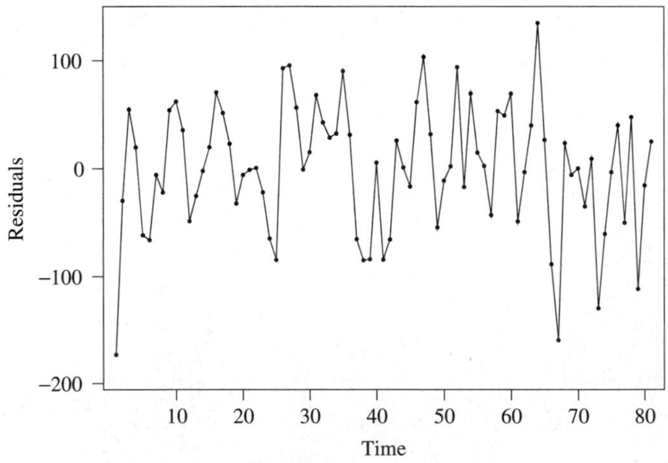

TABLE 10.7 REGRESSION RESULTS OF MODEL (10.47): QUENOUILLE U.S. HOG SERIES

The regression equation is
HogPrice = − 260 − 1.25 HogSupply + 0.491 CornPrice + 0.678 CornSupply + 0.750 FarmWageRate

Predictor	Coef	SE Coef	T	P
Constant	−259.6	102.3	−2.54	0.013
HogSupply	−1.249	0.165	−7.55	0.000
CornPrice	0.491	0.082	5.98	0.000
CornSupply	0.678	0.105	6.47	0.000
FarmWageRate	0.750	0.086	8.71	0.000

S = 61.62   R-Sq = 94.3%   R-Sq(adj) = 94.0%

Analysis of Variance

Source	DF	SS	MS	F	P
Regression	4	4742722	1185681	312.31	0.000
Residual Error	76	288537	3797		
Total	80	5031259			

Durbin-Watson statistic = 1.36

Autocorrelations of residuals

Lag 1	0.268
Lag 2	−0.100
Lag 3	−0.220
Lag 4	0.127
Lag 5	0.075
Lag 6	0.140

TABLE 10.8 REGRESSION RESULTS OF MODEL (10.48): QUENOUILLE U.S. HOG SERIES

The regression equation is
HogPrice =− 210 − 1.10 HogSupply + 0.443 CornPrice + 0.563 CornSupply + 0.780 FarmWageRate + 0.373 Error(t-1)

Predictor	Coef	SE Coef	T	P
Constant	−209.7	123.7	−1.70	0.094
HogSupply	−1.095	0.178	−6.17	0.000
CornPrice	0.443	0.096	4.62	0.000
CornSupply	0.563	0.113	4.99	0.000
FarmWageRate	0.783	0.103	7.57	0.000
MA Parameter	−0.373	0.112	−3.33	0.001

Durbin-Watson statistic = 1.95

The results of the nonlinear estimation are shown in Table 10.8. We have used the computer software EViews, but results from other software packages, such as SCA, are similar. The lag 1 autocorrelation of the errors has now disappeared, and the Durbin–Watson statistic is approximately 2. The changes in the estimates and the standard errors of the regression parameters are minor; major changes were not expected because the autocorrelations of the errors in model (10.47) are small. The standard errors of the coefficients in Table 10.8 are slightly larger because the new model adjusts for the positive autocorrelation of the errors.

Box and Tiao (1977) give this model an interesting interpretation. Notice that the coefficient of hog supply is approximately −1, and that the coefficients of corn supply and corn price are positive and of approximately equal size. Bringing the variable hog supply to the left-hand side of the equation and taking antilogs leads to an equation that relates the product of supply and price of hogs to the product of supply and price of corn. The former measures the return to the farmer, whereas the latter measures the farmer's expenditures.

EXERCISES

10.1. Show the following useful fact. Economic time series often increase proportionally, which means that observations can be expressed as $y_t = y_{t-1}(1 + r_t)$, where $100r_t$ represents the percentage change of the series. Expand $\ln(1 + r_t)$ in a Taylor series around 1, and show that $\ln(y_t) = \ln(y_{t-1}) + \ln(1 + r_t) \cong \ln(y_{t-1}) + r_t$. Hence, the proportional changes in a time series can be expressed as successive differences of the logarithms of the series. That is, $r_t \cong \ln(y_t) - \ln(y_{t-1})$. (Also see Section 6.5).

10.2. Prove that $V^{-1} = (1 - \phi^2)^{-1}L'L$, where V is the correlation matrix in Eq. (10.1), and L is the $n \times n$ matrix in Eq. (10.18).

10.3. Consider the $n = 52$ sales data on thermostat replacement components given here. The data, taken from R. G. Brown, *Smoothing, Forecasting and Prediction of Discrete Time Series* (Englewood Cliffs, NJ: Prentice Hall, 1962), are given in the file **thermostat**.

Sales Thermostat Replacement Components ($n = 52$; Read Across)

206	245	185	169	162	177	207	216	193	230
212	192	162	189	244	209	207	211	210	173
194	234	156	206	188	162	172	210	205	244
218	182	206	211	273	248	262	258	233	255
303	282	291	280	255	312	296	307	281	308
280	345								

a. Construct a time series plot of the data.

b. Regress sales on time, $y_t = \beta_0 + \beta_1 t + \varepsilon_t$, and use this regression to obtain the predictions and 95% prediction intervals for the next three observations. Use the results in Section 4.3.2.

c. Investigate the adequacy of the linear trend model in part (a) by constructing appropriate residual diagnostics. In particular, calculate the autocorrelations of the residuals from this model and compute the Durbin–Watson test statistic. Explain why this model does not appear to be an adequate representation.

d. Consider a random walk for the errors and estimate the model in terms of its differences, $\Delta y_t = \beta_1 + a_t$. Repeat the diagnostics. Obtain the forecasts for the next three periods, and obtain 95% prediction intervals. How do they differ from those obtained in part (b).

e. Consider a noisy random walk for the errors, $\Delta y_t = \beta_1 + a_t - \theta a_{t-1}$, and repeat the analysis. Use standard software (such as the time series procedures in Minitab) to estimate the model.

10.4. The Vienna University of Economics and Business Administration is Austria's top business school. The following table lists the annual enrollments from 1977 through 1997:

Annual Enrollment at Vienna University of Economics ($n = 21$; Read Across)

7,887	8,284	8,916	8,018	8,968	9,863	11,347	13,167
14,800	16,187	17,517	18,202	18,799	19,190	20,086	20,738
20,952	21,150	21,594	21,624	21,531			

a. Study the time trends in enrollment and obtain predictions for the next 3 years. Also calculate 95% prediction intervals. Would a linear trend model be appropriate? Check the model assumptions carefully by considering the appropriate residual diagnostics. Check whether adjacent residuals are autocorrelated.

b. Investigate whether a model with successive differences would be more appropriate.

c. Explore whether a second-order autoregressive model would be more appropriate for answering these questions.

Remarks

One can think of several forecasting approaches. At one extreme, one can consider a (linear) regression on time, $y_t = \beta_0 + \beta_1 t + \varepsilon_t$. Such a model is appropriate if one believes that the time trend is globally constant and does not change over time; that is, the coefficients β_0 and β_1 are fixed. If one wants to use this model, one must make sure that there is no autocorrelation in the residuals.

Alternatively, one could consider a model of the form $y_t = y_{t-1} + \beta_1 + \varepsilon_t$ and r-step-ahead predictions of future y_{t+r} of the form $y_t(r) = y_t + r\beta_1$. This model treats the change from one period to the next (β_1) as stable and uses the (equally weighted) average of historic differences as its estimate. However, this model allows the "intercept" in the linear prediction equation (y_t) to change with time.

Another possibility is to consider the model $y_t = 2y_{t-1} - y_{t-2} + \varepsilon_t$, with predictions $y_t(1) = 2y_t - y_{t-1} = y_t + (y_t - y_{t-1})$; $y_t(2) = 2y_t(1) - y_t = y_t + 2(y_t - y_{t-1})$; $y_t(3) = 2y_t(2) - y_t(1) = y_t + 3(y_t - y_{t-1})$; etc. Here, the predictions are on the linear trend that passes through the two most recent observations; all other observations are ignored. This model is the most adaptive one because the forecasts change very quickly. The model is quite different than the trend regression model, which assumes a stable, nonchanging trend.

Consider a second-order autoregressive model of the form $y_t = \beta_0 + \phi_1 y_{t-1} + \phi_2 y_{t-2} + \varepsilon_t$ to explore these latter two models. Estimate the model, check for autocorrelations in the residuals, compare its mean square error with the mean square error of the linear trend model, and decide which model to use. Keep in mind that we have only few data, and hence the decision may not be very clear.

10.5. The following table contains monthly sales for a center-city bookstore in Vienna, Austria. The last column of the table contains the advertising expenditures of the center-city bookstore. All data are in \$100. The data are given in the file **bookstore**.

Year	Month	Sales	Advertising Expenditures
1990	1	320	5
1990	2	423	1
1990	3	414	3
1990	4	506	3
1990	5	419	3
1990	6	544	2
1990	7	392	7
1990	8	449	4
1990	9	502	4
1990	10	520	9
1990	11	689	11
1990	12	1,377	14
1991	1	375	3
1991	2	318	2

Year	Month	Sales	Advertising Expenditures
1991	3	378	4
1991	4	497	2
1991	5	433	2
1991	6	426	2
1991	7	421	8
1991	8	448	3
1991	9	519	4
1991	10	638	8
1991	11	714	13
1991	12	1,501	15
1992	1	356	3
1992	2	338	1
1992	3	428	4
1992	4	490	8
1992	5	434	2
1992	6	487	4
1992	7	381	7
1992	8	472	4
1992	9	699	6
1992	10	310	2
1992	11	795	18
1992	12	1,605	16
1993	1	292	5
1993	2	295	1
1993	3	408	10
1993	4	480	5
1993	5	460	5
1993	6	576	2
1993	7	436	6
1993	8	436	4
1993	9	437	5
1993	10	556	3
1993	11	747	13
1993	12	1,531	11
1994	1	406	6
1994	2	314	2
1994	3	503	8
1994	4	438	5
1994	5	453	4
1994	6	417	3
1994	7	352	4
1994	8	439	4
1994	9	533	6
1994	10	618	9
1994	11	706	11
1994	12	1,647	5
1995	1	350	4
1995	2	361	0
1995	3	502	7
1995	4	417	4
1995	5	565	5
1995	6	432	2
1995	7	353	3
1995	8	495	4
1995	9	528	6
1995	10	612	4
1995	11	715	20
1995	12	1,487	5
1996	1	447	3
1996	2	364	4
1996	3	437	7
1996	4	478	2
1996	5	504	2
1996	6	438	9
1996	7	429	1
1996	8	439	2
1996	9	544	5
1996	10	671	6
1996	11	848	17
1996	12	1,504	13
1997	1	374	3
1997	2	394	1
1997	3	508	11
1997	4	470	3
1997	5	444	12
1997	6	366	4
1997	7	406	2
1997	8	509	4
1997	9	551	6
1997	10	654	13

a. Construct a time series plot of monthly sales. Confirm that sales in December alone make up a considerable portion of annual sales. Ignore the information on advertising. Consider models with a linear time trend and monthly indicators that account for the strong seasonal pattern. Interpret the model coefficients.

Describe the trend. Describe the seasonal pattern. Use this model to predict the sales for the next 12 months. Obtain 95% prediction intervals. Interpret the findings. Discuss the assumptions that are made by your predictions. Discuss their shortcomings. Check the model assumptions. In particular, investigate whether adjacent residuals are uncorrelated. Calculate the autocorrelations and the Durbin–Watson statistic.

b. Model the autocorrelations in the residuals if autocorrelations are present. Use autoregressive models of low orders (a first-order autoregressive model may be sufficient). Combine the regression model with the model for the autocorrelated errors. Estimate the combined regression/time series model. Check whether this model results in a better representation of the data. Use this model for forecasting.

c. Investigate the impact of advertising on sales. In addition to advertising expenditures, your model should probably also include a time trend and indicator variables for the seasonality. Advertising is also seasonal, but it would be incorrect to assume that seasonality in sales is primarily due to the seasonality in advertising. Also check whether there are lagged effects of advertising (i.e., whether current sales are affected by present as well as past advertising expenditures).

10.6. Generate two independent random walks of length $n = 50$ ($n = 100$). Use your computer program of choice to generate two independent sets of standard normal variables, $\{a_t\}$ and $\{b_t\}$; generate the random walks from $y_t = y_{t-1} + a_t$ and $x_t = x_{t-1} + b_t$; and fit the regression model $y_t = \beta_0 + \beta_1 x_t + \varepsilon_t$ to the data. Examine the behavior of the standard t ratio for testing $\beta_1 = 0$, repeating your simulations 10 times. Discuss your findings. Also examine the coefficient of determination R^2 in each simulation.

10.7. The table on page 338 lists quarterly sales of four different chemical products that are input factors to many chemical and industrial equipment goods. Quarterly production indices for chemicals and industrial equipment are also shown. The data are given in the file **saleschemical**.

a. For each product, explore the relationships between sales and the two indices. Check the residuals for serial correlation.

b. You will find contemporaneous relationships between product sales and the two indices (i.e., a relationship between sales at time t and the indices at the same time period). It is claimed that the indices help you predict future product sales. Assess this claim. Consider regression models that relate the sales at time t to the production indices at previous periods (such as one quarter prior). What is your conclusion?

10.8. **(continuation of Exercise 8.3).** In Exercise 8.3, we analyzed weekly sales and prices of two soft drinks, brand P and brand C. The data are given in the file **softdrink**. In part (a) of this exercise, we regressed the logarithms of the sales of 12-packs on the logarithms of the prices of 6-, 12-, and 24-packs. We noticed autocorrelation in the residuals. Identify a time series model for the errors and combine it with the regression model. Show that the autocorrelations of the residuals suggest an ARIMA(0,1,1) model $(1 - B)\varepsilon_t = (1 - \theta B)a_t$. Estimate the combined regression time series model

$$\begin{aligned} y_t &= \ln \text{SalesP12}_t \\ &= \beta_0 + \beta_1 \ln \text{PriceP6}_t \\ &\quad + \beta_2 \ln \text{PriceP12}_t \\ &\quad + \beta_3 \ln \text{PriceP24}_t + \varepsilon_t \\ (1 - B)\varepsilon_t &= (1 - \theta B)a_t \end{aligned}$$

Table for Exercise 10.7

Year	Quarter	Product1	Product2	Product3	Product4	Chemicals (Index)	Industrial Equipment (Index)
1994	1	391.484	12.1774	185.597	20.0000	101.555	113.257
1994	2	439.438	11.6094	199.984	20.1719	104.844	115.588
1994	3	405.381	11.7619	192.825	22.6349	104.769	117.300
1994	4	417.176	12.2941	209.250	28.7353	103.826	119.648
1995	1	452.143	19.0794	228.095	29.7460	108.015	124.513
1995	2	393.156	15.5938	240.281	28.3281	107.591	125.111
1995	3	403.419	16.5806	237.661	31.0968	104.553	128.057
1995	4	443.677	19.9355	241.339	30.6613	102.020	129.813
1996	1	429.359	11.3438	243.188	31.2656	102.776	130.424
1996	2	437.750	11.4219	264.734	32.9219	104.707	130.592
1996	3	434.258	14.4355	263.919	33.3226	105.529	130.996
1996	4	461.048	19.1774	256.468	34.1935	106.811	132.022
1997	1	439.565	16.1290	251.516	34.6129	110.398	134.551
1997	2	463.077	17.4462	293.646	37.2923	111.482	135.940
1997	3	440.302	17.0794	296.286	35.3333	112.496	137.971
1997	4	449.306	18.2258	282.274	40.1613	113.475	139.388
1998	1	469.242	17.1129	269.726	34.6290	114.492	138.520
1998	2	452.063	13.2188	281.766	42.1563	114.556	138.134
1998	3	399.127	15.0952	261.587	28.0159	110.618	139.660
1998	4	473.066	17.3934	314.852	34.3607	109.197	137.933
1999	1	427.677	14.2154	305.077	35.6615	112.660	135.662
1999	2	476.563	16.6875	312.578	41.9375	117.656	134.934
1999	3	472.661	16.9839	305.419	44.8710	115.985	135.500
1999	4	466.647	20.3971	296.118	40.6176	119.506	136.320
2000	1	480.200	23.3692	317.538	46.2308	119.219	141.640
2000	2	498.031	23.0313	308.703	44.2500	120.355	143.446
2000	3	445.694	23.2419	329.468	49.8548	116.129	146.991
2000	4	417.726	29.1129	287.952	44.7903	112.136	146.144
2001	1	409.969	22.2864	245.285	37.9138	107.045	134.013
2001	2	399.444	9.4418	268.266	35.6807	101.799	127.045
2001	3	386.098	5.1210	247.031	33.5933	101.157	121.842
2001	4	393.686	7.2955	202.202	33.3136	101.037	117.610
2002	1	426.159	10.1111	220.683	32.6984	104.129	113.992
2002	2	470.563	13.1094	243.156	42.8750	108.863	115.461
2002	3	407.349	12.5873	208.143	39.6190	109.395	115.910

and show that the estimate of θ is close to 1. Interpret the result. This model is equivalent to relating differences of log sales to differences of log prices, with moving average errors and a moving average parameter close to 1. Remember that first differences of logs are equivalent to percentage changes.

Note that the data set includes a few weeks with missing data. For simplicity, we suggest that you omit these weeks. For a small number of missing weeks, as in our case, this strategy is reasonable even though it alters the time lag between certain observations.

Note that we have introduced the **backshift notation**, which is standard notation in the time series literature. B is the backshift operator. When applied to a time series, the backshift operator shifts the time index by one unit. That is, $By_t = y_{t-1}$, $B^2 y_t = y_{t-2}$, $B^3 y_t = y_{t-3}$, and so on. Hence, $(1 - B)\varepsilon_t = (1 - \theta B)a_t$ is a short way of writing $\varepsilon_t - \varepsilon_{t-1} = a_t - \theta a_{t-1}$.

The "design" of the price data is interesting because there are periods during which prices are rather flat. Look at the time series graph of (ln) prices. There appears to be a certain "industry price" that stores use as the base when reducing their prices. Occasionally, the industry price changes. One could argue that it is not the actual price but, rather the "unanticipated" price that matters and affects sales. One could measure the unanticipated price by considering the difference between the price p_t and the exponentially weighted average of past prices. That is, one could consider

$$p_t - (1 - \alpha)[p_{t-1} + \alpha p_{t-2} + \alpha^2 p_{t-3} + \cdots] = [(1 - B)/(1 - \alpha B)]p_t$$

as the relevant regressor variable. The parameter α determines how quickly the price information is forgotten. Here, B is the backshift operator, and you should check that the left-hand side of the previous expression can be written this way. For simplicity of exposition, we have considered a single price series. This leads to the model

$$y_t = \beta_0 + \beta_1[(1 - B)/(1 - \alpha B)]p_t + \{\text{unanticipated prices in other pack sizes}\} + \varepsilon_t$$

or

$$(1 - \alpha B)y_t = \beta_0^* + \beta_1(1 - B)p_t + \cdots + (1 - \alpha B)\varepsilon_t$$

Estimate this model, and confirm that α is close to 1. In essence, this goes back to the model with differences in all variables (response as well as regressor variables) and a moving average parameter that is close to 1. It coincides with the regression time series model considered at the beginning of this exercise.

10.9. The following table lists the first few daily stock closing prices at the Vienna Stock Exchange of Lenzing AG, a major producer of viscose fibers. The complete data for all trading days in 2002 are given in the file **lenzingstock**. Construct a time series graph of the data. Assess the stationarity of the data. Calculate first differences, graph the data, and assess the stationarity of the first differences.

Date	Time (Day of Year)	Closing Stock Price, Lenzing AG (EURO)
1/2/2002	2	72.50
1/3/2002	3	75.00
1/4/2002	4	73.00
1/7/2002	7	73.00
1/8/2002	8	71.80
1/9/2002	9	71.10
....		

10.10. The following data are taken from Kadiyala, K. R. Testing for the Independence of Regression Disturbances. *Econometrica,* 38, 97–117, 1970. The data are given in the file **icecreamsales**.

Ice cream consumption (in pints per capita) is measured over consecutive 4-week periods from March 18, 1951, to July 11, 1953. In addition, the data set includes the price of the ice cream (in dollars per pint), the weekly family income of the panel of customers (in dollars), and the mean temperature of the panel city (in °F). The data are taken from an unpublished report by Hildreth and Lu (1960).

Construct a model that relates ice cream consumption to the price, the family income, and the temperature. Pay particular attention to any serial correlation of the error terms. Calculate the Durbin–Watson statistic and the autocorrelations of the residuals. Incorporate any serial correlation into a new error model (by using, for example, a first-order autoregressive model for the errors), and estimate a regression model with serially correlated errors. What are your conclusions with regard to price and income?

Week	Ice Cream Consumption	Price	Family Income	Temperature
1	0.386	0.270	78	41
2	0.374	0.282	79	56
3	0.393	0.277	81	63
4	0.425	0.280	80	68
5	0.406	0.272	76	69
6	0.344	0.262	78	65
7	0.327	0.275	82	61
8	0.288	0.267	79	47
9	0.269	0.265	76	32
10	0.256	0.277	79	24
11	0.286	0.282	82	28
12	0.298	0.270	85	26
13	0.329	0.272	86	32
14	0.318	0.287	83	40
15	0.381	0.277	84	55
16	0.381	0.287	82	63
17	0.470	0.280	80	72
18	0.443	0.277	78	72
19	0.386	0.277	84	67
20	0.342	0.277	86	60
21	0.319	0.292	85	44
22	0.307	0.287	87	40
23	0.284	0.277	94	32
24	0.326	0.285	92	27
25	0.309	0.282	95	28
26	0.359	0.265	96	33
27	0.376	0.265	94	41
28	0.416	0.265	96	52
29	0.437	0.268	91	64
30	0.548	0.260	90	71

10.11. The data are taken from Shaw, N. *Manual of Meteorology,* Vol. 1, p. 284. Cambridge, UK: Cambridge University Press. The following table shows the mean annual level of Lake Victoria Nyanza for the years 1902–1921, relative to a fixed standard, and the number of sunspots in the same years. The data are given in the file **lakelevel**.

Relate the lake levels (y) to the number of sunspots (x). Pay particular attention to any serial correlation of the error terms. Calculate the Durbin–Watson statistic and the autocorrelations of the residuals, and check the adequacy of the fitted model(s).

Year	Y = Lake Level	X = Sunspots
1902	−10	5
1903	13	24
1904	18	42
1905	15	63
1906	29	54
1907	21	62
1908	10	49
1909	8	44
1910	1	19
1911	−7	6
1912	−11	4
1913	−3	1
1914	−2	10
1915	4	47
1916	15	57
1917	35	104
1918	27	81
1919	8	64
1920	3	38
1921	−5	25

10.12. The data are taken from Hanke, J. E., and Reitsch, A. G. *Business Forecasting,* 3rd ed., Needham Heights, MA: Allyn & Bacon, 1989. Annual Sears Roebuck sales (y) and U.S. disposable personal income (in billions of dollars) for the years 1955–1975 are listed in the following table. The data are given in the file **sears**.

a. Regress sales on disposable income, carefully checking for autocorrelations in the errors.

b. Modify the independent error model, and consider a noisy random walk model for the errors in the regression model.

c. Consider a regression of changes in sales ($\text{Sales}_t - \text{Sales}_{t-1}$) on changes in disposable income ($\text{Income}_t - \text{Income}_{t-1}$).

d. Consider a regression of relative changes in sales $(\text{Sales}_t - \text{Sales}_{t-1})/\text{Sales}_{t-1} \cong \ln(\text{Sales}_t) - \ln(\text{Sales}_{t-1})$ on relative changes in disposable income, $(\text{Income}_t - \text{Income}_{t-1})/\text{Income}_{t-1} \cong \ln(\text{Income}_t) - \ln(\text{Income}_{t-1})$.

e. Comment on your findings.

Year	Sears Sales	U.S. Disposable Income
1955	3.307	273.4
1956	3.556	291.3
1957	3.601	306.9
1958	3.721	317.1
1959	4.036	336.1
1960	4.134	349.4
1961	4.268	362.9
1962	4.578	383.9
1963	5.093	402.8
1964	5.716	437.0
1965	6.357	472.2
1966	6.769	510.4
1967	7.296	544.5
1968	8.178	588.1
1969	8.844	630.4
1970	9.251	685.9
1971	10.006	742.8
1972	11.200	801.3
1973	12.500	903.1
1974	13.101	983.6
1975	13.640	1,076.7

10.13. The data are taken from Coen, P. J., Gomme, E. D., and Kendall, M. G. Lagged relationships in economic forecasting. *Journal of the Royal Statistical Society, Series A,* 132, 133–163, 1969. The data are given in the file **cgk**.

The authors analyze quarterly data (from quarter 3 of 1952 to quarter 4 of 1967) on the *Financial Times* (FT) share index, the FT commodity index, and the seasonally adjusted UK car production. One of their models regresses the FT share index (y) on the lagged FT commodity index (with lag 7) and the lagged UK car production (lag 6).

a. Reproduce their regression model, carefully checking the model assumptions. In particular, check whether the errors are uncorrelated. You will find large autocorrelations among the errors that were overlooked by the authors.

b. Consider simple time series models for the errors (such as the random walk or the noisy random walk) and estimate the combined regression time series models.

c. Interpret the findings of your analysis. Discuss how this is related to the discussion of spurious regression in Section 10.2.

Year	Quarter	$y =$ FT Share Index	$x_1 =$ Car Production	$x_2 =$ FT Comm Index
1952	3	112.7	105,761	96.21
1952	4	115.0	121,874	93.74
1953	1	121.4	126,260	91.37
1953	2	118.4	145,248	86.31
1953	3	122.7	160,370	84.98
1953	4	128.8	163,648	86.46
1954	1	135.7	178,195	90.04
1954	2	148.5	187,197	94.74
1954	3	165.4	195,916	92.43
1954	4	178.5	199,253	92.41
1955	1	187.3	227,616	91.65
1955	2	195.9	215,363	89.38
1955	3	205.2	231,728	91.05
1955	4	191.5	231,767	89.89
1956	1	183.0	211,211	90.16
1956	2	184.5	185,200	86.78
1956	3	181.1	152,404	88.45
1956	4	173.7	156,163	90.69
1957	1	185.4	151,567	86.03
1957	2	201.8	213,683	84.85

Year	Quarter	$y =$ FT Share Index	$x_1 =$ Car Production	$x_2 =$ FT Comm Index	Year	Quarter	$y =$ FT Share Index	$x_1 =$ Car Production	$x_2 =$ FT Comm Index
1957	3	198.0	244,543	84.07	1962	4	283.6	312,780	78.13
1957	4	168.0	253,111	81.96	1963	1	295.7	363,336	80.38
1958	1	161.6	266,580	80.03	1963	2	309.3	378,275	81.78
1958	2	170.2	253,543	79.80	1963	3	295.7	414,457	82.81
1958	3	184.5	261,675	80.19	1963	4	342.0	459,158	84.99
1958	4	211.0	249,407	80.13	1964	1	335.1	460,397	86.31
1959	1	218.3	246,248	80.42	1964	2	344.4	462,279	85.95
1959	2	231.7	293,062	82.67	1964	3	360.9	434,255	90.73
1959	3	247.4	285,809	82.78	1964	4	346.5	475,890	92.42
1959	4	301.9	366,265	82.61	1965	1	340.6	439,365	87.18
1960	1	323.8	374,241	82.47	1965	2	340.3	431,666	85.20
1960	2	314.1	375,764	81.86	1965	3	323.3	399,160	85.44
1960	3	321.0	354,411	79.70	1965	4	345.6	449,564	87.85
1960	4	312.9	249,527	77.89	1966	1	349.3	437,555	89.95
1961	1	323.7	206,165	77.61	1966	2	359.7	426,616	90.20
1961	2	349.3	258,410	78.90	1966	3	320.0	399,254	88.89
1961	3	310.4	279,342	79.72	1966	4	299.9	334,587	83.25
1961	4	295.8	264,824	78.08	1967	1	318.5	367,997	81.21
1962	1	301.2	312,983	77.54	1967	2	343.1	393,808	79.90
1962	2	285.8	300,932	76.99	1967	3	360.8	375,968	78.70
1962	3	271.7	323,424	76.25	1967	4	397.8	381,692	81.50

11 Logistic Regression

11.1 THE MODEL

In this chapter, we discuss the regression situation when the response variable is categorical. The simplest situation arises if the response is **binary**, indicating the presence or absence of a characteristic. For such situations, we consider a special model, referred to as the **logistic regression model**. We explain how to estimate and interpret its parameters and show how to check this model for possible misspecification. Furthermore, we extend the analysis to the situation in which the response is categorical with more than two possible outcomes.

We introduce the logistic regression model through the illustrative example in Table 11.1. Part (a) of the table lists the outcomes (y) of individual capital punishment cases, together with the severity of the crimes (x) thought to be a determining factor for getting the death penalty. The outcome variable can take on only two distinct values: $y =$ death sentence, coded as 1, and $y =$ life sentence, coded as 0. The regressor (explanatory) variable x is measured as an index from 1 to 10.

The outcome y_i of case i is assumed to have a Bernouilli distribution with "success" and "failure" probabilities

$$P(y_i = 1) = \pi \quad \text{and} \quad P(y_i = 0) = 1 - \pi, \quad \text{for } i = 1, 2, \ldots, n \qquad (11.1)$$

The parameter π defines the mean of the distribution: $E(y_i) = \pi$. The logistic regression models the success probability as a function of the severity (x). That is, $\pi = \pi(x)$ so that for case i with severity x_i, $\pi_i = \pi(x_i)$.

Table 11.1 considers a single explanatory (regressor) variable, the severity of the crime. In general, we may also want to consider the race of the victim, the race of the defendant, the gender of the defendant, and so on. Then the logistic regression models the probability of getting the death sentence π as a function of p explanatory (regressor) variables, $x_1, x_2, \ldots, x_p$. For case i, with values of the explanatory variables $\boldsymbol{x}_i = (x_{i1}, x_{i2}, \ldots, x_{ip})'$, $\pi_i = \pi(\boldsymbol{x}_i)$.

TABLE 11.1 ILLUSTRATIVE EXAMPLE: DEATH PENALTY DATA

(a) Individual Cases

Case	Severity of Crime (x)	Capital Punishment Outcome (y)
1	5	$y_1 = 1$ (death)
2	4	$y_2 = 0$ (life)
3	5	$y_3 = 0$ (life)
4	7	$y_4 = 1$ (death)
5	5	$y_5 = 0$ (life)
6	5	$y_6 = 1$ (death)
...	...	...
i	x_i	y_i
...	...	...
...	...	...
n	9	$y_n = 1$ (death)

(b) Grouped According to Constellations

Setting Constellation	Severity of Crime (x)	Capital Punishment Outcome (Number of Death Sentences)
1	5	$y_1 = 2$ (death sentences among 4 cases)
2	...	...
3	7	$y_3 = 4$ (death sentences among 8 cases)
...	...	...
i	x_i	y_i death sentences among the n_i cases
...	...	...
m	9	y_m death sentences among the n_m cases

In our illustration, a case refers to a defendant in a death penalty case. In other examples, the case may represent a subject in a medical experiment on the effectiveness of a new drug, a company or a credit customer in studies of credit risk, or a manufactured component in experiments assessing failure.

In many situations, especially if the explanatory variables are categorical, one observes several cases at the same values of the explanatory variables. In the example in Table 11.1, there are several different (and unrelated) cases with the same exact severity index. For example, there are four cases with crime severity 5, there are eight cases with crime severity 7, and so on. In general, there may be n_i cases with a certain crime severity x_i. We call the categories that are characterized by identical levels on all explanatory variables a setting or a constellation. We may have a total of n cases that occur at m constellations ($m \leq n$), with ($n_1, \ldots, n_i, \ldots, n_m$) observations at the m distinct constellations, and $n = n_1 + n_2 + \cdots + n_m$. See the illustration in Table 11.1b. The number of successes y_i at the ith constellation is obtained by summing independent 0/1 Bernouilli outcome variables over all individual cases that make up that constellation. Because of the independence assumption for the individual cases, the number of successes y_i follows a binomial distribution with parameters n_i (the number of cases within the ith constellation) and success probability $\pi_i = \pi(\boldsymbol{x}_i)$. That is,

$$P(Y_i = y_i) = f(y_i) = \binom{n_i}{y_i} [\pi(\boldsymbol{x}_i)]^{y_i} [1 - \pi(\boldsymbol{x}_i)]^{n_i - y_i}, \quad \text{for } i = 1, 2, \ldots, m \tag{11.2}$$

with mean $E(y_i) = n_i \pi(\boldsymbol{x}_i)$. We use the term **case** if we refer to an individual observation, and **constellation** if we refer to the grouped information at distinct levels of the explanatory variables. We use the same symbol y_i to denote both the (Bernouilli) outcome in an individual case i and the (binomial) number of successes at a constellation i consisting of n_i cases. The reference to either a case or a constellation indicates whether y_i is a binary or a binomial response variable.

Fitting a standard linear regression model $y_i = \pi(\boldsymbol{x}_i) + \varepsilon_i = \boldsymbol{x}_i'\boldsymbol{\beta} + \varepsilon_i$ with normal errors to binary response data y_i would be incorrect. A linear representation permits estimates of the response to be outside the range 0 to 1, which is wrong when we model probabilities. Moreover, the normal error distribution is no longer valid because with a binary response only two different errors are possible: $-\pi(\boldsymbol{x}_i)$ if the response is zero and $1 - \pi(\boldsymbol{x}_i)$ if the response is 1.

In the logistic regression model the probabilities $\pi(\boldsymbol{x}_i)$ are parameterized as

$$\pi(\boldsymbol{x}_i) = \frac{e^{\boldsymbol{x}_i'\boldsymbol{\beta}}}{1 + e^{\boldsymbol{x}_i'\boldsymbol{\beta}}} = \frac{1}{1 + e^{-\boldsymbol{x}_i'\boldsymbol{\beta}}} \quad \text{and} \quad 1 - \pi(\boldsymbol{x}_i) = \frac{1}{1 + e^{\boldsymbol{x}_i'\boldsymbol{\beta}}} = \frac{e^{-\boldsymbol{x}_i'\boldsymbol{\beta}}}{1 + e^{-\boldsymbol{x}_i'\boldsymbol{\beta}}} \tag{11.3}$$

where $\boldsymbol{x}_i'\boldsymbol{\beta} = \beta_0 + \beta_1 x_{i1} + \cdots + \beta_p x_{ip}$. The probabilities are nonlinear functions of the parameters $\boldsymbol{\beta}$. It is instructive to look at the probabilities for the case of a single explanatory variable x, as illustrated in the context of the death penalty example in Table 11.1. Another example is the outcome of a purchasing decision y (buy or not buy a product) and the price x of the product. In Figures 11.1 and 11.2, we show the function

$$P(y = 1) = \pi(x) = \frac{e^{\beta_0 + \beta_1 x}}{1 + e^{\beta_0 + \beta_1 x}} \tag{11.4}$$

FIGURE 11.1
Graph of Eq. (11.4) for $\beta_1 > 0$

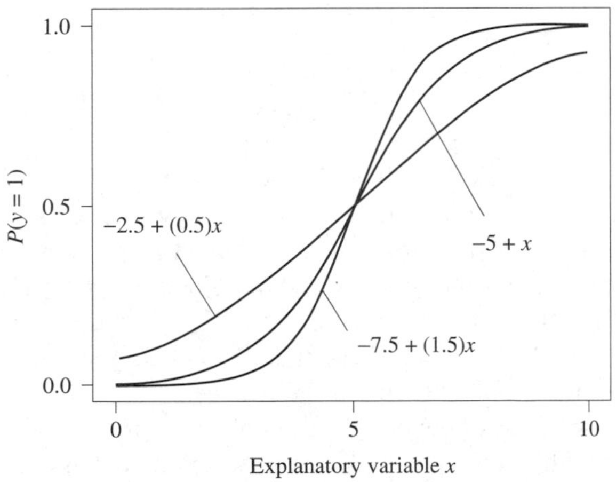

FIGURE 11.2
Graph of Eq. (11.4) for $\beta_1 < 0$

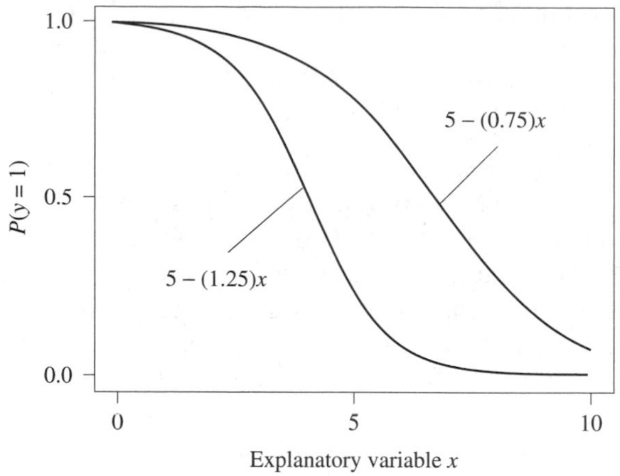

for positive and negative values of β_1. The parameters β_0 and β_1 determine the inflection point and the steepness of the sigmoid-like function. A graph such as the one in Figure 11.1 is appropriate in the death penalty study, in which the likelihood of receiving a death sentence increases with the severity of the crime. A graph such as the one in Figure 11.2 is appropriate in the purchasing study, in which the likelihood of buying decreases with increasing price of the product.

11.2 INTERPRETATION OF THE PARAMETERS

The probabilities in Eq. (11.3) are nonlinear functions of $\boldsymbol{\beta}$. However, a simple transformation results in a linear model. It can be shown that

$$\ln \frac{\pi(\boldsymbol{x}_i)}{1 - \pi(\boldsymbol{x}_i)} = \boldsymbol{x}'_i \boldsymbol{\beta} \tag{11.5}$$

Much of the interpretation of the logistic regression model centers on the ratio

$$\frac{\pi(\boldsymbol{x})}{1 - \pi(\boldsymbol{x})} = \exp(\boldsymbol{x}'\boldsymbol{\beta}) \tag{11.6}$$

This ratio compares the probability of the occurrence of a characteristic to the probability of its nonoccurrence and is usually referred to as the **odds** of occurrence. Whereas probabilities are constrained to lie between 0 and 1, odds can take on values between zero and infinity. For example, for $\pi = 0.5$, the odds are 1 (i.e., even odds, or 1:1 odds). For $\pi = 0.8$, the odds for (in favor of) occurrence are $0.8/0.2 = 4$, or 4:1, implying that the probability of occurrence is four times as large as the probability of nonoccurrence.

The logarithm of the odds in Eq. (11.6), $\ln [\pi/(1 - \pi)]$, is referred to as the **log odds** or the **logit**. The logistic regression model assumes a linear model for the logit; that is,

$$\ln \frac{\pi(\boldsymbol{x})}{1-\pi(\boldsymbol{x})} = \boldsymbol{x}'\beta = \beta_0 + \beta_1 x_1 + \cdots + \beta_p x_p \tag{11.7}$$

This representation shows that the regression coefficients represent changes in the log odds. Let us change one of the regressor variables, for example, variable x, by one unit while keeping the values of all other regressors in the model constant. For simplicity of notation, we have omitted the subscript of the changed regressor and its associated parameter β. The shift of the regressor from value x to the new value $x+1$ changes the log odds by β units. That is,

$$\beta = \ln \frac{\pi(x+1)}{1-\pi(x+1)} - \ln \frac{\pi(x)}{1-\pi(x)} = \ln(a/b) \tag{11.8}$$

where $a = \dfrac{\pi(x+1)}{1-\pi(x+1)}$ are the odds for occurrence if the considered explanatory variable is at level $x+1$, and $b = \dfrac{\pi(x)}{1-\pi(x)}$ are the odds for occurrence at level x. Exponentiation of the regression coefficient provides the odds ratio, a/b. That is,

$$\exp(\beta) = (a/b) \tag{11.9}$$

For example, a regression coefficient $\beta = -0.2$ with $\exp(\beta) = \exp(-0.2) = 0.82$ indicates that a change from x to $x+1$ changes (reduces) the odds of occurrence by the multiplicative factor 0.82; it reduces the odds of occurrence by 18%. A value of $\beta = 0$ and $\exp(\beta) = \exp(0) = 1$ implies that a change in the explanatory variable has no effect on the odds of occurrence. A value of $\beta = 1.5$ and $\exp(\beta) = \exp(1.5) = 4.48$ indicates that a one-unit change in the explanatory variable increases the odds for occurrence by a factor of 4.48, or by $100(4.48-1) = 348\%$.

What do the regression results imply if we change a regressor variable by k units (e.g., $k = 10$, from a value x to the value $x+10$) while keeping the values of all other regressor variables constant? In this case,

$$\ln \frac{\pi(x+k)}{1-\pi(x+k)} - \ln \frac{\pi(x)}{1-\pi(x)} = \beta k$$

and $\exp(\beta k)$measures the odds ratio that is implied by this change. For a meaningful interpretation of the coefficients in logistic regression models, the user needs to decide on the magnitude of the change, k, that is relevant to the variables at hand.

11.2.1 RELATIONSHIP BETWEEN LOGISTIC REGRESSION AND THE ANALYSIS OF 2 × 2 CONTINGENCY TABLES

The interpretation of the regression coefficients in terms of odds ratios has made logistic regression a very powerful tool. The odds ratio is a familiar concept in the analysis of categorical data.

TABLE 11.2 FREQUENCIES IN A 2 $\times$ 2 TABLE[a]

	$x=0$	$x=1$	Row Sum
$y=0$	n_{00} (n_{00}/n_{+0})	n_{01} (n_{01}/n_{+1})	n_{0+}
$y=1$	n_{10} (n_{10}/n_{+0})	n_{11} (n_{11}/n_{+1})	n_{1+}
Column Sum	n_{+0}	n_{+1}	n_{++}

[a] Relative frequencies, based on column sums, are given in parentheses.

If there is just one categorical explanatory variable x with two outcomes (coded as 0 or 1), we can arrange the information in the 2×2 contingency table as shown in Table 11.2. The table lists the cell frequencies n_{ij} and the relative frequencies in parentheses for each outcome of the explanatory variable x. Furthermore, it shows row and column sums and the total number of cases. The response y may indicate the presence or absence of disease, and x may be an indicator for exposure. Or, y may indicate the type of sentence in a death penalty case (death or life), and x may stand for the race of the defendant (White or non-White). Or, the response y may be the outcome of a hiring decision (hired or not hired), and x may be the gender of the applicant (male or female). The odds ratio in the 2×2 table is defined as $(n_{11}/n_{01})/(n_{10}/n_{00}) = (n_{00}n_{11}/n_{01}n_{10})$.

Consider the logistic regression model $P(y=1) = \pi(x) = \dfrac{e^{\beta_0+\beta_1 x}}{1+e^{\beta_0+\beta_1 x}}$, with $\pi(x=0) = \dfrac{e^{\beta_0}}{1+e^{\beta_0}}$ and $\pi(x=1) = \dfrac{e^{\beta_0+\beta_1}}{1+e^{\beta_0+\beta_1}}$. The odds at $x=1$ and $x=0$ are respectively $\pi(x=1)/[1-\pi(x=1)] = \exp(\beta_0+\beta_1)$ and $\pi(x=0)/[1-\pi(x=0)] = \exp(\beta_0)$, and the odds ratio is $\exp(\beta_1)$. This shows that the exponentiation of the slope coefficient in the logistic regression coincides with the odds ratio in the 2×2 table.

The logistic regression model is more flexible than a contingency table because it (i) can adjust odds ratios for the presence of other explanatory variables and (ii) can make the odds ratios depend on other variables. As an illustration, let the response y be the outcome of a hiring decision (hired or not hired), x the gender of the applicant (male or female), and z the experience. Assume that z is an indicator that is 1 for applicants with prior experience and 0 for those with no prior experience.

Consider the following three logistic regression models for the probability of being hired:

$$P(y=1) = \pi(x) = \frac{e^{\beta_0+\beta_1 x}}{1+e^{\beta_0+\beta_1 x}} \tag{11.10a}$$

$$P(y=1) = \pi(x, z) = \frac{e^{\beta_0+\beta_1 x+\beta_2 z}}{1+e^{\beta_0+\beta_1 x+\beta_2 z}} \tag{11.10b}$$

$$P(y=1) = \pi(x, z, xz) = \frac{e^{\beta_0+\beta_1 x+\beta_2 z+\beta_3 xz}}{1+e^{\beta_0+\beta_1 x+\beta_2 z+\beta_3 xz}} \tag{11.10c}$$

The odds ratio computed from the 2×2 contingency table of y (hiring decision) and x (gender), or equivalently computed as $\exp(\beta_1)$ from the results of the logistic regression with success probability (11.10a), is suspect if the omitted variable z affects the probability of being hired [and the model in Eq. (11.10b) is correct], or if the odds ratios vary with the omitted variable [and the model in Eq. (11.10c) is needed].

The second model in Eq. (11.10b) allows the variable z to affect the probability of being hired. However, as simple substitution with Eq. (11.10b) shows, the gender odds ratio for getting hired is still given by $\exp(\beta_1)$. [The odds for getting hired is $\exp(\beta_0 + \beta_1 x + \beta_2 z)$, but the term $\beta_2 z$ cancels when forming the ratio of the odds at $x = 1$ and $x = 0$.] Nevertheless, estimating the model (11.10a) without the needed term $\beta_2 z$ may affect the estimate of β_1 and hence the estimated odds ratio. Omission becomes a problem if the omitted variable z is related to the included variable x and if the gender proportions in the experienced group are much different from the gender proportions in the group without experience. We have seen the same effect in the linear regression model when we discussed the bias that results when omitting from the model an explanatory variable that is correlated with other explanatory variables (see the appendix to Chapter 6).

The third model with success probability given in Eq. (11.10c) includes an "interaction" allowing the gender odds ratio to depend on the explanatory variable z. Simple substitution shows that the gender odds ratio for getting hired is $\exp(\beta_1)$ when $z = 0$ and $\exp(\beta_1 + \beta_3)$ when $z = 1$. Ignoring the interaction and fitting the model in Eq. (11.10b) without the term $\beta_3 xz$ results in an estimated odds ratio that is a combination of the two odds ratios.

This discussion indicates that one should always check whether the analysis needs to be adjusted for other covariates. The logistic regression framework makes such an adjustment straightforward.

11.2.2 ANOTHER ADVANTAGE OF MODELING ODDS RATIOS BY LOGISTIC REGRESSION

A reason for the wide spread use of odds ratios and logistic regression is that it is very easy to adjust the analysis for different sampling schemes. For odds ratios, it does **not** matter whether the data are obtained through a case–control study or through a retrospective study that samples units conditional on the outcome variable y. A **case–control** (or **prospective**) **study** selects subjects on the basis of their explanatory variables (covariates). For example, in an investigation of smoking (x) and lung cancer (y), such a study selects random samples of nonsmokers and smokers and classifies the subjects according to the presence of the disease (y). In a **retrospective study**, on the other hand, the sampling is carried on the response; random samples of disease-free and diseased patients are taken, and subjects are classified according to their smoking history (x). It is clear from looking at Table 11.2 that the odds ratio is the same under both sampling strategies. For retrospective sampling, the odds-ratio is $(n_{11}/n_{10})/(n_{01}/n_{00}) = (n_{00}n_{11}/n_{01}n_{10})$–the same as in the prospective study.

The only coefficient that is affected by the sampling scheme is the intercept in the logistic regression model. We demonstrate this in the following discussion for the general logistic model with success probability

$$P(y=1)=\pi(\boldsymbol{x})=\frac{e^{\beta_0+\boldsymbol{x}'\boldsymbol{\beta}}}{1+e^{\beta_0+\boldsymbol{x}'\boldsymbol{\beta}}}$$

and vector of covariates $\boldsymbol{x}$. Suppose that the data are sampled retrospectively. Let the indicator variable w express whether or not the individual is included in the sample, and denote the sampling proportions by

$$P(w=1\,|\,y=1)=\gamma_1 \quad \text{and} \quad P(w=1\,|\,y=0)=\gamma_0$$

The sampling proportion varies with the outcome but does not depend on the covariate $\boldsymbol{x}$. Applying Bayes theorem, the probability of success in a sampled case with covariates $\boldsymbol{x}$ is given by

$$\begin{aligned}
&P(y=1\,|\,w=1,\boldsymbol{x})\\
&=\frac{P(w=1\,|\,y=1,\boldsymbol{x})P(y=1\,|\,\boldsymbol{x})}{P(w=1\,|\,y=1,\boldsymbol{x})P(y=1\,|\,\boldsymbol{x})+P(w=1\,|\,y=0,\boldsymbol{x})P(y=0\,|\,\boldsymbol{x})}\\
&=\frac{\gamma_1\exp(\beta_0+\boldsymbol{x}'\boldsymbol{\beta})}{\gamma_1\exp(\beta_0+\boldsymbol{x}'\boldsymbol{\beta})+\gamma_0}=\frac{\exp(\beta_0^*+\boldsymbol{x}'\boldsymbol{\beta})}{1+\exp(\beta_0^*+\boldsymbol{x}'\boldsymbol{\beta})}
\end{aligned}$$

where $\beta_0^*=\beta_0+\ln(\gamma_1/\gamma_0)$. The last equality follows from straightforward algebra. Although the data have been sampled retrospectively, the same logistic model continues to apply, with exactly the same slopes $\boldsymbol{\beta}$ but with a different intercept. The methods described in the context of prospective studies can be applied to retrospective studies as well, with the exception of the intercept, which is now affected by the sampling proportions.

11.3 ESTIMATION OF THE PARAMETERS

Under the Bernouilli assumption in Eq. (11.1) (see Table 11.1a), and assuming independence among the n individual cases, the probability density of $y_1, y_2, \ldots, y_n$ is given by

$$p(y_1, y_2, \ldots, y_n\,|\,\boldsymbol{\beta})=\prod_{i=1}^{n}[\pi(\boldsymbol{x}_i)]^{y_i}[1-\pi(\boldsymbol{x}_i)]^{1-y_i}$$

with the parameters $\boldsymbol{\beta}$ entering into the probabilities $\pi_i=\pi(\boldsymbol{x}_i)$. Accordingly, the log-likelihood function is given by

$$\ln L(\boldsymbol{\beta}\,|\,y_1, y_2, \ldots, y_n)=\sum_{i=1}^{n} y_i \ln\pi(\boldsymbol{x}_i)+\sum_{i=1}^{n}(1-y_i)\ln[1-\pi(\boldsymbol{x}_i)] \quad (11.11)$$

Nonlinear optimization procedures such as the Newton–Raphson method of Chapter 9 are employed to determine the estimates that maximize this function.

Assume that the observed values on the p explanatory variables comprise m different settings (constellations); see Table 11.1b. It is this situation that we use in

where

$$\sum_{i=1}^{m} n_i\tilde{\pi}_i(1-\tilde{\pi}_i)\boldsymbol{x}_i z_i = \left[\sum_{i=1}^{m} n_i\tilde{\pi}_i(1-\tilde{\pi}_i)\boldsymbol{x}_i\boldsymbol{x}'_i\right]\tilde{\boldsymbol{\beta}} + \sum_{i=1}^{m}(y_i - n_i\tilde{\pi}_i)\boldsymbol{x}_i$$
$$= G\tilde{\boldsymbol{\beta}} - \boldsymbol{g} \tag{11.21}$$

Hence, the weighted least squares estimate

$$\hat{\boldsymbol{\beta}}^{\text{WLS}} = G^{-1}(G\tilde{\boldsymbol{\beta}} - \boldsymbol{g}) = \tilde{\boldsymbol{\beta}} - G^{-1}\boldsymbol{g} \tag{11.22}$$

is identical to the Newton–Raphson iteration in Eq. (11.18). This shows that one can think of the Newton–Raphson procedure of estimating the coefficients in the logistic regression model as a sequence of weighted least squares steps. The current estimate $\tilde{\boldsymbol{\beta}}$ is used to calculate the "response" vector and the weighting matrix at each step of the iteration.

11.4 INFERENCE

11.4.1 LIKELIHOOD RATIO TESTS

Computation for a given logistic regression model can be carried out with a wide variety of statistical software packages. These packages use the Newton–Raphson method to obtain the maximum likelihood estimates $\hat{\boldsymbol{\beta}}$. Furthermore, they evaluate—through substitution of the estimates into Eq. (11.12)—the maximum value of the likelihood function, $L(\hat{\boldsymbol{\beta}})$, and its logarithm, $\ln L(\hat{\boldsymbol{\beta}})$.

Likelihood ratio tests can be used to compare the maximum likelihood under the current model with success probability $\pi(\boldsymbol{x}_i) = \exp(\boldsymbol{x}'_i\boldsymbol{\beta})/[1+\exp(\boldsymbol{x}'_i\boldsymbol{\beta})]$ (which we call the "full" model), with the maximum likelihood obtained under alternative competing models. Other models of interest have one or more of the explanatory variables omitted (we call these the "restricted" models; see also Section 4.5). Likelihood ratio (LR) tests look at twice the logarithm of the likelihood ratio of the full [L(full)] and the restricted [L(restricted)] models. That is,

$$\text{LR test statistic} = 2\ln\frac{L(\text{full})}{L(\text{restricted})} = 2\,\{\ln L(\text{full}) - \ln L(\text{restricted})\} \tag{11.23}$$

If the constraints imposed by the restricted model are valid, then the LR statistic in Eq. (11.23) follows, in large samples, a chi-square distribution with degrees of freedom given by the number of independent constraints. We reject the imposed restrictions at significance level α if the LR test statistic is larger than the $100(1-\alpha)$ percentile of the applicable chi-square distribution. We retain the restrictions if the test statistic is smaller than this percentile.

LR tests can be used to test the significance of an individual coefficient (a partial test), the significance of two or more coefficients, and the significance of all regressor coefficients (test of overall regression). The **test of overall regression** compares the maximum log-likelihood of the full model, $\ln L(\hat{\boldsymbol{\beta}})$, with the maximum likelihood of the model with constant success probability $\exp(\beta_0)/(1+\exp(\beta_0))$ that is unaffected by all covariates. The maximum likelihood estimate of this constant success probability is given by y/n, where y is the overall number of successes and n is the total number of cases. Substituting this estimate into the log-likelihood function in Eq. (11.12) leads to the maximum log-likelihood

$$\ln L_0 = c + y\ln(y) + (n-y)\ln(n-y) - n\ln(n) \tag{11.24}$$

Hence, the test statistic for testing the overall significance of the regression is given by

$$\text{LR test statistic} = 2\left\{\sum_{i=1}^{m} y_i \ln\hat{\pi}_i + \sum_{i=1}^{m}(n_i - y_i)\ln(1-\hat{\pi}_i) - [y\ln(y) + (n-y)\ln(n-y) - n\ln(n)]\right\} \tag{11.25}$$

A large value of the statistic in Eq. (11.25) indicates that the success probability depends on one or more of the regressors. On the other hand, a small value indicates that none of the regressors in the model influence the success probability.

11.4.2 DEVIANCE

The model with a single success probability unaffected by the covariates, $\exp(\beta_0)/(1+\exp(\beta_0))$, is the simplest one can consider. The most elaborate model is the **saturated model**, where each constellation of the explanatory variables is allowed its own distinct success probability. Such constellation-specific success probabilities are estimated by y_i/n_i, where n_i and y_i are the number of cases and the number of successes at the ith constellation, respectively. The deviance D is defined as twice the log-likelihood ratio between the saturated model and the parameterized (full) model with estimated success probability $\hat{\pi}_i = \hat{\pi}(\boldsymbol{x}_i) = \exp(\boldsymbol{x}_i'\hat{\boldsymbol{\beta}})/[1+\exp(\boldsymbol{x}_i'\hat{\boldsymbol{\beta}})]$. It follows from the log-likelihood function in Eq. (11.12) that

$$D = 2\frac{\ln L(\text{saturated})}{\ln L(\text{full})} = 2\sum_{i=1}^{m}\left[y_i \ln\left(\frac{y_i}{n_i\hat{\pi}_i}\right) + (n_i - y_i)\ln\left(\frac{n_i - y_i}{n_i(1-\hat{\pi}_i)}\right)\right] \tag{11.26}$$

A note on the calculation of the deviance: For $y=0$, the term $y\ln(y/n\hat{\pi}) = 0$. Similarly, $(n-y)\ln[(n-y)/n(1-\hat{\pi})] = 0$ if $y=n$.

The deviance is part of the standard computer output for logistic regression. A large deviance should be viewed as a warning that the parameterization in the full model is not adequate.

The likelihood ratio test statistic in Eq. (11.23) can equivalently be expressed through the deviances,

$$\begin{aligned}\text{LR test statistic} &= 2\{\ln L(\text{full}) - \ln L(\text{restricted})\}\\ &= D(\text{restricted}) - D(\text{full})\end{aligned} \tag{11.27}$$

Comment One can also define the deviance in the linear regression model with normal errors. In the saturated model, each individual observation serves as the estimate of the "mean," and the deviance in the regression model with normal errors is given by $D = \sigma^{-2}\text{SSE} = \sigma^{-2}\sum(y_i - \hat{\mu}_i)^2$. We learned in Chapter 4 that it follows a chi-square distribution with $n - p - 1$ degrees of freedom. With normal errors, the deviance includes the nuisance parameter σ^2 and cannot be calculated directly. Programs such as GLIM (Generalized Linear Interactive Modelling) list the deviance as $\sigma^2 D = \text{SSE} = \sum(y_i - \hat{\mu}_i)^2$ and provide a scale parameter that is an estimate of σ^2, $\hat{\sigma}^2 = D/(n - p - 1)$. Deviance and error sum of squares are equivalent concepts.

11.4.3 STANDARD ERRORS OF THE MAXIMUM LIKELIHOOD ESTIMATES

The matrix of second derivatives of the negative log-likelihood function is given in Eq. (11.17), $G = \sum_{i=1}^{m} n_i \pi_i (1 - \pi_i) \boldsymbol{x}_i \boldsymbol{x}_i'$. The inverse of the matrix G, evaluated at the maximum likelihood estimates $\hat{\boldsymbol{\beta}}$, provides an estimate of the asymptotic covariance matrix of the maximum likelihood estimates,

$$V(\hat{\boldsymbol{\beta}}) \cong \left[\sum_{i=1}^{m} n_i \hat{\pi}_i (1 - \hat{\pi}_i) \boldsymbol{x}_i \boldsymbol{x}_i' \right]^{-1} \tag{11.28}$$

The square roots of the diagonal elements in this matrix provide the asymptotic standard errors s.e.$(\hat{\beta}_0)$, s.e.$(\hat{\beta}_1)$, ..., s.e.$(\hat{\beta}_p)$ of the estimates $\hat{\beta}_0, \hat{\beta}_1, \ldots, \hat{\beta}_p$; off-diagonal elements provide the covariances between different estimates. Of course, this approximation requires that the sample size n is reasonably large.

The standard errors can be used to obtain approximate 95% confidence intervals for β_j,

$$\hat{\beta}_j \pm 1.96\,\text{s.e.}(\hat{\beta}_j) \tag{11.29}$$

and to calculate the t ratios, $\hat{\beta}_j/\text{s.e.}(\hat{\beta}_j)$, to test $H_0\colon \beta_j = 0$ against the alternative hypothesis $H_1\colon \beta_j \neq 0$. The significance is assessed from the percentiles of the normal distribution. These tests are referred to as **Wald tests**.

As indicated previously, the coefficient β_j represents the logarithm of the odds ratio. Hence, the transformed value $\exp(\beta_j)$, the odds ratio, becomes an important quantity for the interpretion. Transforming the estimate and the confidence interval in Eq. (11.29) leads to the estimated odds ratio $\exp(\hat{\beta}_j)$ and the confidence interval from $\exp[\hat{\beta}_j - 1.96\,\text{s.e.}(\hat{\beta}_j)]$ to $\exp[\hat{\beta}_j + 1.96\,\text{s.e.}(\hat{\beta}_j)]$. The confidence interval for the odds ratio is skewed but still has the specified 95% coverage. The point estimate $\exp(\hat{\beta}_j)$ represents the median of its sampling distribution.

The covariance matrix can be used to obtain standard errors of other statistics that are of interest, particularly, estimates of the probabilities, $\hat{\pi}_i = \hat{\pi}(\boldsymbol{x}_i) = \dfrac{e^{\boldsymbol{x}_i'\hat{\boldsymbol{\beta}}}}{1 + e^{\boldsymbol{x}_i'\hat{\boldsymbol{\beta}}}}$. The derivative in Eq. (11.14), $\partial \pi_i / \partial \boldsymbol{\beta} = \pi_i(1 - \pi_i)\boldsymbol{x}_i$, and the expansion $\hat{\pi}_i \approx \pi_i + (\hat{\boldsymbol{\beta}} - \boldsymbol{\beta})' \partial \pi_i / \partial \boldsymbol{\beta} = \pi_i + \pi_i(1 - \pi_i)\boldsymbol{x}_i'(\hat{\boldsymbol{\beta}} - \boldsymbol{\beta})$ can be used to approximate the variance

$$V(\hat{\pi}_i) \approx [\pi_i(1-\pi_i)]^2 \boldsymbol{x}_i' V(\hat{\boldsymbol{\beta}}) \boldsymbol{x}_i$$

$$= [\pi_i(1-\pi_i)]^2 \boldsymbol{x}_i' \left[\sum_{j=1}^{m} n_j \pi_j (1-\pi_j) \boldsymbol{x}_j \boldsymbol{x}_j' \right]^{-1} \boldsymbol{x}_i \qquad (11.30)$$

where the probabilities π_i are evaluated at the maximum likelihood estimates $\hat{\boldsymbol{\beta}}$.

11.5 MODEL-BUILDING APPROACH IN THE CONTEXT OF LOGISTIC REGRESSION MODELS

11.5.1 PRELIMINARY MODEL SPECIFICATION

General regression modeling principles also apply to the logistic regression situation. One seeks to determine parsimonious (simple) models that explain the data. The rationale for preferring simple models is that they tend to be numerically more stable, and they are easier to interpret and generalize. Overfitting a model also leads to unstable estimated coefficients with large standard errors. Although many of the tools for model selection discussed in Chapter 7 are still useful, some modifications need to be made when modeling a binary response.

For situations in which the covariate (the x variable) is categorical with two or more possible outcomes, one may want to start the analysis by performing a chi-square test of homogeneity (independence) in the contingency table of x and y. Rejection of the hypothesis of independence is an indication that the covariate has an influence on the response, and one can proceed to the odds ratios to explore the form of the relationship. In the case of more than one categorical covariate, one can calculate a chi-square test for each covariate separately. Of course, we know from standard regression modeling that such an approach may mask the joint effect. It can be that a collection of covariates, each of which is weakly associated with the outcome, becomes an important predictor if the covariates are considered together. The fact that logistic regression is able to address the effects of more than two covariates on the response is one of its advantages over a pairwise contingency table analysis.

If the covariate x is continuous, one can investigate whether there are differences in the covariate between the success and failure groups. If there are covariate differences among the two groups, one can conclude that the covariate is related to the response and that it should be part of the model. Comparative dot plots of the covariate observations, as well as a two-sample t test, can give us preliminary feedback.

Scatter plots of the response y_i against each explanatory variable x_i are part of the standard model specification tool-kit. With a binary response and very little replication at the covariate constellations ($m \cong n$), these plots need to be modified and replaced by scatter plots of some smoothed response $\tilde{y}_i$ against x_i. Many different smoothing methods can be used. For example, one can calculate smoothed values

$$\tilde{y}_i = \frac{\sum_{j=1}^{n} w_i(x_j) y_j}{\sum_{j=1}^{n} w_i(x_j)}$$

where $w_i(x_j) = \exp[-(x_j - x_i)^2/c]$ is a weight function that gives most of the weight to observations with similar values of x_i, and $c > 0$ is a constant that controls the width of the smoothing window. Of course, if the number of cases in each constellation group is large, then the relative frequencies $\hat{\pi}_i = y_i/n_i$ and their logits, $\text{logit} = \ln(\hat{\pi}_i/1-\hat{\pi}_i)$, can be graphed directly against the explanatory variables. However, adjustments need to be made if there are zero successes in a particular constellation [see Agresti (2002) for a discussion on how to calculate empirical logits and Eno and Terrill (1999) for grayscale graphics].

11.5.2 ASSESSING MODEL FIT

With today's computer power, it is not difficult to fit a large number of different logistic regression models once one has identified the appropriate covariates and thought about whether to include nonlinear components (such as the square of the covariates) and interaction terms. Models can be simplified by looking at appropriate likelihood ratio (or Wald-type) tests, and stepwise modeling procedures can be employed. Since model building is iterative and involves multiple significance tests, the significance levels for these tests should not be set too low; a significance level of $\alpha = 0.10$ is often reasonable. In cases in which the covariates are multicollinear, stepwise procedures may lead to several models with similar fits.

The selection of the "best" model from among all that have been tried depends on the criterion that is chosen. The **deviance** and the **Pearson chi-square** are two commonly used criteria. One prefers models with low values on these statistics.

In the discussion of the standard linear regression model, we showed how replicate observations at each configuration of the explanatory variable(s) allow us to obtain a measure of the pure error sum of squares, which subsequently can be used to test for lack of fit (see Section 6.4). The **deviance** statistic in Eq. (11.26) does exactly this in the context of logistic regression. The deviance is a likelihood ratio statistic that compares the saturated model to a parametric model that explains the success probabilities through fewer parameters.

The goodness of fit is usually assessed over the different constellations of the explanatory variables. We may start with n cases and a model $\pi(\boldsymbol{x}_i) = \exp(\boldsymbol{x}_i'\boldsymbol{\beta})/[1 + \exp(\boldsymbol{x}_i'\boldsymbol{\beta})]$ that includes p explanatory variables and $p + 1$ coefficients. However, there may only be a set of m distinct configurations of the explanatory variables. If the number of cases becomes large, but the number of distinct configurations does not expand fast enough so that the number of replications for each configuration becomes large also, the distribution of the deviance can be approximated by a chi-square distribution with $m - p - 1$ degrees of freedom. The deviance can be compared with the (90th or 95th) percentile of the chi-square distribution with $m - p - 1$ degrees of freedom. Lack of fit and model inadequacy are indicated if the deviance is large and exceeds the percentile. On the other hand, a deviance that is smaller than the percentile indicates that the model is adequate.

The **Pearson chi-square** is a similar statistic for evaluating model adequacy. The logistic regression model specifies probabilities of success and failure at each of the m distinct constellations of the explanatory variables. These lead to the frequencies that are expected under the logistic model; $n_i\hat{\pi}_i$ successes and $n_i(1-\hat{\pi}_i)$ failures at the ith constellation, $i = 1, 2, \ldots, m$. These expected frequencies can be compared with the observed frequencies, y_i and $n_i - y_i$. The Pearson statistic is the chi-square statistic that results from a $2 \times m$ table where the rows correspond to the two outcomes 1 (success) and 0 (failure), and the columns correspond to the m distinct covariate patterns. The Pearson chi-square statistic is given by

$$\chi^2 = \sum_{i=1}^{m}\left[\frac{[y_i - n_i\hat{\pi}_i]^2}{n_i\hat{\pi}_i} + \frac{[(n_i - y_i) - n_i(1-\hat{\pi}_i)]^2}{n_i(1-\hat{\pi}_i)}\right] = \sum_{i=1}^{m}\frac{[y_i - n_i\hat{\pi}_i]^2}{n_i\hat{\pi}_i(1-\hat{\pi}_i)} \tag{11.31}$$

The asymptotic result for the deviance also applies for the Pearson chi-square statistic. The statistic in Eq. (11.31) can be compared with the percentile of a chi-square distribution with $m - p - 1$ degrees of freedom. Large values of this test statistic suggest that the logistic regression model with estimated success probabilities $\hat{\pi}(\boldsymbol{x}_i) = \exp(\boldsymbol{x}_i'\hat{\boldsymbol{\beta}})/[1 + \exp(\boldsymbol{x}_i'\hat{\boldsymbol{\beta}})]$ is inadequate.

One often standardizes the deviance and the Pearson chi-square statistic by dividing them by their degrees of freedom, $m - p - 1$. The adequacy of the model is questioned if these standardized goodness-of-fit statistics are much larger than 1.

The deviance and the Pearson chi-square statistics are useful measures of (lack of) fit, provided that we have replicate observations at each constellation of the explanatory variable(s). If we do not have replicate observations (such is often the case with continuous covariates), their interpretation as measures of model adequacy becomes tenuous. Of course, one can always group continuous covariates into a smaller number of groups.

Methods for grouping the cases if there are no replicates on the explanatory variables (and hence $m = n$) have been suggested in the literature. Hosmer and Lemeshow (1980) and Lemeshow and Hosmer (1982) propose to group the cases on the basis of the estimated probabilities $\hat{\pi}_i = \hat{\pi}(\boldsymbol{x}_i)$. They recommend to rank the estimated probabilities from the smallest to the largest and to use this ranking to categorize the cases into g groups of equal size. Usually, $g = 10$, in which case the groups are referred to as "deciles of risk." For each group k , $k = 1, 2, \ldots, g$, one calculates the number of successes o_k and the number of failures $n_k - o_k$ that are associated with the n_k cases in the group. The observed frequencies are compared with the expected frequencies $n_k\bar{\pi}_k$ and $n_k(1-\bar{\pi}_k)$, where $\bar{\pi}_k = \frac{\sum_{i \in \text{Group } k} \hat{\pi}_i}{n_k}$ is the average estimated success probability in the kth group. The Pearson chi-square statistic is calculated from the resulting $2 \times g$ table,

$$\mathrm{HL} = \sum_{k=1}^{g} \frac{[o_k - n_k \bar{\pi}_k]^2}{n_k \bar{\pi}_k (1 - \bar{\pi}_k)} \tag{11.32}$$

It is referred to as the Hosmer–Lemeshow statistic and is standard output of most software packages. Hosmer and Lemeshow study this grouping in the case in which there are no replicates on the explanatory variables. Assuming that the fitted logistic regression is the correct model, they show that the distribution of HL is well approximated by a chi-square distribution with $g - 2$ degrees of freedom. Large values of the Hosmer–Lemeshow statistic indicate lack of fit.

11.5.3 MODEL DIAGNOSTICS

The Pearson chi-square statistic and the deviance are overall goodness-of-fit measures. Each of these statistics is made up of m squared residuals, which are known as **Pearson residuals** and **deviance residuals**, respectively. The Pearson residual for covariate pattern i is defined as

$$r_i = r(y_i, \hat{\pi}_i) = \frac{y_i - n_i \hat{\pi}_i}{\sqrt{n_i \hat{\pi}_i (1 - \hat{\pi}_i)}} \tag{11.33}$$

The deviance residual is given by

$$d_i = d(y_i, \hat{\pi}_i) = \pm \left\{ 2 \left[y_i \ln \left(\frac{y_i}{n_i \hat{\pi}_i} \right) + (n_i - y_i) \ln \left(\frac{n_i - y_i}{n_i (1 - \hat{\pi}_i)} \right) \right] \right\}^{1/2} \tag{11.34}$$

where the sign is determined by the sign of $y_i - n_i \hat{\pi}_i$. For covariate patterns with $y_i = 0$, the deviance residuals are $d(y_i, \hat{\pi}_i) = -\sqrt{-2n_i \ln(1 - \hat{\pi}_i)}$, and for $y_i = n_i$, the deviance residuals are $d(y_i, \hat{\pi}_i) = \sqrt{-2n_i \ln \hat{\pi}_i}$.

The deviance D and the Pearson chi-square χ^2 are sums of squares of the respective residuals. They provide single numbers that summarize the agreement of observed and fitted values. However, it is important to examine the fit at individual constellations and check whether the fit is supported over the entire range of covariate patterns or whether there are certain covariate patterns for which there is lack of fit. Case diagnostics for logistic regressions have been developed and are part of most computer packages. Case diagnostics include leverage and measures of influence of a given covariate constellation on the deviance, the Pearson chi-square, and the parameter estimates.

Starting from the iteratively reweighted least squares representation of logistic regression, Pregibon (1981) obtains a linear approximation to the fitted values that yields a "hat" matrix for logistic regression, $\hat{\mu} = (n_1\hat{\pi}_1, n_2\hat{\pi}_2, \ldots, n_m\hat{\pi}_m)' \cong H\mathbf{y}$. This matrix is given by

$$H = V^{-1/2} X (X'V^{-1}X)^{-1} X'V^{-1/2} \tag{11.35}$$

where V^{-1} is the diagonal matrix containing the weights $n_i\hat{\pi}_i(1 - \hat{\pi}_i)$ in the iteratively reweighted least squares algorithm. The diagonal elements of the hat matrix, h_{ii}, can be used to standardize the Pearson residuals,

$$r_i^s = \frac{r_i}{\sqrt{1-h_{ii}}} = \frac{y_i - n_i\hat{\pi}_i}{\sqrt{n_i\hat{\pi}_i(1-\hat{\pi}_i)}\sqrt{1-h_{ii}}} \tag{11.36}$$

Absolute values larger than 2 or 3 (roughly) indicate lack of fit at that particular covariate constellation. Pregibon extends Cook's influence measure to the logistic regression and approximates the change in the estimates from omitting the ith covariate constellation as

$$\Delta\hat{\beta}_i = \left(\hat{\beta} - \hat{\beta}_{(i)}\right)'(X'V^{-1}X)^{-1}\left(\hat{\beta} - \hat{\beta}_{(i)}\right) \approx \left(r_i^s\right)^2 \frac{h_{ii}}{1-h_{ii}} \tag{11.37}$$

where $\hat{\beta}_{(i)}$ is the estimate with the ith constellation dropped from the estimation.

Residual diagnostic statistics are usually graphed against the estimated (success) probabilities. Quite often, the square of the standardized residuals is considered because large residuals—irrespective of their signs—indicate poorly fitted points. When working with the squares, one can use the value "4" to judge the magnitude of the residual. It should be noted that the visual assessment of diagnostics in logistic regression is more challenging than in the normal regression situation because the discreteness of binary data makes the interpretation of the resulting displays more difficult.

Pregibon (1981) and Landwehr *et al.* (1984) give a detailed discussion of case diagnostics for logistic regression. Many useful suggestions on model checking are also given by Cook and Weisberg (1999) and Hosmer and Lemeshow (2000).

11.6 EXAMPLES

The previous sections gave a detailed theoretical discussion of logistic regression. In this section, we give several examples that illustrate the technique.

11.6.1 EXAMPLE 1: DEATH PENALTY AND RACE OF THE VICTIM

George Woodworth and Dave Baldus of the University of Iowa collected extensive data on death penalty sentencing in Georgia—information they presented for defendant McClesky in the 1988 law case, *McClesky v. Zant*. Here, we look at a subset of their data. Table 11.3 provides information on 362 death penalty cases. For each case, we consider the outcome (death penalty: yes/no), the race of the victim (White/Black), and the aggravation level of the crime. The cases with the lowest aggravation level (level 1) involve bar room brawls, liquor-induced arguments, and lovers' quarrels. Level 6 comprises the most vicious, cruel, cold-blooded, unprovoked crimes. There are a total of 362 cases ($n = 362$) and a total of 12 different covariate patterns ($m = 12$).

Table 11.3 also lists the race of victim odds ratios for receiving the death penalty at each aggravation level as well as the odds ratio for the summary information when collapsing the tables of frequencies over all aggravation levels. The odds of receiving the death penalty when a White victim is involved are 8.24 times higher than when the victim is Black.

TABLE 11.3 DEATH PENALTY DATA[a]

Aggravation Level	Race of Victim	Death Penalty Yes	Death Penalty No	Odds Ratio
1	White	2	60	(2/60)/(1/181) = 6.03
	Black	1	181	
2	White	2	15	2.80
	Black	1	21	
3	White	6	7	3.86
	Black	2	9	
4	White	9	3	6.00
	Black	2	4	
5	White	9	0	Cannot be calculated
	Black	4	3	
6	White	17	0	Cannot be calculated
	Black	4	0	
Summary All levels	White	45	85	(45/85)/(14/218) = 8.24
	Black	14	218	

[a] The data are given in the file **deathpenalty**.

For the logistic regression, we code the response y as 1 if the sentence is death and 0 otherwise, and we code the covariate "race of victim" x_1 as 1 if the victim is White and 0 if the victim is Black. The aggravation index x_2 is treated as a continuous measurement variable.

We fit the logistic regression model with probability

$$P(y=1)=\pi(x_1,x_2)=\frac{\exp(\beta_0+\beta_1x_1+\beta_2x_2)}{1+\exp(\beta_0+\beta_1x_1+\beta_2x_2)} \tag{11.38a}$$

and logit

$$\ln\frac{P(y=1)}{1-P(y=1)}=\beta_0+\beta_1x_1+\beta_2x_2 \tag{11.38b}$$

Most computer packages include routines for logistic regression, and using these procedures is as easy as running a standard regression model. The partial output from the Minitab logistic regression function in Table 11.4 shows the maximum likelihood estimates and their standard errors. The hypothesis that neither covariate is important, H_0: $\beta_1=\beta_2=0$, is soundly rejected. The likelihood ratio test statistic $=208.4$ [calculated from Eq. (11.25), and denoted by G in Minitab] is highly significant when compared to percentiles of the chi-square distribution with 2 degrees of freedom. The probability of receiving the death sentence is affected by one or both of the covariates.

The maximum likelihood estimate of the effect of race of victim is $\hat{\beta}_1=1.81$, leading to the odds ratio $\exp(\hat{\beta}_1)=\exp(1.81)=6.11$. The estimate is statistically different from zero; the probability value of the standardized test statistic is 0.001

TABLE 11.4 REGRESSION OUTPUT OF MODEL (11.38). DEATH PENALTY DATA[a]

Logistic Regression Table

Predictor	Coef	SE Coef	Z	P	Odds Ratio	95% CI Lower	Upper
Constant	−6.6760	0.7574	−8.81	0.000			
RaceVictim	1.8106	0.5361	3.38	0.001	6.11	2.14	17.49
Aggravation	1.5397	0.1867	8.25	0.000	4.66	3.23	6.72

Log-Likelihood = −56.738
Test that all slopes are zero: G = 208.402, DF = 2, P-Value = 0.000

Goodness-of-Fit Tests

Method	Chi-Square	DF	P
Pearson	3.094	9	0.960
Deviance	3.882	9	0.919

[a] Note that $Z = \text{Coef}/\text{SE Coef}$ is the usual t ratio of the estimated coefficient.

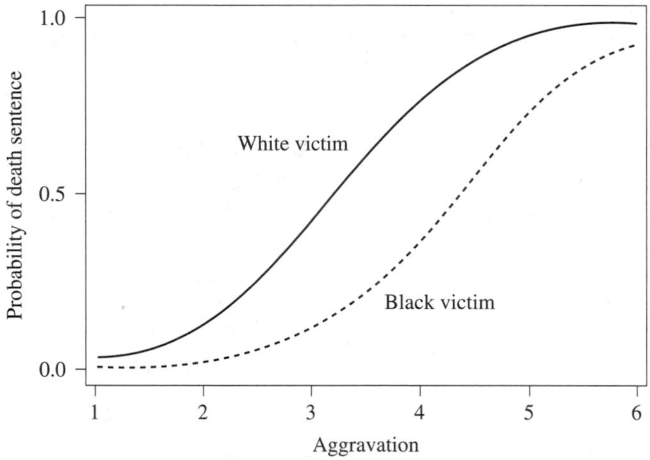

FIGURE 11.3 Estimated Probabilities. Death Penalty Data

and much smaller than any reasonable significance level. A 95% confidence interval for β_1 is given by $1.81 \pm (1.96)(0.5361)$, and the 95% confidence interval for the odds-ratio extends from $\exp(0.759) = 2.14$ to $\exp(2.861) = 17.49$. Conditioning on the severity of the crime, the odds of receiving death when the victim is White are 6.11 times the odds of getting death when the victim is Black. There is no doubt that the victim's race makes a major difference.

The aggravation of the crime also has a significant impact. The partial t statistic is 8.25, and its probability value is very small. The severity of the crime increases the odds of receiving death. Each extra unit on the aggravation scale multiplies the odds of receiving the death penalty by a factor of 4.66. This effect can be seen very clearly from Figure 11.3, in which the estimated probabilities of receiving death as a function of aggravation are graphed separately for a White and Black victim.

TABLE 11.5 OBSERVED AND EXPECTED FREQUENCIES AND PEARSON AND DEVIANCE RESIDUALS: DEATH PENALTY DATA

Aggravation 1–6	Race: 0, Black; 1, White	Number of cases	Probability (Death)	Observed (Death)	Expected (Death)	Observed (Life)	Expected (Life)	Pearson Residual	Pearson Residual (Standardized)	Deviance Residual
1	1	62	0.035	2	2.15	60	59.85	−0.105	−0.139	−0.106
1	0	182	0.006	1	1.06	181	180.94	−0.062	−0.080	−0.063
2	1	17	0.144	2	2.44	15	14.56	−0.305	−0.351	−0.313
2	0	22	0.027	1	0.59	21	21.41	0.546	0.588	0.497
3	1	13	0.439	6	5.70	7	7.30	0.166	0.202	0.165
3	0	11	0.113	2	1.25	9	9.75	0.716	0.800	0.666
4	1	12	0.785	9	9.42	3	2.58	−0.293	−0.353	−0.287
4	0	6	0.373	2	2.24	4	3.76	−0.203	−0.237	−0.205
5	1	9	0.944	9	8.50	0	0.50	0.728	0.781	1.014
5	0	7	0.735	4	5.15	3	1.85	−0.984	−1.229	−0.936
6	1	17	0.988	17	16.79	0	0.21	0.463	0.488	0.653
6	0	4	0.928	4	3.71	0	0.29	0.556	0.589	0.771

Table 11.5 lists the observed and expected frequencies for death and life at each of the 12 covariate constellations as well as the Pearson and the deviance residuals. You may want to check the calculations that are involved for at least one of the covariate constellations. For example, consider the case of a White victim ($x_1 = 1$) and aggravation level $3(x_2 = 3)$. The probability of receiving the death penalty is

$$\hat{\pi} = \frac{\exp(-6.676 + 1.81 + (3)(1.54))}{1 + \exp(-6.676 + 1.81 + (3)(1.54))} = 0.439$$

Six of 13 defendants received the death penalty. Hence, the Pearson residual is given by

$$r(y, \hat{\pi}) = \frac{6 - (13)(0.439)}{\sqrt{(13)(0.439)(0.561)}} = 0.166$$

The deviance residual is given by

$$d(y, \hat{\pi}) = +\left\{2\left[(6)\ln\frac{6}{(13)(0.439)} + (7)\ln\frac{7}{(13)(1-0.439)}\right]\right\}^{1/2} = 0.165$$

The two residuals are very similar. This is also true for the residuals at the other constellations.

The Pearson chi-square from Eq. (11.31) is 3.094, with $m - p - 1 = 12 - 3 = 9$ degrees of freedom; see Table 11.4. It is much smaller than the 95th percentile of the chi-square distribution with 9 degrees of freedom, and it indicates that we have found an adequate model. The deviance, calculated from Eq. (11.26), is 3.822; it, too, leaves little doubt that the model is adequate. There is no need to calculate the Hosmer–Lemeshow statistic because the number of different covariate constellations ($m = 12$) is much smaller than the number of cases ($n = 362$).

There is good evidence that we have found an adequate model. Nevertheless, we extend the model in various directions, always checking whether such extensions are really necessary. The aggravation index entered the logits in a linear fashion. One should check whether this is appropriate or whether a quadratic component is needed. We revise the model to include a quadratic component but find that the estimate of β_3 in

$$\ln \frac{P(y=1)}{1-P(y=1)} = \beta_0 + \beta_1 x_1 + \beta_2 x_2 + \beta_3 (x_2)^2$$

is insignificant.

Our model in Eq. (11.38) implies that the aggravation index has an impact on the probability of receiving the death sentence. However, this is different from having the odds ratio depend on the aggravation. Odds ratios depend on the aggravation if there is a significant interaction effect of race and aggravation. We check this by adding the product $x_1 x_2$ to the model,

$$\ln \frac{P(y=1)}{1-P(y=1)} = \beta_0 + \beta_1 x_1 + \beta_2 x_2 + \beta_3 (x_1 x_2)$$

but find that the coefficient β_3 is insignificant, given that both x_1 and x_2 are already in the model.

A plot of the squared standardized Pearson residuals against the estimated probabilities for model (11.38) is shown in Figure 11.4a. All points are considerably smaller than 4, our usual cutoff value. Figure 11.4b plots the influence measures in Eq. (11.37). It shows that the constellation of aggravation level 5 and Black victim, where four of seven defendants receive death, has the largest influence on the regression estimates. Omitting this constellation from the regression reduces the odds ratio for race from 6.11 to 4.82, and the odds ratio for aggravation changes from 5.21 to 4.66. However, these estimates fall well within the confidence intervals that are obtained with the complete data set, and they remain highly significant.

11.6.2 EXAMPLE 2: HAND WASHING IN BATHROOMS AT THE UNIVERSITY OF IOWA

During the fall semester of 2001, a group of undergraduate students in the introductory statistics course collected data on hand washing after using the restroom. Students took observations at several locations around campus. They developed ingenious ways of "hiding" in bathrooms so that students being observed were not aware of their presence. They collected data on whether users of the bathroom washed their hands (0, no; 1, yes), gender (0, male; 1, female), whether users carried a backpack (0, no; 1, yes), and whether others were present (0, no; 1, yes). The data are listed in Table 11.6.

At the preliminary model specification stage, we looked at pairwise cross-classification tables of washing vs. gender, washing vs. backpack, and washing vs. others present (Table 11.7). Gender seems to matter most, with women washing hands more frequently.

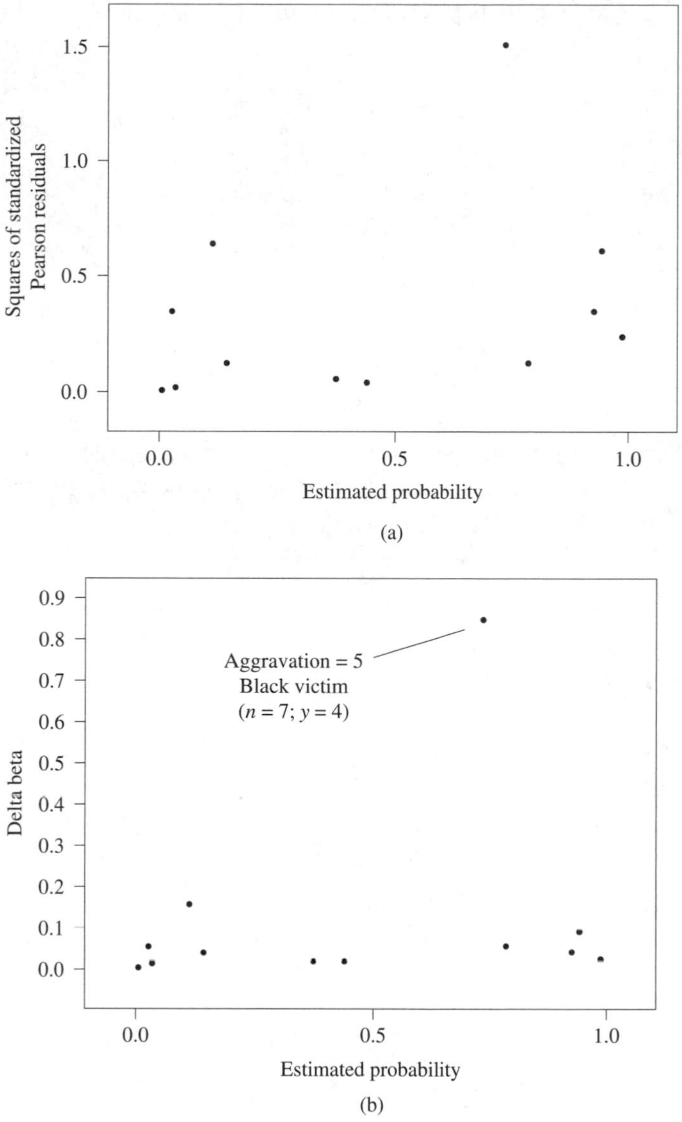

FIGURE 11.4
(a) Plot of Squared Standardized Pearson Residuals against the Estimated Probabilities. Death Penalty Data.
(b) Plot of Delta Beta in Eq. (11.37) against the Estimated Probability. Death Penalty Data

We start out with a model that includes all three covariates. We find that "Others Present" is not statistically significant, which leads us to the logit

$$\ln \frac{P(y=1)}{1-P(y=1)} = \beta_0 + \beta_1 \text{Gender} + \beta_2 \text{Backpack} \tag{11.39}$$

The estimation and goodness-of-fit results for this model in Table 11.8 are satisfactory. The standardized goodness-of-fit statistics (Pearson and deviance) do not exceed the 95th percentile. All terms in the model are significant. Women are more likely to wash hands; the odds for females are 4.27 times the odds for males. The presence of a backpack reduces the odds that people wash hands.

TABLE 11.6 UNIVERSITY OF IOWA WASHROOM DATA[a]

Gender	Backpack	Others Present	Washing: No	Washing: Yes
Male	No	Alone	27	26
Male	No	Others	5	5
Male	Yes	Alone	13	8
Male	Yes	Others	3	2
Female	No	Alone	4	28
Female	No	Others	0	5
Female	Yes	Alone	7	7
Female	Yes	Others	4	6

[a] The data are given in the file **washroom**.

TABLE 11.7 PRELIMINARY MODEL SPECIFICATION: UNIVERSITY OF IOWA WASHROOM DATA. CHI-SQUARE TEST OF INDEPENDENCE (HOMOGENEITY) IN A 2 × 2 CONTINGENCY TABLE.

```
Rows:     Washing  Columns:  Gender

             MALE  FEMALE     All
0 (NO)        48     15        63
1 (YES)       41     46        87
All           89     61       150

Chi-Square = 12.792, DF = 1, P-Value = 0.000

Rows:     Washing  Columns:  Backpack

               NO  YES        All
0 (NO)         36   27         63
1 (YES)        64   23         87
All           100   50        150

Chi-Square = 4.433, DF = 1, P-Value = 0.035

Rows:     Washing  Columns:  People Present

               NO  YES        All
0 (NO)         51   12         63
1 (YES)        69   18         87
All           120   30        150
Chi-Square = 0.062, DF = 1, P-Value = 0.804
```

11.6.3 EXAMPLE 3: SURVIVAL OF THE DONNER PARTY

Ramsey and Schafer (2002) analyze the survival of members of the Donner party. The Donner party left Springfield, Illinois, in 1846. They tried to reach the Sacramento Valley but were stranded in the eastern Sierra Nevada Mountains when they tried to use an untested route. By the time they were rescued in April 1847, many of their members had died. Data on this group have been used by

TABLE 11.8 ESTIMATION RESULTS: UNIVERSITY OF IOWA WASHROOM DATA

Model with Gender, Backpack, Others

Response Information

Variable	Value	Count	
Wash	1	87	(Event)
	0	63	
	Total	150	

Logistic Regression Table

Predictor	Coef	SE Coef	Z	P	Odds Ratio	95% CI Lower	95% CI Upper
Constant	0.0998	0.2465	0.40	0.685			
Sex	1.4433	0.3880	3.72	0.000	4.23	1.98	9.06
Backpack	−1.0149	0.3914	−2.59	0.010	0.36	0.17	0.78
People	0.1398	0.4500	0.31	0.756	1.15	0.48	2.78

Log-Likelihood = −91.915
Test that all slopes are zero: G = 20.258, DF = 3, P-Value = 0.000

Goodness-of-Fit Tests

Method	Chi-Square	DF	P
Pearson	4.160	4	0.385
Deviance	4.890	4	0.299

Model with Gender, Backpack

Logistic Regression Table

Predictor	Coef	SE Coef	Z	P	Odds Ratio	95% CI Lower	95% CI Upper
Constant	0.1184	0.2391	0.50	0.620			
Sex	1.4533	0.3871	3.75	0.000	4.28	2.00	9.13
Backpack	−0.9958	0.3860	−2.58	0.010	0.37	0.17	0.79

Log-Likelihood = −91.963
Test that all slopes are zero: G = 20.161, DF = 2, P-Value = 0.000

Goodness-of-Fit Tests

Method	Chi-Square	DF	P
Pearson	3.461	1	0.063
Deviance	3.507	1	0.061

anthropologists to test the hypothesis that females are better able to withstand harsh conditions than males. The covariates include categorical information (gender) as well as a continuous variable (age) (Table 11.9).

The 2×2 cross-classification table with respect to gender and survival shows that 10 of 15 females survived compared to only 10 of 30 males. The odds ratio is $(10/5)/(10/20) = 4$, implying that the female odds of survival are four times better than the male odds. However, this analysis does not adjust for the age of the person. The dot plots in Figure 11.5 show that the age of survivors is slightly younger than the age of those who perished.

TABLE 11.9 GENDER, AGE, FAMILY STATUS, AND SURVIVAL OF THE DONNER PARTY MEMBERS, 15 YEARS AND OLDER[a]

Name	MultiFamily	Gender	Age (years)	Survival
Antonio	0	0	23	0
Breen, Mary	1	1	40	1
Breen, Patrick	1	0	40	1
Burger, Charles	0	0	30	0
Denton, John	0	0	28	0
Dolan, Patrick	0	0	40	0
Donner, Elizabeth	1	1	45	0
Donner, George	1	0	62	0
Donner, Jacob	1	0	65	0
Donner, Tamsen	1	1	45	0
Eddy, Eleanor	1	1	25	0
Eddy, William	1	0	28	1
Elliot, Milton	0	0	28	0
Fosdick, Jay	1	0	23	0
Fosdick, Sarah	1	1	22	1
Foster, Sarah	1	1	23	1
Foster, William	1	0	28	1
Graves, Eleanor	1	1	15	1
Graves, Elizabeth	1	1	47	0
Graves, Franklin	1	0	57	0
Graves, Mary	1	1	20	1
Graves, William	1	0	18	1
Halloran, Luke	0	0	25	0
Hardkoop, Mr.	0	0	60	0
Herron, William	0	0	25	1
Noah, James	0	0	20	1
Keseberg, Lewis	1	0	32	1
Keseberg, Phillipine	1	1	32	1
McCutcheon, Amanda	1	1	24	1
McCutcheon, William	1	0	30	1
Murphy, John	1	0	15	0
Murphy, Lavin	1	1	50	0
Pike, Harriet	1	1	21	1
Pike, William	1	0	25	0
Reed, James	1	0	46	1
Reed, Margaret	1	1	32	1
Reinhardt, Joseph	0	0	30	0
Shoemaker, Samuel	0	0	25	0
Smith, James	0	0	25	0
Snyder, John	0	0	25	0
Spitzer, Augustus	0	0	30	0
Stanton, Charles	0	0	35	0
Trubode, J.B.	0	0	23	1
Williams, Baylis	1	0	24	0
Williams, Eliza	1	1	25	1

[a] The data are given in the file **donner**. Multifamily: 0, single; 1, other members present. Gender: 0, male; 1, female. Survival: 0, no; 1, yes.

FIGURE 11.5
Dot plot of Age, Separate for Each Group

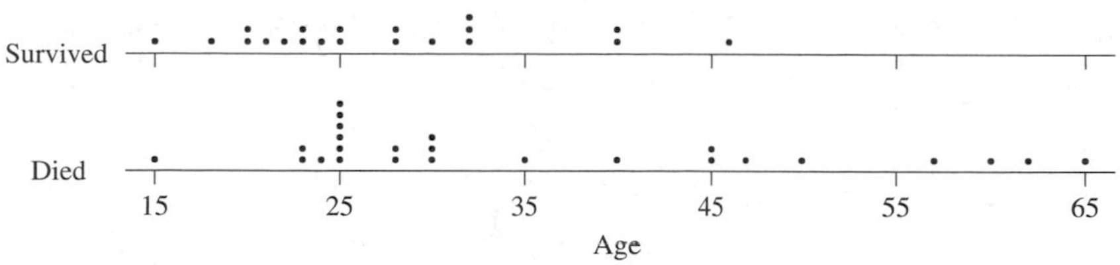

FIGURE 11.6
Estimated Probability of Survival: Males and Females. Donner Party

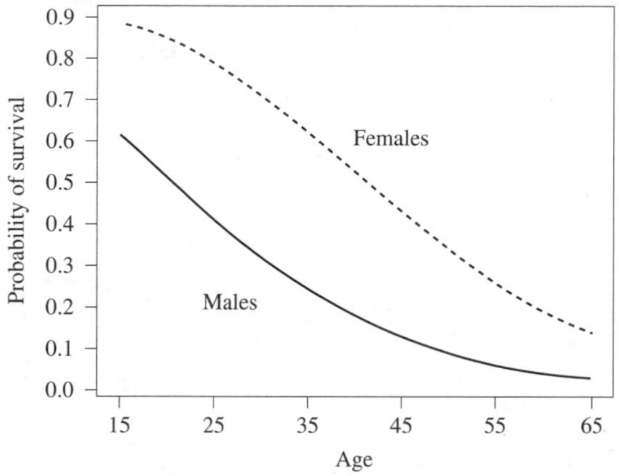

The results of the logistic regression with gender and age are given in Table 11.10. The partial tests for both variables, age and gender, are statistically significant. Older age reduces one's chance of survival. For example, a 10-year increase in age changes the odds of survival by a factor of $\exp[(-0.0782)(10)] = 0.46$; it reduces the odds of survival by 54%. Gender is important; the female odds of survival are 4.94 times higher than the male odds of survival. Graphs of the estimated survival probabilities are shown in Figure 11.6. Table 11.10 also shows the results of the logistic regression model with an added interaction between gender and age. However, the interaction is insignificant and can be dropped from the model (t ratio, -1.71; $p = 0.086$).

In this data set, we have $n = 45$ subjects and a total of $m = 28$ different covariate constellations. The number of subjects in most constellation groups is 1. Hence, the assumption of a chi-square distribution for the deviance and the Pearson statistic is suspect, and not much weight should be given to these statistics. It is preferable to consider the Hosmer–Lemeshow statistic, which aggregates the frequencies into fewer groups. Here, we use the commonly suggested aggregation into decile groups. The Hosmer–Lemeshow statistic, 10.95 with 8 degrees of freedom from its chi-square approximation, does not indicate serious overall lack of fit.

Residual diagnostics are shown in Figure 11.7. The graph of the squared standardized Pearson residuals against the estimated probabilities in Figure 11.7a shows one worrisome residual that is somewhat larger than expected. The residual comes from a 46-year-old male who survived. Survival in this case is unusual because the model indicates a low probability for survival. The graph in Figure 11.7b shows that two other constellations have a major impact on the estimates.

TABLE 11.10 ESTIMATION RESULTS: SURVIVAL OF THE DONNER PARTY

Model with Gender and Age

Response Information

Variable	Value	Count	
survival	1	20	(Event)
	0	25	
	Total	45	

Logistic Regression Table

					Odds	95% CI	
Predictor	Coef	SE Coef	Z	P	Ratio	Lower	Upper
Constant	1.633	1.110	1.47	0.141			
Gender	1.5973	0.7555	2.11	0.034	4.94	1.12	21.72
age	−0.07820	0.03729	−2.10	0.036	0.92	0.86	0.99

Log-Likelihood = −25.628
Test that all slopes are zero: G = 10.570, DF = 2, P-Value = 0.005

Goodness-of-Fit Tests

Method	Chi-Square	DF	P
Pearson	24.353	25	0.499
Deviance	26.441	25	0.384
Hosmer-Lemeshow	10.952	8	0.204

Table of Observed and Expected Frequencies:
(See Hosmer-Lemeshow Test for the Pearson Chi-Square Statistic)

	Group										
Value	1	2	3	4	5	6	7	8	9	10	Total
1											
Obs	0	3	1	2	1	1	3	3	4	2	20
Exp	0.2	1.0	1.3	1.8	2.9	2.7	2.2	2.9	3.3	1.7	
0											
Obs	4	2	3	3	6	5	1	1	0	0	25
Exp	3.8	4.0	2.7	3.2	4.1	3.3	1.8	1.1	0.7	0.3	
Total	4	5	4	5	7	6	4	4	4	2	45

Model with Gender. Age, and Gender*Age

Logistic Regression Table

					Odds	95% CI	
Predictor	Coef	SE Coef	Z	P	Ratio	Lower	Upper
Constant	0.318	1.131	0.28	0.778			
Gender	6.928	3.399	2.04	0.042	1020.50	1.30	7.98E+05
age	−0.03248	0.03527	−0.92	0.357	0.97	0.90	1.04
gender*age	−0.16160	0.09426	−1.71	0.086	0.85	0.71	1.02

Log-Likelihood = −23.673
Test that all slopes are zero: G = 14.480, DF = 3, P-Value = 0.002

Goodness-of-Fit Tests

Method	Chi-Square	DF	P
Pearson	20.781	24	0.652
Deviance	22.532	24	0.548
Hosmer-Lemeshow	5.169	7	0.639

Model With Family Only

Logistic Regression Table

					Odds	95% CI	
Predictor	Coef	SE Coef	Z	P	Ratio	Lower	Upper
Constant	−1.4663	0.6405	−2.29	0.022			
Family	1.8146	0.7432	2.44	0.015	6.14	1.43	26.35

Log-Likelihood = −27.389
Test that all slopes are zero: G = 7.048, DF = 1, P-Value = 0.008

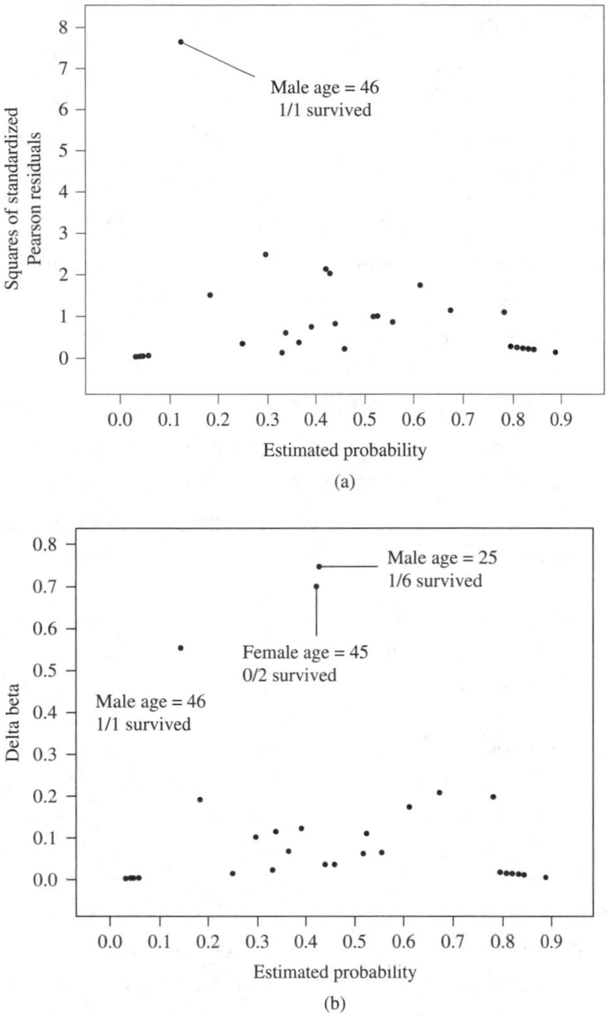

FIGURE 11.7 (a) Squared Standardized Pearson Residuals against Estimated Probabilities. Donner party. (b) Plot of Delta Beta in Eq. (11.37) against the Estimated Probability. Donner Party

Finally, we put forward another possible interpretation of the data. We create a new covariate that indicates whether a subject has other family members present or whether the subject travels alone. One could argue that family support structure enhances the chance of survival. The indicator gets value 1 if other members with the same family name are present and 0 otherwise. We use the family name as an indicator of the relationship. This may be a crude approximation because there may be relationships among members with different family names. The information in Table 11.9 indicates that none of 16 members without family ties survived, whereas 15 of 29 members with family ties survived. Because of "zero" survivors in the first group, we cannot calculate a "crude" odds ratio for survival. However, we can fit a logistic regression model to the data. The results in Table 11.10 show that the odds of survival for subjects with other family members present are 6.14 times higher than the odds for members traveling as individuals.

11.7 OVERDISPERSION

When estimating logistic regression models on grouped data, sometimes the model does not fit. The deviance and the Pearson chi-square statistic are too large, relative to their degrees of freedom. This lack of fit can arise for one (or all) of the following reasons:

- The choice of the logit may be inappropriate. In such a situation, one should explore other functional forms, such as the probit or the complementary log–log models discussed in the next section.
- The specification of covariate terms in the logit may be incorrect. In this case, one should check whether added nonlinear terms (such as terms involving the square of the covariates and interactions) help improve the fit.
- Outliers may be present, and this may explain why the observed frequencies at some of the covariate constellations are not explained by the model.
- The lack of fit may arise because of **extra binomial variation**; one refers to this as **overdispersion**. This issue is discussed in the current section.

Overdispersion is quite common if one has grouped data. With grouped data, the response is the number of occurrences at the various covariate constellations. It is often the case that the variability of the response exceeds the variance $n_i\pi_i(1-\pi_i)$ that is implied by the binomial distribution. A common root cause for this is **unobserved heterogeneity**. Let us explain this in more detail by considering a single covariate constellation. Assume that n subjects are involved at this particular constellation, and assume that their success probabilities (the $\pi's$) are not identical but vary around mean μ_π with variance σ_π^2. The number of successes among the n subjects is subject to two sources of variability: the variability in the outcomes of each of the n cases that is due to the Bernouilli variability (occurrence or nonoccurrence) and the variability in the success probabilities that change from case to case. Results from mathematical statistics allow us to calculate the mean and the variance of a random variable (here, y) in terms of the mean and variance of the conditional distribution of y given a second random variable (here, π). The number of successes y among the n subjects has mean

$$E(y) = E_\pi[E_{y|\pi}(y\,|\,\pi)] = nE_\pi[\pi] = n\mu_\pi$$

and variance

$$\begin{aligned} V(y) &= V_\pi\,[E(y\,|\,\pi)] + E_\pi\,[V(y\,|\,\pi)] = V_\pi(n\pi) + E_\pi\,[n\pi(1-\pi)] \\ &= n^2\sigma_\pi^2 + nE[\pi] - nE[\pi^2] = n^2\sigma_\pi^2 + n\mu_\pi - n\left[\sigma_\pi^2 + \mu_\pi^2\right] \\ &= n\mu_\pi(1-\mu_\pi) + n(n-1)\sigma_\pi^2 \\ &= \phi n\mu_\pi(1-\mu_\pi) \end{aligned} \tag{11.40}$$

where the scale parameter $\phi = 1 + \dfrac{(n-1)\sigma_\pi^2}{\mu_\pi(1-\mu_\pi)} > 1$ accounts for the overdispersion.

Overdispersion can also be expected if the independence assumption among the subjects within the ith covariate constellation group is violated. In a medical-type study, a certain constellation may involve animals from the same litter, and the responses of animals from the same litter may be correlated. Similarly, the constellation may involve subjects from the same region, firms that belong to the same family, and so on. Of course, one could incorporate this information into the logit specification, but often this is not possible because one does not know about these factors.

Overdispersion implies a true variance $V(y_i) = \phi n_i \pi_i(1-\pi_i)$, where $\phi > 1$, and this affects the variance of the estimated regression coefficients. The covariance matrix $V(\hat{\beta})$ in Eq. (11.28) needs to be adjusted for the overdispersion. The corrected variance of the estimates

$$V(\hat{\boldsymbol{\beta}}) = \phi \left(\sum_{i=1}^{m} n_i \pi_i (1-\pi_i) \boldsymbol{x}_i \boldsymbol{x}_i' \right)^{-1} \tag{11.41}$$

is a multiple $\phi > 1$ of the variance that ignores the overdispersion. Consequently, the standard errors that ignore overdispersion are too small, t ratios are too large, probability values are too small, and estimates appear "too" significant. Ignoring the overdispersion may lead the investigator to find "spurious" regression relationships.

In grouped logistic regression in which the sample sizes in each group are large, it is reasonable to estimate ϕ in the previous equation by the standardized deviance, $D/(m-p-1)$. Alternatively, one can use the scaled Pearson chi-square statistic, $\chi^2/(m-p-1)$, as an estimate of the scale parameter ϕ. Also, the likelihood ratio tests need to be adjusted for the overdispersion; the expressions in Eqs. (11.23) and (11.25) need to be divided by the standardized deviance.

11.8 OTHER MODELS FOR BINARY RESPONSES

In the analysis of binary responses, the probability of sucess $\pi = \pi(\boldsymbol{x})$ needs to be restricted between 0 and 1. The logistic regression model achieves this restriction by transforming π into the **logit** $\ln\left(\dfrac{\pi}{1-\pi}\right)$, which is between $-\infty$ and $+\infty$, and modeling the logit as a linear function of the covariates,

$$\text{logit}(\pi) = \ln \frac{\pi}{1-\pi} = \boldsymbol{x}'\boldsymbol{\beta} \quad \text{or} \quad \pi = \frac{\exp(\boldsymbol{x}'\boldsymbol{\beta})}{1+\exp(\boldsymbol{x}'\boldsymbol{\beta})} \tag{11.42a}$$

Functions other than the logit can also be used to transform a probability into a quantity on $(-\infty, +\infty)$. An alternate function that transforms a probability π into a quantity on $(-\infty, +\infty)$ is the **probit** $\Phi^{-1}(\pi)$, the inverse of the distribution

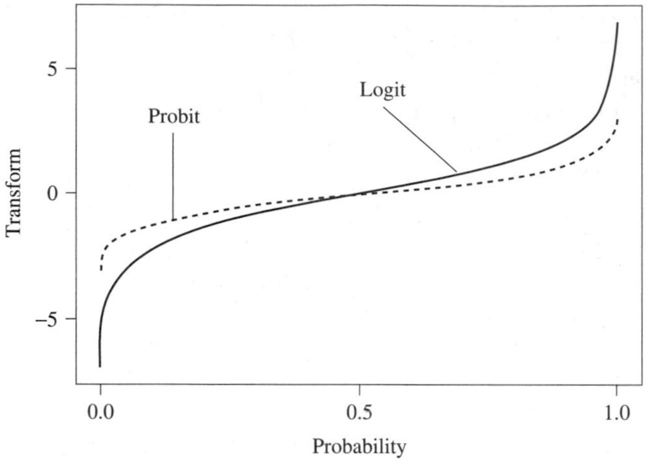

FIGURE 11.8 Comparison of Logit and Probit Transformations

function Φ of the standard normal distribution. A linear regression can be used to model the effects of covariates on the probit,

$$\text{probit}(\pi) = \Phi^{-1}(\pi) = \boldsymbol{x}'\boldsymbol{\beta} \quad \text{or} \quad \pi = \Phi(\boldsymbol{x}'\boldsymbol{\beta}) \tag{11.42b}$$

The comparison of the two transformations for a single regressor x with $\boldsymbol{x}'\boldsymbol{\beta} = \beta_0 + \beta_1 x$ in Figure 11.8 shows that the logit and probit transformations are very similar, except in situations in which the probabilities are very close to 0 or 1. Furthermore, both models achieve success probability $\pi = 1/2$ when x takes the value $x_0 = -\beta_0/\beta_1$. The function $\pi(x)$ exhibits "symmetry" around this value in the sense that $\pi(x_0 + k) = 1 - \pi(x_0 - k)$. One rarely obtains qualitative differences in the conclusions from the logistic and the probit models.

Another possible transformation of π is provided by $\log[-\log(1-\pi)]$. The model with

$$\log[-\log(1 - \pi(\boldsymbol{x}))] = \boldsymbol{x}'\boldsymbol{\beta} \quad \text{or} \quad \pi = 1 - \exp(-\exp(\boldsymbol{x}'\boldsymbol{\beta})) \tag{11.42c}$$

is referred to as the **complementary log–log model**. It lacks the symmetry of the logit and probit models and is not used as often as the other two models.

11.9 MODELING RESPONSES WITH MORE THAN TWO CATEGORICAL OUTCOMES

Logistic regression models the relationship between a binary (or dichotomous) outcome and a set of covariates. The model can be extended to the **polytomous** situation in which the response has more than two outcomes. A distinction needs to be made depending on whether the polytomous response is on an ordinal (ordered) or a nominal (not ordered) scale.

An extension of logistic regression for a **nominal response** when there is no particular ordering among the possible outcomes is discussed first. Assume that we have $k+1$ possible outcomes, which we list as $0, 1, 2, \ldots, k$. As before, we assume that we have several covariates and that their values for the ith case are collected in the vector $\boldsymbol{x}_i$.

The probabilities that the response on the ith case takes on one (and only one) of the $k+1$ possible outcomes are modeled by

$$\begin{aligned} P(y_i = 0) &= \frac{1}{1 + \sum_{j=1}^{k} \exp\left[\boldsymbol{x}'_i \boldsymbol{\beta}^{(j)}\right]} \\ P(y_i = 1) &= \frac{\exp\left[\boldsymbol{x}'_i \boldsymbol{\beta}^{(1)}\right]}{1 + \sum_{j=1}^{k} \exp\left[\boldsymbol{x}'_i \boldsymbol{\beta}^{(j)}\right]} \\ &\ldots\ldots \\ P(y_i = k) &= \frac{\exp\left[\boldsymbol{x}'_i \boldsymbol{\beta}^{(k)}\right]}{1 + \sum_{j=1}^{k} \exp\left[\boldsymbol{x}'_i \boldsymbol{\beta}^{(j)}\right]} \end{aligned} \tag{11.43}$$

The probabilities are nonnegative, and they sum to 1. The model in Eq. (11.43) includes k parameter vectors, $\boldsymbol{\beta}^{(1)}, \boldsymbol{\beta}^{(2)}, \ldots, \boldsymbol{\beta}^{(k)}$, and with p covariates each of these k vectors includes $p+1$ parameters.

Pairing each response category with the baseline group 0 (a choice that is arbitrary) leads to logits

$$\ln \frac{P(y_i = 1)}{P(y_i = 0)} = \boldsymbol{x}'_i \boldsymbol{\beta}^{(1)},\ \ln \frac{P(y_i = 2)}{P(y_i = 0)} = \boldsymbol{x}'_i \boldsymbol{\beta}^{(2)}, \ldots,\ \ln \frac{P(y_i = k)}{P(y_i = 0)} = \boldsymbol{x}'_i \boldsymbol{\beta}^{(k)} \tag{11.44}$$

that are linear functions of the parameters. Similarly, the logits of group r with respect to group r^* other than 0 are given by

$$\ln \frac{P(y_i = r)}{P(y_i = r^*)} = \boldsymbol{x}'_i \left(\boldsymbol{\beta}^{(r)} \quad \boldsymbol{\beta}^{(r^*)}\right) \tag{11.45}$$

The distribution of y_i is multinomial with the probability vector given in Eq. (11.43). Maximum likelihood estimates of the parameters and their approximate standard errors can be obtained by extending the analysis in Sections 11.3 and 11.4. The extensions involved are fairly straightforward mathematically.

The interpretation is not always easy because the model involves many parameters. Changing the explanatory variable x_s by one unit (from a value x to a value $x+1$) and keeping all other covariates fixed, changes the odds of outcome group r relative the base group 0 by the factor $\exp(\beta_s^{(r)})$. If we are interested in the effect of such a change on the odds of outcome group r relative to a group r^* other than 0, then the odds are changed by the factor $\exp(\beta_s^{(r)} - \beta_s^{(r^*)})$. If a single continuous covariate x is involved, the interpretation simplifies because the probabilities in Eq. (11.43) can be graphed as functions of x.

For a binary 0/1 covariate x_s (indicating the absence/presence of a condition), an alternative but equivalent approach calculates the odds ratios from the 2×2

table of frequencies for groups r and 0 on the response, and conditions 0 and 1 on the covariate x_s. However, an advantage of modeling the probabilities with Eq. (11.43) and calculating the odds ratios from $\exp(\beta_s^{(r)})$ is the fact that all other explanatory variables are included in this analysis, and the resulting odds ratios are adjusted for the other covariates.

Next, we discuss the case in which the response variable is **ordinal**. For example, a bond may be rated as junk bond, B grade, or A grade; of course, finer groups, such as triple A, double A, etc., can be considered. Or, mental impairment may be rated as fully impaired, with moderate symptoms of impairment, with mild symptoms, and fully functional. Let us order these outcomes into $k+1$ groups from the smallest (least inclusive) group, coded 0, to the largest (most inclusive) outcome group, coded k. In the bond example, junk bond rating is coded as 0, B-rated bonds as 1, and A-rated bonds as 2. In the mental health example, fully impaired is coded as 0, moderate symptoms as 1, mild symptoms as 2, and fully functional as 3.

For ordinal data such as these, a parsimonious model representation considers the logits of the **cumulative probabilities**. That is, for subject i with covariate vector $\boldsymbol{x}_i$ we model

$$\ln \frac{P(y_i \le r)}{1 - P(y_i \le r)} = \alpha_r + \boldsymbol{x}'_i\boldsymbol{\beta}, \quad \text{for } r = 0, 1, 2, \ldots, k-1 \tag{11.46}$$

We can express the cumulative probabilities as

$$P(y_i \le r) = \frac{\exp(\alpha_r + \boldsymbol{x}'_i\boldsymbol{\beta})}{1 + \exp(\alpha_r + \boldsymbol{x}'_i\boldsymbol{\beta})}, \quad \text{for } r = 0, 1, 2, \ldots, k-1 \tag{11.47}$$

The intercepts are increasing ($\alpha_0 \le \alpha_1 \le \ldots \le \alpha_{k-1}$) as the cumulative probabilities in Eq. (11.47) increase in r for fixed $\boldsymbol{x}_i$. Observe that the model specification assumes constant covariate coefficients $\boldsymbol{\beta}$ for all r. Also note that for the last group k, $P(y_i \le k) = 1$.

It is instructive to examine the cumulative probability for the case of a single continuous covariate x. For $r > s$, we can write

$$\begin{aligned} P(y_i \le r \,|\, x) &= \frac{\exp(\alpha_r + x\beta)}{1 + \exp(\alpha_r + x\beta)} = \frac{\exp\left(\alpha_s + \left[x + \frac{\alpha_r - \alpha_s}{\beta}\right]\beta\right)}{1 + \exp\left(\alpha_s + \left[x + \frac{\alpha_r - \alpha_s}{\beta}\right]\beta\right)} \\ &= P\left(y_i \le s \,|\, x + \frac{\alpha_r - \alpha_s}{\beta}\right) \end{aligned} \tag{11.48}$$

For $r > s$, the curve $P(y_i \le r)$ is the curve for $P(y_i \le s)$, but translated by $(\alpha_r - \alpha_s)/\beta$ in the x direction. This shows that the response curves (for $r = 0$, $1, 2, \ldots, k-1$) share the same rate of increase or decrease but are horizontally displaced from each other.

The ratio of the odds of cumulative probabilities, $P(y_i \le r)/[1 - P(y_i \le r)]$, is called the **cumulative odds ratio**. The model in Eq. (11.46) implies that the logarithm of the cumulative odds ratio is given by

$$\ln \frac{P(y_i \le r \mid \boldsymbol{x})/[1 - P(y_i \le r \mid \boldsymbol{x})]}{P(y_i \le r \mid \boldsymbol{x}^*)/[1 - P(y_i \le r \mid \boldsymbol{x}^*)]} = (\boldsymbol{x} - \boldsymbol{x}^*)'\boldsymbol{\beta} \tag{11.49}$$

The log odds ratio is proportional to the distance between $\boldsymbol{x}$ and $\boldsymbol{x}^*$. The odds of observing a response $y \le r$ at $\boldsymbol{x}$ are $\exp((\boldsymbol{x} - \boldsymbol{x}^*)'\boldsymbol{\beta})$ times the odds at $\boldsymbol{x}^*$. The same proportionality constant applies to all groups r; hence, one refers to this model as the **proportional odds model**.

If one wants to learn how changes in a (single) continuous covariate x affect the probabilities of each group, $P(y = r \mid x)$, one needs to calculate these probabilities recursively from Eqs. (11.47). Starting with $P(y = 0 \mid x) = P(y \le 0 \mid x)$, one calculates $P(y = 1 \mid x)$ from $P(y \le 1 \mid x)$ and $P(y = 0 \mid x)$; $P(y = 2 \mid x)$ from $P(y \le 2 \mid x)$ and $P(y \le 1 \mid x)$; and so on. Graphs of these probabilities as a function of x indicate how the covariate affects $P(y = r \mid x)$.

EXERCISES

11.1. This exercise deals with the production of a certain viscose fiber, a wood-based product (made from beech trees) that is an important component of modern textiles. The production process is described in Ledolter (2002). The second phase of the production process involves liquid viscose being pressed through very fine nozzles into a certain spin bath solution.

Here, we consider weekly production data (51 consecutive weeks in 1998) on two spinning machines (called "streets" 5 and 6) that are being fed by the same batch of liquid viscose. Our interest is to study the presence of a quality problem that is related to the presence of **"long fibers."** The presence of long fibers is thought to be a major contributor to downstream production problems when fibers are spun into textile fabric. For each of the 51 consecutive weeks, we record the total number of samples taken and the number of positive samples that contain fibers. Since the two spinning machines are not identical, data are taken on both machines (streets).

It is thought that the amount of stretch that is applied by the spinning machine is an important factor, with a lower **stretch reduction** increasing the odds for quality problems. The two spinning machines are also somewhat different (different age) and they use slightly different **production processes**. The production processes have to do with the number of direction changes when fibers are being pulled through the spin bath. Street 5 uses only the standard process (process 1), whereas street 6 uses both process 1 and process 2. Process 2 is thought to be slightly "tougher" on the fiber, increasing the odds of quality problems. It is also thought that the **throughput** may have an impact on the presence of quality problems; hence, data on total production (sum of both streets) are given.

Analyze the data.

a. Calculate weekly proportions of long fibers for each street. For each street separately, construct time series graphs of these proportions and graph the proportions against stretch reduction, total throughput, and the type of process being used (street 6 only).

b. For each street separately, conduct a logistic regression with stretch reduction, throughput, and production process (street 6 only) as explanatory variables. Interpret the output and draw conclusions. What can you say about the effects of stretch reduction, throughput, and type of process?

Week	Samples Street6	Positive Street6	Process Street6	Stretch ReductionStreet6	Samples Street5	Positive Street5	Process Street5	Stretch ReductionStreet5	Production Combined
1	57	1	1	58.3810	51	0	1	58.0000	396.767
2	55	0	1	58.0000	51	0	1	58.0000	394.729
3	52	0	1	58.0000	56	2	1	57.8025	398.143
4	57	1	1	58.0000	55	1	1	57.7664	388.000
5	55	2	1	58.0000	51	1	1	57.9057	392.329
6	65	3	1	57.8865	59	2	1	57.6273	398.171
7	57	1	1	57.1304	54	1	1	57.1286	395.643
8	55	1	1	57.0000	59	2	1	57.3630	394.871
9	60	1	1	57.0000	50	0	1	57.2203	386.371
10	54	3	1	57.8879	56	0	1	57.0819	395.386
11	55	1	1	57.7857	60	2	1	57.7091	392.786
12	62	1	1	57.5720	56	3	1	57.2397	383.086
13	50	1	1	57.0000	50	2	1	57.1805	366.900
14	55	0	1	57.0000	55	0	1	57.0749	384.157
15	62	1	1	56.9416	55	1	1	57.3986	392.371
16	61	2	1	56.6290	49	0	1	57.0000	389.657
17	56	2	1	57.7587	55	0	1	57.5532	386.371
18	56	1	1	56.9901	62	2	1	58.0000	393.029
19	58	1	1	56.5204	63	2	1	57.0440	396.686
20	58	3	1	57.2875	48	0	1	57.0000	391.200
21	57	1	1	57.5397	59	1	1	57.1817	387.217
22	58	2	1	57.9437	66	3	1	57.9328	391.343
23	55	0	1	57.6205	56	1	1	57.2621	389.486
24	42	1	1	57.0313	34	1	1	57.2671	377.860
25	49	1	1	57.0962	46	0	1	55.2997	411.767
26	58	1	1	57.2647	54	1	1	57.0894	408.757
27	60	1	1	57.8860	58	0	1	55.6831	404.029
28	75	6	1	56.0348	83	5	1	56.0000	397.643
29	61	3	1	55.9564	63	3	1	56.1179	406.686
30	63	3	1	57.0203	69	4	1	57.2488	385.700
31	57	1	1	57.0186	57	3	1	54.6700	392.886
32	63	3	1	54.6460	73	2	1	57.0900	369.971
33	69	5	1	55.4447	61	2	1	56.0000	370.357
34	121	6	1	54.1579	87	5	1	55.3558	383.357
35	72	4	1	53.4398	55	1	1	55.1995	381.843
36	70	4	1	54.1363	69	5	1	54.9039	378.571
37	70	2	1	53.9795	76	5	1	53.2166	380.214
38	71	5	1	53.8262	79	4	1	54.6508	368.729
39	57	2	1	53.2414	57	3	1	53.1061	361.571
40	67	5	1	53.4392	64	3	1	52.4849	388.900
41	58	1	1	53.0000	67	3	1	52.1213	404.171
42	62	2	1	53.4306	56	2	1	52.7922	402.700
43	87	5	2	52.0000	77	6	1	52.6179	410.714
44	113	7	2	52.0538	90	6	1	52.2571	409.043
45	61	6	2	51.3803	82	6	1	52.8258	382.333
46	122	7	2	51.6279	73	6	1	52.0000	385.214
47	100	7	2	51.0000	86	5	1	52.0000	382.971
48	118	7	2	51.1126	71	4	1	51.0594	398.014
49	67	6	2	51.0099	55	1	1	51.0000	389.514
50	85	5	2	51.0000	69	3	1	51.7356	405.543
51	56	4	2	51.0000	46	4	1	52.0000	387.560

c. Check your models by considering the appropriate diagnostics. In particular, obtain the Pearson and deviance residuals. Check for the presence of serial correlation among the residuals.

The data are given in the file **lenzing**.

11.2. The data used in this exercise derive from a study by the Cranfield Network on European Human Resource Management (Cranet-E), a research network dedicated to analyzing developments in the area of human resource management in public and private organizations with more than 200 employees (Brewster & Hegewisch, 1994).

Organizations are surveyed on various aspects of human resource management through extensive questionnaires. Here, we analyze the 1999 UK data on private sector for-profit organizations. We study the relationship between the profitability of the firm and firm-specific factors such as the size of the firm, the sector of the firm (primary, secondary, and tertiary), and two management practices, the number of incentives and the presence of a formal evaluation procedure, that are thought to have an influence on the success of the firm. **Profitability**, the response variable, is defined as an indicator that is 1 if gross revenues are "well in excess of costs." **Size** is a categorical variable with three groups: 1 for firms with 200–500 employees, 2 for firms with 501–1,000 employees, and 3 for firms with more than 1,000 employees. Two different **sectors** are distinguished: the primary/secondary sector and the tertiary sector. Firms are asked whether incentives (tied to share options, profit sharing, group bonus, and merit pay) apply to the following four groups of employees: management, professional, clerical, and manual. An **incentives** index ranging from 0 to 16 is calculated, where 0 indicates no incentives and 16 expresses the fact that all four types of incentives are offered to all four groups. Firms are also asked whether their performance is regularly evaluated through a formal process. **Evaluation** is an indicator that is set 1 if such special evaluation takes place and 0 otherwise.

Data on 429 firms are listed in the file **cranfield**. Information on the first 10 firms is given here.

Sector	Size	Incentives	Evaluation	Profitable	Sector Recoded
2	1	7	0	0	2
2	2	4	0	0	2
3	3	2	0	0	3
2	3	13	0	1	2
2	2	8	1	0	2
2	1	0	0	1	2
3	2	5	0	1	3
2	3	8	0	1	2
2	3	0	0	0	2
3	1	10	0	1	3
...					

a. Construct a plot of the success proportions against the incentive index. Furthermore, plot the success proportions against the size of the firm, the sector, and the evaluation indicator.

b. Consider a logistic regression model with the following regressors: size (two parameters for the three groups), sector (one parameter), evaluation, and a linear component for incentive index. Can you simplify the model by omitting certain regressor variables? If so, estimate the simplified model. Check the fitted model by considering appropriate model diagnostics. Interpret the results.

11.3. The following data are taken from Higgins, J. E., and Koch, G. G. Variable selection and generalized chi-square analysis of categorical data applied to a large cross-sectional occupational health survey. *International Statistical Review,* 45, 51–62, 1977. The data are given in the file **byssinosis**.

The data derive from an extensive survey of workers in the cotton industry. The variable of interest is the presence of lung byssinosis. Byssinosis, also called brown lung disease, is a chronic, asthma-like narrowing of the airways resulting from inhaling particles of cotton, flax, hemp, or jute. It has been recognized as an occupational hazard for textile workers. More than 35,000 textile workers, mostly from textile-producing regions of North and South Carolina, have been disabled by byssinosis and 183 have died between 1979 and 1992.

Numbers of workers suffering from (yes) and not suffering from (no) byssinosis for different categories of workers are given here. The covariates are race (1, White; 2, other), gender (1, male; 2, female), smoking history (1, smoker; 2, nonsmoker), length of employment in the cotton industry (1, less than 10 years; 2, between 10 and 20 years; 3, more than 20 years), and the dustiness of the workplace (1, high; 2, medium; 3, low).

Analyze the data. In particular,

Yes	No	Number	Dust	Race	Sex	Smoking	Employment Length
3	37	40	1	1	1	1	1
0	74	74	2	1	1	1	1
2	258	260	3	1	1	1	1
25	139	164	1	2	1	1	1
0	88	88	2	2	1	1	1
3	242	245	3	2	1	1	1
0	5	5	1	1	2	1	1
1	93	94	2	1	2	1	1
3	180	183	3	1	2	1	1
2	22	24	1	2	2	1	1
2	145	147	2	2	2	1	1
3	260	263	3	2	2	1	1
0	16	16	1	1	1	2	1
0	35	35	2	1	1	2	1
0	134	134	3	1	1	2	1
6	75	81	1	2	1	2	1
1	47	48	2	2	1	2	1
1	122	123	3	2	1	2	1
0	4	4	1	1	2	2	1
1	54	55	2	1	2	2	1
2	169	171	3	1	2	2	1
1	24	25	1	2	2	2	1
3	142	145	2	2	2	2	1
4	301	305	3	2	2	2	1
8	21	29	1	1	1	1	2
1	50	51	2	1	1	1	2
1	187	188	3	1	1	1	2
8	30	38	1	2	1	1	2
0	5	5	2	2	1	1	2
0	33	33	3	2	1	1	2
0	0	0	1	1	2	1	2
1	33	34	2	1	2	1	2
2	94	96	3	1	2	1	2
0	0	0	1	2	2	1	2
0	4	4	2	2	2	1	2
0	3	3	3	2	2	1	2
2	8	10	1	1	1	2	2
1	16	17	2	1	1	2	2
0	58	58	3	1	1	2	2
1	9	10	1	2	1	2	2
0	0	0	2	2	1	2	2
0	7	7	3	2	1	2	2
0	0	0	1	1	2	2	2
0	30	30	2	1	2	2	2
1	90	91	3	1	2	2	2
0	0	0	1	2	2	2	2
0	4	4	2	2	2	2	2
0	4	4	3	2	2	2	2
31	77	108	1	1	1	1	3
1	141	142	2	1	1	1	3
12	495	507	3	1	1	1	3
10	31	41	1	2	1	1	3
0	1	1	2	2	1	1	3
0	45	45	3	2	1	1	3
0	1	1	1	1	2	1	3
3	91	94	2	1	2	1	3
3	176	179	3	1	2	1	3
0	1	1	1	2	2	1	3
0	0	0	2	2	2	1	3
0	2	2	3	2	2	1	3
5	47	52	1	1	1	2	3
0	39	39	2	1	1	2	3
3	182	185	3	1	1	2	3
3	15	18	1	2	1	2	3
0	1	1	2	2	1	2	3
0	23	23	3	2	1	2	3
0	2	2	1	1	2	2	3
3	187	190	2	1	2	2	3
2	340	342	3	1	2	2	3
0	0	0	1	2	2	2	3
0	2	2	2	2	2	2	3
0	3	3	3	2	2	2	3

a. Discuss whether (and which of) the covariates have an influence on the presence of byssinosis.

b. Explore the presence of interactions.

c. Check the adequacy of the logistic regression model. Interpret the deviance and the Pearson statistic. Investigate whether your model(s) is adequate in explaining the presence of byssinosis.

11.4. The following data are taken from Brown, P. J., Stone, J., and Ord-Smith, C. Toxaemic signs during pregnancy. *Applied Statistics*, 32, 69–72, 1983. The data are given in the file **toxemia**.

Data collected in Bradford, United Kingdom, between 1968 and 1977 relate to 13,384 women giving birth to their first child. The data set includes information on toxaemic signs exhibited by the mother during pregnancy: hypertension only, proteinurea (i. e., the presence of protein in urine) only, both hypertension and proteinurea, and neither hypertension nor proteinurea. The aim of the study was to determine if the level of smoking is related to the incidence of toxaemic signs and how this might depend on social class. The two covariates are social class (1–5) and the number of cigarettes smoked (1, none; 2, 1–19 cigarettes per day; 3, 20 or more cigarettes per day).

Analyze the data. Discuss whether (and which of) the covariates have an influence on the presence of toxaemic signs. Consider each symptom group separately. In particular,

a. Suppose hypertension is the response in a logistic model. Find a suitable logistic model using the covariates social class and smoking history, and interpret the model and the associations it implies.

b. Suppose proteinurea is the response in a logistic model. Repeat the analysis.

c. Suppose the occurence of both hypertension and proteinurea is the response. Repeat the analysis.

d. Suppose that the occurence of either hypertension or proteinurea is the response. Repeat the analysis.

Check the adequacy of your models.

Class	Smoking	Both Hypertension and Proteinurea	Proteinurea Only	Hypertension Only	Neither Problem Exhibited	Total
1	1	28	82	21	286	417
1	2	5	24	5	71	105
1	3	1	3	0	13	17
2	1	50	266	34	785	1,135
2	2	13	92	17	284	406
2	3	0	15	3	34	52
3	1	278	1,101	164	3,160	4,703
3	2	120	492	142	2,300	3,054
3	3	16	92	32	383	523
4	1	63	213	52	656	984
4	2	35	129	46	649	859
4	3	7	40	12	163	222
5	1	20	78	23	245	366
5	2	22	74	34	321	451
5	3	7	14	4	65	90

12 Generalized Linear Models and Poisson Regression

12.1 THE MODEL

The logistic regression model of Chapter 11 is a special case of the generalized linear model considered by Nelder and Wedderburn (1972). Generalized linear models are extensions of traditional linear models that allow (i) the mean of a population to depend on a linear predictor through a nonlinear **link function** and (ii) the response probability distribution to be any member of a special set of distributions referred to as the **exponential family**.

Generalized linear models consist of three major building blocks:

- Response variates $y_1, y_2, \ldots, y_n$, which are assumed to share the same distribution from the exponential family. For a definition of the exponential family, consult any text on mathematical statistics. The normal, the Poisson, and the binomial distributions are all members of this general class.
- A set of parameters $\boldsymbol{\beta}$ and p explanatory variables $x_1, x_2, \ldots, x_p$.
- A monotone link function that relates a transform of the mean $\mu_i = E(y_i)$ linearly to the explanatory variables. That is, a function g such that

$$g(\mu_i) = \beta_0 + \beta_1 x_{i1} + \cdots + \beta_p x_{ip} = \boldsymbol{x}_i' \boldsymbol{\beta} \tag{12.1}$$

In the standard linear regression model, the link function is the identity $g(\mu) = \mu$, and the response follows a normal distribution. In the logistic regression model of Chapter 11, the link function is the logit, $g(\mu) = \ln \dfrac{\mu}{1-\mu}$, and the response follows a Bernoulli (or binomial) distribution. In the Poisson regression model considered in this chapter, the link function is the logarithm $g(\mu) = \ln \mu$, and the response follows a Poisson distribution. All three link functions are monotone functions of μ. That is, $g(\mu) \geq g(\mu^*)$ for $\mu > \mu^*$.

The link function notation g is standard in the generalized linear models literature. The link g should not be confused with the derivative of the log-likelihood function in nonlinear least squares (see Chapter 9).

Discussion of generalized linear models, in all their generality and details, can be found in the 1972 paper by Nelder and Wedderburn and in books by Agresti (2002), Dobson (2002), Fahrmeir and Tutz (2001), McCullagh and Nelder (1987), and Myers *et al.* (2002). A detailed discussion of the logistic regression model was given in Chapter 11. In this section, we describe the Poisson regression model.

The Poisson regression model arises when the response represents count data. The Poisson distribution is commonly used when modeling variables such as the number of daily equipment failures, the number of weekly traffic fatalities, the monthly number of insurance claims, the incidence of rare diseases in a certain locality, the counts of certain organisms in an area, and the yearly number of coup d'etats in South America. The Poisson distribution is described by the probabilities

$$P(Y=y)=\frac{\mu^y}{y!}e^{-\mu},\quad y=0,1,2,\ldots \tag{12.2}$$

Its mean and its variance are given by $\mu > 0$.

The mean of the Poisson distribution may depend on explanatory variables, but the relationship cannot be linear because this could lead to negative values for μ. However, a logarithmic link and a linear model for

$$g(\mu)=\ln\mu=\beta_0+\beta_1x_1+\cdots+\beta_px_p \tag{12.3}$$

satisfies the nonnegativity constraint. This parameterization implies that the mean is

$$\mu=\exp(\beta_0+\beta_1x_1+\cdots+\beta_px_p) \tag{12.4}$$

The interpretation of the coefficients in the Poisson regression model is as follows. Assume that we change one of the explanatory variables, for example, the first one, by one unit from x to $x+1$ while keeping all other regressors $x_2,\ldots,x_p$ fixed. This change affects the mean of the Poisson response by

$$100\frac{\exp[\beta_0+\beta_1(x+1)+\beta_2x_2+\cdots+\beta_px_p]-\exp[\beta_0+\beta_1x+\beta_2x_2+\cdots+\beta_px_p]}{\exp[\beta_0+\beta_1x+\beta_2x_2+\cdots+\beta_px_p]}$$

$$=100[\exp(\beta_1)-1] \tag{12.5}$$

percent.

12.2 ESTIMATION OF THE PARAMETERS IN THE POISSON REGRESSION MODEL

It is straightforward to write down the likelihood function for this model. It is given by

$$L(\beta\mid y_1,y_2,\ldots,y_n)=\prod_{i=1}^{n}\frac{(\mu_i)^{y_i}}{y_i!}\exp(-\mu_i) \tag{12.6}$$

The parameters β enter through the means $\mu_i=\exp(\beta_0+\beta_1x_{i1}+\cdots+\beta_px_{ip})$. The factor $\prod_{i=1}^{n}y_i!$ does not involve the parameters. Hence, the log-likelihood function is given by

$$\ln L(\boldsymbol{\beta} \mid y_1, y_2, \ldots, y_n) = c + \sum_{i=1}^{n} y_i \ln \mu_i - \sum_{i=1}^{n} \mu_i \tag{12.7}$$

where $c = -\sum_{i=1}^{n} \ln(y_i!)$ is a constant that does not depend on the parameters. The Newton–Raphson method (see Section 9.4) can be used to determine the estimates that maximize this function. The steps here are very similar to the ones we used for the estimation in the logistic regression model.

The derivative of the log-likelihood function in Eq. (12.7) with respect to μ_i is

$$\frac{\partial \ln L}{\partial \mu_i} = \sum_{i=1}^{n} \left[\frac{y_i}{\mu_i} - 1 \right] = \sum_{i=1}^{n} \left[\frac{y_i - \mu_i}{\mu_i} \right] \tag{12.8}$$

The $(p+1)$ vector of first derivatives of $\mu_i = \exp(\beta_0 + \beta_1 x_{i1} + \cdots + \beta_p x_{ip})$ is given by $\partial \mu_i / \partial \boldsymbol{\beta} = \mu_i \boldsymbol{x}_i$, where $\boldsymbol{x}_i = (1, x_{i1}, \ldots, x_{ip})'$ is a $(p+1)$ column vector. Hence, we find that

$$\frac{\partial \ln L}{\partial \boldsymbol{\beta}} = \frac{\partial \ln L}{\partial \mu_i} \frac{\partial \mu_i}{\partial \boldsymbol{\beta}} = \sum_{i=1}^{n} \left[\frac{y_i - \mu_i}{\mu_i} \right] \mu_i \boldsymbol{x}_i = \sum_{i=1}^{n} (y_i - \mu_i) \boldsymbol{x}_i \tag{12.9}$$

Setting the vector of first derivatives equal to zero leads to the $p+1$ maximum likelihood score equations

$$X'(\boldsymbol{y} - \boldsymbol{\mu}) = \boldsymbol{0} \tag{12.10}$$

where X is the design matrix with the n rows $\boldsymbol{x}_1', \ldots, \boldsymbol{x}_i', \ldots, \boldsymbol{x}_n'$ representing the levels of the explanatory variables at the n cases, $\boldsymbol{y} = (y_1, y_2, \ldots, y_n)'$ is the $n \times 1$ column vector of the responses at these levels, and $\boldsymbol{\mu} = (\mu_1, \mu_2, \ldots, \mu_n)'$ is the vector of means. The Eqs. (12.10) resemble the normal equations in the standard linear model, but as in the logistic regression model, the mean vector $\boldsymbol{\mu}$ is a nonlinear function of the parameters $\boldsymbol{\beta}$.

One component of the Newton–Raphson procedure in Section 9.4 is the negative first derivative of the log-likelihood function, $\boldsymbol{g} = -\dfrac{\partial \ln L}{\partial \boldsymbol{\beta}} = -\sum_{i=1}^{n} (y_i - \mu_i) \boldsymbol{x}_i$, evaluated at the current value $\tilde{\boldsymbol{\beta}}$. The derivative is easy to calculate; all that is needed is an evaluation of the means μ_i in Eq. (12.4) at the current value $\tilde{\boldsymbol{\beta}}$.

The Hessian matrix G in the Newton–Raphson procedure is the matrix of second derivatives of the negative log-likelihood function, evaluated at the current value $\tilde{\boldsymbol{\beta}}$. Taking another derivative of the first derivative in Eq. (12.9), we obtain the second derivative with respect to the two scalar elements β_j and β_{j*} of the parameter vector $\boldsymbol{\beta}$,

$$-\frac{\partial^2 \ln L}{\partial \beta_j \partial \beta_{j^*}} = -\frac{\partial}{\partial \beta_j} \left\{ \sum_{i=1}^{n} (y_i - \mu_i) x_{ij^*} \right\} = \sum_{i=1}^{n} \mu_i x_{ij} x_{ij^*} \tag{12.11}$$

The Hessian matrix

$$G = \sum_{i=1}^{n} \mu_i \boldsymbol{x}_i \boldsymbol{x}_i' \tag{12.12}$$

is evaluated at the current value $\tilde{\boldsymbol{\beta}}$, and the Newton–Raphson iteration is given by

$$\boldsymbol{\beta}^* = \tilde{\boldsymbol{\beta}} - [G(\tilde{\boldsymbol{\beta}})]^{-1}\boldsymbol{g}(\tilde{\boldsymbol{\beta}}) \tag{12.13}$$

The Hessian matrix $G(\tilde{\boldsymbol{\beta}})$ does not depend on random variables. Hence, the scoring method and the Newton–Raphson method reduce to the same algorithm. Let us denote the final estimate by $\hat{\boldsymbol{\beta}}$.

Comment As in the logistic regression model, this procedure can be implemented through an iteratively reweighted least squares (IRLS) algorithm. A weighted linear regression of the iteratively computed response

$$z_i = \boldsymbol{x}_i'\tilde{\boldsymbol{\beta}} + \frac{1}{\tilde{\mu}_i}(y_i - \tilde{\mu}_i), \qquad \text{for } i = 1, 2, \ldots, n \tag{12.14}$$

on the explanatory vector $\boldsymbol{x}_i$, with weights $w_i = \tilde{\mu}_i$, is equivalent to the Newton–Raphson iteration in Eq. (12.13).

This is easy to show. The weighted least squares (WLS) estimate is

$$\hat{\boldsymbol{\beta}}^{\text{WLS}} = \left[\sum_{i=1}^{n} w_i \boldsymbol{x}_i \boldsymbol{x}_i'\right]^{-1} \left[\sum_{i=1}^{n} w_i \boldsymbol{x}_i z_i\right] = \left[\sum_{i=1}^{n} \tilde{\mu}_i \boldsymbol{x}_i \boldsymbol{x}_i'\right]^{-1} \left[\sum_{i=1}^{n} \tilde{\mu}_i \boldsymbol{x}_i z_i\right] \tag{12.15}$$

with

$$\left[\sum_{i=1}^{n} \tilde{\mu}_i \boldsymbol{x}_i z_i\right] = \sum_{i=1}^{n} \tilde{\mu}_i \boldsymbol{x}_i \left[\boldsymbol{x}_i'\tilde{\boldsymbol{\beta}} + \frac{1}{\tilde{\mu}_i}(y_i - \tilde{\mu}_i)\right] = \left[\sum_{i=1}^{n} \tilde{\mu}_i \boldsymbol{x}_i \boldsymbol{x}_i'\right]\tilde{\boldsymbol{\beta}} + \sum_{i=1}^{n}(y_i - \tilde{\mu}_i)\boldsymbol{x}_i$$

$$= G\tilde{\boldsymbol{\beta}} - \boldsymbol{g}$$

Hence, the weighted least squares estimate

$$\hat{\boldsymbol{\beta}}^{\text{WLS}} = G^{-1}(G\tilde{\boldsymbol{\beta}} - \boldsymbol{g}) = \tilde{\boldsymbol{\beta}} - G^{-1}\boldsymbol{g}$$

is identical to the Newton–Raphson iteration in Eq. (12.13).

12.3 INFERENCE IN THE POISSON REGRESSION MODEL

12.3.1 LIKELIHOOD RATIO TESTS

Likelihood ratio (LR) tests compare the log-likelihoods of competing models (the full and the restricted models),

$$\text{LR test statistic} = 2\ln\frac{L(\text{full})}{L(\text{restricted})} = 2\{\ln L(\text{full}) - \ln L(\text{restricted})\} \tag{12.16}$$

If the constraints imposed by the restricted model are valid, then the LR statistic in Eq. (12.16) follows, in large samples, a chi-square distribution with degrees of freedom given by the number of independent constraints. We reject the imposed restrictions at significance level α if the LR test statistic is larger than the $100(1-\alpha)$ percentile of the appropriate chi-square distribution. We retain the restrictions if the test statistic is smaller than the percentile. LR tests can be used to test the significance of an individual coefficient (a partial test), the joint significance of two or more coefficients, and the significance of all regression coefficients (test of overall regression).

12.3.2 STANDARD ERRORS OF THE MAXIMUM LIKELIHOOD ESTIMATES AND WALD TESTS

The inverse of the matrix of second derivatives of the negative log-likelihood function, $G = \sum_{i=1}^{n} \hat{\mu}_i \boldsymbol{x}_i \boldsymbol{x}_i'$, evaluated at the maximum likelihood estimates $\hat{\boldsymbol{\beta}}$, provides an estimate of the covariance matrix of the maximum likelihood estimates,

$$V(\hat{\boldsymbol{\beta}}) \cong G^{-1} = \left[\sum_{i=1}^{n} \hat{\mu}_i \boldsymbol{x}_i \boldsymbol{x}_i' \right]^{-1} \tag{12.17}$$

Of course, this approximation requires that the sample size n is reasonably large. The standard errors s.e.$(\hat{\beta}_j)$, the square roots of the diagonal elements of the matrix in Eq. (12.17), can be used to obtain Wald confidence intervals for individual regression parameters, $\hat{\beta}_j \pm 1.96$ s.e.$(\hat{\beta}_j)$.

Standard Errors of the Estimated Mean

The estimate of the ith mean $\mu_i = \exp(\beta_0 + \beta_1 x_{i1} + \cdots + \beta_p x_{ip}) = \exp(\boldsymbol{x}_i'\boldsymbol{\beta})$ is given by $\hat{\mu}_i = \exp(\boldsymbol{x}_i'\hat{\boldsymbol{\beta}})$. Its variance can be approximated after expanding $\hat{\mu}_i$ in a Taylor series around $\boldsymbol{\beta}$, $\hat{\mu}_i \cong \mu_i + (\hat{\boldsymbol{\beta}} - \boldsymbol{\beta})' \left.\frac{\partial \hat{\mu}_i}{\partial \hat{\boldsymbol{\beta}}}\right|_{\hat{\boldsymbol{\beta}}=\boldsymbol{\beta}} = \mu_i + \mu_i \boldsymbol{x}_i'(\hat{\boldsymbol{\beta}} - \boldsymbol{\beta})$. Hence, the estimated variance of $\hat{\mu}_i$ is

$$V(\hat{\mu}_i) \cong \mu_i^2 V(\boldsymbol{x}_i'(\hat{\boldsymbol{\beta}} - \boldsymbol{\beta})) = \mu_i^2 \boldsymbol{x}_i' V(\hat{\boldsymbol{\beta}}) \boldsymbol{x}_i \cong \hat{\mu}_i^2 \boldsymbol{x}_i' \left[\sum_{j=1}^{n} \hat{\mu}_j \boldsymbol{x}_j \boldsymbol{x}_j' \right]^{-1} \boldsymbol{x}_i \tag{12.18}$$

after using the covariance matrix in Eq. (12.17) and replacing the unknown parameters by their estimates. Approximate 95% confidence intervals can be obtained from

$$\hat{\mu}_i \pm 1.96 \hat{\mu}_i \sqrt{\boldsymbol{x}_i' \left[\sum_{j=1}^{n} \hat{\mu}_j \boldsymbol{x}_j \boldsymbol{x}_j' \right]^{-1} \boldsymbol{x}_i} \tag{12.19}$$

Deviance, Goodness of Fit, and Residual Diagnostics

The **saturated model**, in which a separate parameter is estimated for each observation, is the most elaborate model one can consider. The contribution of the ith observation to the log-likelihood, $y_i \ln(\mu_i) - \mu_i$, is maximized for $\mu_i = y_i$. Hence, the log-likelihood function in Eq. (12.7) for the saturated model is given by $c + \sum_{i=1}^{n} [y_i \ln(y_i) - y_i]$. The deviance of the estimated model with parameterization $\hat{\mu}_i = \exp(\hat{\beta}_0 + \hat{\beta}_1 x_{i1} + \cdots + \hat{\beta}_p x_{ip})$ is

$$\begin{aligned} D &= 2\{\ln L(\text{saturated}) - \ln L(\text{parametrized})\} \\ &= 2\left\{ \sum_{i=1}^{n} [y_i \ln(y_i) - y_i] - \sum_{i=1}^{n} [y_i \ln(\hat{\mu}_i) - \hat{\mu}_i] \right\} \\ &= 2\sum_{i=1}^{n} \left[y_i \ln\left(\frac{y_i}{\hat{\mu}_i}\right) - (y_i - \hat{\mu}_i) \right] = 2\sum_{i=1}^{n} \left[y_i \ln\left(\frac{y_i}{\hat{\mu}_i}\right) \right] \end{aligned} \tag{12.20}$$

since $\sum_{i=1}^{n} [y_i - \hat{\mu}_i] = 0$ as long as an intercept is included in the model; see Eq. (12.10). For $y = 0$, $y \ln(y) = 0$.

Expanding $y \ln\left(\frac{y}{\mu}\right)$ in a second-order Taylor series around $y = \mu$, $y \ln\left(\frac{y}{\mu}\right) \cong (y - \mu) + \frac{1}{2\mu}(y - \mu)^2$, and substituting this result into Eq. (12.20) leads to

$$D \cong 2\sum_{i=1}^{n}\left[(y_i - \hat{\mu}_i) + \frac{1}{2\hat{\mu}_i}(y_i - \hat{\mu}_i)^2 - (y_i - \hat{\mu}_i)\right] = \sum_{i=1}^{n}\frac{(y_i - \hat{\mu}_i)^2}{\hat{\mu}_i} = \chi^2 \tag{12.21}$$

This is known as the Pearson chi-square statistic, which is also part of the standard output of Poisson regression programs.

The deviance and the Pearson chi-square statistics are useful measures of (lack of) fit. One often standardizes the deviance and the Pearson chi-square statistic by dividing them by their degrees of freedom, $n - p - 1$. If the standardized goodness-of-fit statistics are much larger than 1, one starts to question the adequacy of the model.

Residuals can be defined from the expressions in Eqs. (12.20) and (12.21). **Deviance residuals** are defined as

$$d_i = \pm\sqrt{2\left[y_i \ln\left(\frac{y_i}{\hat{\mu}_i}\right) - (y_i - \hat{\mu}_i)\right]} \tag{12.22}$$

where the sign is positive if $y_i - \hat{\mu}_i > 0$ and negative if $y_i - \hat{\mu}_i < 0$. **Pearson residuals** are given by

$$r_i = \frac{y_i - \hat{\mu}_i}{\sqrt{\hat{\mu}_i}} \tag{12.23}$$

12.4 OVERDISPERSION

Large standardized deviances (and large standardized Pearson chi-square statistics) indicate that the fitted model is not adequate. One possible reason for lack of fit is an incorrect functional specification of the mean vector. However, deviances of richly parameterized models may still be large. In this case, overdispersion may be at work.

The Poisson model restricts the mean and the variance to be the same. In some cases, the variability is larger than what is allowed by the Poisson distribution. "Extra" Poisson variation may be present because of unmeasured effects, certain clustering of events, or other contaminating influences that combine to produce more variation than what is predicted by the Poisson model. For small count data, overdispersion is usually not much of an issue. For large count data, it may be a problem. One approach to working with large count data is to transform the data; in many cases, the square root transformation stabilizes the variance and also achieves normality. The transformed data are then modeled by standard regressions with normal errors and a variance parameter that is not tied to the mean.

Overdispersion in the Poisson model implies a true variance $V(y_i) = \phi\mu_i$, where $\phi > 1$ is the overdispersion parameter. Overdispersion affects the variance of the estimates in Eq. (12.17). We saw that the estimates are the weighted least squares estimates in the regression of $z_i = \boldsymbol{x}_i'\tilde{\boldsymbol{\beta}} + \frac{1}{\tilde{\mu}_i}(y_i - \tilde{\mu}_i)$ on $\boldsymbol{x}_i$, given in Eq. (12.15). With overdispersion $V(y_i) = \phi\mu_i$,

$$V(z_i) = V\left[\boldsymbol{x}_i'\boldsymbol{\beta} + \frac{1}{\mu_i}(y_i - \mu_i)\right] = \frac{1}{\mu_i^2}V(y_i) = \frac{1}{\mu_i^2}\phi\mu_i = \frac{1}{\mu_i}\phi = \frac{1}{w_i}\phi$$

where $w_i = \mu_i$ are the weights in the regression. Hence,

$$V(\hat{\boldsymbol{\beta}}^{\text{WLS}}) = V\left\{\left[\sum_{i=1}^{n} w_i \boldsymbol{x}_i \boldsymbol{x}_i'\right]^{-1}\left[\sum_{i=1}^{n} w_i \boldsymbol{x}_i z_i\right]\right\}$$

$$= \left[\sum_{i=1}^{n} w_i \boldsymbol{x}_i \boldsymbol{x}_i'\right]^{-1}\left[\sum_{i=1}^{n} w_i^2 \boldsymbol{x}_i \boldsymbol{x}_i' \frac{1}{w_i}\phi\right]\left[\sum_{i=1}^{n} w_i \boldsymbol{x}_i \boldsymbol{x}_i'\right]^{-1} = \phi\left[\sum_{i=1}^{n} w_i \boldsymbol{x}_i \boldsymbol{x}_i'\right]^{-1} \tag{12.24}$$

The true variance of the estimates is a multiple $\phi > 1$ of the variance that ignores the overdispersion. Consequently, the standard errors that ignore overdispersion are too small, t ratios are too large, probability values are too small, and estimates appear "too" significant. Ignoring the overdispersion may lead the investigator to find "spurious" regression relationships.

The solution to this problem is rather simple and involves a correction to the variance of the estimates. The standardized deviance (or equivalently, the standardized Pearson chi-square statistic) provides an estimate of ϕ, and $\hat{\phi} = D/(n - p - 1)$ can be substituted into Eq. (12.24). Also, the likelihood ratio tests need to be adjusted for the overdispersion; the expressions in Eq. (12.16) need to be divided by $\hat{\phi}$.

A Comment on the Analysis of Correlated Data

The analysis of correlated data arising from time series measurements when the measurements follow the standard regression model with normal errors has been studied in Chapter 10. However, the normal assumption is not reasonable when the responses are discrete, as in the logistic or Poisson regression models. A different methodology must be used in the data analysis when the responses are discrete and correlated. **Generalized estimating equations** (GEEs) provide a practical method with reasonable statistical efficiency to analyze such data, and this approach has been implemented in the SAS procedure GENMOD. We do not discuss this here but refer the interested reader to the book by Diggle *et al.* (2002).

12.5 Examples

We illustrate the Poisson regression analysis on two data sets. The first one deals with the mating success of African elephants. The second example concerns the number of organisms and its relationship to a certain chemical in the environment.

The SAS PROC GENMOD routine is used for the calculations. This is a very general routine that allows the estimation of a wide variety of generalized linear models, such as logistic and Poisson regressions. Alternatively, one can use routines in other packages, such as S-Plus.

12.5.1 EXAMPLE 1: MATING SUCCESS OF AFRICAN ELEPHANTS

The data are taken from Poole (1989) and are given in Table 12.1. This data set is also analyzed in Ramsey and Schafer (2002). The paper reports on a study of 41 male African elephants over a period of 8 years. The age of the elephant at the beginning of the study and the number of successful matings during the 8 years were recorded. The objective was to learn whether older animals are more successful at mating or whether they have diminished success after reaching a certain age.

A graph of the numbers of successful matings y against the ages x of the 41 male elephants is shown in Figure 12.1. We notice an exponential-like increase in matings with age, which is expected with the link $\mu = \exp(\beta_0 + \beta_1 x)$. The fitted values $\hat{\mu} = \exp(\hat{\beta}_0 + \hat{\beta}_1 x)$, where the estimates are taken from the Poisson regression model that is discussed next, are shown as the solid line in Figure 12.1. The fit from the model with $\mu = \exp(\beta_0 + \beta_1 x + \beta_2 x^2)$ is shown as the dashed line.

How do we get the estimates, how can we decide whether the quadratic component of age is needed, and how do we judge the adequacy of the model? Here is where the theory discussed in this chapter comes into play, and where we use statistical software such as the SAS PROC GENMOD routine. The slightly edited output from fitting the Poisson regression with $\mu = \exp(\beta_0 + \beta_1 x)$ is shown in Table 12.2. It lists the estimates calculated by the iteratively reweighted least squares algorithm, the covariance matrix of the estimates in Eq. (12.17), and the correlation matrix. The square roots of the diagonal elements of the covariance matrix are the standard errors of the estimates, s.e.$(\hat{\beta}_j)$, and they are listed immediately to the right of the estimates. Approximate 95% confidence intervals are obtained from $\hat{\beta}_j \pm 1.96$ s.e.$(\hat{\beta}_j)$; SAS calls these the Wald confidence intervals. For the coefficient on age β_1, the 95% confidence interval is given by 0.0687 ± (1.96)(0.0137); it extends from 0.0418 to 0.0956. The likelihood ratio statistic for testing $\beta_1 = 0$ (the coefficient on age) is given next to the confidence interval. The likelihood ratio test statistic is twice the difference of the log-likelihood of the full model (here, the model with intercept and age) and the log-likelihood of the reduced model (the model with just the intercept); it is given by 24.97. The p value is the probability that a chi-square random variable with one degree of freedom exceeds 24.97, and it is less than 0.0001. Since it is small—smaller than

TABLE 12.1 AGE AND MATING SUCCESS OF AFRICAN ELEPHANTS[a]

Age of Elephant	Number of Matings
27	0
28	1
28	1
28	1
28	3
29	0
29	0
29	0
29	2
29	2
29	2
30	1
32	2
33	4
33	3
33	3
33	3
33	2
34	1
34	1
34	2
34	3
36	5
36	6
37	1
37	1
37	6
38	2
39	1
41	3
42	4
43	0
43	2
43	3
43	4
43	9
44	3
45	5
47	7
48	2
52	9

[a] The data are given in the file **elephants**.

the usual significance level 0.05—we reject the hypothesis that $\beta_1 = 0$. Age is indeed an important predictor of mating success. The estimate $\hat{\beta}_1 = 0.0687$ with $\exp(\hat{\beta}_1) - 1 = \exp(0.0687) - 1 = 0.071$ implies that each additional year of age increases the number of matings by 7.1%, at least over the considered age range.

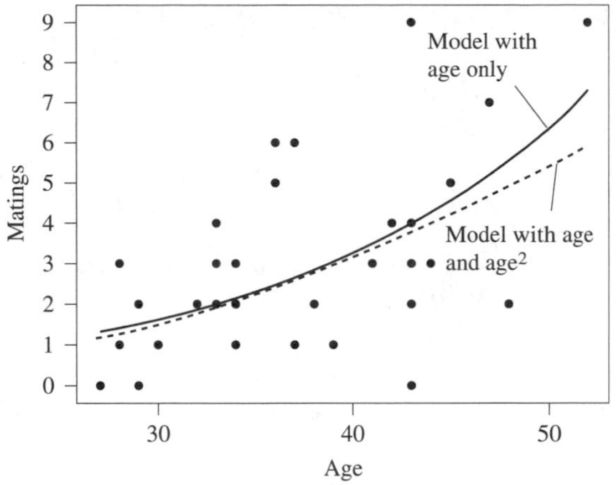

FIGURE 12.1 Scatter Plot of the Number of Successful Matings y against Age x

Also note the large correlation among the estimates of the intercept and the coefficient of age. The magnitude of this correlation is to a large extent a consequence of the data design.

The deviance and the Pearson chi-square statistic are calculated from Eqs. (12.20) and (12.21); they are $D = 51.0116$ and $\chi^2 = 45.136$. These statistics use the observations and the fitted means $\hat{\mu}_i = \exp(\hat{\beta}_0 + \hat{\beta}_1 x_i)$. Their degrees of freedom are given by $n - p - 1 = 41 - 2 = 39$. The 95th percentile of a chi-square distribution with 39 degrees of freedom is 54.57. The model is adequate because the deviance (as well as the Pearson statistic) is smaller than this percentile. Division of the goodness-of-fit statistics by their degrees of freedom leads to the standardized values, also shown in Table 12.2. They are not much larger than 1, indicating that there is little evidence of overdispersion. Hence, we see little need to adjust the standard errors and the likelihood ratio tests.

The estimates can be used to obtain the mean number of matings of an elephant with a certain age, for example, $x = 40$ years. The estimate is given by $\hat{\mu}_{x=40} = \exp[-1.5820 + (0.0687)(40)] = 3.21$. Its standard error from Eq. (12.18) can be calculated as follows. With the covariance matrix of the parameter estimates in Table 12.2, we find

$$[1 \quad 40]V(\hat{\beta})\begin{bmatrix} 1 \\ 40 \end{bmatrix} = [1 \quad 40]\begin{bmatrix} 0.29661 & -0.007371 \\ -0.007371 & 0.0001889 \end{bmatrix}\begin{bmatrix} 1 \\ 40 \end{bmatrix} = 0.00917$$

Hence,

$$\text{s.e.}(\hat{\mu}_{x=40}) \cong \hat{\mu}_{x=40}\sqrt{0.00917} = (3.21)(0.096) = 0.31$$

Deviance and Pearson residuals can also be calculated, and they should be graphed against the explanatory variables (here, just age) and the fitted values $\hat{\mu}$. One should see no patterns in these graphs; furthermore, the variance should be roughly constant. The residual plots in Figure 12.2 show no major inadequacies.

Another issue that needs to be settled is whether the model can be improved by fitting a more elaborate model that involves the square of age.

TABLE 12.2 OUTPUT FOR THE POISSON REGRESSION MODEL WITH $\mu = \exp(\beta_0 + \beta_1 x)$

```
                 The GENMOD Procedure
                  Model Information

          Data Set                      WORK.ELEPHANTS
          Distribution                         Poisson
          Link Function                            Log
          Dependent Variable                    mating
          Observations Used                         41

                Parameter Information

                Parameter     Effect

                Prm1          Intercept
                Prm2          age

         Criteria For Assessing Goodness Of Fit

   Criterion                DF        Value     Value/DF

   Deviance                 39      51.0116       1.3080
   Scaled Deviance          39      51.0116       1.3080
   Pearson Chi-Square       39      45.1360       1.1573
   Scaled Pearson X2        39      45.1360       1.1573
   Log Likelihood                   10.7400

             Estimated Covariance Matrix

                        Prm1             Prm2

           Prm1      0.29661        -0.007371
           Prm2    -0.007371        0.0001889

             Estimated Correlation Matrix

                        Prm1            Prm2

             Prm1     1.0000         -0.9846
             Prm2    -0.9846          1.0000

                 The GENMOD Procedure
             Analysis Of Parameter Estimates

                                 Standard    Wald 95% Confidence      Chi-
Parameter    DF    Estimate       Error           Limits            Square   Pr > ChiSq

Intercept     1     -1.5820      0.5446     -2.6494     -0.5146       8.44       0.0037
age           1      0.0687      0.0137      0.0418      0.0956      24.97       <.0001
Scale         0      1.0000      0.0000      1.0000      1.0000

NOTE: The scale parameter was held fixed.
```

The slightly edited output from fitting the Poisson regression with $\mu = \exp(\beta_0 + \beta_1 x + \beta_2 x^2)$ is shown in Table 12.3. The log-likelihood of the full model (the model with intercept, age, and age^2) is $\ln L(\text{full}) = 10.8327$ from Table 12.3. The log-likelihood of the reduced model (the model with just age) is $\ln L(\text{reduced}) = 10.7400$ from Table 12.2. The likelihood ratio test statistic for testing $\beta_2 = 0$ is

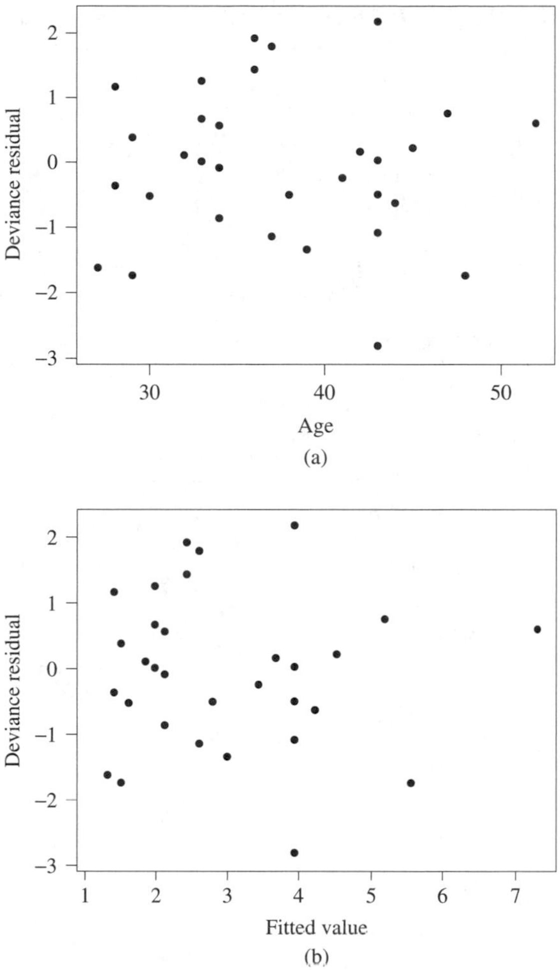

FIGURE 12.2 Residual Plots for the Poisson Regression Model with $\mu = \exp(\beta_0 + \beta_1 x)$

2(10.8327 – 10.7400) = 0.185. The associated probability value, 0.6693 in Table 12.3 in the line corresponding to age2 = age^2, is the probability that a chi-square random variable with 1 degree of freedom exceeds this statistic. The probability value is much larger than the usual significance level of 0.05 (note that the LR test statistic is much smaller than the 95th percentile of the chi-square with one degree of freedom). Hence, we cannot reject the null hypothesis that $\beta_2 = 0$. We conclude that $\beta_2 = 0$, which indicates that the square of age is not needed. The same conclusion is reached from the 95% confidence interval of β_2. The interval from −0.0048 to 0.0031 in Table 12.3 includes zero, indicating that age^2 is not significant. The confidence interval uses the standard error s.e.$(\hat{\beta}_2) = 0.0020 = \sqrt{0.0000040497}$, the square root of the corresponding diagonal element in the covariance matrix. The two tests, the likelihood ratio test and the test using the Wald confidence intervals, are equivalent.

TABLE 12.3 OUTPUT FOR THE POISSON REGRESSION MODEL WITH $\mu = \exp(\beta_0 + \beta_1 x + \beta_2 x^2)$

The GENMOD Procedure

Model Information

Data Set	WORK.ELEPHANTS
Distribution	Poisson
Link Function	Log
Dependent Variable	mating
Observations Used	41

Parameter Information

Parameter	Effect
Prm1	Intercept
Prm2	age
Prm3	age2

Criteria For Assessing Goodness Of Fit

Criterion	DF	Value	Value/DF
Deviance	38	50.8262	1.3375
Scaled Deviance	38	50.8262	1.3375
Pearson Chi-Square	38	44.6574	1.1752
Scaled Pearson X2	38	44.6574	1.1752
Log Likelihood		10.8327	

Estimated Covariance Matrix

	Prm1	Prm2	Prm3
Prm1	9.21510	−0.47737	0.006006
Prm2	−0.47737	0.02497	−0.000317
Prm3	0.006006	−0.000317	4.0497E-6

Estimated Correlation Matrix

	Prm1	Prm2	Prm3
Prm1	1.0000	−0.9952	0.9831
Prm2	−0.9952	1.0000	−0.9960
Prm3	0.9831	−0.9960	1.0000

The GENMOD Procedure

Analysis Of Parameter Estimates

Parameter	DF	Estimate	Standard Error	Wald 95% Confidence Limits		Chi-Square	Pr > ChiSq
Intercept	1	−2.8574	3.0356	−8.8071	3.0923	0.89	0.3466
age	1	0.1360	0.1580	−0.1737	0.4456	0.74	0.3896
age2	1	−0.0009	0.0020	−0.0048	0.0031	0.18	0.6693
Scale	0	1.0000	0.0000	1.0000	1.0000		

NOTE: The scale parameter was held fixed.

12.5.2 EXAMPLE 2: REPRODUCTION OF CERIODAPHNIA ORGANISMS

The second data set is taken from Myers, *et al.* (2002). The data are listed in Table 12.4. This example deals with the number of Ceriodaphnia organisms that are counted in a controlled environment in which reproduction occurs among

TABLE 12.4 REPRODUCTION OF CERIODAPHNIA ORGANISMS[a]

Ceriodaphnia	Concentration	Strain
82	0	1
106	0	1
63	0	1
99	0	1
101	0	1
45	0.50	1
34	0.50	1
26	0.50	1
44	0.50	1
42	0.50	1
31	0.75	1
22	0.75	1
16	0.75	1
30	0.75	1
29	0.75	1
22	1	1
14	1	1
10	1	1
21	1	1
20	1	1
15	1.25	1
8	1.25	1
6	1.25	1
14	1.25	1
13	1.25	1
10	1.50	1
8	1.50	1
11	1.50	1
10	1.50	1
10	1.50	1
8	1.75	1
8	1.75	1
3	1.75	1
8	1.75	1
1	1.75	1
58	0	2
58	0	2
62	0	2
58	0	2
73	0	2

(*Continued*)

TABLE 12.4 (Continued)

Ceriodaphnia	Concentration	Strain
27	0.50	2
28	0.50	2
31	0.50	2
28	0.50	2
38	0.50	2
19	0.75	2
20	0.75	2
22	0.75	2
20	0.75	2
28	0.75	2
14	1	2
14	1	2
15	1	2
14	1	2
21	1	2
9	1.25	2
10	1.25	2
12	1.25	2
10	1.25	2
16	1.25	2
7	1.50	2
3	1.50	2
1	1.50	2
8	1.50	2
7	1.50	2
4	1.75	2
3	1.75	2
2	1.75	2
8	1.75	2
4	1.75	2

[a] The data are given in the file **ceriodaphnia**.

the organisms. Two different strains of organisms are involved, and the environment is changed by adding varying amounts of a chemical component intended to impair reproduction.

Scatter plots of counts against concentration, for each of the two strains of organisms, are shown in Figure 12.3. The counts decrease with concentration, and the decrease is roughly exponential. Looking at the two graphs, one notices differences among the two strains.

The scatter plots suggest a Poisson regression with link $\mu = \exp(\beta_0 + \beta_1 x_1 + \beta_2 x_2) = \exp(\beta_0 + \beta_1 \text{Conc} + \beta_2 \text{Strain})$, where Strain is an indicator that is 0 for strain 1 and 1 for strain 2. The output from the Poisson regression is shown in Table 12.5. The estimates, the standard errors, and 95% Wald confidence intervals are shown in the last part of this table. All estimates are significantly different from zero; the confidence intervals do not cover zero and the probability values of the

FIGURE 12.3 Scatter Plot of Counts *y* against Concentration, for Organism Strains 1 and 2

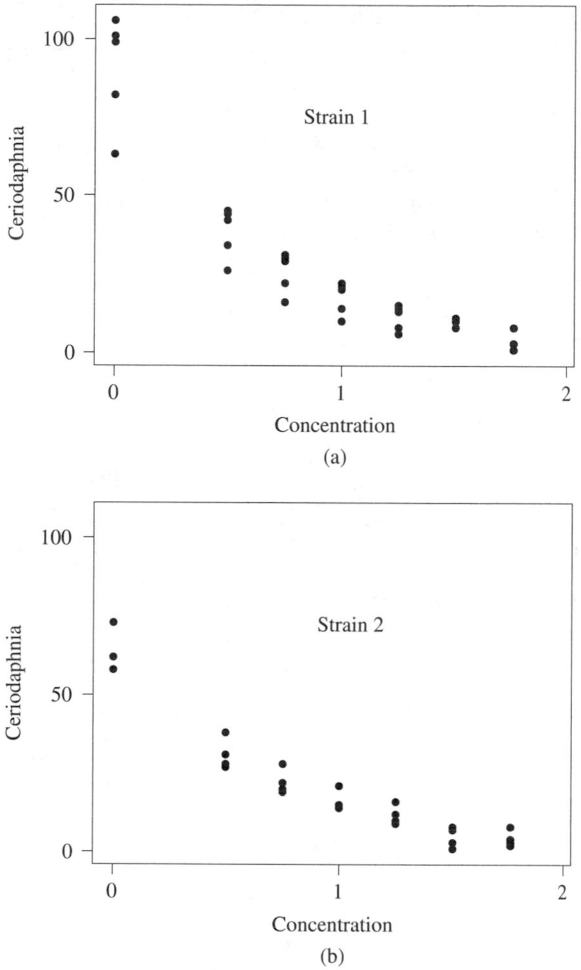

partial likelihood ratio tests are smaller than 0.05. Hence, we cannot simplify the model. The estimate $\hat{\beta}_1 = -1.54$ and $\exp(\hat{\beta}_1) - 1 = \exp(-1.54) - 1 = -0.786$ implies that for both strains each additional unit of concentration reduces the mean count by 78.6%. The estimate $\hat{\beta}_2 = -0.275$ and $\exp(\hat{\beta}_2) - 1 = \exp(-0.275) - 1 = -0.240$ implies that the mean count for strain 2 is 24% smaller than that for strain 1.

The deviance and the Pearson chi-square statistic in Table 12.5 (86.38 and 79.83, respectively) indicate that the fitted model is adequate; the statistics do not exceed the 95th percentile (87.11) of the chi-square distribution with $70 - 2 - 1 = 67$ degrees of freedom. The standardized deviance and Pearson chi-square statistic are not much larger than 1. Residual plots (deviance residuals against concentration, against strain, and against fitted values; not shown) indicate no unusual patterns.

TABLE 12.5 OUTPUT FOR THE POISSON REGRESSION MODEL WITH $\mu = \exp(\beta_0 + \beta_1 x_1 + \beta_2 x_2)$

The GENMOD Procedure

Model Information

Data Set	WORK.CERIODAPHNIA
Distribution	Poisson
Link Function	Log
Dependent Variable	count
Observations Used	70

Parameter Information

Parameter	Effect
Prm1	Intercept
Prm2	concent
Prm3	strain

Criteria For Assessing Goodness Of Fit

Criterion	DF	Value	Value/DF
Deviance	67	86.3765	1.2892
Scaled Deviance	67	86.3765	1.2892
Pearson Chi-Square	67	79.8301	1.1915
Scaled Pearson X2	67	79.8301	1.1915
Log Likelihood		4493.8023	

Estimated Covariance Matrix

	Prm1	Prm2	Prm3
Prm1	0.001532	−0.001064	−0.001010
Prm2	−0.001064	0.002172	1.525E-18
Prm3	−0.001010	1.525E-18	0.002340

Estimated Correlation Matrix

	Prm1	Prm2	Prm3
Prm1	1.0000	−0.5836	−0.5335
Prm2	−0.5836	1.0000	0.0000
Prm3	−0.5335	0.0000	1.0000

The GENMOD Procedure

Analysis Of Parameter Estimates

Parameter	DF	Estimate	Standard Error	Wald 95% Confidence Limits		Chi-Square	Pr > ChiSq
Intercept	1	4.4546	0.0391	4.3779	4.5313	12954.8	<.0001
concent	1	−1.5431	0.0466	−1.6344	−1.4517	1096.34	<.0001
strain	1	−0.2750	0.0484	−0.3689	−0.1802	32.31	<.0001
Scale	0	1.0000	0.0000	1.0000	1.0000		

NOTE: The scale parameter was held fixed.

Finally, we check whether it is necessary to include an interaction term in the model and we fit the Poisson regression model with $\mu = \exp(\beta_0 + \beta_1 x_1 + \beta_2 x_2 + \beta_3 x_1 x_2)$. The partial likelihood ratio statistic for testing $\beta_3 = 0$ is given by 1.78, and its probability value is 0.18 (output is not shown). The probability value is considerably larger than the significance level of 0.05 (equivalently, the 95% Wald confidence interval of β_3 covers zero). Hence, we do not reject $\beta_3 = 0$ and find no reason to include an interaction term in the model.

What if the standardized deviance (or the Pearson chi-square statistic) were large and it were not possible to decrease these statistics to acceptable levels by considering more elaborate specifications of the mean function? In such a case, one would conclude that overdispersion is present, and one would adjust the standard errors and the likelihood ratio tests. PROC GENMOD in SAS does this if one asks for this option. It uses either the standardized deviance or Pearson chi-square statistic as the estimate of ϕ in the adjustment in Eq. (12.24). One could use this adjustment in Example 2 because the deviance 86.38 is close to the 95th percentile of 87.11. Although this adjustment increases the standard errors slightly, our conclusions about the very strong effects of concentration and strain are unaffected.

EXERCISES

12.1. The data are taken from P. McCullagh and J. A. Nelder. *Generalized Linear Models*. New York: Chapman & Hall, 1983. The data are given in the file **damage**.

Consider the number of reported damage incidents (response y) and the aggregate months of service by ship type, year of construction, and period of operation. Data on five different types of ships (1–5), four different years of construction (1–4, with 1960–64 coded 1, 1965–1969 coded 2, 1970–1974 coded 3, and 1975–1979 coded 4), and two periods of operation (1 and 2, with 1960–1974 coded 1, and 1975–1979 coded 2) are available.

$X =$ Type of Ship	$Z =$ Year of Construction	$W =$ Period of Service	MS = Months of Service	$Y =$ Number of Damage Incidents
1	1	1	127	0
1	1	2	63	0
1	2	1	1,095	3
1	2	2	1,095	4
1	3	1	1,512	6

$X =$ Type of Ship	$Z =$ Year of Construction	$W =$ Period of Service	MS = Months of Service	$Y =$ Number of Damage Incidents
1	3	2	3,353	18
1	4	1	*	*
1	4	2	2,244	11
2	1	1	44,882	39
2	1	2	17,176	29
2	2	1	28,609	58
2	2	2	20,370	53
2	3	1	7,064	12
2	3	2	13,099	44
2	4	1	*	*
2	4	2	7,117	18
3	1	1	1,179	1
3	1	2	552	1
3	2	1	781	0
3	2	2	676	1
3	3	1	783	6
3	3	2	1,948	2
3	4	1	*	*
3	4	2	274	1
4	1	1	251	0
4	1	2	105	0

$X =$ Type of Ship	$Z =$ Year of Construction	$W =$ Period of Service	MS = Months of Service	$Y =$ Number of Damage Incidents
4	2	1	288	0
4	2	2	192	0
4	3	1	349	2
4	3	2	1,208	11
4	4	1	*	*
4	4	2	2,051	4
5	1	1	45	0
5	1	2	*	*
5	2	1	789	7
5	2	2	437	7
5	3	1	1,157	5
5	3	2	2,161	12
5	4	1	*	*
5	4	2	542	1

a. The first three variables are categorical. A program such as SAS GENMOD allows you to enter categorical variables into the model and it creates the associated indicator variables automatically. The program creates indicator variables $X1$–$X5$ for the type of ship, $Z1$–$Z4$ for the year of construction, and $W1$ and $W2$ for the period of operation. Consider the Poisson regression model with link

$$\ln \mu = \beta_0 + \beta_1 \ln(\mathrm{MS}) + \beta_2 X2 + \cdots + \beta_5 X5 + \beta_6 Z2 + \cdots + \beta_8 Z4 + \beta_9 W2$$

and estimate this model using an available software program such as SAS GENMOD.

b. It seems reasonable to suppose that the number of damage incidents is directly proportional to MS, the months of service, and one can expect that the coefficient β_1 is 1. Test whether $\beta_1 = 1$. The literature refers to the term ln(MS) as an "offset."

c. Set $\beta_1 = 1$ and consider the model with link

$$\ln \mu = \beta_0 + \ln(\mathrm{MS}) + \beta_2 X2 + \cdots + \beta_5 X5 + \beta_6 Z2 + \cdots + \beta_8 Z4 + \beta_9 W2$$

SAS GENMOD allows you to fit models with an offset. Interpret the estimates of the regression coefficients $\beta_2, \ldots, \beta_9$.

d. Different ships are involved in this study, and one can expect intership variability in the proneness for accidents. Check for overdispersion. Adjust for overdispersion if necessary.

e. The model in (c) treats the factor effects as additive. Explore whether interaction components are needed.

f. Summarize your findings.

12.2. Aitkin *et al.* (1989) model insurance claims that are classified according to two factors: age group A (two levels) and car type T (three levels). The number of insurance claims (Y) and the number of insurance policyholders (H) are listed for the six claim groups. The number of policy holders can be treated as an "offset" because one can expect that the number of claims should be strictly proportional to the number of policies [see Exercise 12.1(b) for discussion].

$H =$ Number of Policies	$Y =$ Number of Claims	$A =$ Age Group	$T =$ Car Type
500	42	1	Small
1,200	37	1	Medium
100	1	1	Large
400	101	2	Small
500	73	2	Medium
300	14	2	Large

Consider the Poisson regression model with link

$$\ln \mu = \beta_0 + \ln(H) + \beta_1 A_2 + \beta_2 T_2 + \beta_3 T_3$$

where A_1, A_2 and T_1, T_2, T_3 are the corresponding indicator variables. Estimate and interpret the model.

12.3. The data are taken from Jorgenson, D. W. Multiple regression analysis of a Poisson process. *Journal of the American Statistical Association,* 56, 235–245, 1961. The data are given in the file **failures**.

The following table lists the number of failures for a complex piece of electronic equipment that is operating under cycles of operation (weeks) in which each cycle is divided into two operating regimes. The number of failures and the time spent in regime 1 and regime 2 during the cycle are listed.

$T1 =$ Time in Regime 1	$T2 =$ Time in Regime 2	$Y =$ Number of Failures
33.3	25.3	15
52.2	14.4	9
64.7	32.5	14
137.0	20.5	24
125.9	97.6	27
116.3	53.6	27
131.7	56.6	23
85.0	87.3	18
91.9	47.8	22

Under each regime of operation, the number of failures should be proportional to the time spent in this regime, but the constants of proportionality λ_1 and λ_2 may not be the same. The number of failures should follow a Poisson regression model with link

$$\ln \mu = \lambda_1 T_1 + \lambda_2 T_2$$

Obtain estimates of the parameters λ_1 and λ_2, and interpret the results.

12.4. The following data are from Hand *et al. A Handbook of Small Data Sets,* Chapman and Hall, 1994. They show the incidence of nonmelanoma skin cancer among women in Minneapolis–St. Paul (coded 0) and Dallas–Ft. Worth (coded 1). The data are given in the file **skincancer**.

Number of Cases	Town	Age Group	Population
1	0	15–24	172,675
16	0	25–34	123,065
30	0	35–44	96,216
71	0	45–54	92,051
102	0	55–64	72,159
130	0	65–74	54,722
133	0	75–84	32,185
40	0	85+	8,328
4	1	15–24	181,343
38	1	25–34	146,207
119	1	35–44	121,374
221	1	45–54	111,353
259	1	55–64	83,004
310	1	65–74	55,932
Missing	1	75–84	Missing
65	1	85+	7,583

a. Cancer incidence should be directly proportional to the size of the population (POP). Hence, it is reasonable to consider ln(POP) as an offset. Age is a categorical variable. Use indicator variables for the eight age groups ($X1$–$X8$) and consider the Poisson regression with link

$$\ln \mu = \beta_0 + \ln(\text{POP}) + \beta_2 X2 + \cdots + \beta_8 X8 + \beta_9 \text{Town}$$

Fit the model and interpret the regression coefficients $\beta_2, \ldots, \beta_8$ (measuring the age effect) and β_9 (measuring the effect of location). One would expect that sun exposure is greater in Texas than in Minnesota.

b. Estimate the more general model

$$\ln \mu = \beta_0 + \beta_1 \ln(\text{POP}) + \beta_2 X2 + \cdots + \beta_8 X8 + \beta_9 \text{Town}$$

and test whether $\beta_1 = 1$.

12.5. The data are from G. Rodríguez (Princeton University). The following table gives information on the number of lung cancer deaths by age and smoking status. The data are given in the file **lungcancer**. The four columns represent

Age: in 5-year age groups coded 1–9, for age brackets 40–44, 45–49, 50–54, 55–59, 60–64, 65–69, 70–74, 75–79, and 80+;

Smoking status: coded 1, does not smoke; 2, smokes cigars or pipe only; 3, smokes cigarettes and cigar/pipe; and 4, smokes cigarettes only;

Population: in hundreds of thousands; and Deaths: number of lung cancer deaths in a year.

Age	Smoking Status	Population (100,000)	Lung Cancer Deaths
1	1	656	18
2	1	359	22
3	1	249	19
4	1	632	55
5	1	1,067	117
6	1	897	170
7	1	668	179
8	1	361	120
9	1	274	120
1	2	145	2
2	2	104	4
3	2	98	3
4	2	372	38
5	2	846	113
6	2	949	173
7	2	824	212
8	2	667	243
9	2	537	253
1	3	4,531	149
2	3	3,030	169
3	3	2,267	193
4	3	4,682	576
5	3	6,052	1001
6	3	3,880	901
7	3	2,033	613
8	3	871	337
9	3	345	189
1	4	3,410	124
2	4	2,239	140
3	4	1,851	187
4	4	3,270	514
5	4	3,791	778
6	4	2,421	689
7	4	1,195	432
8	4	436	214
9	4	113	63

Age and smoking status are categorical variables. A program such as SAS GENMOD creates the associated indicator variables—$X1$–$X9$ for the nine age groups and $Z1$–$Z4$ for smoking status—automatically. Population size (POP) can be treated as an offset because one can expect that the incidence of lung cancer is directly proportional to population.

Consider the Poisson regression model with link

$$\ln \mu = \beta_0 + \beta_1 \ln(\text{POP}) + \beta_2 X2 + \cdots + \beta_9 X9 + \beta_{10} Z2 + \beta_{11} Z3 + \beta_{12} Z4$$

and estimate the model using an available software program such as GENMOD. Interpret the regression coefficients and summarize your findings. Check the model for overdispersion and explore whether interaction terms are needed in the model.

12.6. Hill *et al.* (2001, Section 10.4) discuss the results of a survey that assessed the number of household visits to Lake Keepit per year, the distance of the household from the lake, the household income, and the number of household members. The data are given in the file **lake**. Data for the first 5 of the 250 households are listed here.

DIST = Distance	INC = Family Income	SIZE = Family Members	Y = Number of Visits
27	4.45	5	1
72	7.69	4	1
44	10.04	4	5
23	8.97	4	4
89	9.15	5	2

Consider the Poisson regression with link

$$\ln \mu = \beta_0 + \beta_1 \text{DIST} + \beta_2 \text{INC} + \beta_3 \text{SIZE}$$

Fit the model and interpret your findings.

Brief Answers to Selected Exercises

CHAPTER 1

1.1 Data from experiments are usually more informative because one can control the conditions under which the experimental runs are carried out. Experimentation is probably not possible in case (f). The relative humidity conditions in the plant cannot be varied according to a fixed experimental plan. Instead, one takes measurements in the plant on the relative humidity and at the same time on the output (performance) of the process. A danger with such data is that the relative humidity in the plant may be affected by factors that also affect the output.

1.2 A graph of ln(Payout) against the product of interest rate and maturity indicates a linear relationship. This is expected from the model Payout $= Pe^{RT}$. Taking the logarithm on both sides of the equation leads to ln(Payout) $=$ $\ln(P) + RT$. The intercept changes with the logarithm of the invested principle; the regression coefficient of RT is one.

1.3 Selected examples:

Exercise 2.9: MBA grade point average and GMAT score: observational study

Exercise 4.14: Survival of bull semen: experimental data

Exercise 8.2: Height and weight of boys: observational study

1.6 Usually, it is not easy to spot relationships from three-dimensional graphs. The bivariate scatter plots for the silkworm data are easier to interpret.

1.7 The polynomial model (with $p > 1$)

$$y = \beta_0 + \beta_1 x + \beta_2 x^2 + \cdots + \beta_p x^p + \varepsilon$$

is nonlinear in the explanatory variable x but linear in the parameters.

1.8 The model $y = \alpha + (0.49 - \alpha)\exp[-\beta(x - 8)] + \varepsilon$ is nonlinear in the parameters. Trace out the mean response for changing levels of x. Take $\alpha = 0.39$ and $\beta = 0.10$, and consider x values between 8 and 40. This particular model is studied in Chapter 9; x is the age of a chemical product in weeks, and the response y is its remaining chlorine.

1.9 Think in terms of percentages. The variability (expressed as standard deviation) may be $\pm 10\%$. If sales are at level 10, this implies an uncertainty of ± 1 units. For level 1000, the uncertainty is ± 100 units. If the standard deviation is proportional to the level, one should analyze the logarithm of sales; see chapter 6.

1.10 Poverty of a school district affects the number of students in subsidized lunch programs, with poorer districts having more children in these nationally subsidized programs. Poverty also affects the scholastic test scores in these districts. The strong positive correlation between the number of children in subsidized lunch programs and test achievement scores in these districts does not imply that there is a causal connection between subsidized lunch and test scores. It is poverty that is the driving causal factor.

1.12 a. Absolute raise (raise in terms of dollars earned) can be written as

$$\text{AbsoluteRaise} = (R)(\text{PreviousSalary}) = (\beta\text{PreviousSalary})\text{Performance}$$

A graph of AbsoluteRaise against Performance does not exhibit a perfect linear association because the slope depends on the previous salary that changes from person to person. A regression of AbsoluteRaise on Performance may not provide the correct estimate of the parameter β. Take two workers; the previous salary of the first worker is half the salary of the second one, but the first worker is twice as productive. Their absolute raises are the same. The slope in the plot of AbsoluteRaise against Performance is zero and not the desired parameter β.

b. Let $R =$ RelativeRaise, where R is a small number such as 0.03 (3%). The ratio CurrentSalary/PreviousSalary = $[(1+R)\text{PreviousSalary}]/\text{PreviousSalary} = 1+R$. A first-order Taylor series expansion of $\ln(1+R) \approx R$ is valid for small R. Hence, $\ln(\text{CurrentSalary}/\text{PreviousSalary}) = \ln(1+R) \approx R = \beta\text{Performance}$ is linearly related to Performance. A regression of ln(CurrentSalary/PreviousSalary) on Performance provides an estimate of β.

CHAPTER 2

2.1 a. 95th percentile = 14.93; 99th percentile = 16.98

b. $t(0.95;10) = 1.812$; $t(0.95;25) = 1.708$; $t(0.99;10) = 2.764$; $t(0.99;25) = 2.485$

c. $\chi^2(0.95;1) = 3.841$; $\chi^2(0.95;4) = 9.488$; $\chi^2(0.95;10) = 18.307$
$\chi^2(0.99;1) = 6.635$; $\chi^2(0.99;4) = 13.277$; $\chi^2(0.99;10) = 23.209$

d. $F(0.95;2,10) = 4.10$; $F(0.95;4,10) = 3.48$; $F(0.99;2,10) = 7.56$; $F(0.99;4,10) = 5.99$

2.3 Correlation = 0.816; $R^2 = 0.667$; Estimated equation: $\hat{\mu} = 3 + 0.5x$

Same (linear regression) results for all four data sets. However, scatter plots show that linear regression is only appropriate for the first data set.

2.6 c. Estimated equation: $\hat{\mu} = 49.268 + 0.478x$; $R^2 = 0.998$; $s = \sqrt{\text{MSE}} = 0.402$. Model is appropriate.

d. (i) $\hat{\beta}_1 = 0.478$; s.e.$(\hat{\beta}_1) = 0.0040$; 95% confidence interval: (0.470 , 0.486)

(ii) $\hat{\mu}(x = 100\ln(25)) = 203.19$; 95% confidence interval: (202.95, 203.44)

e. Estimates and standard errors of β_0 and β_1 change by a factor of 5/9.

2.8 a. Estimated equation: $\hat{\mu} = 31.9114 + 0.2625x$; t ratio$(\hat{\beta}_1) = 0.2625/0.0393 = 6.68$; p value= 0.0002; number of cars sold is a significant predictor variable.

b. 95% confidence interval for β_1: (0.172, 0.353)

c. 90% confidence interval for β_1: (0.189, 0.336)

d. $R^2 = 0.848$

e. $s = \sqrt{\text{MSE}} = 264.0$, compared with $s_y = 638.4$

f. $\hat{\mu}(x = 1187) = 343.5$

2.10 a. Approximate 95% prediction interval: (2.104, 3.111)

b. Approximate 95% prediction interval: (1.592, 2.599)

2.11 $R^2 = \left[1 + \dfrac{n-p-1}{pF}\right]^{-1}$

2.13 a. Estimated equation: $\hat{\mu} = 0.520x$; $s^2 = 46.2/16 = 2.89$; $\hat{\beta}_1 = 0.520$; s.e.$(\hat{\beta}_1) = 0.0132$; 95% confidence interval: (0.492, 0.548)

b. Estimated equation: $\hat{\mu} = 0.725 + 0.498x$; $\hat{\beta}_0 = 0.725$; s.e.$(\hat{\beta}_0) = 1.549$; $\hat{\beta}_0/\text{s.e.}(\hat{\beta}_0) = 0.725/1.549 = 0.47$; p value = 0.65; conclude $\beta_0 = 0$.

2.15 a. Estimated equation: $\hat{\mu} = 68.45 - 0.41x$; $R^2 = 0.677$; $s = 4.563$; F statistic = 10.47; p value = 0.023; reject $\beta_1 = 0$.

b. s.e.$(\hat{\beta}_0) = 12.93$; $\hat{\beta}_0/\text{s.e.}(\hat{\beta}_0) = 68.45/12.93 = 5.29$; p value = 0.003; s.e.$(\hat{\beta}_1) = 0.127$; $\hat{\beta}_1/\text{s.e.}(\hat{\beta}_1) = -0.41/0.127 = -3.23$;

p value $= 0.023$; reject $\beta_1 = 0$.
99% confidence interval for β_1: $(-0.92, 0.11)$

c. $\hat{\mu}(x = 100) = 27.41$; s.e.$(\hat{\mu}(x = 100)) = 1.73$; 95% confidence interval: (22.97, 31.86)

d. $\hat{\mu}(x = 84) = 33.98$; s.e.$(\hat{\mu}(x = 84)) = 2.76$; 95% confidence interval: (26.88, 41.07)

2.17 b. Estimated equation: $\hat{\mu} = -46.44 + 1.792x$;
95% confidence interval for β_0: $(-77.37, -15.50)$
95% confidence interval for β_1: (1.48, 2.11)

c. Good fit; $R^2 = 0.956$

d. $\hat{\mu}(x = 100) = 132.8$; 95% prediction interval: (122.39,143.22)

e. Strong linear relationship

2.18 b. Estimated equation: $\hat{\mu} = 33.31 + 2.168x$

c.

```
Analysis of Variance
Source          DF      SS      MS      F       P
Regression       1 4361.5 4361.5 14.58 0.002
Residual Error 13 3889.4  299.2
Total           14 8250.9
```

d. $F = 14.58$; p value $= 0.002$; reject $\beta_1 = 0$.

e. s.e.$(\hat{\beta}_1) = 0.568$; $\hat{\beta}_1/$s.e.$(\hat{\beta}_1) = 2.168/0.568 = 3.82$; same p value $= 0.002$; reject $\beta_1 = 0$.

f. Individual with $x = 63$ and $y = 220$ unusual. Estimates and standard errors change; R^2 increases.

2.19 a. Estimated equation: $\hat{\mu} = 3.707 - 0.0123x$; $R^2 = 0.950$

b. F statistic $= 75.41$; p value $= 0.001$; reject $\beta_1 = 0$ at the 0.01 significance level.

c. Response is average of three observations. Use of individual values would improve the sensitivity of the analysis.

d. No; molecular weight 200 is far outside the region of experimentation; one does not know whether the linear relationship will continue to hold.

2.21 Estimated equation: $\hat{\mu} = 8.957 + 0.587x$; $R^2 = 0.537$; $F = 59.21$; reject $\beta_1 = 0$.
Significant linear relationship between the methods. However, variability is large and predictive power low.

2.22 Plot of y (memory retention) against x (time) shows a nonlinear (exponentially decaying) pattern. Graphs of $\ln(y)$ against x and $\ln(y)$ against $\ln(x)$ show similar patterns. Plot of y against $\ln(x)$ shows a linear pattern.
Estimated equation: $\hat{\mu} = 0.846 - 0.079\ln(x)$; $R^2 = 0.990$; good model.

2.27

$$\begin{aligned} y = \text{Takeup(kg)}: \hat{\mu} &= -9.896 + 0.0753x;\ R^2 = 0.986; \\ F &= 1530.3;\ \text{reject } \beta_1 = 0. \\ y = \text{Takeup(\%)}: \hat{\mu} &= 4.737 + 0.00162x;\ R^2 = 0.703; \\ F &= 52.07;\ \text{reject } \beta_1 = 0. \end{aligned}$$

CHAPTER 3

3.1 a. $A' = \begin{bmatrix} 2 & 3 & 2 \\ 0 & 2 & 1 \\ 1 & 2 & 4 \end{bmatrix}$ b. $A'A = \begin{bmatrix} 17 & 8 & 16 \\ 8 & 5 & 8 \\ 16 & 8 & 21 \end{bmatrix}$

c. $\text{tr}(A) = 8$; $\text{tr}(A'A) = 43$

d. $\det(A) = 11$; $\det(A'A) = 121$

3.2 a.

$$X'X = \begin{bmatrix} 4 & 0 & 0 \\ 0 & 4 & 0 \\ 0 & 0 & 4 \end{bmatrix};\quad (X'X)^{-1} = \begin{bmatrix} 1/4 & 0 & 0 \\ 0 & 1/4 & 0 \\ 0 & 0 & 1/4 \end{bmatrix};$$

$$X'\mathbf{y} = \begin{bmatrix} 19 \\ 1 \\ 5 \end{bmatrix};\quad (X'X)^{-1}X'\mathbf{y} = \begin{bmatrix} 4.75 \\ 0.25 \\ 1.25 \end{bmatrix}$$

3.5 a. $\det(A) = 10$; $A^{-1} = \begin{bmatrix} 0.4 & 0 & -0.2 \\ 0 & 0.5 & -0.5 \\ -0.2 & -0.5 & 1.1 \end{bmatrix}$

b. Eigenvalues: 5.8951, 2.3973, 0.7076. Eigenvectors are the columns in

$$P = \begin{bmatrix} -0.4317 & 0.8857 & 0.1706 \\ -0.7526 & -0.4579 & 0.4732 \\ -0.4973 & -0.0759 & -0.8643 \end{bmatrix}$$

c.

$$A = P\Lambda P' == \begin{bmatrix} -0.4317 & 0.8857 & 0.1706 \\ -0.7526 & -0.4579 & 0.4732 \\ -0.4973 & -0.0759 & -0.8643 \end{bmatrix}$$

$$\times \begin{bmatrix} 5.8951 & 0 & 0 \\ 0 & 2.3973 & 0 \\ 0 & 0 & 0.7076 \end{bmatrix} \begin{bmatrix} -0.4317 & 0.8857 & 0.1706 \\ -0.7526 & -0.4579 & 0.4732 \\ -0.4973 & -0.0759 & -0.8643 \end{bmatrix}'$$

d. Eigenvalues are positive. Correlation matrix

$$\begin{bmatrix} 1 & 0.289 & 0.408 \\ 0.289 & 1 & 0.707 \\ 0.408 & 0.707 & 1 \end{bmatrix}$$

3.7 a. $\det(A) = 0$

b. Eigenvalues: 9, 1, and 0. Eigenvectors are the columns in

$$P = \begin{bmatrix} 1/\sqrt{6} & 1/\sqrt{2} & -1/\sqrt{3} \\ 1/\sqrt{6} & -1/\sqrt{2} & -1/\sqrt{3} \\ 2/\sqrt{6} & 0 & 1/\sqrt{3} \end{bmatrix}$$

c. One eigenvalue is zero. The linear combination $-y_1 - y_2 + y_3$ has variance zero.

3.8 a. $AB = \begin{bmatrix} 16 & 17 \\ 18 & 13 \end{bmatrix}$; b. $BA = \begin{bmatrix} 7 & 17 & 10 \\ 8 & 10 & 8 \\ 14 & 12 & 12 \end{bmatrix}$

3.11 a. Bivariate normal, with mean vector $\begin{bmatrix} 2 \\ 6 \end{bmatrix}$ and covariance matrix $\begin{bmatrix} 1 & 0 \\ 0 & 2 \end{bmatrix}$.

b. Bivariate normal, with mean vector $\begin{bmatrix} 7/3 \\ 17/3 \end{bmatrix}$ and covariance matrix $\begin{bmatrix} 2/3 & 1/3 \\ 1/3 & 5/3 \end{bmatrix}$.

3.17 Quadratic form with $A = \begin{bmatrix} 1 & 0 & 0 \\ 0 & 0.5 & 0.5 \\ 0 & 0.5 & 0.5 \end{bmatrix}$.

Matrix A is idempotent, with determinant 0 and rank 2. Distribution of the (normalized) quadratic form $(y_1^2 + 0.5y_2^2 + 0.5y_3^2 + y_2y_3)/\sigma^2$ is chi-square with 2 degrees of freedom.

CHAPTER 4

4.1

$$X'X = \begin{bmatrix} 10 & 55 \\ 55 & 385 \end{bmatrix}; \quad (X'X)^{-1} = \begin{bmatrix} 0.4667 & -0.0667 \\ -0.0667 & 0.0121 \end{bmatrix};$$

$$V(\hat{\beta}) = \sigma^2 \begin{bmatrix} 0.4667 & -0.0667 \\ -0.0667 & 0.0121 \end{bmatrix}$$

$$V(\hat{\beta}_0) = (0.4667)\sigma^2; \quad V(\hat{\beta}_1) = (0.0121)\sigma^2$$

4.5 a. $V(\hat{\beta}_1) = 18$

b. $\text{Cov}(\hat{\beta}_1, \hat{\beta}_3) = 1.2$

c. $\text{Corr}(\hat{\beta}_1, \hat{\beta}_3) = 0.0943$

d. $V(\hat{\beta}_1 - \hat{\beta}_3) = 24.6$

4.7 a. $R^2 = 0.9324$

b. F statistic = 110.35; p value = 0.000; reject $\beta_1 = \beta_2 = \beta_3 = 0$.

c. 95% confidence interval for β_{taxes} : (0.074, 0.306); reject $\beta_{\text{taxes}} = 0$; cannot simplify model.

95% confidence interval for β_{baths}: (−16.83, 180.57); cannot reject $\beta_{\text{baths}} = 0$; can simplify model by dropping "baths."

4.10 a. Estimated equation: $\hat{\mu} = 3.453 + 0.496x_1 + 0.0092x_2$; $s^2 = 4.7403$; s.e.$(\hat{\beta}_0) = 2.431$, s.e.$(\hat{\beta}_1) = 0.00605$, s.e.$(\hat{\beta}_2) = 0.00097$

b. $\hat{\beta}_1/\text{s.e.}(\hat{\beta}_1) = 81.98$; p value = 0.000; reject $\beta_1 = 0$. $\hat{\beta}_2/\text{s.e.}(\hat{\beta}_2) = 9.48$; p value = 0.000; reject $\beta_2 = 0$.

4.12 Output (R software, using the function "lm"):

Coefficients	Estimate	Std. Error	t value	Pr(>|t|)
Intercept	39.437054	12.110986	3.256	0.00765
TEMP	0.084067	0.060469	1.390	0.19194
PROD	0.001876	0.000607	3.091	0.01027
DAYS	0.131704	0.289800	0.454	0.65833
PAYR	−0.215677	0.098810	−2.183	0.05162
HOUR	−0.014475	0.030052	−0.482	0.63949

$R^2 = 0.645$; adjusted $R^2 = 0.483$; $s = 3.213$; F statistic = 3.99; p value = 0.026

b. Using additional SS, $F = [(205.956 - 183.48)/2]/10.322 = 1.09$; p value = $P(F(2, 11) > 1.09) = 0.37$; cannot reject H_0: $\beta_1 = \beta_3 = \beta_5 = 0$

c. Prefer reduced model $\hat{\mu} = 46.02 + 0.00204\text{PROD} - 0.216\text{PAYR}$; $R^2 = 0.574$

d. PROD has smallest p value

e. Water usage as linear function of PROD and PAYR

4.14 c. Estimated equation: $\hat{\mu} = 39.482 + 1.0092x_1 - 1.873x_2 - 0.367x_3$

d. (i) (22.802, 25.653); (ii) (20.109, 28.346)

e. F statistic = 30.08; p value = 0.000; reject H_0: $\beta_1 = \beta_2 = \beta_3 = 0$

4.16 a. SST = 60; SSR = 57.9725; SSE = 2.0275

Source	DF	SS	MS	F	P
Regression	3	57.9725	19.3242	47.66	2.129815e-05
Residual	5	2.0275	0.4055		
Total	8	60.0000			

F statistic = 47.66; reject $\beta_1 = \beta_2 = \beta_3 = 0$.

b. Estimated equation: $\hat{\mu} = -1.16346 + 0.13527x_1 + 0.01995x_2 + 0.12195x_3$; $s^2 = 0.4055$; s.e.$(\hat{\beta}_0) = 1.974$; s.e.$(\hat{\beta}_1) = 0.45474$; s.e.$(\hat{\beta}_2) = 0.23784$; s.e.$(\hat{\beta}_3) = 0.01225$; $t(\hat{\beta}_1) = 0.295$; p value = 0.78; cannot reject $\beta_1 = 0$. $t(\hat{\beta}_2) = 0.084$; p value = 0.94; cannot reject $\beta_2 = 0$. $t(\hat{\beta}_3) = 9.955$; p value = 0.000; reject $\beta_3 = 0$.

4.20 a. $\hat{\beta}^{\text{WLS}} = \sum y_i / \sum x_i$; $V(\hat{\beta}^{\text{WLS}}) = \sigma^2 / \sum x_i$

b. $\hat{\beta}^{\text{WLS}} = 30/2 = 15$; $V(\hat{\beta}^{\text{WLS}}) = \sigma^2/150$

4.22 a. Linear model not appropriate.

b. Fitted equation: TensileStrength = −6.674 + 11.764 Hardwood − 0.635 $(\text{Hardwood})^2$

Source	DF	SS	MS	F	P
Regression	2	3104.2	1552.1	79.43	0.000
Residual Error	16	312.6	19.5		
Total	18	3416.9			

Model adequate; quadratic term needed; increases R^2 from 0.305 to 0.909

95% confidence interval: (38.14, 44.00)

95% prediction interval: (31.25, 50.88)

4.23 Quadratic model. Estimated equation: $\hat{\mu} = 82.385 - 38.310x + 4.703x^2$
Regression significant; adequate fit
Stars with ln(surface temperature) <4 appear different and should be investigated separately. Without these stars, a linear model is appropriate.

Predictor	Coef	SE Coef	T	P
Constant	82.385	9.581	8.60	0.000
x	−38.310	4.790	−8.00	0.000
x2	4.7025	0.5939	7.92	0.000

S = 0.3667   R-Sq = 60.6%   R-Sq(adj) = 58.8%

Analysis of Variance

Source	DF	SS	MS	F	P
Regression	2	9.0945	4.5472	33.82	0.000
Residual Error	44	5.9165	0.1345		
Total	46	15.0110			

CHAPTER 5

5.2 a. \$3000; b. \$900

5.4 $\text{VIF}_1 = 2.5$; $\text{VIF}_2 = 5$; $\text{VIF}_3 = 10$; evidence of multicollinearity

5.6 F statistic = 22.33; p value < 0.0001; reject H_0: $\mu_1 = \mu_2 = \mu_3$.

Source	DF	Sum of Squares	Mean Square	F Value	Pr > F
Model	2	31.112	15.556	22.33	<.0001
Error	42	29.261	0.697		
Corrected Total	44	60.372			

5.8 a. Expected difference in systolic blood pressure for females versus males who drink the same number of cups of coffee, excercise the same, and are of the same age

b. Represents variation due to measurement error and omitted factors

c. Association, but not causation

d. Represents interaction between gender and coffee consumption

5.9 a. $E(y_t) = \begin{cases} \beta_0 + \beta_1 t, & t = 1, 2, \ldots, 7 \\ \beta_2 + \beta_3 t, & t = 8, 9, \ldots, 14 \end{cases}$

Intersecting lines at $t = 8: \beta_2 = \beta_0 + 8(\beta_1 - \beta_3)$, and

$$E(y_t) = \begin{cases} \beta_0 + \beta_1 t, & t = 1, 2, \ldots, 7 \\ \beta_0 + \beta_1 8 + \beta_3 (t-8), & t = 8, 9, \ldots, 14 \end{cases}$$

b. $E(y_t) = \beta_0 + \beta_1 t, \quad t = 1, 2, \ldots, 14$

c. $F = 55.95$; p value $= P(F(1, 11) > 55.95) = 0.0000$; model in (a) is preferable.

5.12 b. $\hat{\mu} = -0.9122 + 0.1607x_1 + 0.2198x_2 + 0.0112x_3 + 0.1020x_4$; $R^2 = 0.692$; $s = 0.8365$

(i) $t(\hat{\beta}_1) = 2.43$; p value = 0.023; reject $\beta_1 = 0$.

(ii) $F = 3.90$; p value $= 0.034$; reject hypothesis $\beta_3 = \beta_4 = 0$

(iii) $F = 14.07$; p value < 0.0001; reject hypothesis $\beta_1 = \beta_2 = \beta_3 = \beta_4 = 0$

c. $\hat{\mu} = -1.462 + 0.1536x_1 + 0.3221x_2 + 0.0166x_3 + 0.0571x_4 - 0.00087x_2x_3 + 0.00599x_2x_4$; $F = 0.40$; p value $= 0.67$; interactions are not important.

5.13 b. $z = 0$ (protein-rich); $z = 1$ (protein-poor); $\hat{\mu} = 50.324 + 16.009x + 0.918z - 7.329xz$ $F = 107.24$; p value < 0.0001; reject $\beta_2 = \beta_3 = 0$; linear relationship between height and age is not the same for the two diets.

5.15 Weight (x_1); $x_2 = 0$ (type A engine); $x_2 = 1$ (type B engine);

a. $\mu = \beta_0 + \beta_1 x_1 + \beta_2 x_2$;

b. $\mu = \beta_0 + \beta_1 x_1 + \beta_2 x_2 + \beta_3 x_1 x_2$

CHAPTER 6

6.2 a. Linear model: $\hat{\mu} = 23.35 + 1.045x$; $R^2 = 0.955$; $s = 0.737$; F(lack of fit) $=$ 10.01; p value $= 0.002$; lack of fit

b. Quadratic model: $\hat{\mu} = 22.56 + 1.67x - 0.068x^2$; $R^2 = 0.988$; $s = 0.394$; $t(\hat{\beta}_2) = -0.06796/0.01031 = -6.59$; reject $\beta_2 = 0$; F(lack of fit) $= 2.30$; p value $= 0.13$; no lack of fit

6.7 a. True. For a correct model, $\text{Cov}(\boldsymbol{e}, \hat{\boldsymbol{\mu}}) = O$, and a plot of the residuals e_i against the fitted values $\hat{\mu}_i$ should show no association. However, $\text{Cov}(\boldsymbol{e}, \boldsymbol{y}) = \sigma^2(I - H)$; the correlation makes the interpretation of the plot of e_i against y_i difficult.

b. Not true. Outliers should be scrutinized but not necessarily rejected.

c. True.

6.8 a. 5; b. 2; c. 4; d. 1

6.11 a. No; b. No; c. No; d. No; e. True

6.12 A (Palm Beach); B (Broward); C (Dade); D (Pasco)

6.15 Estimate of transformation parameter in Box–Cox family $\hat{\lambda} \approx 0$, indicating a logarithmic transformation of the response y.

Regression of $\ln(y)$ on x: $\hat{\mu} = 2.436 + 0.000567x$; $R^2 = 0.986$; $s = 0.0845$.

The first case is quite influential ($x = 574$; $y = 21.9$; Cook $= 0.585$).

6.21 Scatter plot of $\ln(y)$ against $\ln(x)$ shows a linear association with three outlying observations (brachiosaurus, diplodocus, and triceratops). Omitting these three cases and fitting the linear model to the reduced data set leads to an adequate fit.

Estimated equation: $\hat{\mu} = 2.15 + 0.752\ln(x)$; $R^2 = 0.922$; $s = 0.726$. The two observations with the largest positive residuals and the largest Cook influence are human (stand. residual $=$ 2.72; Cook $= 0.174$) and rhesus monkey (stand. residual $= 2.25$; Cook $= 0.119$).

6.22 Estimated equation: $\hat{\mu} = 74.319 - 2.089\text{Conc} + 0.430\text{Ratio} - 0.372\text{Temp}$; $R^2 = 0.939$; $s = 0.74$; F(lack of fit) $= 7.44$; p value $= 0.036$; indication of lack of fit

```
Analysis of Variance
Source            DF      SS      MS      F      P
Regression         3  92.304  30.768  56.17  0.000
Residual Error    11   6.026   0.548
  Lack of Fit      7   5.596   0.799   7.44  0.036
  Pure Error       4   0.430   0.108
Total             14  98.329
```

Run #2 (Conc $= 1$, Ratio $= -1$, Temp $= -1$; Yield $= 73.9$) is influential, with a large Cook's distance. This run should be investigated. Without this run, there is no lack of fit.

6.26 Linear model: $\hat{\mu} = 0.131 + 0.241x$, with $R^2 = 0.874$, is not appropriate.

Quadratic model: $\hat{\mu} = -1.16 + 0.723x - 0.0381x^2$, with $R^2 = 0.968$, is a possibility.
90% confidence interval: (1.972, 2.102).

Reciprocal transformation on x: $\hat{\mu} = 2.98 - 6.93(1/x)$, with $R^2 = 0.980$, is better.
90% confidence interval: (1.951, 2.026).

CHAPTER 7

7.1 a. Backward elimination: Drop x_3 (step 1); drop x_4 (step 2); next candidate x_2 for elimination cannot be dropped. Model with x_1 and x_2.

b. Forward selection: Enter x_4 (step 1); enter x_1 (step 2); enter x_2 (step 3); next candidate x_3

for selection cannot be entered. Model with x_1, x_2, and x_4.

c. Stepwise regression: Steps 1, 2, and 3 of forward selection; x_4 can be dropped from model containing x_1, x_2, and x_4; no reason to add x_3 to the model with x_1 and x_2. Model with x_1 and x_2.

d. Model with x_1 and x_2 : $C_p = 2.68$, close to desired value 3. Full model: $C_p = 5$. Prefer model with x_1 and x_2.

e. x_2 and x_4 highly correlated.

f. $F = 68.6$; p value <0.001; reject $\beta_1 = \beta_3 = 0$.

7.2 a. C_p: Model with x_1 and x_2 ($C_p = 2.7$). R^2: Model with x_1 and x_2, or model with x_1 and x_4. Small gain by going to more complicated models.

b. Backward elimination ($\alpha_{\text{drop}} = 0.1$): Model with x_1 and x_2.
Forward selection ($\alpha_{\text{enter}} = 0.1$): Model with x_1, x_2, and x_4.
Stepwise regression ($\alpha_{\text{drop}} = \alpha_{\text{enter}} = 0.1$): Model with x_1 and x_2.

7.7 $\hat{\mu} = -5.0359 + 0.0671\text{AirFlow} + 0.1295\text{CoolTemp}$; $R^2 = 0.909$; $C_p = 2.9$
Last case (AirFlow = 70; CoolTemp = 20; StackLoss = 1.5) is an influential observation and should be scrutinized. Without this case.

$$\hat{\mu} = -5.1076 + 0.0863\text{AirFlow} + 0.0803\text{CoolTemp};$$
$$R^2 = 0.946$$

7.8 Stepwise regression ($\alpha_{\text{drop}} = \alpha_{\text{enter}} = 0.15$) : $\hat{\mu} = -62.60 + 7.427\%\text{ASurf} + 6.828\%\text{ABase} - 5.2685\text{Run}$; $R^2 = 0.724$; $R^2_{\text{adj}} = 0.693$; $C_p = 1.3$
Similar model: $\hat{\mu} = -23.00 + 5.975\%\text{AsphSurf} - 5.4058\text{Run}$; $R^2 = 0.695$; $R^2_{\text{adj}} = 0.673$; $C_p = 1.9$
Cases 13 and 15 with large Cook's influence. Second set of runs with considerably smaller change in rut depth.

CHAPTER 8

8.1 b. Extremely small sample sizes ($n = 13$ and $n = 11$) make it difficult to detect violations of the independence assumption.

c. Root mean square errors of the fitted models range from 1.5 to 2 percentage points. Their size implies that half-widths of 95% prediction intervals are at least 3 or 4 percentage points without considering the estimation error. Incorporating the uncertainty from the estimation with small samples makes the actual prediction intervals even wider. Predictions are "within-sample" predictions (the case being predicted is part of the data used for estimation). Prediction errors for "out-of-sample" predictions (the case being predicted is not part of the data used for estimation) are larger; see part d.

d. Leaving out case i, running the regression on the reduced data set, and predicting the response of the case that has been left out using the estimates from the reduced data set lead to PRESS residuals $e_{(i)} = y_{(i)} - \hat{y}_{(i)} = e_i/(1 - h_{ii})$; see Chapter 6.

e. The four prediction models studied in this exercise are no better and no worse than the models proposed by Fair and Lewis-Beck/Tien. Although they give some indication about the winner of presidential elections, their large uncertainty makes them only useful in the rather uninteresting situation in which there is little doubt about the outcome of the election.

8.2 1a. Models with a linear component of Age provide an adequate representation of the relationships. The addition of Age^2 is not needed. The models lead to R^2 of approximately 60% for height and 45% for weight. Addition of birth weight to the regression of weight at referral on age at referral increases the R^2 from 45.9 to 48.3%. Birth weight is marginally significant (estimate 2.26, with p value 0.064). Each extra pound at birth increases the weight at referral by 2.26 pounds.

1c. Models with a linear component of AgeCo provide an adequate representation of the relationship between HeightCo and AgeCo. For weight, the addition of the quadratic component AgeCo^2 becomes necessary. The scatter plot of weight against age suggests that the variability increases with the level. The scatter plot of the logarithm of weight against age indicates that the variability is stabilized by this transformation. No major lack of fit can be detected in the residuals from the regression of ln(WeightCo) on AgeCo.

The estimate of the transformation parameter λ in the Box–Cox family of transformations results in an estimate that is close to zero, confirming the appropriateness of the logarithmic transformation.

2b. The correlation between the height of mothers and the height of fathers is small (0.077). The correlation between the weight of mothers and the weight of fathers is larger (0.242). There is (weak) evidence that both partners tend to be above or below the average weight. The scatter plot shows three unusual cases. In one case, the father is quite heavy, whereas the mother is of average weight. In the other two cases, the fathers are of average weight, whereas the mothers are heavy. However, the omission of these three cases does not change the correlation coefficient ($r = 0.243$).

CHAPTER 9

9.1 Scatter plot of leaf area against age indicates a nonlinear relationship.

Model with a quadratic component of age: $\hat{\mu} = -0.123 + 2.15\text{Age} - 0.096(\text{Age})^2$. Quadratic term needed (estimate -0.096, with t ratio -4.95); $R^2 = 0.966$.

Gompertz model: $\hat{\mu} = 12.5 \exp[-2.5 \exp(-0.36\text{Age})]$. Least squares estimates are statistically significant. Some correlation among the estimates. Coefficient of determination $R^2 = 0.961$ similar to the one for quadratic regression. Quadratic regression (linear in the parameters) and the nonlinear Gompertz model lead to similar fitted curves.

9.2 Large R^2 for all four models. Michaelis–Menton ($R^2 = 0.954$) and its modification ($R^2 = 0.979$). The modification leads to a significant improvement. Modified Michaelis–Menton and modified exponential rise models perform similarly; fitted values are virtually indistinguishable. Reject the null hypothesis $\alpha_1 = \alpha_2 = \alpha_3 = 0$ in the quadratic Michaelis–Menton model with day indicator; $F = [(43{,}222{,}556 - 32{,}022{,}591)/3]/[32{,}022{,}591/42] = 4.90$, with probability value $P[F(3,42) \geq 4.90] = 1 - 0.995 = 0.005$.

9.3 Logarithmic transformations of the response and the regressors in model 1; reciprocal transformation of the response in model 2; $\ln[(1/y) - 1]$ in model 3.

CHAPTER 10

10.3 b. Regression of sales on time: $\text{Sales}_t = 166.396 + (2.325)t$; $R^2 = 0.566$

Predictions and 95% prediction intervals (from Minitab):

$$y_{52}(1) = 289.60; \quad 224.65 \text{ to } 354.56$$
$$y_{52}(2) = 291.93; \quad 226.84 \text{ to } 357.02$$
$$y_{52}(3) = 294.25; \quad 229.02 \text{ to } 359.49$$

c. Significant autocorrelations in residuals, especially at lag 1 (0.41); Durbin–Watson = 1.09. Regression of sales on time is not an appropriate forecasting model.

d. $\Delta y_t = 2.7255 + a_t . \hat{\sigma}_a = 32.51$. Lag one autocorrelation (-0.37) exceeds twice its standard error 0.14. Not an appropriate forecasting model.

Predictions and 95% prediction intervals:

$$y_{52}(1) = 347.73; \quad 347.73 \pm 63.72$$
$$y_{52}(2) = 350.46; \quad 350.46 \pm 90.11$$
$$y_{52}(3) = 353.19; \quad 353.19 \pm 110.37$$

e. $\Delta y_t = y_t - y_{t-1} = 2.252 + a_t - 0.72a_{t-1}$. Autocorrelations of the residuals are small; acceptable model.

10.4 a. Regression on time: $\text{Enrollment}_t = 6527 + (830)t$. Positive autocorrelations and unacceptable Durbin–Watson test statistic (0.26).

c. Regression of enrollment on previous two enrollments:

$$\text{Enrollment}_t = 914 + 1.469 \text{ Enrollment}_{t-1} - 0.506 \text{ Enrollment}_{t-2}$$

Autocorrelations of the residuals are small. Durbin–Watson = 2.32. Appropriate forecasting method. Predictions: $y_{21}(1) = 21{,}536$; $y_{21}(2) = 21{,}592$; $y_{21}(3) = 21{,}670$.

10.5 a. Regression model:

$$\begin{aligned}\text{Sales} = 1500 &+ 0.449\ \text{Time} - 1154\ \text{IndJan}\\ &- 1169\ \text{IndFeb} - 1073\ \text{IndMar}\\ &- 1049\ \text{IndApr} - 1057\ \text{IndMay}\\ &- 1061\ \text{IndJun} - 1126\ \text{IndJul}\\ &- 1062\ \text{IndAug} - 984\ \text{IndSep}\\ &- 951\ \text{IndOct} - 776\ \text{IndNov}\end{aligned}$$

b. Autocorrelation function of residuals with spike at lag one, suggesting a first-order moving average error model:

$$\begin{aligned}\text{Sales}_t = 1501 &+ 0.44\text{Time} - 1155\text{IndJan}_t\\ &- 1169\text{IndFeb}_t + \cdots - 779\text{IndNov}_t\\ &+ a_t - 0.27a_{t-1}\end{aligned}$$

Residuals are uncorrelated. Trend coefficient small and negligible for all practical purposes. Strong seasonal component.

c. Model with current advertising:

$$\begin{aligned}\text{Sales}_t = \beta_0 &+ \beta_1 t + \beta_2\text{IndJan}_t + \beta_3\text{IndFeb}_t\\ &+ \cdots + \beta_{12}\text{IndNov}_t + \beta_{13}\text{Adv}_t\\ &+ a_t - \theta a_{t-1}\end{aligned}$$

Insignificant advertising effect ($\hat{\beta}_{13} = 2.04$ with t ratio $= -0.88$). Residuals are uncorrelated.

10.10 Regression model: Cons $= 0.197 - 1.04$Price $+$ 0.00331Income $+ 0.00346$Temp. Independent error model inadequate. Durbin–Watson statistic $= 1.02$. Lag one autocorrelation (0.32) large compared to its standard error 0.18. Regression model with MA(1) errors:

$$\begin{aligned}\text{Cons}_t = 0.33 &- 1.39\text{Price}_t + 0.0029\text{Inc}_t\\ &+ 0.0034\text{Temp}_t + a_t + 0.50a_{t-1}\end{aligned}$$

Regression coefficients for income and temperature significant (t ratios exceed two). Income and temperature with positive regression coefficients; ice cream sales increase with increasing income and rising temperature. Coefficient of price is negative and not very significant (t ratio $= -1.77$).

10.13 a. Regression model: FTEShares$_t = 595 +$ (0.000514)CarProd$_{t-6}$ $-$ (5.54)FTECom$_{t-7} + \varepsilon_t$

Significant t ratios: $t(\hat{\beta}_1) = 15.10$ and $t(\hat{\beta}_2) = -8.24$. Durbin–Watson (0.87) unacceptable. Significant autocorrelations in residuals.

b. Noisy random walk model for errors. Regression in first differences with moving average errors

$$\begin{aligned}\Delta\text{FTEShares}_t = &\ (0.0001)\Delta\text{CarProd}_{t-6}\\ &- (0.69)\Delta\text{FTECom}_{t-7} + a_t\\ &+ 0.15a_{t-1}\end{aligned}$$

No autocorrelation in the residuals.

c. Moving average parameter close to zero. Regression model with random walk errors

$$\begin{aligned}\Delta\text{FTEShares}_t = 3.71 &+ (0.00014)\Delta\text{CarProd}_{t-6}\\ &- (0.79)\Delta\text{FTECom}_{t-7} + a_t\end{aligned}$$

No autocorrelation in the residuals. Regressors not statistically significant (p values of 0.085 and 0.51), implying the random walk model

$$\begin{aligned}\Delta\text{FTEShares}_t &= \text{FTEShares}_t - \text{FTEShares}_{t-1}\\ &= a_t\end{aligned}$$

Expected result: "Significant" regression in (a) is spurious, implied by the incorrect model for the error terms.

CHAPTER 11

11.1 Scatter plots of weekly proportions of long fibers against time and the explanatory variables show an increase during the second half of the year and a relationship to the amount of stretch reduction.

Logistic regressions of the proportions of long fibers on stretch reduction, throughput, and the type of process show that stretch reduction remains the only significant variable. An increase in the stretch reduction of one unit (percent) reduces the odds for the occurrence of long fibers on machine (street) 6 by 15%. A small stretch reduction increases the odds for quality problems. The model is adequate, considering the small Hosmer–Lemeshow statistic and the absence of serial correlation in the residuals.

The fitted proportion of long fibers π at a certain stretch reduction x is given by

$$\hat{\pi}(x) = \frac{\exp(5.928 - 0.1662x)}{1 + \exp(5.928 - 0.1662x)}$$

For example, $\hat{\pi}(x=52)=0.062$, $\hat{\pi}(x=53)=0.053$, and $\hat{\pi}(x=57)=0.028$.

```
Minitab Logistic Regression Output
                                              Odds    95% CI
Predictor        Coef  SE Coef       Z      P Ratio  Lower Upper
Constant       5.928    1.818    3.26 0.001
stretch6   -0.16619  0.03359  -4.95 0.000    0.85   0.79  0.90

Log-Likelihood =-566.554
Test that all slopes are zero: G = 24.891,
DF = 1, P-value = 0.000

Goodness-of-Fit Tests
Method              Chi-Square   DF       P
Pearson                 27.801   41   0.943
Deviance                26.951   41   0.955
Hosmer-Lemeshow          6.938    7   0.435
```

11.3 Logistic regressions relating the proportions of affected workers to the five factors—dust, race, sex, smoking, and length of employment—show that race and sex can be omitted from the model.

Smoking is an important contributor to byssinosis. Everything else equal, not smoking reduces the odds of contracting byssinosis by 46%.

The length of employment in the cotton industry matters. The odds that a worker with 10 to 20 years of employment contracts byssinosis are 1.66 times the odds of a worker with less than 10 years in the industry. The odds for a worker with more than 20 years are twice (1.96) the odds of a worker with less than 10 years in the industry.

Dustiness of the workplace matters. The odds of contracting byssinosis at workplaces with medium and low levels of dustiness are considerably less than the odds for workplaces with a high level of dustiness.

The evidence for interaction terms is not particularly strong.

CHAPTER 12

12.1 a. The Poisson regression on ln(MS) and the three factors—type of ship, year of construction, and period of service—involves the estimation of 10 parameters. All three factors are needed in the model. Type 3 ships report the smallest number of damage incidents. Ships constructed in years 2 (1965–1969) and 3 (1970–1974) experience the highest number of reported damage incidents. The second period of operation (1975–1979) is associated with a higher number of reported damage incidents.

b. The parameter estimate for ln(MS) is $\hat{\beta}_1=0.9027$. The 95% confidence interval $0.70 \le \beta_1 \le 1.10$ includes one, which makes the offset interpretation plausible.

c. The estimation results of the model with offset are similar to the results in (a).

d. The deviance $D=37.80$, with probability value $P(\chi^2(24) \ge 37.80)=0.0363$, indicates some overdispersion. Adjusting the analysis for overdispersion (by using the SCALE = DEVIANCE option in SAS GENMOD) does not change the findings.

e. Interaction terms indicate that interaction components are not needed.

12.4 a. The Poisson regression of lung cancer incidence with offset ln(POP) and the two factors—age group and town—involves nine parameters. Both age and town are significant. The estimate of the town effect is $\hat{\beta}_9=0.85$, with standard error 0.06. Women in Texas have a 100[exp(0.85) − 1] = 134% higher incidence of skin cancer. The deviance and the Pearson chi-square statistics are approximately one and indicate no problem with over- or underdispersion.

b. The estimation of the general model without assuming an offset leads to the estimate $\hat{\beta}_1=1.96$. The 95% confidence interval $0.73 \le \beta_1 \le 3.18$ includes one, which makes the offset interpretation plausible.

In a subsequent analysis, age is introduced as a continuous variable. Both age and town are significant. Every 10 years, the cancer rate (deaths per population) increases by 100[exp(0.6133) − 1] = 85%.

12.6 Treating size, with groups 1–5, as a continuous variable leads to the Poisson link $\ln \mu = \beta_0 + \beta_1 \text{DIST} + \beta_3 \text{SIZE}$. Income is not significant and can be omitted from the model. A change in distance by 10 miles reduces the mean number of visits by 19%.

A change in the family size by one unit increases the mean number of visits by 14.5%.

A test whether size needs to be treated as a class factor (with five groups requiring four indicators) or as a continuous variable was carried out. The log-likelihood ratio test statistic, $2(10.7564 - 9.8849) = 1.74$, is small compared to cutoffs of the chi square distribution with 3 degrees of freedom. It is appropriate to treat size as a continuous variable.

Statistical Tables

TABLE A: CUMULATIVE PROBABILITIES $P(Z \leq z)$ FOR THE STANDARD NORMAL DISTRIBUTION

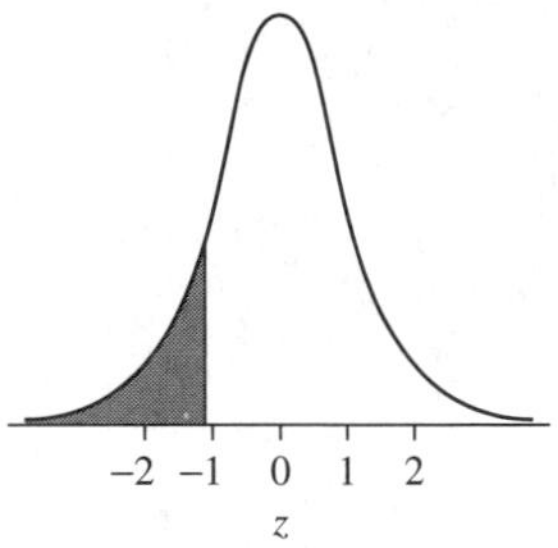

Normal Distribution Table (Each entry is the total area under the standard normal curve to the left of z, which is specified to two decimal places by joining the row value to the column value.)

z	0.00	0.01	0.02	0.03	0.04	0.05	0.06	0.07	0.08	0.09
−3.9	0.0000	0.0000	0.0000	0.0000	0.0000	0.0000	0.0000	0.0000	0.0000	0.0000
−3.8	0.0001	0.0001	0.0001	0.0001	0.0001	0.0001	0.0001	0.0001	0.0001	0.0001
−3.7	0.0001	0.0001	0.0001	0.0001	0.0001	0.0001	0.0001	0.0001	0.0001	0.0001
−3.6	0.0002	0.0002	0.0001	0.0001	0.0001	0.0001	0.0001	0.0001	0.0001	0.0001
−3.5	0.0002	0.0002	0.0002	0.0002	0.0002	0.0002	0.0002	0.0002	0.0002	0.0002
−3.4	0.0003	0.0003	0.0003	0.0003	0.0003	0.0003	0.0003	0.0003	0.0003	0.0002
−3.3	0.0005	0.0005	0.0005	0.0004	0.0004	0.0004	0.0004	0.0004	0.0004	0.0003
−3.2	0.0007	0.0007	0.0006	0.0006	0.0006	0.0006	0.0006	0.0005	0.0005	0.0005
−3.1	0.0010	0.0009	0.0009	0.0009	0.0008	0.0008	0.0008	0.0008	0.0007	0.0007
−3.0	0.0013	0.0013	0.0013	0.0012	0.0012	0.0011	0.0011	0.0011	0.0010	0.0010
−2.9	0.0019	0.0018	0.0018	0.0017	0.0016	0.0016	0.0015	0.0015	0.0014	0.0014
−2.8	0.0026	0.0025	0.0024	0.0023	0.0023	0.0022	0.0021	0.0021	0.0020	0.0019
−2.7	0.0035	0.0034	0.0033	0.0032	0.0031	0.0030	0.0029	0.0028	0.0027	0.0026
−2.6	0.0047	0.0045	0.0044	0.0043	0.0041	0.0040	0.0039	0.0038	0.0037	0.0036
−2.5	0.0062	0.0060	0.0059	0.0057	0.0055	0.0054	0.0052	0.0051	0.0049	0.0048
−2.4	0.0082	0.0080	0.0078	0.0075	0.0073	0.0071	0.0069	0.0068	0.0066	0.0064
−2.3	0.0107	0.0104	0.0102	0.0099	0.0096	0.0094	0.0091	0.0089	0.0087	0.0084
−2.2	0.0139	0.0136	0.0132	0.0129	0.0125	0.0122	0.0119	0.0116	0.0113	0.0110
−2.1	0.0179	0.0174	0.0170	0.0166	0.0162	0.0158	0.0154	0.0150	0.0146	0.0143
−2.0	0.0228	0.0222	0.0217	0.0212	0.0207	0.0202	0.0197	0.0192	0.0188	0.0183
−1.9	0.0287	0.0281	0.0274	0.0268	0.0262	0.0256	0.0250	0.0244	0.0239	0.0233
−1.8	0.0359	0.0351	0.0344	0.0336	0.0329	0.0322	0.0314	0.0307	0.0301	0.0294
−1.7	0.0446	0.0436	0.0427	0.0418	0.0409	0.0401	0.0392	0.0384	0.0375	0.0367
−1.6	0.0548	0.0537	0.0526	0.0516	0.0505	0.0495	0.0485	0.0475	0.0465	0.0455
−1.5	0.0668	0.0655	0.0643	0.0630	0.0618	0.0606	0.0594	0.0582	0.0571	0.0559
−1.4	0.0808	0.0793	0.0778	0.0764	0.0749	0.0735	0.0721	0.0708	0.0694	0.0681
−1.3	0.0968	0.0951	0.0934	0.0918	0.0901	0.0885	0.0869	0.0853	0.0838	0.0823
−1.2	0.1151	0.1131	0.1112	0.1093	0.1075	0.1056	0.1038	0.1020	0.1003	0.0985
−1.1	0.1357	0.1335	0.1314	0.1292	0.1271	0.1251	0.1230	0.1210	0.1190	0.1170
−1.0	0.1587	0.1562	0.1539	0.1515	0.1492	0.1469	0.1446	0.1423	0.1401	0.1379
−0.9	0.1841	0.1814	0.1788	0.1762	0.1736	0.1711	0.1685	0.1660	0.1635	0.1611
−0.8	0.2119	0.2090	0.2061	0.2033	0.2005	0.1977	0.1949	0.1921	0.1894	0.1867
−0.7	0.2420	0.2389	0.2358	0.2327	0.2296	0.2266	0.2236	0.2206	0.2177	0.2148
−0.6	0.2743	0.2709	0.2676	0.2643	0.2611	0.2578	0.2546	0.2514	0.2483	0.2451
−0.5	0.3085	0.3050	0.3015	0.2981	0.2946	0.2812	0.2877	0.2843	0.2810	0.2776
−0.4	0.3446	0.3409	0.3372	0.3336	0.3300	0.3264	0.3228	0.3192	0.3156	0.3121
−0.3	0.3821	0.3783	0.3745	0.3707	0.3669	0.3632	0.3594	0.3557	0.3520	0.3483
−0.2	0.4207	0.4168	0.4129	0.4090	0.4052	0.4013	0.3974	0.3936	0.3897	0.3859
−0.1	0.4602	0.4562	0.4522	0.4483	0.4443	0.4404	0.4364	0.4325	0.4286	0.4247
−0.0	0.5000	0.4960	0.4920	0.4880	0.4840	0.4801	0.4761	0.4721	0.4681	0.4641

TABLE A: (CONTINUED) CUMULATIVE PROBABILITIES $P(Z \leq z)$ FOR THE STANDARD NORMAL DISTRIBUTION

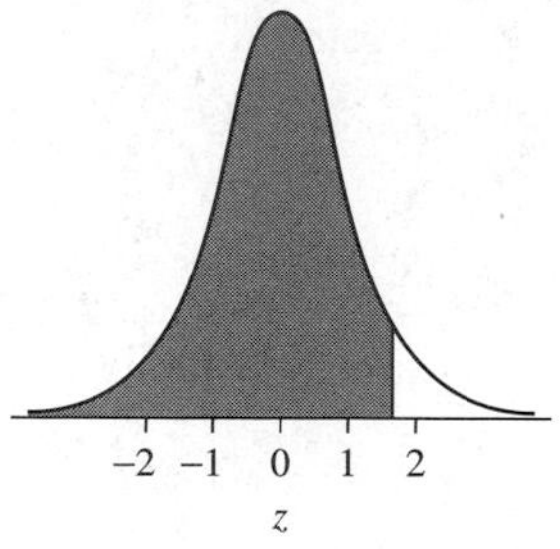

Normal Distribution Table (Each entry is the total area under the standard normal curve to the left of z, which is specified to two decimal places by joining the row value to the column value.)

z	0.00	0.01	0.02	0.03	0.04	0.05	0.06	0.07	0.08	0.09
0.0	0.5000	0.5040	0.5080	0.5120	0.5160	0.5199	0.5239	0.5279	0.5319	0.5359
0.1	0.5398	0.5438	0.5478	0.5517	0.5557	0.5596	0.5636	0.5675	0.5714	0.5753
0.2	0.5793	0.5832	0.5871	0.5910	0.5948	0.5987	0.6026	0.6064	0.6103	0.6141
0.3	0.6179	0.6217	0.6255	0.6293	0.6331	0.6368	0.6406	0.6443	0.6480	0.6517
0.4	0.6554	0.6591	0.6628	0.6664	0.6700	0.6736	0.6772	0.6808	0.6844	0.6879
0.5	0.6915	0.6950	0.6985	0.7019	0.7054	0.7088	0.7123	0.7157	0.7190	0.7224
0.6	0.7257	0.7291	0.7324	0.7357	0.7389	0.7422	0.7454	0.7486	0.7517	0.7549
0.7	0.7580	0.7611	0.7642	0.7673	0.7704	0.7734	0.7764	0.7794	0.7823	0.7852
0.8	0.7881	0.7910	0.7939	0.7967	0.7995	0.8023	0.8051	0.8078	0.8106	0.8133
0.9	0.8159	0.8186	0.8212	0.8238	0.8264	0.8289	0.8315	0.8340	0.8365	0.8389
1.0	0.8413	0.8438	0.8461	0.8485	0.8508	0.8531	0.8554	0.8577	0.8599	0.8621
1.1	0.8643	0.8665	0.8686	0.8708	0.8729	0.8749	0.8770	0.8790	0.8810	0.8830
1.2	0.8849	0.8869	0.8888	0.8907	0.8925	0.8944	0.8962	0.8980	0.8997	0.9015
1.3	0.9032	0.9049	0.9066	0.9082	0.9099	0.9115	0.9131	0.9147	0.9162	0.9177
1.4	0.9192	0.9207	0.9222	0.9236	0.9251	0.9265	0.9279	0.9292	0.9306	0.9319
1.5	0.9332	0.9345	0.9357	0.9370	0.9382	0.9394	0.9406	0.9418	0.9429	0.9441
1.6	0.9452	0.9463	0.9474	0.9484	0.9495	0.9505	0.9515	0.9525	0.9535	0.9545
1.7	0.9554	0.9564	0.9573	0.9582	0.9591	0.9599	0.9608	0.9616	0.9625	0.9633
1.8	0.9641	0.9649	0.9656	0.9664	0.9671	0.9678	0.9686	0.9693	0.9699	0.9706
1.9	0.9713	0.9719	0.9726	0.9732	0.9738	0.9744	0.9750	0.9756	0.9761	0.9767
2.0	0.9772	0.9778	0.9783	0.9788	0.9793	0.9798	0.9803	0.9808	0.9812	0.9817
2.1	0.9821	0.9826	0.9830	0.9834	0.9838	0.9842	0.9846	0.9850	0.9854	0.9857
2.2	0.9861	0.9864	0.9868	0.9871	0.9875	0.9878	0.9881	0.9884	0.9887	0.9890
2.3	0.9893	0.9896	0.9898	0.9901	0.9904	0.9906	0.9909	0.9911	0.9913	0.9916
2.4	0.9918	0.9920	0.9922	0.9925	0.9927	0.9929	0.9931	0.9932	0.9934	0.9936
2.5	0.9938	0.9940	0.9941	0.9943	0.9945	0.9946	0.9948	0.9949	0.9951	0.9952
2.6	0.9953	0.9955	0.9956	0.9957	0.9959	0.9960	0.9961	0.9962	0.9963	0.9964
2.7	0.9965	0.9966	0.9967	0.9968	0.9969	0.9970	0.9971	0.9972	0.9973	0.9974
2.8	0.9974	0.9975	0.9976	0.9977	0.9977	0.9978	0.9979	0.9979	0.9980	0.9981
2.9	0.9981	0.9982	0.9982	0.9983	0.9984	0.9984	0.9985	0.9985	0.9986	0.9986
3.0	0.9987	0.9987	0.9987	0.9988	0.9988	0.9989	0.9989	0.9989	0.9990	0.9990
3.1	0.9990	0.9991	0.9991	0.9991	0.9992	0.9992	0.9992	0.9992	0.9993	0.9993
3.2	0.9993	0.9993	0.9994	0.9994	0.9994	0.9994	0.9994	0.9995	0.9995	0.9995
3.3	0.9995	0.9995	0.9995	0.9996	0.9996	0.9996	0.9996	0.9996	0.9996	0.9997
3.4	0.9997	0.9997	0.9997	0.9997	0.9997	0.9997	0.9997	0.9997	0.9997	0.9998
3.5	0.9998	0.9998	0.9998	0.9998	0.9998	0.9998	0.9998	0.9998	0.9998	0.9998
3.6	0.9998	0.9998	0.9999	0.9999	0.9999	0.9999	0.9999	0.9999	0.9999	0.9999
3.7	0.9999	0.9999	0.9999	0.9999	0.9999	0.9999	0.9999	0.9999	0.9999	0.9999
3.8	0.9999	0.9999	0.9999	0.9999	0.9999	0.9999	0.9999	0.9999	0.9999	0.9999
3.9	1.0000	1.0000	1.0000	1.0000	1.0000	1.0000	1.0000	1.0000	1.0000	1.0000

TABLE B: PERCENTILES OF CHI-SQUARE DISTRIBUTIONS: α IS THE UPPER TAIL AREA AND $\chi^2(1-\alpha; v)$ IS THE $100(1-\alpha)$ PERCENTILE

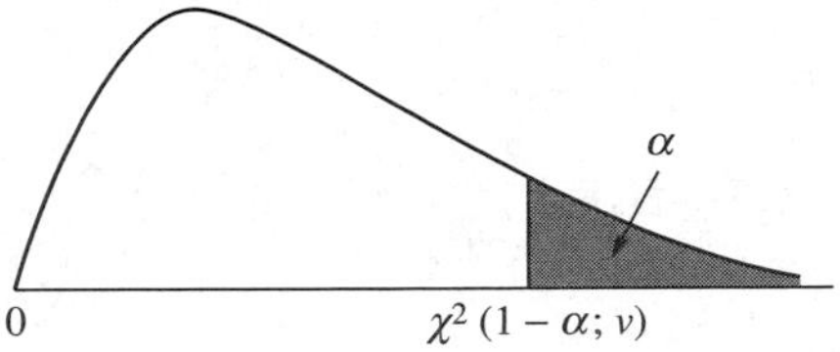

v	$\alpha = 0.995$	$\alpha = 0.99$	$\alpha = 0.975$	$\alpha = 0.95$	$\alpha = 0.05$	$\alpha = 0.025$	$\alpha = 0.01$	$\alpha = 0.005$	v
1	0.0^4393	0.0^3157	0.0^3982	0.00393	3.841	5.024	6.635	7.879	1
2	0.0100	0.0201	0.0506	0.103	5.991	7.378	9.210	10.597	2
3	0.0717	0.115	0.216	0.352	7.815	9.348	11.345	12.838	3
4	0.207	0.297	0.484	0.711	9.488	11.143	13.277	14.860	4
5	0.412	0.554	0.831	1.145	11.070	12.832	15.086	16.750	5
6	0.676	0.872	1.237	1.635	12.592	14.449	16.812	18.548	6
7	0.989	1.239	1.690	2.167	14.067	16.013	18.475	20.278	7
8	1.344	1.646	2.180	2.733	15.507	17.535	20.090	21.955	8
9	1.735	2.088	2.700	3.325	16.919	19.023	21.666	23.589	9
10	2.156	2.558	3.247	3.940	18.307	20.483	23.209	25.188	10
11	2.603	3.053	3.816	4.575	19.675	21.920	24.725	26.757	11
12	3.074	3.571	4.404	5.226	21.026	23.337	26.217	28.300	12
13	3.565	4.107	5.009	5.892	22.362	24.736	27.688	29.819	13
14	4.075	4.660	5.629	6.571	23.685	26.119	29.141	31.319	14
15	4.601	5.229	6.262	7.261	24.996	27.488	30.578	32.801	15
16	5.142	5.812	6.908	7.962	26.296	28.845	32.000	34.267	16
17	5.697	6.408	7.564	8.672	27.587	30.191	33.409	35.718	17
18	6.265	7.015	8.231	9.390	28.869	31.526	34.805	37.156	18
19	6.844	7.633	8.907	10.117	30.144	32.852	36.191	38.582	19
20	7.434	8.260	9.591	10.851	31.410	34.170	37.566	39.997	20
21	8.034	8.897	10.283	11.591	32.671	35.479	38.932	41.401	21
22	8.643	9.542	10.982	12.338	33.924	36.781	40.289	42.796	22
23	9.260	10.196	11.688	13.091	35.172	38.076	41.638	44.181	23
24	9.886	10.856	12.401	13.848	36.415	39.364	42.980	45.558	24
25	10.520	11.524	13.120	14.611	37.652	40.646	44.314	46.928	25
26	11.160	12.198	13.844	15.379	38.885	41.923	45.642	48.290	26
27	11.808	12.879	14.573	16.151	40.113	43.194	46.963	49.645	27
28	12.461	13.565	15.308	16.928	41.337	44.461	48.278	20.993	28
29	13.121	14.256	16.047	17.708	42.557	45.722	49.588	52.336	29
30	13.787	14.953	16.791	18.493	43.773	46.979	50.892	53.672	30

Source: Reproduced from Table 8 of E. S. Pearson and H. O. Hartley, *Biometrika Tables for Statisticians,* Vol. 1 (Cambridge, UK: Cambridge University Press, 1954). Reproduced with permission of the Biometrika Trustees.

TABLE C: PERCENTILES OF STUDENT *t* DISTRIBUTIONS: α IS THE UPPER TAIL AREA AND $t(1-\alpha; v)$ IS THE $100(1-\alpha)$ PERCENTILE

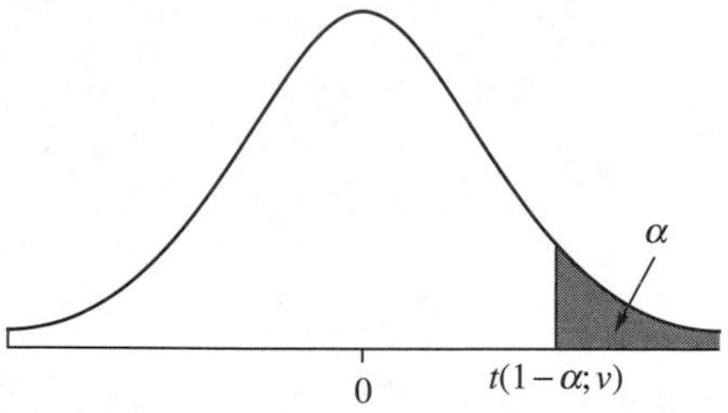

v	$\alpha = 0.10$	$\alpha = 0.05$	$\alpha = 0.025$	$\alpha = 0.01$	$\alpha = 0.005$
1	3.078	6.314	12.706	31.821	63.657
2	1.886	2.920	4.303	6.965	9.925
3	1.638	2.353	3.182	4.541	5.841
4	1.533	2.132	2.776	3.747	4.604
5	1.476	2.015	2.571	3.365	4.032
6	1.440	1.943	2.447	3.143	3.707
7	1.415	1.895	2.365	2.998	3.499
8	1.397	1.860	2.306	2.896	3.355
9	1.383	1.833	2.262	2.821	3.250
10	1.372	1.812	2.228	2.764	3.169
11	1.363	1.796	2.201	2.718	3.106
12	1.356	1.782	2.179	2.681	3.055
13	1.350	1.771	2.160	2.650	3.012
14	1.345	1.761	2.145	2.624	2.977
15	1.341	1.753	2.131	2.602	2.947
16	1.337	1.746	2.120	2.583	2.921
17	1.333	1.740	2.110	2.567	2.898
18	1.330	1.734	2.101	2.552	2.878
19	1.328	1.729	2.093	2.539	2.861
20	1.325	1.725	2.086	2.528	2.845
21	1.323	1.721	2.080	2.518	2.831
22	1.321	1.717	2.074	2.508	2.819
23	1.319	1.714	2.069	2.500	2.807
24	1.318	1.711	2.064	2.492	2.797
25	1.316	1.708	2.060	2.485	2.787
26	1.315	1.706	2.056	2.479	2.779
27	1.314	1.703	2.052	2.473	2.771
28	1.313	1.701	2.048	2.467	2.763
29	1.311	1.699	2.045	2.462	2.756
30	1.310	1.697	2.042	2.457	2.750
40	1.303	1.684	2.021	2.423	2.704
60	1.296	1.671	2.000	2.390	2.660
120	1.289	1.658	1.980	2.358	2.617
∞	1.282	1.645	1.960	2.326	2.576

Source: Reproduced from Table 12 of E. S. Pearson and H. O. Hartley, *Biometrika Tables for Statisticians,* Vol. 1 (Cambridge, UK: Cambridge University Press, 1954). Reproduced with permission of the Biometrika Trustees.

TABLE D: 95TH PERCENTILES OF F DISTRIBUTIONS: $\alpha = 0.05$ IS THE UPPER TAIL AREA AND $F(0.95; v, w)$ IS THE 95TH PERCENTILE

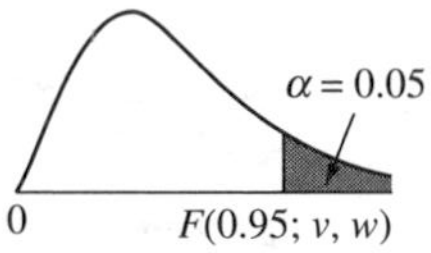

w = Degrees of Freedom for Denominator	v = Degrees of Freedom for Numerator																		
	1	2	3	4	5	6	7	8	9	10	12	15	20	24	30	40	60	120	∞
1	161.4	199.5	215.7	224.6	230.2	234.0	236.8	238.9	240.5	241.9	243.9	245.9	248.0	249.1	250.1	251.1	252.2	253.3	254.3
2	18.51	19.00	19.16	19.25	19.30	19.33	19.35	19.37	19.38	19.40	19.41	19.43	19.45	19.45	19.46	19.47	19.48	19.49	19.50
3	10.13	9.55	9.28	9.12	9.01	8.94	8.89	8.85	8.81	8.79	8.74	8.70	8.66	8.64	8.62	8.59	8.57	8.55	8.53
4	7.71	6.94	6.59	6.39	6.26	6.16	6.09	6.04	6.00	5.96	5.91	5.86	5.80	5.77	5.75	5.72	5.69	5.66	5.63
5	6.61	5.79	5.41	5.19	5.05	4.95	4.88	4.82	4.77	4.74	4.68	4.62	4.56	4.53	4.50	4.46	4.43	4.40	4.36
6	5.99	5.14	4.76	4.53	4.39	4.28	4.21	4.15	4.10	4.06	4.00	3.94	3.87	3.84	3.81	3.77	3.74	3.70	3.67
7	5.59	4.74	4.35	4.12	3.97	3.87	3.79	3.73	3.68	3.64	3.57	3.51	3.44	3.41	3.38	3.34	3.30	3.27	3.23
8	5.32	4.46	4.07	3.84	3.69	3.58	3.50	3.44	3.39	3.35	3.28	3.22	3.15	3.12	3.08	3.04	3.01	2.97	2.93
9	5.12	4.26	3.86	3.63	3.48	3.37	3.29	3.23	3.18	3.14	3.07	3.01	2.94	2.90	2.86	2.83	2.79	2.75	2.71
10	4.96	4.10	3.71	3.48	3.33	3.22	3.14	3.07	3.02	2.98	2.91	2.85	2.77	2.74	2.70	2.66	2.62	2.58	2.54
11	4.84	3.98	3.59	3.36	3.20	3.09	3.01	2.95	2.90	2.85	2.79	2.72	2.65	2.61	2.57	2.53	2.49	2.45	2.40
12	4.75	3.89	3.49	3.26	3.11	3.00	2.91	2.85	2.80	2.75	2.69	2.62	2.54	2.51	2.47	2.43	2.38	2.34	2.30
13	4.67	3.81	3.41	3.18	3.03	2.92	2.83	2.77	2.71	2.67	2.60	2.53	2.46	2.42	2.38	2.34	2.30	2.25	2.21
14	4.60	3.74	3.34	3.11	2.96	2.85	2.76	2.70	2.65	2.60	2.53	2.46	2.39	2.35	2.31	2.27	2.22	2.18	2.13
15	4.54	3.68	3.29	3.06	2.90	2.79	2.71	2.64	2.59	2.54	2.48	2.40	2.33	2.29	2.25	2.20	2.16	2.11	2.07
16	4.49	3.63	3.24	3.01	2.85	2.74	2.66	2.59	2.54	2.49	2.42	2.35	2.28	2.24	2.19	2.15	2.11	2.06	2.01
17	4.45	3.59	3.20	2.96	2.81	2.70	2.61	2.55	2.49	2.45	2.38	2.31	2.23	2.19	2.15	2.10	2.06	2.01	1.96
18	4.41	3.55	3.16	2.93	2.77	2.66	2.58	2.51	2.46	2.41	2.34	2.27	2.19	2.15	2.11	2.06	2.02	1.97	1.92
19	4.38	3.52	3.13	2.90	2.74	2.63	2.54	2.48	2.42	2.38	2.31	2.23	2.16	2.11	2.07	2.03	1.98	1.93	1.88
20	4.35	3.49	3.10	2.87	2.71	2.60	2.51	2.45	2.39	2.35	2.28	2.20	2.12	2.08	2.04	1.99	1.95	1.90	1.84
21	4.32	3.47	3.07	2.84	2.68	2.57	2.49	2.42	2.37	2.32	2.25	2.18	2.10	2.05	2.01	1.96	1.92	1.87	1.81
22	4.30	3.44	3.05	2.82	2.66	2.55	2.46	2.40	2.34	2.30	2.23	2.15	2.07	2.03	1.98	1.94	1.89	1.84	1.78
23	4.28	3.42	3.03	2.80	2.64	2.53	2.44	2.37	2.32	2.27	2.20	2.13	2.05	2.01	1.96	1.91	1.86	1.81	1.76
24	4.26	3.40	3.01	2.78	2.62	2.51	2.42	2.36	2.30	2.25	2.18	2.11	2.03	1.98	1.94	1.89	1.84	1.79	1.73
25	4.24	3.39	2.99	2.76	2.60	2.49	2.40	2.34	2.28	2.24	2.16	2.09	2.01	1.96	1.92	1.87	1.82	1.77	1.71
26	4.23	3.37	2.98	2.74	2.59	2.47	2.39	2.32	2.27	2.22	2.15	2.07	1.99	1.95	1.90	1.85	1.80	1.75	1.69
27	4.21	3.35	2.96	2.73	2.57	2.46	2.37	2.31	2.25	2.20	2.13	2.06	1.97	1.93	1.88	1.84	1.79	1.73	1.67
28	4.20	3.34	2.95	2.71	2.56	2.45	2.36	2.29	2.24	2.19	2.12	2.04	1.96	1.91	1.87	1.82	1.77	1.71	1.65
29	4.18	3.33	2.93	2.70	2.55	2.43	2.35	2.28	2.22	2.18	2.10	2.03	1.94	1.90	1.85	1.81	1.75	1.70	1.64
30	4.17	3.32	2.92	2.69	2.53	2.42	2.33	2.27	2.21	2.16	2.09	2.01	1.93	1.89	1.84	1.79	1.74	1.68	1.62
40	4.08	3.23	2.84	2.61	2.45	2.34	2.25	2.18	2.12	2.08	2.00	1.92	1.84	1.79	1.74	1.69	1.64	1.58	1.51
60	4.00	3.15	2.76	2.53	2.37	2.25	2.17	2.10	2.04	1.99	1.92	1.84	1.75	1.70	1.65	1.59	1.53	1.47	1.39
120	3.92	3.07	2.68	2.45	2.29	2.17	2.09	2.02	1.96	1.91	1.83	1.75	1.66	1.61	1.55	1.50	1.43	1.35	1.25
∞	3.84	3.00	2.60	2.37	2.21	2.10	2.01	1.94	1.88	1.83	1.75	1.67	1.57	1.52	1.46	1.39	1.32	1.22	1.00

Source: Reproduced from Table 18 of E. S. Pearson and H. O. Hartley, *Biometrika Tables for Statisticians,* Vol. 1 (Cambridge, UK: Cambridge University Press, 1954). Reproduced with permission of the Biometrika Trustees.

TABLE D: (CONTINUED) 99TH PERCENTILES OF *F* DISTRIBUTIONS: $\alpha = 0.01$ IS THE UPPER TAIL AREA AND $F(0.99; v, w)$ IS THE 99TH PERCENTILE

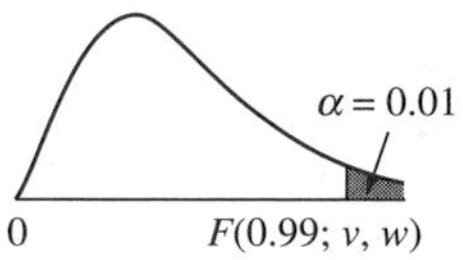

w = Degrees of Freedom for Denominator	v = Degrees of Freedom for Numerator																		
	1	2	3	4	5	6	7	8	9	10	12	15	20	24	30	40	60	120	∞
1	4052	4999.5	5403	5625	5764	5859	5928	5982	6022	6056	6106	6157	6209	6235	6261	6287	6313	6339	6366
2	98.50	99.00	99.17	99.25	99.30	99.33	99.36	99.37	99.39	99.40	99.42	99.43	99.45	99.46	99.47	99.47	99.48	99.49	99.50
3	34.12	30.82	29.46	28.71	28.24	27.91	27.67	27.49	27.35	27.23	27.05	26.87	26.69	26.60	26.50	26.41	26.32	26.22	26.13
4	21.20	18.00	16.69	15.98	15.52	15.21	14.98	14.80	14.66	14.55	14.37	14.20	14.02	13.93	13.84	13.75	13.65	13.56	13.46
5	16.26	13.27	12.06	11.39	10.97	10.67	10.46	10.29	10.16	10.05	9.89	9.72	9.55	9.47	9.38	9.29	9.20	9.11	9.02
6	13.75	10.92	9.78	9.15	8.75	8.47	8.26	8.10	7.98	7.87	7.72	7.56	7.40	7.31	7.23	7.14	7.06	6.97	6.88
7	12.25	9.55	8.45	7.85	7.46	7.19	6.99	6.84	6.72	6.62	6.47	6.31	6.16	6.07	5.99	5.91	5.82	5.74	5.65
8	11.26	8.65	7.59	7.01	6.63	6.37	6.18	6.03	5.91	5.81	5.67	5.52	5.36	5.28	5.20	5.12	5.03	4.95	4.86
9	10.56	8.02	6.99	6.42	6.06	5.80	5.61	5.47	5.35	5.26	5.11	4.96	4.81	4.73	4.65	4.57	4.48	4.40	4.31
10	10.04	7.56	6.55	5.99	5.64	5.39	5.20	5.06	4.94	4.85	4.71	4.56	4.41	4.33	4.25	4.17	4.08	4.00	3.91
11	9.65	7.21	6.22	5.67	5.32	5.07	4.89	4.74	4.63	4.54	4.40	4.25	4.10	4.02	3.94	3.86	3.78	3.69	3.60
12	9.33	6.93	5.95	5.41	5.06	4.82	4.64	4.50	4.39	4.30	4.16	4.01	3.86	3.78	3.70	3.62	3.54	3.45	3.36
13	9.07	6.70	5.74	5.21	4.86	4.62	4.44	4.30	4.19	4.10	3.96	3.82	3.66	3.59	3.51	3.43	3.34	3.25	3.17
14	8.86	6.51	5.56	5.04	4.69	4.46	4.28	4.14	4.03	3.94	3.80	3.66	3.51	3.43	3.35	3.27	3.18	3.09	3.00
15	8.68	6.36	5.42	4.89	4.56	4.32	4.14	4.00	3.89	3.80	3.67	3.52	3.37	3.29	3.21	3.13	3.05	2.96	2.87
16	8.53	6.23	5.29	4.77	4.44	4.20	4.03	3.89	3.78	3.69	3.55	3.41	3.26	3.18	3.10	3.02	2.93	2.84	2.75
17	8.40	6.11	5.18	4.67	4.34	4.10	3.93	3.79	3.68	3.59	3.46	3.31	3.16	3.08	3.00	2.92	2.83	2.75	2.65
18	8.29	6.01	5.09	4.58	4.25	4.01	3.84	3.71	3.60	3.51	3.37	3.23	3.08	3.00	2.92	2.84	2.75	2.66	2.57
19	8.18	5.93	5.01	4.50	4.17	3.94	3.77	3.63	3.52	3.43	3.30	3.15	3.00	2.92	2.84	2.76	2.67	2.58	2.49
20	8.10	5.85	4.94	4.43	4.10	3.87	3.70	3.56	3.46	3.37	3.23	3.09	2.94	2.86	2.78	2.69	2.61	2.52	2.42
21	8.02	5.78	4.87	4.37	4.04	3.81	3.64	3.51	3.40	3.31	3.17	3.03	2.88	2.80	2.72	2.64	2.55	2.46	2.36
22	7.95	5.72	4.82	4.31	3.99	3.76	3.59	3.45	3.35	3.26	3.12	2.98	2.83	2.75	2.67	2.58	2.50	2.40	2.31
23	7.88	5.66	4.76	4.26	3.94	3.71	3.54	3.41	3.30	3.21	3.07	2.93	2.78	2.70	2.62	2.54	2.45	2.35	2.26
24	7.82	5.61	4.72	4.22	3.90	3.67	3.50	3.36	3.26	3.17	3.03	2.89	2.74	2.66	2.58	2.49	2.40	2.31	2.21
25	7.77	5.57	4.68	4.18	3.85	3.63	3.46	3.32	3.22	3.13	2.99	2.85	2.70	2.62	2.54	2.45	2.36	2.27	2.17
26	7.72	5.53	4.64	4.14	3.82	3.59	3.42	3.29	3.18	3.09	2.96	2.81	2.66	2.58	2.50	2.42	2.33	2.23	2.13
27	7.68	5.49	4.60	4.11	3.78	3.56	3.39	3.26	3.15	3.06	2.93	2.78	2.63	2.55	2.47	2.38	2.29	2.20	2.10
28	7.64	5.45	4.57	4.07	3.75	3.53	3.36	3.23	3.12	3.03	2.90	2.75	2.60	2.52	2.44	2.35	2.26	2.17	2.06
29	7.60	5.42	4.54	4.04	3.73	3.50	3.33	3.20	3.09	3.00	2.87	2.73	2.57	2.49	2.41	2.33	2.23	2.14	2.03
30	7.56	5.39	4.51	4.02	3.70	3.47	3.30	3.17	3.07	2.98	2.84	2.70	2.55	2.47	2.39	2.30	2.21	2.11	2.01
40	7.31	5.18	4.31	3.83	3.51	3.29	3.12	2.99	2.89	2.80	2.66	2.52	2.37	2.29	2.20	2.11	2.02	1.92	1.80
60	7.08	4.98	4.13	3.65	3.34	3.12	2.95	2.82	2.72	2.63	2.50	2.35	2.20	2.12	2.03	1.94	1.84	1.73	1.60
120	6.85	4.79	3.95	3.48	3.17	2.96	2.79	2.66	2.56	2.47	2.34	2.19	2.03	1.95	1.86	1.76	1.66	1.53	1.38
∞	6.63	4.61	3.78	3.32	3.02	2.80	2.64	2.51	2.41	2.32	2.18	2.04	1.88	1.79	1.70	1.59	1.47	1.32	1.00

References

CHAPTER 1

Poincare, H. *The Foundations of Science*. New York: The Science Press, 1913 (reprinted 1929), p. 169.

CHAPTER 2

Anscombe, F. J. Graphs in statistical analysis. *The American Statistician,* 27, 17–21, 1973.

Atkinson, A. C. *Plots, Transformations, and Regression*. Oxford, UK: Clarendon Press, 1985.

Bissell, A. F. Lines through the origin—IS NO INT the answer? *Journal of Applied Statistics,* 19, 193–210, 1992.

Gilchrist, W. *Statistical Modelling*. Chichester, UK: Wiley, 1984.

Mosteller, F., Rourke, R. E. K., and Thomas, G. B. *Probability with Statistical Applications* (2nd ed.). Reading, MA: Addison-Wesley, 1970.

Risebrough, R. W. Effects of environmental pollutants upon animals other than man. In *Proceedings of the 6th Berkeley Symposium on Mathematics and Statistics*. Berkeley: University of California Press, pp. 443–463, 1972.

Roberts, H. V., and Ling, R. F. *Conversational Statistics with IDA*. New York: Scientific Press/McGraw-Hill, 1982.

Wallach, D., and Goffinet, B. Mean square error of prediction in models for studying ecological and agronomic systems. *Biometrics,* 43, 561–573, 1987.

Weisberg, S. *Applied Linear Regression,* New York, Wiley, 1980. [2nd ed., 1985.]

CHAPTER 4

Davies, O. L., and Goldsmith, P. L. (Eds.). *Statistical Methods in Research and Production* (4th ed.). Edinburgh, UK: Oliver & Boyd, 1972.

Davison, A. C., and Hinkley, D. V. *Bootstrap Methods and Their Applications*. New York: Cambridge University Press, 1997.

Efron, B. Bootstrap methods: Another look at the jackknife. *Annals of Statistics,* 7, 1–26, 1979.

Efron, B., and Tibshirani, R. J. *An Introduction to the Bootstrap*. New York: Chapman & Hall, 1993.

Horowitz, J. L. The bootstrap. In *Handbook of Econometrics* (Vol. 6). Amsterdam: North Holland, 1999.

Humphreys, R. M. Studies of luminous stars in nearby galaxies. I. Supergiants and O stars in the Milky Way. *Astrophysics Journal, Supplementary Series,* 38, 309–350, 1978.

Joglekar, G., Schuenemeyer, J. H., and LaRiccia, V. Lack-of-fit testing when replicates are not available. *The American Statistician,* 43, 135–143, 1989.

CHAPTER 5

Latter, O. H. The Cuckoo's egg. *Biometrika,* 1, 164–176, 1901.

Mazess, R. B., Peppler, W. W., and Gibbons, M. Total body composition by dualphoton (^{153}Gd) absorptiometry. *American Journal of Clinical Nutrition,* 40, 834–839, 1983.

CHAPTER 6

Abraham, B., and Ledolter, J. *Statistical Methods for Forecasting*. New York: Wiley, 1983.

Bennett, G. W. Determination of anaerobic threshold. *Canadian Journal of Statistics,* 16, 307–310, 1988.

Box, G. E. P., and Cox, D. R. An analysis of transformations. *Journal of the Royal Statistical Society, Series B,* 26, 211–243, 1964; discussion, 244–252.

Brown, B. M., and Maritz, J. S. Distribution-free methods in regression. *Australian Journal of Statistics,* 24, 318–331, 1982.

Cook, R. D., and Weisberg, S. *Residuals and Influence in Regression*. London: Chapman & Hall, 1982.

Dempster, A. P. *Elements of Continuous Multivariate Analysis*. Reading, MA: Addison-Wesley, 1969.

Hill, W. J., and Wiles, R. A. Plant experimentation (PLEX). *Journal of Quality Technology,* 7, 115–122, 1975.

Jerison, H. J. *Evolution of the Brain and Intelligence*. New York: Academic Press, 1973.

Joglekar, G., Schuenemeyer, J. H., and LaRiccia, V. Lack-of-fit testing when replicates are not available. *The American Statistician,* 43, 135–143, 1989.

Robertson, J. D., and Armitage, P. Comparison of two hypertensive agents. *Anaesthesia,* 14, 53–64, 1959.

Ryan, T. A., Joiner, B. L., and Ryan, B. F. *The Minitab Student Handbook*. Boston: Duxbury, 1985.

Snapinn, S. M., and Small, R. D. Tests of significance using regression models for ordered categorical data. *Biometrics,* 42, 583–592, 1966.

Snee, R. D. An alternative approach to fitting models when re-expression of the response is useful. *Journal of Quality Technology,* 18, 211–225, 1986.

Weiner, B. *Discovering Psychology*. Chicago: Science Research Association, 1977.

Williams, E. J. *Regression Analysis*. New York: Wiley, 1959.

CHAPTER 7

Brownlee, K. A. *Statistical Theory and Methodology in Science and Engineering* (2nd ed.). London: Wiley, 1965.

Cox, D. R., and Snell, E. J. *Applied Statistics: A Handbook of BMDP Analyses*. London: Chapman & Hall, 1987.

Gorman, J. W., and Toman, R. J. Selection of variables for fitting equations to data. *Technometrics,* 8, 27–51, 1966.

Kieschnick, R., and McCullough, B. D. Regression analysis of variates observed on (0,1): Percentages, proportions, and fractions. *Statistical Modelling: An International Journal,* 3, 193–213, 2003.

Mallows, C. L. Some comments on C_p. *Technometrics,* 15, 661–675, 1973.

Vandaele, W. Participation in illegitimate activities: Erlich revisited. In *Deterrence and Incapacitation* (A. Blumstein, J. Cohen, and D. Nagin, Eds.). Washington, DC: National Academy of Sciences, pp. 270–335, 1978.

Woodley, W. L., Biondini, R., and Berkeley, J. Rainfall results 1970–75: Florida Area Cumulus Experiment. *Science,* 195, 735–742, 1977.

CHAPTER 8

Ashenfelter, O., Ashmore, D., and Lalonde, R. Bordeaux wine vintage quality and the weather. *Chance Magazine,* 8, 7–14, 1995.

Campbell, J. E., and Garand, J. C. *Before the Vote: Forecasting American National Elections*. Thousand Oaks, CA: Sage, 2000.

Fair, R. C. *Predicting Presidential Elections and Other Things*. Stanford, CA: Stanford University Press, 2002.

Hunter, W. G. Some ideas about teaching design of experiments, with 2^5 examples of experiments conducted by students. *The American Statistician,* 31, 12–17, 1977.

Kramer, J. R., Loney, J., Ponto, L. B., Roberts, M. A., and Grossman, S. Predictors of adult height and weight in boys treated with methylphenidate for childhood behavior problems. *Journal of the American Academy of Child and Adolescent Psychiatry,* 39, 517–524, 2000.

Ledolter, J. Projects in introductory statistics courses. *The American Statistician,* 49, 364–367, 1995.

CHAPTER 9

Bates, D. M., and Watts, D. G. *Nonlinear Regression Analysis and Its Applications*. New York: Wiley, 1988.

Gallant, A. R. *Nonlinear Statistical Models*. New York: Wiley, 1987.

Huet, S., Bouvier, A., Gruet, M., and Jolivet, E. *Statistical Tools for Nonlinear Regression: A Practical Guide with S-Plus Examples*. New York: Springer, 1996.

Morgan, P. H., Mercer, L. P., and Flodin, N. W. General model for nutritional responses of higher organisms. *Proceedings of the National Academy of Sciences,* 72, 4327–4331, 1975.

Rasch, D. The robustness against parameter variation of exact locally optimum designs in nonlinear regression—A case study. *Computational Statistics and Data Analysis,* 20, 441–453, 1995.

Richards, F. J. A flexible growth function for empirical use. *Journal of Experimental Botany,* 10, 290–300, 1959.

Seber, G. A. F., and Wild, C. J. *Nonlinear Regression*. New York: Wiley, 1989.

Singh, M., Kanji, G. K., and El-Bizri, K. S. A note on inverse estimation in non-linear models. *Journal of Applied Statistics,* 19, 473–477, 1992.

Smith, H., and Dubey, S. D. Some reliability problems in the chemical industry. *Industrial Quality Control,* 21, 64–70, 1964.

Sedlacek, G. Zur Quantifizierung und Analyse der Nichtlinearitaet von Regressionsmodellen. *Austrian Journal of Statistics,* 27, 171–190, 1998.

CHAPTER 10

Abraham, B., and Ledolter, J. *Statistical Methods for Forecasting*. New York: Wiley, 1983.

Blattberg, R. C., and Jeuland, A. P. A micromodeling approach to investigate the advertising–sales relationship. *Management Science,* 27, 988–1005, 1981.

Box, G. E. P., Jenkins, G. M., and Reinsel, G. C. *Time Series Analysis, Forecasting and Control* (3rd ed.). New York: Prentice Hall, 1994.

Box, G. E. P., and Newbold, P. Some comments on a paper by Coen, Gomme and Kendall. *Journal of the Royal Statistical Society, Series A,* 134, 229–240, 1971.

Box, G. E. P., and Tiao, G. C. A canonical analysis of multiple time series. *Biometrika,* 64, 355–365, 1977.

Brown, R. G. *Smoothing, Forecasting and Prediction of Discrete Time Series*. Englewood Cliffs, NJ: Prentice Hall, 1962.

Coen, P. J., Gomme, E. D., and Kendall, M. G. Lagged relationships in economic forecasting. *Journal of the Royal Statistical Society, Series A,* 132, 133–163, 1969.

Granger, C. W. J., and Newbold, P. Spurious regressions in econometrics. *Journal of Econometrics,* 2, 111–120, 1974.

Granger, C. W. J., and Newbold, P. *Forecasting Economic Time Series* (2nd ed.). New York: Academic Press, 1978.

Hanke, J. E., and Reitsch, A. G. *Business Forecasting* (3rd ed.). Needham Heights, MA: Allyn & Bacon, 1989.

Kadiyala, K. R. Testing for the independence of regression disturbances. *Econometrica,* 38, 97–117, 1970.

Newbold, P., and Davies, N. Error mis-specification and spurious regressions. *International Economic Review,* 19, 513–519, 1978.

Quenouille, M. H. *The Analysis of Multiple Time Series*. New York: Hafner, 1968

Shaw, N. *Manual of Meteorology* (Vol. 1). Cambridge, UK: Cambridge University Press, p. 284, 1926.

CHAPTER 11

Agresti, A. *Categorical Data Analysis* (2nd ed.). New York: Wiley, 2002.

Brewster, C., and Hegewisch, A. *Policy and Practice in European Human Resource Management. The Price Waterhouse Cranfield Survey*. London: Routledge, 1994.

Brown, P. J., Stone, J., and Ord-Smith, C. Toxaemic signs during pregnancy. *Applied Statistics,* 32, 69–72, 1983.

Cook, R. D., and Weisberg, S. *Applied Regression including Computing and Graphics*. New York: Wiley, 1999.

Eno, D. R., and Terrell, G. R. Scatterplots for logistic regression (with discussion). *Journal of Computational and Graphical Statistics,* 8, 413–430, 1999.

Higgins, J. E., and Koch, G. G. Variable selection and generalized chi-square analysis of categorical data applied to a large cross-sectional occupational health survey. *International Statistical Review,* 45, 51–62, 1977.

Hosmer, D. W., and Lemeshow, S. A goodness-of-fit test for the multiple logistic regression model. *Communications in Statistics, Series A,* 10, 1043–1069, 1980.

Hosmer, D. W., and Lemeshow, S. *Applied Logistic Regression* (2nd ed.). New York: Wiley, 2000.

Landwehr, J. M., Pregibon, D., and Shoemaker, A. C. Graphical methods for assessing logistic regression models (with discussion). *Journal of the American Statistical Association,* 79, 61–83, 1984.

Ledolter, J. A case study in design of experiments: Improving the manufacture of viscose fiber. *Quality Engineering,* 15, 311–322, 2002.

Lemeshow, S., and Hosmer, D. W. The goodness-of-fit statistics in the development of logistic regression models. *American Journal of Epidemiology,* 115, 92–106, 1982.

McCullagh, P., and Nelder, J. A. *Generalized Linear Models*. London: Chapman & Hall, 1983.

Montgomery, D. C., Peck, E. A., and Vining, G. G. *Introduction to Linear Regression Analysis* (3rd ed.). New York: Wiley, 2001.

Myers, R. H., Montgomery, D. C., and Vining, G. G. *Generalized Linear Models, with Applications in Engineering and the Sciences*. New York: Wiley, 2002.

Pregibon, D. Logistic regression diagnostics. *Annals of Statistics,* 9, 705–724, 1981.

Ramsey, F. L., and Schafer, D. W. *The Statistical Sleuth* (2nd ed.). Pacific Grove, CA: Duxbury, 2002.

CHAPTER 12

Agresti, A. *Categorical Data Analysis* (2nd ed.). New York: Wiley, 2002.

Aitkin, M., Anderson, D., Francis, B., and Hinde, J. *Statistical Modeling in GLIM*. Oxford, UK: Clarendon, 1989.

Diggle, P., Heagerty, P., Liang, K. Y., and Zeger, S. *Analysis of Longitudinal Data*. New York: Oxford University Press, 2002.

Dobson, A. J. *An Introduction to Generalized Linear Models* (2nd ed.). London: Chapman & Hall, 2002.

Fahrmeir, L., and Tutz, G. *Multivariate Statistical Modeling Based on Generalized Linear Models* (2nd ed.). New York: Springer, 2001.

Hand, D. J., Daly, F., Lunn, A. D., McConway, K. J., and Ostrowski, E. *A Handbook of Small Data Sets*. London: Chapman & Hall, 1994.

Hill, R. C., Griffiths, W. E. and Judge, G. G. *Undergraduate Econometrics* (2nd ed.). New York: Wiley, 2001.

Jorgenson, D. W. Multiple regression analysis of a Poisson process. *Journal of the American Statistical Association,* 56, 235–245, 1961.

McCullagh, P., and Nelder, J. A. *Generalized Linear Models* (2nd ed.). London: Chapman & Hall, 1987.

Myers, R. H., Montgomery, D. C., and Vining, G. G. *Generalized Linear Models with Applications in Engineering and the Sciences*. New York: Wiley, 2002.

Nelder, J. A., and Wedderburn, R. W. M. Generalised linear models. *Journal of the Royal Statistical Society, Series A,* 135, 370–384, 1972.

Poole, J. H. Mate guarding, reproductive success and female choice in African elephants. *Animal Behavior,* 37, 842–849, 1989.

Ramsey, F. L., and Shafer, D. W. *The Statistical Sleuth* (2nd ed.). Pacific Grove, CA: Duxbury, 2002.

List of Data Files

The following data sets are available on the enclosed data disk as well as on the Web site **http://statistics.brookscole.com/abraham_ledolter/**. All files are in plain text format and in the format of frequently-used statistical software packages.

CHAPTER 1

contraceptive
gasconsumption
hardness
payout
uffi

CHAPTER 2

absorption
anscombe
bloodpressure
boiling
distance
energyrequirement
hooker
iowatest
ironcontent
memory
overhead
pelicaneggs
sriver
turtles

CHAPTER 3

No data file

CHAPTER 4

abrasion
bsemen
hospital
lightintensity
sales
silkw
water
woodstrength

CHAPTER 5

bodyfat
catalyst
cellulose
cuckoo
diet
pizza
shrinkage
weightloss

CHAPTER 6

brainweight
chemreaction
chemyield
cloud
florida2000
forestry
kidney
liver
phosphorus
recovery
rigidity
stopping
tearpaper
ventilation
vocabulary
volumeeggs
volumetrees
windmill

CHAPTER 7

achievement
asphalt
crimerate

election2000
hald
market
powerplant
rainseeding
stackloss

CHAPTER 8

abramowitz
campbell
cows
election(Beck&Tien)
election(Fair)
height&weight
holbrook
iowastudent
lockerbie
softdrink
wine

CHAPTER 9

chlorine
nitrate
palmtrees
shootlength

CHAPTER 10

bookstore
cgk
hogs
icecreamsales
lakelevel
lenzingstock
salesadvert
saleschemical
sears
thermostat

CHAPTER 11

byssinosis
cranfield
deathpenalty
donner
lenzing
toxemia
washroom

CHAPTER 12

ceriodaphnia
damage
elephants
failures
lake
lungcancer
skincancer

Index